MyEngineeringLab™

...ht now, in your course, there are young men and women whose engineering ...ievements could revolutionize, improve, and sustain future generations.

...n't Let Them Get Away.

...king Like an Engineer, 2nd edition, together with MyEngineeringLab, is a complete ...tion for providing an engaging in-class experience that will inspire your students to ... in engineering, while also giving them the practice and scaffolding they need to keep ...nd be successful in the course.

...n more at **www.myengineeringlab.com**

...YS LEARNING

PEARSON

THINKING LIKE AN ENGINEER

AN ACTIVE LEARNING APPROACH

Second Edition

Elizabeth A. Stephan
Clemson University

David R. Bowman
Clemson University

William J. Park
Clemson University

Benjamin L. Sill
Clemson University

Matthew W. Ohland
Purdue University

PEARSON

Upper Saddle River Boston Columbus San Francisco New York
Indianapolis London Toronto Sydney Singapore Tokyo Montreal
Dubai Madrid Hong Kong Mexico City Munich Paris Amsterdam Cape Town

Vice President and Editorial Director, ECS: *Marcia J. Horton*
Executive Editor: *Holly Stark*
Editorial Assistant: *Carlin Heinle*
Vice President, Production: *Vince O'Brien*
VP/Director of Marketing: *Patrice Jones*
Executive Marketing Manager: *Tim Galligan*
Marketing Assistant: *Jon Bryant*
Permissions Project Manager: *Karen Sanatar*
Senior Managing Editor: *Scott Disanno*
Production Project Manager: *Clare Romeo*
Director of Operations: *Nick Sklitsis*

Operations Specialist: *Lisa McDowell*
Cover Designer: *Kenny Beck*
Cover Photos: *Blue guitar/Evgeny Guityaev/ Shutterstock; X-ray of guitar/Gustoimages/ Science Photo Library*
Composition/Full-Service Project Management: *Jouve India Private Limited*
Full-Service Project Management: *Pavithra Jayapaul, Jouve India Private Limited*
Cover Printer: *Lehigh-Phoenix Color/Hagerstown*
Printer/Binder: *Quad Graphics*
Typeface: *10/12 Times Ten LT STD Roman*

Credits and acknowledgments borrowed from other sources and reproduced, with permission, in this textbook appear on appropriate page within text.

Library of Congress Cataloging-in-Publication Data

Thinking like an engineer : an active learning approach / Elizabeth A. Stephan . . . [et al.].—2nd ed.
  p. cm.
 Includes bibliographical references and index.
 ISBN-13: 978-0-13-276671-5 (alk. paper)
 ISBN-10: 0-13-276671-X (alk. paper)
  1. Engineering—Study and teaching (Higher) 2. Active learning.   I. Stephan, Elizabeth A.
TA147.T45 2012
620.0071'1—dc23

                                                                    2011037789

10 9 8 7 6 5 4 3 2 1

ISBN-13: 978-0-13-276671-5
ISBN-10:   0-13-276671-X

CONTENTS

PREFACE

At our university, all students who wish to major in engineering begin in the General Engineering Program, and after completing a core set of classes, they can declare a specific engineering major. Within this core set of classes, students are required to take math, physics, chemistry, and a two-semester engineering sequence. Over the past 12 years, our courses have evolved to address not only the changing qualities of our students, but also the changing needs of our customers. The material taught in our courses is the foundation upon which the upper level courses depend for the skills necessary to master more advanced material. It was for these freshman courses that this text was created.

We didn't set out to write a textbook: we simply set out to find a better way to teach our students. Our philosophy was to help students move from a mode of learning, where everything was neatly presented as lecture and handouts where the instructor was looking for the "right" answer, to a mode of learning driven by self-guided inquiry. We wanted students to advance beyond "plug-and-chug" and memorization of problem-solving methods—to ask themselves if their approaches and answers make sense in the physical world. We couldn't settle on any textbooks we liked without patching materials together—one chapter from this text, four chapters from another—so we wrote our own notes. Through them, we tried to convey that engineering isn't always about having the answer—sometimes it's about asking the right questions, and we want students to learn how to ask those sorts of questions. Real-world problems rarely come with all of the information required for their solutions. Problems presented to engineers typically can't be solved by looking at how someone else solved the exact same problem. Part of the fun of engineering is that every problem presents a unique challenge and requires a unique solution. Engineering is also about arriving at an answer and being able to justify the "why" behind your choice, and equally important, the "why not" of the other choices.

We realized quickly, however, that some students are not able to learn without sufficient scaffolding. Structure and flexibility must be managed carefully. Too much structure results in rigidity and unnecessary uniformity of solutions. On the other hand, too much flexibility provides insufficient guidance, and students flounder down many blind alleys, thus making it more difficult to acquire new knowledge. The tension between these two must be managed constantly. We are a large public institution, and our student body is very diverse. Our hope is to provide each student with the amount of scaffolding they need to be successful. Some students will require more background work than others. Some students will need to work five problems, and others may need to work 50. We talk a great deal to our students about how each learner is unique. Some students need to listen to a lecture; some need to read the text over three times, and others just need to try a skill and make mistakes to discover what they still don't understand. We have tried to provide enough variety for each type of learner throughout.

Over the years, we have made difficult decisions on exactly what topics, and how much of each topic, to teach. We have refined our current text to focus on mastering four areas, each of which is introduced below.

PART 1: ENGINEERING ESSENTIALS

There are three threads that bind the first six chapters in Engineering Essentials together. The first is expressed in the part title: all are essential for a successful career in engineering. The second is communications. Part 1 concludes with an introduction to a problem-solving methodology.

First, as an aspiring engineer, it is important that students attempt to verify that engineering is not only a career that suits their abilities but also one in which they will find personal reward and satisfaction.

Second, practicing engineers often make decisions that will affect not only the lives of people but also the viability of the planetary ecosystem that affects all life on Earth. Without a firm grounding in making decisions based on ethical principles, there is an increased probability that undesirable or even disastrous consequences may occur.

Third, most engineering projects are too large for one individual to accomplish alone; thus, practicing engineers must learn to function effectively as a team, putting aside their personal differences and combining their unique talents, perspectives, and ideas to achieve the goal.

Finally, communications bind it all together. Communication, whether written, graphical, or spoken, is essential to success in engineering.

This part ends off where all good problem solving should begin—with estimation and a methodology. It's always best to have a good guess at any problem before trying to solve it more precisely. SOLVEM provides a framework for solving problems that encourages creative observation as well as methodological rigor.

PART 2: UBIQUITOUS UNITS

The world can be described using relatively few dimensions. We need to know what these are and how to use them to analyze engineering situations. Dimensions, however, are worthless in allowing engineers to find the numeric solution to a problem. Understanding units is essential to determine the correct numeric answers to problems. Different disciplines use different units to describe phenomena (particularly with respect to the properties of materials such as viscosity, thermal conductivity, density and so on). Engineers must know how to convert from one unit system to another. Knowledge of dimensions allows engineers to improve their problem-solving abilities by revealing the interplay of various parameters.

PART 3: SCRUPULOUS WORKSHEETS

When choosing an analysis tool to teach students, our first pick is Excel™. Students enter college with varying levels of experience with Excel. To allow students who are novice users to learn the basics without hindering more advanced users, we have placed the basics of Excel in the Appendix material, which is available online. To help students determine if they need to review the Appendix material, an activity has been included in the introductions to Chapter 10 (Worksheets), Chapter 11 (Graphing), and Chapter 12 (Trendlines) to direct students to Appendices B, C, and D, respectively.

Once students have mastered the basics, each chapter in this part provides a deeper usage of Excel in each category. Some of this material extends beyond a simple introduction to Excel, and often, we teach the material in this unit by jumping around, covering half of each chapter in the first semester, and the rest of the material in the second semester course.

Chapter 12 introduces students to the idea of similarities among the disciplines, and how understanding a theory in one application can often aid in understanding a similar theory in a different application. We also emphasize the understanding of models (trendlines) as possessing physical meaning. Chapter 13 discusses a process for determining a mathematical model when presented with experimental data and some advanced material on dealing with limitations of Excel.

Univariate statistics and statistical process control wrap up this part of the book by providing a way for engineering students to describe both distributions and trends.

PART 4: PUNCTILIOUS PROGRAMMING

Part 4 (Punctilious Programming) covers a variety of topics common to any introductory programming textbook. In contrast to a traditional programming textbook, this part approaches each topic from the perspective of how each can be used in unison with the others as a powerful engineering problem-solving tool. The topics presented in Part 4 are introduced as if the student has no prior programming ability and are continually reiterated throughout the remaining chapters.

For this textbook we chose MATLAB™ as the programming language because it is commonly used in many engineering curricula. The topics covered provide a solid foundation of how computers can be used as a tool for problem solving and provide enough scaffolding for transfer of programming knowledge into other languages commonly used by engineers (such as C/C++/Java).

THE "OTHER" STUFF WE'VE INCLUDED...

Throughout the book, we have included sections on surviving engineering, time management, goal setting, and study skills. We did not group them into a single chapter, but have scattered them throughout the part introductions to assist students on a topic when they are most likely to need it. For example, we find students are much more open to discuss time management in the middle of the semester rather than the beginning.

In addition, we have called upon many practicing and aspiring engineers to help us explain the "why" and "what" behind engineering. They offer their "Wise Words" throughout this text. We have included our own set of "Wise Words" as the introduction to each topic here as a glimpse of what inspired us to include certain topics.

NEW TO THIS EDITION

The 2nd edition of *Thinking Like an Engineer: An Active Learning Approach* (TLAE) contains new material and revisions based off of the comments from faculty teaching with our textbook, the recommendations of the reviewers of our textbook, and most importantly, the feedback from our students. We have added approximately 30% new questions and six projects that intend to engage students to create meaningful learning experiences. In addition, we have added new material that reflects the constant changing face of engineering education because many of our upperclassman teaching assistants frequently comment to us "I wish I had ___ when I took this class."

New to This Edition, by chapter:

- Chapter 1: Everyday Engineering

  - New section on Engineering Grand Challenges.

- Chapter 2: Ethics

  - New sections on Plagiarism and Social Responsibility.

- Chapter 3: Design and Teamwork

  - New section on Sustainability.
  - Improved section on Teamwork by linking to CATME team evaluation system.

- Chapter 5: Estimation

  - New sections on Estimation by Analogy, Estimation by Aggregation, Establishing Upper and Lower Bounds, and Estimation Using Modeling.

- Chapter 11: Graphical Solutions

  - Combined material from Chapter 7: Graphing Guidelines in TLAE 1e.

- Chapter 14: Statistics

  - Combined material from Chapters 9 and 14 in TLAE 1e to make a single unified chapter on Statistics.

- Chapter 15: Algorithms

  - Selected material from Chapter 17 in TLAE 1e.
  - New material on algorithm variables, user interaction, calculations and conversions, referencing other algorithms (or functions), error checking and prevention, different types of errors to look for in algorithms, iteration, and testing algorithms.

- Chapter 16: Programs and Functions

  - Selected material from Chapter 17 in TLAE 1e. New separate section on programming basics.
  - New material on efficient creation of matrices and advanced matrix addressing techniques.

- Chapter 17: Input/Output in MATLAB

  - New material on creating error and warning messages.
  - New examples on creating proper plots in MATLAB using advanced graphing properties.
  - New section on use of POLYFIT for adding trendlines to graphs.

- Chapter 18: Logic and Conditionals
  - New sections on Applications of Loops: Cell Arrays and Cell Matrices, Applications of Loops: Excel I/O, and Applications of Loops: Graphical User Interfaces.
- Back materials:
  - New section, Umbrella Projects added to the end of the textbook.
  - New quick reference pages: MATLAB special characters and MATLAB graphing properties.
- Online Appendix Materials
  - New material in Appendix A: Interpolation (previously Chapter 8 in TLAE 1e). www.pearsonhighered.com/TLAE

HOW TO USE

As we have alluded to previously, this text contains many different types of instruction to address different types of learners. There are two main components to this text: hard copy and online.

In the hardcopy, the text is presented topically rather than sequentially, but hopefully with enough autonomy for each piece to stand alone. For example, we routinely discuss only part of the Excel material in our first semester course, and leave the rest to the second semester. We hope this will give you the flexibility to choose how deeply into any given topic you wish to dive, depending on the time you have, the starting abilities of your students, and the outcomes of your course. More information about topic sequence options can be found in the instructor's manual.

Within the text, there are several checkpoints for students to see if they understand the material. Within the reading are **Comprehension Checks**, with the answers provided in the back of the book. Our motivation for including Comprehension Checks within the text rather than include them as end of part questions is to maintain the active spirit of the classroom within the reading, allowing the students to self-evaluate their understanding of the material in preparation for class—to enable students to be self-directed learners, we must encourage them to self-evaluate regularly. At the end of each chapter, **In-Class Activities** are given to reinforce the material in each chapter. In-Class Activities exist to stimulate active conversation within pairs and groups of students working through the material. We generally keep the focus on student effort, and ask them to keep working the problem until they arrive at the right answer. This provides them with a set of worked out problems, using their own logic, before they are asked to tackle more difficult problems. The **Review** sections provide additional questions, often combining skills in the current chapter with previous concepts to help students climb to the next level of understanding. By providing these three types of practice, students are encouraged to reflect on their understanding in preparing for class, during class, and at the end of each chapter as they prepare to transfer their knowledge to other areas. Finally at the end of the text, we have provided a series of **Umbrella Projects** to allow students to apply skills that they have mastered to larger-scope problems. We have found the use of these problems extremely helpful in providing context for the skills that they learn throughout a unit.

Understanding that every student learns differently, we have included several media components in addition to traditional text. Each section within each chapter has an accompanying set of **video lecture slides** ▶. Within these slides, the examples presented are unique from those in the text to provide another set of sample solutions. The slides are presented with **voiceover**, which has allowed us to move away from traditional in-class lecture. We expect the students to listen to the slides outside of class, and then in class we typically spend time working problems, reviewing assigned problems, and providing **"wrap-up" lectures**, which are mini-versions of the full lectures to summarize what they should have gotten from the assignment. We expect the students to come to class with questions from the reading and lecture that we can then help clarify. We find with this method, the students pay more attention, as the terms and problems are already familiar to them, and they are more able to verbalize what they don't know. Furthermore, they can always go back and listen to the lectures again to reinforce their knowledge as many times as they need.

Some sections of this text are difficult to lecture, and students will learn this material best by **working through examples**. This is especially true with Excel and MATLAB, so you will notice that many of the lectures in these sections are shorter than previous material. The examples are scripted the first time a skill is presented, and students are expected to have their laptop open and work through the examples (not just read them). When students ask us questions in this section, we often start the answer by asking them to "show us your work from Chapter XX." If the student has not actually worked the examples in that chapter, we tell them to do so first; often, this will answer their questions.

After the first few basic problems, in many cases where we are discussing more advanced skills than data entry, we have **provided starting worksheets and code** ▣ in the online version by "hanging" the worksheets within the online text. Students can access the starting data through the online copy of the book. In some cases, though, it is difficult to explain a skill on paper, or even with slides, so for these instances we have included **videos** ▶.

Finally, for the communication section, we have provided **templates** ▣ ▣ for several types of reports and presentations. These can also be accessed in the Pearson eText version, available with adoption of MyEngineeringLab™. Visit www.pearsonhighered.com/TLAE for more information.

MyEngineeringLab™

Thinking Like an Engineer, 2nd edition, together with MyEngineeringLab provides an engaging in-class experience that will inspire your students to stay in engineering, while also giving them the practice and scaffolding they need to keep up and be successful in the course. It's a complete digital solution featuring:

- A Customized Study Plan for each student with remediation activities provides an opportunity for self paced learning for students at all different levels of preparedness.
- Automatically graded Homework Review Problems from the book and Self Study Quizzes give immediate feedback to the student and provide comprehensive gradebook tracking for instructors.
- Interactive Tutorials with additional algorithmically generated exercises provide opportunity for point-of-use help and for more practice.
- "Show My Work" feature allows instructors to see the entire solution, not only the graded answer.
- Learning Objectives mapped to ABET outcomes provide comprehensive reporting tools.
- Available with or without the full eText.

If adopted, access to MyEngineeringLab can be bundled with the book or purchased separately. For a fully digital offering, learn more at www.myengineeringlab.com or www.pearsonhighered.com/TLAE.

ADDITIONAL RESOURCES FOR INSTRUCTORS

Instructor's Manual—Available to all adopters, this provides a complete set of solutions for all activities and review exercises. For the In-Class Activities, suggested guided inquiry questions along with time frame guidelines are included. Suggested content sequencing and descriptions of how to couple assignments to the umbrella projects are also provided.

PowerPoints—A complete set of lecture PowerPoint slides, available with voiceover or as standard slides, make course planning as easy as possible.

Sample Exams—Available to all adopters, these will assist in creating tests and quizzes for student assessment.

MyEngineeringLab—Provides web-based assessment, tutorial, homework and course management. www.myengineeringlab.com

All requests for instructor resources are verified against our customer database and/or through contacting the requestor's institution. Contact your local Pearson/Prentice Hall representative for additional information.

WHAT DOES THINKING LIKE AN ENGINEER MEAN?

We are often asked about the title of the book. We thought we'd take a minute and explain what this means, to each of us. Our responses are included in alphabetical order.

For me, thinking like an engineer is about creatively finding a solution to some problem. In my pre-college days, I was very excited about music. I began my musical pursuits by learning the fundamentals of music theory by playing in middle school band and eventually worked my way into different bands in high school (orchestra, marching and, jazz band) and branching off into teaching myself how to play guitar. I love playing and listening to music because it gives me an outlet to create and discover art. I pursued engineering for the same reason; as an engineer, you work in a field that creates or improves designs or processes. For me, thinking like an engineer is exactly like thinking like a musician—through my fundamentals, I'm able to be creative, yet methodical, in my solutions to problems.

D. Bowman, Computer Engineer

Thinking like an engineer is about solving problems with whatever resources are most available—or fixing something that has broken with materials that are just lying around. Sometimes, it's about thinking ahead and realizing what's going to happen before something breaks or someone gets hurt—particularly in thinking about what it means to fail safe—to design how something will fail when it fails. Thinking like an engineer is figuring out how to communicate technical issues in a way that anyone can understand. It's about developing an instinct to protect the public trust—an integrity that emerges automatically.

M. Ohland, Civil Engineer

To me, understanding the way things work is the foundation on which all engineering is based. Although most engineers focus on technical topics related to their specific discipline, this understanding is not restricted to any specific field, but applies to everything! One never knows when some seemingly random bit of knowledge, or some pattern discerned in a completely disparate field of inquiry, may prove critical in solving an engineering problem. Whether the field of investigation is Fourier analysis, orbital mechanics, Hebert boxes, personality types, the Chinese language, the life cycle of mycetozoans, or the evolution of the music of Western civilization, the more you understand about things, the more effective an engineer you can be. Thus, for me, thinking like an engineer is intimately, inextricably, and inexorably intertwined with the Quest for Knowledge. Besides, the world is a truly fascinating place if one bothers to take the time to investigate it.

W. Park, Electrical Engineer

Engineering is a bit like the game of golf. No two shots are ever exactly the same. In engineering, no two problems or designs are ever exactly the same. To be successful, engineers need a bag of clubs (math, chemistry, physics, English, social studies) and then need to have the training to be able to select the right combination of clubs to move from the tee to the green and make a par (or if we are lucky, a birdie). In short, engineers need to be taught to THINK.

B. Sill, Aerospace Engineer

I like to refer to engineering as the color grey. Many students enter engineering because they are "good at math and science." I like to refer to these disciplines as black and white—there is one way to integrate an equation and one way to balance a chemical reaction. Engineering is grey, a blend of math and science that does not necessarily have one clear answer. The answer can change depending on the criteria of the problem. Thinking like an engineer is about training your mind to conduct the methodical process of problem solving. It is examining a problem from many different angles, considering the good, the bad and the ugly in every process or product. It is thinking creatively to discover ways of solving problems, or preventing issues from becoming problems. It's about finding a solution in the grey and presenting it in black and white.

E. Stephan, Chemical Engineer

Lead author note: When writing this preface, I asked each of my co-authors to answer this question. As usual, I got a wide variety of interpretations and answers. This is typical of the way we approach everything we do, except that I usually try and mesh the responses into one voice. In this instance, I let each response remain unique. As you progress throughout this text, you will (hopefully) see glimpses of each of us interwoven with the one voice. We hope that through our uniqueness, we can each reach a different group of students and present a balanced approach to problem solving, and, hopefully, every student can identify with at least one of us.

—Beth Stephan
Clemson University
Clemson, SC

ACKNOWLEDGMENTS

When we set out to formalize our instructional work, we wanted to portray engineering as a reality, not the typical flashy fantasy portrayed by most media forums. We called on many of our professional and personal relationships to help us present engineering in everyday terms. During a lecture to our freshman, Dr. Ed Sutt [PopSci's 2006 Inventor of the Year for the HurriQuake Nail] gave the following advice: *A good engineer can reach an answer in two calls: the first, to find out who the expert is; the second, to talk to the expert.* Realizing we are not experts, we have called on many folks to contribute articles. To our experts who contributed articles for this text, we thank: Dr. Lisa Benson, Dr. Neil Burton, Jan Comfort, Jessica (Pelfrey) Creel, Solange Dao, Jason Huggins, Leidy Klotz, and Troy Nunmaker.

To Dr. Lisa Benson, thank you for allowing us to use "Science as Art" for the basis of many photos that we have chosen for this text. To explain "Science as Art": *Sometimes, science and art meet in the middle. For example, when a visual representation of science or technology has an unexpected aesthetic appeal, it becomes a connection for scientists, artists and the general public. In celebration of this connection, Clemson University faculty and students are challenged to share powerful and inspiring visual images produced in laboratories and workspaces for the "Science as Art" exhibit.* For more information, please visit www.scienceasart.org. To the creators of the art, thank you for letting us showcase your work in this text: Martin Beagley, Dr. Caye Drapcho, Eric Fenimore, Dr. Scott Husson, Dr. Jaishankar Kutty, Dr. Kathleen Richardson, and Dr. Ken Webb. A special thanks to Kautex Machines and Chuck Flammer, and to Russ Werneth for getting us the great Hubble teamwork photo.

To the Rutland Institute for Ethics at Clemson University: The four-step procedure outlined in Chapter 2 on Ethics is based on the toolbox approach presented in the Ethics Across the Curriculum Seminar. Our thanks to Dr. Daniel Wueste, Director, and the other Rutlanders (Kelly Smith, Stephen Satris and Charlie Starkey) for their input into this chapter.

To Jonathan Feinberg and all the contributors to the Wordle (http://www.wordle.net) project, thank you for the tools to create for the Wordle images in the introduction sections. We hope our readers enjoy this unique way of presenting information, and are inspired to create their own Wordle!

To our friends and former students who contributed their Wise Words: Tyler Andrews, Corey Balon, Ed Basta, Sergey Belous, Brittany Brubaker, Tim Burns, Ashley Childers, Jeremy Comardelle, Matt Cuica, Jeff Dabling, Christina Darling, Ed D'Avignon, Brian Dieringer, Lauren Edwards, Andrew Flowerday, Stacey Forkner, Victor Gallas Cervo, Lisa Gascoigne, Khadijah Glast, Tad Hardy, Colleen Hill, Tom Hill, Becky Holcomb, Beth Holloway, Selden Houghton, Allison Hu, Ryan Izard, Lindy Johnson, Darryl Jones, Maria Koon, Rob Kriener, Jim Kronberg, Rachel Lanoie, Mai Lauer, Jack Meena, Alan Passman, Mike Peterson, Candace Pringle, Derek Rollend,

Eric Roper, Jake Sadie, Janna Sandel, Ellen Styles, Adam Thompson, Kaycie (Smith) Timmons, Devin Walford, Russ Werneth, and Aynsley Zollinger.

To our fellow faculty members, for providing inspiration, ideas, and helping us find countless mistakes: Dr. Lisa Benson, Dr. Steve Brandon, Dr. Ashley Childers, Dr. Jonathan Maier, John Minor, and Dr. Julie Trenor. You guys are the other half of this team that makes this the best place on earth to work! We could not have done this without you.

To the staff of the GE program, we thank you for your support of us and our students: Kelli Blankenship, Lib Crockett, Chris Porter, and all of our terrific advising staff both past and present. To Chuck Heck, Linda Nilson, Barbara Weaver, and the rest of the Teaching with Technology group for helping us discover better ways to teach, spark our innovation, and create such wonderful classroom environments. To the administration at Clemson, we thank you for your continued support of our program: Associate Dean Dr. Randy Collins, Interim Director Dr. Don Beasley, Dean Dr. Esin Gulari, Provost Dr. Dori Helms, Former Dean Dr. Tom Keinath, and Former Dean Dr. Steve Melshimer. Special thanks to President Jim Barker for his inspirational leadership of staying the course and giving meaning to "One Clemson."

To the thousands of students who used this text in various forms over the years—thanks for your patience, your suggestions, and your criticism. You have each contributed not only to the book, but to our personal inspirations to keep doing what we do.

To all the reviewers who provided such valuable feedback to help us improve. We appreciate the time and energy needed to review this material, and your thoughtful comments have helped push us to become better.

To the great folks at Prentice Hall—this project would not be a reality without all your hard work. To Eric Hakanson, without that chance meeting this project would not have begun! Thanks to Holly Stark for her belief in this project and in us! Thanks to Scott Disanno for keeping us on track and having such a great vision to display our hard work. You make us look great! Thanks to Dan Sandin and his team for the work on the electronic text. Thanks to Tim Galligan and the fabulous Pearson sales team all over the country for promoting our book to other schools and helping us allow so many students to start "Thinking Like Engineers"! A special thanks to Clare Romeo for editing the text—both editions—about a zillion times and putting up with all of our seemingly endless questions! Your patience with us for the second edition has been amazing. We would not have made it through this without all of the Pearson team efforts and encouragement!

FINALLY, ON A PERSONAL NOTE

DRB: Thanks to my parents and sister for supporting my creative endeavors with nothing but encouragement and enthusiasm. To my grandparents, who value science, engineering, and education to be the most important fields of study. To my co-authors, who continue to teach me to think like an engineer. To Dana, you are the glue that keeps me from falling to pieces. Thank you for your support, love, laughter, inspiration, and determination, among many other things. You are entirely too rad. I love you.

MWO: My wife Emily has my love, admiration, and gratitude for all she does, including holding the family together. For my children, who share me with my students—Charlotte, whose "old soul" touches all who take the time to know her; Carson, who is quietly inspiring; and Anders, whose love of life and people endears him to all. I acknowledge my father Theodor, who inspired me to be an educator; my mother Nancy, who helped me understand people; my sister Karen, who lit a pathway in engineering; my brother Erik, who showed me that one doesn't need to be loud to be a leader; and my mother-in-law Nancy Winfrey, who shared the wisdom of a long career. I recognize those who helped me create an engineering education career path: Fred Orthlieb, Civil and Coastal Engineering at the University of Florida, Marc Hoit, Duane Ellifritt, Cliff Hays, Mary Grace Kantowski, and John Lybas, the NSF's SUCCEED Coalition, Tim Anderson, Clemson's College of Engineering and Science and General Engineering, Steve Melsheimer, Ben Sill, and Purdue's School of Engineering Education.

WJP: Choosing only a few folks to include in an acknowledgment is a seriously difficult task, but I have managed to reduce it to five. First, Beth Stephan has been the guiding force behind this project, without whom it would never have come to fruition. In addition, she has shown amazing patience in putting up with my shenanigans and my weird perspectives. Next, although we exist in totally different realities, my parents have always supported me, particularly when I was a newly married, destitute graduate student fresh off the farm. Third, my son Isaac, who has the admirable quality of being willing to confront me with the truth when I am behaving badly, and for this I am grateful. Finally, and certainly most importantly, to Lila, my partner of more than one-third century, I owe a debt beyond anything I could put into words. Although life with her has seldom been easy, her influence has made me a dramatically better person.

BLS: To my amazing family, who always picked up the slack when I was off doing "creative" things, goes all my gratitude. To Anna and Allison, you are wonderful daughters who both endured and "experienced" the development of many "in class, hands on" activities—know that I love you and thank you. To Lois who has always been there with her support and without whining for over 40 years, all my love. Finally, to my co-authors who have tolerated my eccentricities and occasional tardiness with only minimum grumbling, you make great teammates.

EAS: To my co-authors, for tolerating all my strange demands, my sleep-deprived ravings and the occasional "I need this now" hysteria—and it has gotten worse with the second edition—you guys are the best! I am beginning to understand that it does take a village, so to all those "villagers" who care for my family while this project consumes my every waking minute, I am eternally indebted. To my mom, Kay and Denny—thanks for your love and support. To Khadijah, wishes for you to continue to conquer the world! To Brock and Katie, I love you both a bushel and a peck. You are the best kids in the world, and the older you get the more you inspire me to be great at my job. Finally, to Sean . . . for all the times you held it all together, kept things running, and brought me a Diet Coke—I love you more than I can say—and know that even when I forget to say it, I still believe in us. I know this round has been rougher than the first . . . but I think we are finally finding our way. Thanks for continuing to remind me of what really matters most. "Show a little faith, there's magic in the night . . ."

Part 1

ENGINEERING ESSENTIALS

LEARNING OBJECTIVES

The overall learning objectives for this unit include the following:

- Communicate technical information effectively by composing clear and concise oral presentations and written descriptions of experiments and projects.
- Conduct research on ethical issues related to engineering; formulate and justify positions on these issues.
- Demonstrate an ability to design a system, component, or process to meet desired needs.
- Demonstrate an ability to function on multidisciplinary teams.
- Identify process variability and measurement uncertainty associated with an experimental procedure, and interpret the validity of experimental results.
- Use "practical" skills, such as visualizing common units and conducting simple measurements, calculations, and comparisons to make estimations.

As the reader of this text, you are no doubt in a situation where you have an idea you want to be an engineer. Someone or something put into your head this crazy notion—that you might have a happy and successful life working in the engineering profession. Maybe you are good at math or science, or you want a job where creativity is as important as technical skill. Maybe someone you admire works as an engineer. Maybe you are looking for a career that will challenge you intellectually, or maybe you like to solve problems.

You may recognize yourself in one of these statements from practicing engineers on why they chose to pursue an engineering degree.

"The National Academy of Engineering (NAE) is an independent, non-profit institution that serves as an adviser to government and the public on issues in engineering and technology. Its members consist of the nation's premier engineers, who are elected by their peers for their distinguished achievements. Established in 1964, NAE operates under the congressional charter granted to the National Academy of Sciences."
http://www.nae.edu/About.aspx

I chose to pursue engineering because I enjoyed math and science in school, and always had a love for tinkering with electronic and mechanical gadgets since I was old enough to hold a screwdriver.

S. Houghton, Computer Engineer

I chose to pursue engineering because I always excelled in science and math and I really enjoy problem solving. I like doing hands-on activities and working on "tangible" projects.

M. Koon, Mechanical Engineer

I wanted to pursue engineering to make some kind of positive and (hopefully) enduring mark on the world.

J. Kronberg, Electrical Engineer

I was good at science and math, and I loved the environment; I didn't realize how much I liked stream and ground water movement until I look at BioSystems Engineering.

C. Darling, Biosystem

My parents instilled a responsibility to our community in us kids. As an engineer, I can serve my community through efficient and responsible construction while still satisfying my need to solve challenging problems.

J. Meena, Civil Engineer

I asked many different majors one common question: "What can I do with this degree?" The engineering department was the only one that could specifically answer my question. The other departments often had broad answers that did not satisfy my need for a secure job upon graduating.

L. Johnson, Civil Engineer

I am a first-generation college student and I wanted to have a strong foundation when I graduated from college.

C. Pringle, Industrial Engineer

Engineering is a highly regarded and often highly compensated profession that many skilled high-school students choose to enter for the challenge, engagement, and ultimately the reward of joining the ranks of the esteemed engineers of the world. But what, exactly, does an engineer do? This is one of the most difficult questions to answer because of the breadth and depth of the engineering field. So, how do the experts define engineering?

The National Academy of Engineering (NAE) says:

"Engineering has been defined in many ways. It is often referred to as the "application of science" because engineers take abstract ideas and build tangible products from them. Another definition is "design under constraint," because to "engineer" a product means to construct it in such a way that it will do exactly what you want it to, without any unexpected consequences."

According to the Merriam-Webster online dictionary:

Engineering is the application of science and mathematics by which the properties of matter and the sources of energy in nature are made useful to people.

More or less, engineering is a broad, hard-to-define field requiring knowledge of science and mathematics and other fields to turn ideas into reality. The ideas and problems posed to engineers often do not require a mastery-level knowledge of any particular scientific field, but instead require the ability to put together all of the pieces learned in those fields.

Because engineers solve real-life problems, their ultimate motivation is to work toward making life better for everyone. In "The Heroic Engineer" (*Journal of Engineering Education*, January 1997) by Taft H. Broome (Howard University), and Jeff Peirce (Duke University), those authors claimed:

Engineers who would deem it their professional responsibility to transcend self-interests to help non-experts advance their own interests may well prove indispensable to free societies in the twenty-first century.

Broome and Peirce go on to explain that the traits and behaviors of engineers can be compared to those of a hero. The motivation of any hero is to save someone's life; engineers create products, devices, and methods to help save lives. Heroes intervene to protect from danger; engineers devise procedures, create machines, and improve processes to protect people and the planet from danger. While learning an engineering discipline can be challenging, the everyday engineer does not see it as an obstacle: it is merely an opportunity to be a hero.

Scattered throughout this text, you will find quotes from practicing engineers. As a good engineering team would, we recognize we (the authors) are not experts at all things, and request input and advice when needed. We asked engineers we know who work at "everyday engineering" jobs to reflect on the choices they made in school and during their careers. We hope you benefit from their collective knowledge. When asked for advice to give to an incoming freshman, one gave the following reply, summing up this section better than we ever could have imagined.

[A career in engineering] is rewarding both financially and personally. It's nice to go to work and see some new piece of technology—to be on the cutting edge. It's also a great feeling to know that you are helping improve the lives of other people. Wherever there has been a great discovery, an engineer is to thank. That engineer can be you.

A. Thompson, Electrical Engineer

ENGINEERING IS AN . . . ITCH!

Contributed by: Dr. Lisa Benson, Assistant Professor of Engineering and Science Education, Clemson University

There are a lot of reasons why you are majoring in engineering. Maybe your goal is to impress someone, like your parents, or to defy all those who said you would never make it, or simply to prove to yourself that you have it in you. Maybe your goal is to work with your hands as well as your mind. Maybe you have no idea why you are here, but you know you like cars. There are about as many goals as there are students, and they serve to motivate students to learn. Some goals are better motivators than others.

Lots of experts have studied goals and how they affect what students do in school. Not surprisingly, there are as many ideas and theories about goals as there are experts. But most experts agree on the idea that there are mastery goals and performance goals.

*Students who are **mastery oriented** try to do things well because they want to do their best. They are not driven by external factors like grades or praise, but instead they seek to learn things because they want to really understand them and not just get the correct answer.*

*Students who are **performance oriented** seek to earn good grades to reflect how hard they've worked. They study because they know it will get them something—a scholarship, an above-average grade, or praise from their parents. Since grades are tied to their sense of achievement, students who are performance oriented tend to feel discouraged and anxious when they earn low grades. They tend to want to memorize and pattern-match to solve problems, rather than learn the underlying concepts and methods.*

Most students have been performance oriented throughout high school. In college, you will be more successful if you start thinking in terms of mastery. If you seek to really understand what you are learning in your classes, performance (i.e., good grades) will follow. But performance (a grade, an award, or praise) is not everything, and it will not be enough to keep you motivated when projects and coursework are challenging.

There is nothing like the feeling when you finally understand something that you did not get before. Sometimes it is an "aha!" moment, and sometimes it is a gradual dawning. The feeling is like an itch you can't scratch—you will want to keep at it once you get it. *When you are motivated to understand and master something, you're taking pride in your achievement of conquering the material, not just getting a good grade on an exam.* And you are going to keep scratching that itch. Keep "scratching" at the material—working, practicing, drilling skills if you need to, whatever it takes—to master that . . . itch!

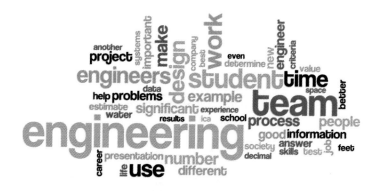

WISE WORDS: WHAT WAS THE HARDEST ADJUSTMENT FROM HIGH SCHOOL TO COLLEGE?

The biggest adjustment was the overwhelming amount of responsibility that I had to take on. There was no longer anybody there to tell me what to do or when to do it. I had to rely on myself to get everything done. All the things I took for granted when I was at home—not having to do my own laundry, not preparing all of my meals, not having to rely on my alarm clock to wake me up, etc.—quickly became quite apparent to me after coming to college. I had to start managing my time better so that I would have time to get all of those things done.

T. Andrews, CE

For me, the most difficult adjustment from high school to college has been unlearning some of the study habits adopted early on. In high school, you can easily get by one semester at a time and just forget what you "learned" when you move into a new semester or a new chapter of the text. College is just a little bit different. To succeed, you have to really make an effort to keep up with your studies—even the classes you have finished already. If you do not, chances are that a topic mentioned in a prerequisite course is going to reappear in a later class, which requires mastery of the previous material in order to excel.

R. Izard, CpE

The hardest adjustment was learning how to study. I could no longer feel prepared for tests by simply paying attention in class. I had to learn to form study groups and begin studying for tests well in advance. You can't cram for engineering tests.

M. Koon, ME

The hardest adjustment was taking full personal responsibility for everything from school work, to social life, and to finances. Life becomes a lot more focused when you realize that you are paying for your education and that your decisions will greatly impact your future. The key is to manage your time between classes, studying, having fun, and sleeping.

S. Belous, CpE

Studying, networking, talking to my professors about my strengths and weaknesses, taking responsibility for my actions, just the whole growing up into an adult was tough.

C. Pringle, IE

The hardest adjustment I had to make going from high school to college was realizing that I was on my own—and not just for academics, either. I was responsible for making sure I remembered to eat dinner, for not eating candy bars for lunch everyday, for balancing my social life with my studies, for managing my money . . . for everything.

J. Sandel, ME

The hardest adjustment from high school to college was changing my study habits. In high school, teachers coordinated their tests so we wouldn't have several on the same day or even in the same week. I had to learn how to manage my time more efficiently. Moreover, it was difficult to find a balance between both the social and academic aspects of college.

D. Walford, BioE

Since the tests cover more material and have more weight in college, I had to alter my study habits to make myself start studying more than a day in advance. It was overwhelming my first semester because there was always something that I could be studying for or working on.

A. Zollinger, CE

CHAPTER 1
EVERYDAY ENGINEERING

Most students who start off in a technical major know very little about their chosen field. This is particularly true in engineering, which is not generally present in the high-school curriculum. Students commonly choose engineering and science majors because someone suggested them. In this section, we help you ask the right questions about your interests, skills, and abilities; we then show you how to combine the answers with what you learn about engineering and science in order to make the right career decision.

1.1 CHOOSING A CAREER

In today's society, the careers available to you upon your graduation are numerous and diverse. It is often difficult as a young adult to determine exactly what occupation you want to work at for the rest of your life because you have so many options. As you move through the process, there are questions that are appropriate to ask. You cannot make a good decision without accurate information. No one can (or should) make the decision for you: not your relatives, professors, advisors, or friends. Only you know what feels right and what does not. You may not know all the answers to your questions right away. That means you will have to get them by gathering more information from outside resources and through your personal experience. Keep in mind that choosing your major and ulti-mately your career is a process. You constantly evaluate and reevaluate what you learn and experience. A key component is whether you feel challenged or overwhelmed. True success in a profession is not measured in monetary terms; it is measured in job satisfac-tion . . . enjoying what you do, doing what you enjoy. As you find the answers, you can choose a major that leads you into a successful career path that you enjoy.

Before you decide, answer the following questions about your tentative major choice. Start thinking about the questions you cannot answer and look for ways or resources to get the information you need. It may take a long time before you know, and that is okay!

- What do I already know about this major?
- What courses will I take to earn a degree in this major?
- Do I have the appropriate academic preparation to complete this major? If not, what will I have to do to acquire it?
- Am I enjoying my courses? Do I feel challenged or stressed?
- What time demands are involved? Am I willing to spend the time it takes to complete this major?
- What kinds of jobs will this major prepare me for? Which sounds most interesting?
- What kinds of skills will I need to do the job I want? Where can I get them?

This process will take time. Once you have the information, you can make a choice. Keep in mind, nothing is set in stone—you can always change your mind!

1.2 CHOOSING ENGINEERING AS A CAREER

In the previous section, we gave several examples of why practicing engineers wanted to pursue a career in engineering. Here are a few more:

- I was always into tinkering with things and I enjoyed working with computers from a young age. Math, science, and physics came very natural to me in high school. For me it was an easy choice.

    J. Comardelle, Computer Engineer

- My initial instinct for a career path was to become an engineer. I was the son of a mechanical engineer, performed well in science and mathematics during primary education, and was always "tinkering" with mechanical assemblies.

    M. Ciuca, Mechanical Engineer

- I chose engineering for a lot of the same reasons that the "typical" entering freshman does—I was good at math and science. I definitely did not know that there were so many types of engineering and to be honest, was a little overwhelmed by the decision I needed to make of what type of engineering was for me.

    L. Edwards, Civil Engineer

- I wasn't really sure what I wanted to do. My parents were not college graduates so there was not a lot of guidance from them, so my high school teachers influenced me a lot. I was taking advanced math and science classes and doing well in them. They suggested that I look into engineering, and I did.

    S. Forkner, Chemical Engineer

- I was a night time/part time student while I worked full time as a metallurgical technician. I was proficient in math and science and fortunate to have a mentor who stressed the need for a bachelor's degree.

    E. Basta, Materials Engineer

- Coming into college, I knew I wanted to pursue a career in medicine after graduation. I also knew that I did not want to major in chemistry, biology, etc. Therefore, bioengineering was a perfect fit. It provides a challenging curriculum while preparing me for medical school at the same time. In addition, if pursuing a career in medicine does not go according to plan, I know that I will also enjoy a career as a bioengineer.

    D. Walford, BioE

Table 1-1 describes the authors' perspective on how various engineering and science disciplines might contribute to different industries or innovations. This table is only an interpretation by a few engineers and does not handle every single possibility of how an engineer might contribute toward innovation. For example, an industrial engineer might be called into work on an energy to share a different perspective on energy efficiency. The broad goal of any engineering discipline is to solve problems, so there is often a need for a different perspective to possibly shed new light toward an innovative solution.

Table 1-1 Sample career paths and possible majors. Shaded boxes indicate a good starting point for further exploration

Careers	Aerospace	Biomedical	BioSystems	Civil	Chemical	Materials	Electric/computer	Environmental	Industrial	Mechanical	Chemistry	Computer Science	Geology	Mathematics	Physics
GENERAL															
Energy industry				■		■	■			■	■		■		■
Machines		■					■		■	■					
Manufacturing					■	■			■	■					
Materials	■	■		■	■					■	■				
Structures	■			■						■			■		
Technical sales		■	■		■	■						■			■
SPECIFIC															
Rocket/airplane	■									■					
Coastal engineering			■	■				■					■		
Computing							■					■		■	
Cryptography							■					■		■	
Defense	■			■		■	■			■	■				■
Environment					■			■					■		■
Fiber optics						■							■		
Forensics					■						■				
Groundwater				■				■					■		
Healthcare		■							■		■				
Human factors	■								■			■			
Industrial sensors					■		■								
Intelligent systems							■					■		■	
Management	■	■							■						
Operations research									■			■		■	
Outdoor work			■	■				■					■		
Pharmaceutical		■			■						■				
Plastics					■	■				■	■				
Robotics	■					■	■					■			
Semiconductors	■					■	■								■
Telecommunications							■					■			
Transportation				■				■	■	■			■		
Waste management			■	■	■			■							

MORE WISE WORDS: HOW DID YOU CHOOSE A MAJOR IN COLLEGE?

Since I knew I wanted to design computers, I had a choice between electrical and computer engineering. I chose computer engineering, so I could learn about both the hardware and software. It was my interests in computers and my high school teachers that were the biggest influence in my decision.

E. D'Avignon, CpE

My first choice in majors was Mechanical Engineering. I changed majors after taking a drafting class in which I did well enough to get a job teaching the lab portion, but I did not enjoy the work. After changing to Electrical and Computer Engineering, I took a Statics and Dynamics course as part of my required coursework and that further confirmed my move as I struggled with that material.

A. Flowerday, EE

Some people come into college, knowing exactly what they want their major and career to be. I, on the other hand, was not one of those people. I realized that I had a wide spectrum of interests, and college allows you to explore all those options. I wanted a major that was innovative and would literally change the future of how we live. After looking through what I loved and wanted to do, my choice was Computer Engineering.

S. Belous, CpE

1.3   NAE GRAND CHALLENGES FOR ENGINEERING

History (and prehistory) is replete with examples of technological innovations that forever changed the course of human society: the mastery of fire, the development of agriculture, the wheel, metallurgy, mathematics of many flavors, the printing press, the harnessing of electricity, powered flight, nuclear power, and many others. The NAE has established a list of 14 challenges for the twenty-first century, each of which has the potential to transform the way we live, work, and play. Your interest in one or more of the Grand Challenges for Engineering may help you select your engineering major. For more information, visit the NAE website at http://www.engineeringchallenges.org/. In case this address changes after we go to press, you can also type "NAE Grand Challenges for Engineering" into your favorite search engine.

A burgeoning planetary population and the technological advances of the last century are exacerbating many current problems as well as engendering a variety of new ones, for example:

- Relatively inexpensive and rapid global travel make it possible for diseases to quickly span the globe whereas a century ago, they could spread, but much more slowly.
- The reliance of the developed world on computers and the Internet makes the fabric of commerce and government vulnerable to cyber terrorism.
- Increased demand for limited resources not only drives up prices for those commodities, but also fosters strain among the nations competing for them.

These same factors can also be a force for positive change in the world.

- Relatively inexpensive and rapid global travel allows even people of modest means to experience different cultures and hopefully promote a more tolerant attitude toward those who live by different sets of social norms.

- Modern communications systems—cell phones, the Internet, etc.—make it essentially impossible for a government to control the flow of information to isolate the members of a population or to isolate that population from the political realities in other parts of the world. An excellent example was the rapid spread of rebellion in the Middle East and Africa in early 2011 against autocratic leaders who had been in power for decades.
- Increased demand for, and rising prices of limited resources is driving increased innovation in alternatives, particularly in meeting the world's energy needs.

As should be obvious from these few examples, technology not only solves problems, but also creates them. A significant portion of the difficulty in the challenges put forth by the NAE to solve critical problems in the world lies in finding solutions that do not create other problems. Let us consider a couple of the stated challenges in a little more detail. You probably already have some familiarity with several of them, such as "make solar energy economical," "provide energy from fusion," "secure cyberspace," and "enhance virtual reality," so we will begin with one of the NAE Grand Challenges for Engineering that is perhaps less well known.

The Nitrogen Cycle

Nitrogen is an element required for all known forms of life, being part of every one of the 20 amino acids that are combined in various ways to form proteins, all five bases used to construct RNA and DNA, and numerous other common biological molecules such as chlorophyll and hemoglobin. Fortunately, the supply of nitrogen is—for all practical purposes—inexhaustible, constituting over 75% of the Earth's atmosphere. However, nitrogen is mostly in the molecular form N_2, which is chemically unavailable for uptake in biological systems since the two nitrogen atoms are held together by a very strong triple bond.

For atmospheric nitrogen to be available to biological organisms, it must be converted, or "fixed," by the addition of hydrogen, into ammonia, NH_3, that may then be used directly or converted by other microorganisms into other reactive nitrogen compounds for uptake by microorganisms and plants. The term nitrogen fixation includes conversion of N_2 into both ammonia and these other reactive compounds, such as the many oxides of nitrogen. Eventually the cycle is completed when these more readily available forms of nitrogen are converted back to N_2 by microorganisms, a process called denitrification.

Prior to the development of human technology, essentially all nitrogen fixation was performed by bacteria possessing an enzyme capable of splitting N_2 and adding hydrogen to form ammonia, although small amounts of fixed nitrogen are produced by lightning and other high-energy processes. In the early twentieth century, a process called the Haber-Bosch process was developed that would allow conversion of atmospheric nitrogen into ammonia and related compounds on an industrial scale. Today, slightly more than a century later, approximately one-third of all fixed nitrogen is produced using this process.

The ready availability of relatively inexpensive nitrogen fertilizers has revolutionized agriculture, allowing people to increase yields dramatically and to grow crops on previously unproductive lands. However, the widespread use of synthetic nitrogen has caused numerous problems, including water pollution, air pollution, numerous human health problems, and disruption of marine and terrestrial ecosystems to the extent that entire populations of some organisms have died off.

Deliberate nitrogen fixation is only one part of the nitrogen cycle problem, however. Many human activities, especially those involving the combustion of fossil fuels, pump huge quantities of various nitrogen compounds into the atmosphere. Nitrous oxide (N_2O), also of some notoriety as the dissociative anesthetic commonly known as "laughing gas," is particularly problematic since it is about 200 times more effective than carbon dioxide as a greenhouse gas, and persists in the atmosphere for over a century.

Altogether, human-caused conversion of nitrogen into more reactive forms now accounts for about half of all nitrogen fixation, meaning that there is twice as much nitrogen fixed today than there was a little more than a century ago. However, we have done little to augment the natural denitrification process, so the deleterious effects of excessive fixed nitrogen continue to increase. We have overwhelmed the natural nitrogen cycle. If we are to continue along this path, we must learn to manage the use of these products more efficiently and plan strategies for denitrification to bring the cycle back into balance.

Reverse-Engineering the Brain

The development of true artificial intelligence (AI) holds possibly the most overall potential for positive change in the human race as well as the most horrendous possible negative effects. This is reflected in science fiction, where the concept of thinking machines is a common plot device, ranging from Isaac Asimov's benevolent R. Daneel Olivaw to the malevolent Skynet in the Terminator movies. If history is any guide, however, the potential for disastrous consequences seldom deters technological advances, so let us consider what is involved in the development of AI.

Although great strides have been made in creating machines that seem to possess "intelligence," almost all such systems that have come to the public notice either rely on brute-force calculations, such as the chess-playing computer, Deep Blue, that defeated world champion Garry Kasparov in 1997, or reliance on incredibly fast access to massive databases, such as the Jeopardy-playing computer, Watson, that defeated both the highest money winner, Brad Rutter, and the record holder for longest winning streak, Ken Jennings, in 2011. Perhaps needless to say, these are oversimplifications, and there are many more aspects to both of these systems. However, one would be hard-pressed to argue that these computers are truly intelligent—that they are self-aware and contain the unexplainable spark of creativity, which is the hallmark of humans, and arguably other highly intelligent creatures on Earth.

Today's robots perform many routine tasks, from welding and painting vehicles to vacuuming our homes and cutting our grass. However, all of these systems are programmed to perform within certain restrictions and have serious limitations when confronted with unexpected situations. For example, if your school utilized vacuuming robots to clean the floors in the classrooms, it would probably be unable to handle the situation effectively if someone became nauseous and regurgitated on the carpet. If we could endow such robots with more human-like intelligence, the range of tasks that they could successfully accomplish would increase by orders of magnitude, thus increasing their utility tremendously.

To date, we have almost exclusively attempted merely to construct intelligent systems that mimic behavior and thought, not design systems that actually store and process information in a manner analogous to that of a biologically based computer (a brain). The human brain utilizes a network of interconnections between specialized subsections that makes even the most advanced computers look like a set of children's building blocks. Although some understanding has been gained, the means of encoding information and its transfer in the brain is almost completely a mystery.

Gaining even a basic understanding of brain function might allow us to develop prosthetic limbs that actually function as well as the originals, restore sight to the blind, repair brain damage, or even enhance human intelligence.

1.4 CHOOSING A SPECIFIC ENGINEERING FIELD

The following paragraphs briefly introduce several different types of engineering majors. By no means is this list completely inclusive.

Bioengineering or Biomedical Engineering

Bioengineering (BioE) and biomedical engineering (BME) apply engineering principles to the understanding and solution of medical problems. Bioengineers are involved in research and development in all areas of medicine, from investigating the physiological behavior of single cells to designing implants for the replacement of diseased or traumatized body tissues. Bioengineers design new instruments, devices, and software, assemble knowledge from many scientific sources to develop new procedures, and conduct research to solve medical problems.

Typical bioengineers work in such areas as artificial organs, automated patient monitoring, blood chemistry sensors, advanced therapeutic and surgical devices, clinical laboratory design, medical imaging systems, biomaterials, and sports medicine.

Bioengineers are employed in universities, industry, hospitals, research facilities, and government. In industry, they may be part of a team serving as a liaison between engineers and clinicians. In hospitals, they select appropriate equipment and supervise equipment performance, testing, and maintenance. In government agencies, they are involved in safety standards and testing.

Biosystems Engineering

Biosystems engineering (BE) is the field of engineering most closely allied with advances in biology. BE emphasizes two main areas: (1) bioprocess engineering, with its basis in microbiology, and (2) ecological engineering, with its basis in ecology. The field focuses on the sustainable production of biorefinery compounds (biofuels, bioactive molecules, and biomaterials) using metabolic pathways found in nature and green processing technologies. Further, BE encompasses the design of sustainable communities utilizing low-impact development strategies (bioretention basins, rainwater harvesting) for stormwater retention and treatment—and ecologically sound food and energy-crop production. Scientific emphasis is shifting toward the biosciences. Biosystems engineers apply engineering design and analysis to biological systems and incorporate fundamental biological principles to engineering designs to achieve ecological balance.

Here are some activities of biosystems engineers:

- Design bioprocesses and systems for biofuels (biodiesel, hydrogen, ethanol), biopharmaceutical, bioplastics, and food processing industries
- Develop ecological designs (permeable pavement, bioswales, green infrastructure) to integrate water management into the landscape
- Integrate biological sustainability concepts into energy, water, and food systems
- Provide engineering expertise for agriculture, food processing, and manufacturing
- Pursue medical or veterinary school or graduate school in the fields of BE, BME, or ecological engineering

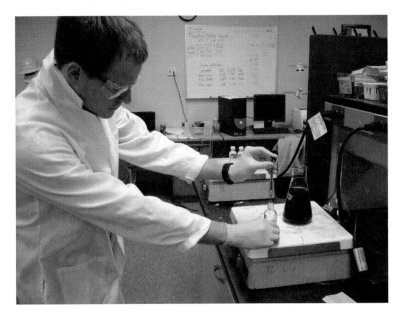

My research is part of a Water Research Foundation project, which is investigating the formation of emerging disinfection byproducts (DBPs) in drinking water treatment.

DBPs are undesirable, toxic compounds that are formed when water is chlorinated. I am investigating the effects of pH, bromide and iodide concentrations, and preoxidants on the formation of a specific family of DBPs.

D. Jones, BE

I am a project manager for new product development.

I oversee and coordinate the various activities that need to be completed in order to get a new product approved and manufactured, and ultimately in the hands of our consumers.

S. Forkner, ChE

Chemical Engineering

Chemical engineering (ChE) incorporates strong emphasis on three sciences: chemistry, physics, and mathematics. Chemical engineers are involved in the research and development, manufacture, sales, and use of chemicals, pharmaceuticals, electronic components, food and consumer goods, petroleum products, synthetic fibers and films, pulp and paper, and many other products. They work on environmental remediation and pollution prevention, as well as in medical and health-related fields. Chemical engineers:

- Conduct research and develop new products
- Develop and design new manufacturing processes
- Earn additional degrees to practice medicine or patent, environmental, or corporate law
- Sell and provide technical support for sophisticated chemical products to customers
- Solve environmental problems; work in biotechnology
- Troubleshoot and solve problems in chemical manufacturing facilities

My team is responsible for implementing the engineered design in the field.

We install, tune, test, and accept into operations all of the electronics that allow customers to use our state of the art fiber optic network to run voice, video, and data for their residential needs.

L. Gascoigne, CE

My current responsibilities include:

- Analysis of traffic signal operations and safety for municipal and private clients;
- Preparation of traffic impact studies;
- Review of plans and traffic studies for municipalities and counties;
- Design of traffic signal installations and traffic signing projects.

C. Hill, CE

Civil Engineering

CE involves the planning, design, construction, maintenance, and operation of facilities and systems to control and improve the environment for modern civilizations. This includes projects of major importance such as bridges, transportation systems, buildings, ports, water distribution systems, and disaster planning.

Here are just a few of many opportunities available for civil engineers:

- Design and analyze structures ranging from small buildings to skyscrapers to offshore oil platforms
- Design dams and building foundations
- Develop new materials for pavements, buildings, and bridges
- Design improved transportation systems
- Design water distribution and removal systems
- Develop new methods to improve safety, reduce cost, speed construction, and reduce environmental impact
- Provide construction and project management services for large engineered projects throughout the world

I develop, manage, and support all software systems. I also deal with system scalability, customer satisfaction, and data management.

J. Comardelle, CpE

I am responsible for assisting in the management of commercial and healthcare projects for Brasfield & Gorrie. Working closely with the owner and architect, I maintain open lines of communication and aim to provide exceptional service to the entire project team from the preconstruction phase of the project through construction. I assist in establishing and monitoring procedures for controlling the cost, schedule, and quality of the work in accordance with the construction contract.

L. Edwards, CE

I am a digital designer and work on the read channel for hard disk drives. The read channel is the portion of the controller SOC (system on a chip) that decodes the analog signal read from the hard disk and converts it to digital data.

I am responsible for writing Verilog RTL code, verification, synthesis into gates, and meeting timing requirements of my blocks.

E. D'Avignon, CpE

Computer Engineering

Computer engineering (CpE) spans the fields of computer science and engineering, giving a balanced view of hardware, software, hardware-software trade-offs, and basic modeling techniques that represent the computing process involving the following technologies:

- Communication system design
- Computer interface design
- Computer networking
- Digital signal processing applications
- Digital system design
- Embedded computer design
- Process instrumentation and control
- Software design

As a Radiation Effects Engineer I test the performance of electronic components in a specific application exposed to different types of radiation. Responsibilities include interfacing with design and system engineers, creating test plans, performing testing and data analysis, and authoring test reports.

A. Passman, EE

I manage global programs that help develop leadership capabilities and skills of our current and future leaders. I am a consultant, a coach, a mentor, and a guide. If leaders are interested in improving how they lead and the impact they have on their employees and on company results, we work with them to identify the best ways for them to continue their development.

A. Hu, EE

Electrical Engineering

Electrical engineering (EE) ranges from the generation and delivery of electrical power to the use of electricity in integrated circuits. The rapid development of technology, based on integrated circuit devices, has enabled the pervasive use of computers in command, control, communication, and computer-aided design. Some systems electrical engineers work on include the following:

- Communication system design
- Control systems—from aircraft to automotive
- Electrical power generation and distribution
- Electromagnetic waves
- Integrated circuit design
- Process instrumentation and control
- Robotic systems design
- Telecommunications

My group supports [a major automotive manufacturer's] decisions pertaining to where to put new plants around the world, what products to build in them, and at what volumes.

In particular, my work involves understanding what the other auto manufacturers are planning for the future (footprint, capacity, technology, processes, etc.), so that information can be used to affect decisions about how to compete around the globe.

M. Peterson, EE

I work with scientists and engineers to protect their innovations by writing patent applications describing their inventions and presenting the applications before the United States Patent & Trademark Office. I also assist clients in determining whether another party is infringing their patents and help my clients to avoid infringing other's patents.

M. Lauer, EnvE

Environmental Engineering

Environmental engineering (EnvE) is an interdisciplinary field of engineering that is focused on cleaning up environmental contamination, as well as designing sustainable approaches to prevent future contamination. Environmental engineers apply concepts from basic sciences (including chemistry, biology, mathematics, and physics) to develop engineered solutions to complex environmental problems.

Environmental engineers design, operate, and manage both engineered and natural systems to protect the public from exposure to environmental contamination and to develop a more sustainable use of our natural resources. These activities include the following:

- Production of safe, potable drinking water
- Treatment of wastewater so that it is safe to discharge to surface water or reuse in applications such as landscape irrigation
- Treatment of air pollutants from mobile (e.g., automobiles) and stationary (e.g., power plants) sources
- Characterization and remediation of sites contaminated with hazardous wastes (e.g., polychlorinated biphenyls, or PCBs)
- Disposal of municipal solid wastes
- Management of radioactive wastes, including characterization of how radioactive materials move through the environment and the risks they pose to human health
- Evaluation of methods to minimize or prevent waste production and inefficient use of energy by manufacturing facilities
- Reduce human health risks by tracking contaminants as they move through the environment
- Design a more sustainable future by understanding our use of resources

With a BS degree in EnvE, students will find employment with consulting engineering firms, government agencies involved in environmental protection, and manufacturing industries.

Industrial Engineering

Industrial engineering (IE) deals with the design and improvement of systems, rather than with the objects and artifacts that other engineers design. A second aspect of IE is the involvement of people in these systems—from the people involved in the design and production to the people who are ultimate end users. A common theme is the testing and evaluation of alternatives that may depend on random events. Industrial engineers use mathematical, physical, social sciences, and engineering combined with the analytical and design methods to design, install, and improve complex systems that provide goods and services to our society. Industrial engineers are called upon to:

- Analyze and model complex work processes to evaluate potential system improvements
- Analyze how combinations of people and machines work together
- Analyze how the surroundings affect the worker, and design to reduce the negative effects of this environment
- Develop mathematical and computer models of how systems operate and interact
- Improve production and service processes from the perspectives of quality, productivity, and cost
- Work on teams with other professionals in manufacturing, service industries, government agencies

My primary job responsibilities include maintaining, upgrading, and designing all the computer systems and IT infrastructure for the Vermont Railroad. I handle all the servers and take care of network equipment. When needed, I also program customized applications and websites for customers or our own internal use. I also serve as a spare conductor and locomotive engineer when business needs demand.

S. Houghton, CpE

As the Business Leader for Central Florida at a major power company, I develop and manage a $42 million budget. I ensure that our engineering project schedule and budget match, and report on variances monthly. I also conduct internal audits and coach employees on Sarbanes-Oxley compliance requirements.

R. Holcomb, IE

Currently I am working on a project to determine patient priorities for evacuations from healthcare facilities during emergencies. The assumption of an evacuation is that there will be enough time to transfer all of the patients, but in the event of limited resources, there may not be enough time to move all of the patients to safety. Further—and depending on the emergency type—it may be an increased risk to transport some patient types. Based on certain objectives, we are developing guidelines to most ethically determine a schedule for choosing patients for emergency evacuations.

A. Childers, IE

Material Science and Engineering

Material science and engineering (MSE) focuses on the properties and production of materials. Nature supplied only 92 naturally occurring elements to serve as building blocks to construct all modern conveniences. A materials engineer works to unlock the relationship between atomic, molecular, and larger-scale structures and the resultant properties. This category includes majors such as ceramic engineering, metallurgical engineering, and polymer science and engineering.

Here is a partial list of products designed and manufactured by MSE:

- Brick, tile, and whitewares research and manufacturing for the home and workplace
- Ceramic spark plugs, oxygen sensors, and catalytic converters that optimize engine performance
- Metal and ceramic materials that enable biomedical implants and prosthetics
- Microwave responsive ceramics that stabilize and filter cellular phone reception
- Nanotechnology, including silver nanoparticles used as antibacterial agents in socks and t-shirts and carbon nanotubes used to reinforce the fork of racing bicycles
- Plastics found in bulletproof vests, replacement heart valves, and high tension wires on bridges

- Superconducting metals that are used in medical imaging devices like magnetic resonance imaging (MRI)
- Ultrapure glass optical fibers that carry telephone conversations and Internet communications

As a Metallurgical Engineer, my duties include

- Consulting firm management/administration
- Failure analysis
- Subcontracted metals testing services
- Metallurgical quality systems design/auditing
- Metallurgical expert in litigation cases
- Materials selection and design consultant, in-process and final inspection and testing services

E. Basta,
Metallurgical Engineer

Mechanical Engineering

I implement technology to protect national assets against adversaries.

J. Dabling, ME

Mechanical engineering (ME) involves areas related to machine design, manufacturing, energy production and control, materials, and transportation. Areas supported by mechanical engineers include:

- Construction
- Energy production and control
- Environmental systems
- Food production
- Management
- Materials processing
- Medicine
- Military service
- Propulsion and transportation systems
- Technical sales

In my job as a management consultant, I address CEO-level management decisions as part of a project team by helping clients identify, analyze, and solve business-related problems. My responsibilities include generating hypotheses, gathering, and analyzing data, conducting benchmarking and best practices assessments, recommending actions, and working with clients to develop implementation plans.

M. Ciuca, ME

As Plant Engineering Manager I report directly to the Plant Manager. My primary responsibilities are managing all the capital investments; providing technical support and expertise to the plant leadership team; and mentoring and developing the plant's engineering staff and technical resources.

J. Huggins, ME

I am a salesman, so at the end of the day I'm looking to grow my market share while trying to protect the market share I already have. I help companies maintain a safe, reliable, and efficient steam and condensate system by utilizing the many products and services that we have to offer. This is mostly done by designing and installing upgrades and improving my customer's existing steam systems.

T. Burns, ME

I am an aerospace engineering manager responsible for developing unique astronaut tools and spacewalk procedures and for testing and training for NASA's Hubble Space Telescope servicing missions. My job ranges from tool and procedure design and development to underwater scuba testing to real-time, on-console support of Space Shuttle missions.

R. Werneth, ME

WISE WORDS: WHAT DID YOU DO YESTERDAY MORNING AT WORK?

I worked on completing a failure analysis report for an industrial client.

E. Basta, CME

I reviewed the results of the overnight simulation runs. There were several failures, so I analyzed the failures and devised fixes for the problems.

E. D'Avignon, CpE

On any given day, my morning might be spent this way: in meetings, at my computer (e-mail, drafting documents/reports), making phone calls, talking to other project members, running a test on the manufacturing lines. Not glamorous, but necessary to solve problems and keep the project moving forward.

S. Forkner, ChE

I continued to design a warehouse/office building on a nuclear expansion project.

T. Hill, CE

Yesterday I designed a spreadsheet to assist in more precisely forecasting monthly expenditures.

R. Holcomb, IE

I attended the Plant Morning Meeting and the Boardmill Leadership Team Meeting, followed by the Plant Budget Meeting. In between meetings I returned e-mails and project-related phone calls. Typically I spend about 50% of my time in meetings. I use the information I receive at these meetings to direct and focus the efforts of the engineering staff.

J. Huggins, ME

Testing some failed prototype biostimulators returned by a trial user, to determine why and how they failed and how to prevent it from happening in coming production versions.

J. Kronberg, EE

Yesterday, I worked on a patent infringement opinion involving agricultural seeding implements, a Chinese patent office response for a component placement and inspection machine used in circuit board manufacturing, and a U.S. patent office response for database navigation software.

M. Lauer, EnvE

In my current position, I spend much of my time reading technical manuals and interface control documents. I attended a meeting detailing lightning protection for the Ares rocket.

E. Styles, EE

1.5   GATHERING INFORMATION

You will need to gather a lot of information in order to answer your questions about engineering or any other major. Many resources are available on your campus and through the Internet.

The Career Center

Most universities have a centralized campus career center. The staff specializes in helping students explore various occupations and make decisions. They offer testing and up-to-date information on many career fields. Professional counselors are available by appointment to assist students with job and major selection decisions.

Career Websites

To learn more about engineering and the various engineering fields, you can find a wealth of information from engineering professional societies. Each engineering field has a professional society dedicated to promoting and disseminating knowledge about that particular discipline. Table 1-2 provides a list of most major engineering fields and the professional society in the United States with which it is associated. In some cases, more than one society is connected with different subdisciplines. Other regions of the world may have their own professional societies.

Perusing the various societies' websites can provide you with information invaluable in helping you decide on a future career. We have not given URLs for the societies, since these sometimes change. To find the current address, simply use an online search engine with the name of the society.

In addition, a few engineering societies are not specific to a discipline, but to their membership;

- National Society of Black Engineers (NSBE)
- Society of Hispanic Professional Engineers (SHPE)
- Society of Women Engineers (SWE)
- Tau Beta Pi, The Engineering Honor Society (TBP).

Most engineering schools have student chapters of the relevant organizations on campus. These organizations provide an excellent opportunity for you to learn more about your chosen discipline or the ones you are considering, and they also help you meet other students with similar interests. Student membership fees are usually nominal, and the benefits of membership far outweigh the small cost.

Active participation in these societies while in school not only gives you valuable information and experience, but also helps you begin networking with professionals in your field and enhances your résumé.

Table 1-2 **Website research starting points**

Society	Abbreviation
American Ceramic Society	ACerS
American Institute of Aeronautics and Astronautics	AIAA
American Institute of Chemical Engineers	AIChE
American Nuclear Society	ANS
American Society of Agricultural and Biological Engineers	ASABE
American Society of Civil Engineers	ASCE
American Society for Engineering Education	ASEE
American Society of Mechanical Engineers	ASME
American Society of Metals International	ASM Int'l.
Association for Computing Machinery	ACM
Audio Engineering Society	AES
Biomedical Engineering Society	BMES
Engineers Without Borders	EWB
Institute of Biological Engineering	IBE
Institute of Electrical and Electronics Engineers	IEEE
Institute of Industrial Engineers	IIE
Institute of Transportation Engineers	ITE
Materials Research Society	MRS
National Academy of Engineering	NAE
National Society of Black Engineers	NSBE
National Society of Professional Engineers	NSPE
Society of Automotive Engineers International	SAE Int'l.
Society of Hispanic Professional Engineers	SHPE
Society of Petroleum Engineers	SPE Int'l.
Society of Plastics Engineers	SPE
Society of Women Engineers	SWE
Tau Beta Pi, The Engineering Honor Society	TBP

WISE WORDS: ADVICE ABOUT SOCIETY PARTICIPATION

Get involved! It is so much fun! Plus, you're going to meet a ton of cool people doing it!

T. Andrews, CE

I am a very involved person, and I love it. I definitely recommend participating in professional societies because not only do they look good on a résumé but they also provide you with useful information for your professional life. It also allows you to network with others in your field which can be helpful down road. Also, do something fun as it is a nice stress relief and distraction when life seems to become really busy.

C. Darling, Biosystem

My advice to students willing to participate on student activities is for them to not be shy when going to a student organization for the first couple of times. It takes time to get well known and feel comfortable around new people, but don't let that prevent you from being part of a student organization that can bring many benefits to you. Always have a positive attitude, be humble, and learn to listen to others; these last advices are traits which you will use in your professional life.

V. Gallas Cervo, ME

My advice to first-year students is to not get involved with too many organizations all at once. It is easy to get distracted from your class work with all the activities on campus. Focus on a couple and be a dedicated officer in one of them. This way you have some-thing to talk about when employers see it on your résumé. I would recommend that you are involved with one organization that you enjoy as a hobby and one organization that is a professional organization.

D. Jones, BE

The most important thing to do when joining any group is to make sure you like the people in it. This is probably even more important than anything the group even does. Also make sure that if the group you're joining has a lot of events they expect you to be at, you have the time to be at those events.

R. Kriener, EE

The field of engineering is a collaborative project; therefore, it is important to develop friendships within your major.

S. Belous, CpE

Make the time to participate in student activities. If possible, try to get a leadership position in one of the activities because it will be useful in interviews to talk about your involvement. While employers and grad schools may not be impressed with how you attended meetings occasionally as a general member, they will be interested to hear about the projects that you worked on and the challenges that you faced in a leadership position.

K. Smith, ChE

Get involved! College is about more than just academics. Participating in student activities is a lot of fun and makes your college experience more memorable. I've made so many friends not just at my school, but all over the country by getting involved. It is also a great way to develop leadership and interpersonal skills that will become beneficial in any career.

A. Zollinger, CE

1.6   PURSUING STUDENT OPPORTUNITIES

In addition to the traditional educational experience, many students seek experience outside of the classroom. Many engineering colleges and universities have special depart-ments that help place students in programs to gain real engineering work experience or provide them with a culturally rich study environment. Ask a professor or advisor if your university provides experiences similar to those described in this section.

Cooperative Education

Contributed by: Dr. Neil Burton, Director of Cooperative Education Program, Clemson University

People learn things in many different ways. Some people learn best by reading; others by listening to others; and still others by participating in a group discussion. One very effective form of learning is called **experiential learning**, also referred to as engaged learning in some places. As the name suggests, experiential learning means learning through experience, and there is a very good chance you used this method to learn how to ride a bike, bake a cake, change a flat tire, or perform any other complex process that took some practice to perfect. The basic assumption behind experiential learning is that you learn more by doing than by simply listening or watching.

Becoming a good engineer is a pretty challenging process, so it seems only natural that experiential learning would be especially useful to an engineering student. In 1906, Herman Schneider, the Dean of Engineering at the University of Cincinnati, developed an experiential learning program for engineering students because he felt students would understand the material in their engineering classes much better if they had a chance to put that classroom knowledge into practice in the workplace. Schneider called this program **Cooperative Education**, and more than 100 years later, colleges and universities all over the world offer cooperative (co-op) education assignments to students in just about every major, although engineering students remain the primary focus of most co-op programs.

There are many different kinds of co-op programs, but all of them offer engineering students the chance to tackle real-world projects with the help and guidance of experienced engineers. One common model of cooperative education allows students to alternate semesters of co-op with semesters of school. In this model, students who accept co-op assignments spend a semester working full-time with a company, return

On my co-op, I worked on the hazard analysis for the new Ares I launch vehicle that NASA is designing to replace the Space Shuttles when they retire in 2010.

Basically, we looked at the design and asked: what happens if this part breaks, how likely is it that this part will break, and how can we either make it less likely for the part to break or give it fault tolerance so the system can withstand a failure?

J. Sandel, on co-op

The value of the cooperative education was to apply the classroom material to real-world applications, develop an understanding of the expectations post-grad, and provide the opportunity for a trial run for a future career path in a low-risk environment.

M. Ciuca, ME

I wanted some practical experience, and I wasn't exactly sure what career I wanted to pursue when I graduated. My experience at a co-op set me on a completely different career path than I had been on previously.

K. Smith, ChE

to school for the following academic term, go back out for a second co-op **rotation**, return to school the following term, and continue this pattern until they have spent enough time on assignment to complete the co-op program.

Students learn a lot about engineering during their co-op assignments, but there are many other benefits as well. Engineering is a tough discipline, and a co-op assignment can often help a student determine if he or she is in the right major. It is a lot better to figure out that you do not want to be an engineer before you have to take thermodynamics or heat transfer! Students who participate in co-op also have a chance to develop some great professional contacts, and these contacts are very handy when it comes time to find a permanent job at graduation. The experience students receive while on a co-op assignment is also highly valued by employers who want to know if a student can handle the challenges and responsibilities of a certain position. In fact, many students receive full-time job offers from their co-op employers upon graduation.

Perhaps the most important benefit a co-op assignment can provide is improved performance in the classroom. By putting into practice the theories you learn about in class, you gain a much better understanding of those theories. You may also see something on your co-op assignment that you will cover in class the following semester, putting you a step ahead of everyone else in the class. You will also develop time management skills while on a co-op assignment, and these skills should help you complete your school assignments more efficiently and effectively when you return from your assignment.

Companies that employ engineers often have cooperative education programs because co-op provides a number of benefits to employers as well as students. While the money companies pay co-op students may be double or even triple than what those students would earn from a typical summer job, it is still much less than companies would pay full-time engineers to perform similar work. Many employers also view cooperative education as a recruiting tool—what better way to identify really good employees than to bring aboard promising students and see how they perform on co-op assignments!

I did my research, and it really made sense to pursue a co-op—you get to apply the skills you learn in class, which allows you to retain the information much better, as well as gain an increased understanding of the material.

As for choosing a co-op over an internship, working for a single company for an extended period of time allows students to learn the ropes and then progress to more intellectually challenging projects later in the co-op. And, if you really put forth your best effort for the duration of your co-op, you could very well end up with a job offer before you graduate!

R. Izard, on co-op

WISE WORDS: WHAT DID YOU GAIN FROM YOUR CO-OP OR INTERNSHIP EXPERIENCE?

I was able to learn how to practically apply the knowledge I was gaining from college. Also, the pay allowed me to fund my schooling.

B. Dieringer, ME

The best part about that experience was how well it meshed with my courses at the time. My ability to apply what I was learning in school every day as well as to take skills and techniques I was learning from experienced engineers and use them toward the projects I was working on in school was invaluable.

A. Flowerday, CpE

My internships were a great introduction to the professional work place—the skills and responsibilities that are expected; the relationships and networks that are needed.

S. Forkner, ChE

I decided to pursue a co-op because I had trouble adapting to the school environment in my first years and taking some time off to work at a company seemed a good way to rethink and reorganize myself. I made a good decision taking some time off, since it allowed me to learn a lot more about myself, how I work, how I learn, and how I operate. I learned that the biggest challenges were only in my mind and believing in me was, and still is, the hardest thing.

V. Gallas Cervo, ME

My internship taught me that I could be an engineer—and a good one too. I had a lot of self-doubt before that experience, and I learned that I was better than I thought I was.

B. Holloway, ME

Being able to immediately apply the things I learned in school to real-world applications helped reinforce a lot of the concepts and theories. It also resulted in two job offers after graduation, one of which I accepted.

J. Huggins, ME

This was an amazing opportunity to get a real taste for what I was going to be doing once I graduated. I began to realize all the different types of jobs I could have when I graduated, all working in the same field. In addition, my work experience made my résumé look 100 times more appealing to potential employers. The experience proved that I could be a team player and that I could hit the ground running without excessive training.

L. Johnson, CE

It gave me a chance to see how people work together in the "real world" so that I could learn how to interact with other people with confidence, and also so that I could learn what kind of worker or manager I wanted to be when I "grew up."

M. Peterson, ME

Without a doubt, the best professional decision of my life. After my co-op rotation finished, I approached school as more like a job. Furthermore, cooping makes school easier! Imagine approaching something in class that you have already seen at work!

A. Thompson, EE

Internship

Contributed by: Mr. Troy Nunamaker, Director of Graduate and Internship Programs, Michelin Career Center, Clemson University

Internships offer the unique opportunity to gain career-related experience in a variety of settings. Now, more than ever, employers look to hire college graduates with internship experience in their field.

Employers indicate that good grades and participation in student activities are not always enough to help students land a good, full-time job. In today's competitive job market, the students with career-related work experience are the students getting the best interviews and job offers. As an added bonus, many companies report that over 70% of full-time hires come directly from their internship program.

WHY CHOOSE AN INTERNSHIP?

■ Bridge classroom applications to the professional world
■ Build a better résumé
■ Possibly receive higher full-time salary offers upon graduation
■ Gain experience and exposure to an occupation or industry
■ Network and increase marketability
■ Potentially fulfill academic requirements and earn money

Searching for an Internship

Although a number of students will engage in an internship experience during their freshman and sophomore years, most students pursue an internship during their junior and senior years. Some students might participate in more than one internship during their college career. Allow plenty of time for the search process to take place and be sure to keep good records of all your applications and correspondences.

■ **Figure out what you are looking for.** You should not start looking for an internship before you have answered the following questions:

  • What are my interests, abilities, and values?
  • What type of organization or work environment am I looking for?
  • Are there any geographical constraints, or am I willing to travel anywhere?

■ **Start researching internship opportunities.** Start looking one to two semesters before your desired start date. Many students find that the search process can take anywhere from 3 to 4 weeks up to 5 to 6 months before securing an internship. You should utilize as many resources as possible in order to have the broadest range of options.

The biggest project I worked on was the Athena model. The Athena is one of NASA's launch platforms.

Before I came here to the contract, several other interns had taken and made Solid Works parts measured from the actual Athena. My task was to take their parts, make them dimensionally correct and put it together in a large assembly.

I took each individual part (~ 300 of them!) and made them dimensionally correct, then put them together into an assembly. After I finished the assembly I animated the launcher and made it move and articulate.

E. Roper, ME

- Visit your campus's career center office to do the following: meet with a career counselor; attend a workshop on internships; find out what positions and resources are available; and look for internship postings through the career center's recruiting system and website resources.
- Attend a career fair on your campus or in your area. Career fairs typically are not just for full-time jobs, but are open to internship applicants as well. In addition, if there are specific companies where you would like to work, contact them directly and find out if they offer internships.
- Network. Network. Network. Only about a quarter of internship opportunities are actually posted. Talk to friends, family, and professors and let them know that you are interested in an internship. Networking sites like LinkedIn and Facebook are also beginning to see more use by employers and students. However, be conscious what images and text are associated with your profile.

■ **Narrow down the results and apply for internships.** Look for resources on your campus to help with developing a résumé and cover letter. *Each résumé and cover letter should then be tailored for specific applications.* As part of the application process, do not be surprised if a company requests additional documents such as references, transcripts, writing samples, and formal application packets.

■ **Wait for responses.** It may take up to a month to receive any responses to your applications. One to 2 weeks after you have submitted your application, call the organization to make sure they received all the required documents from you.

■ **Interview for positions.** Once you have your interviews scheduled, stop by the career center to see what resources they have available to help you prepare for the interview. Do not forget to send a thank-you note within 24 hours of the interview, restating your interest in the position.

Accepting an Internship

Once you have secured an internship, look to see if academic internship coursework is available on your campus so that the experience shows up on your transcripts. If you were rejected from any organizations, take it as a learning experience and try to determine what might have made your application stronger.

POINTS TO PONDER

Am I eligible for an internship?
Most companies look to hire rising juniors and seniors, but a rising sophomore or even a freshman with relevant experience and good grades can be a strong candidate.

Will I be paid for my internship?
The pay rate will depend on your experience, position, and the individual company. However, most engineering interns receive competitive compensation; averages are $14–$20 per hour.

When should I complete an internship?
Contrary to some popular myths, an internship can be completed not only during summers, but also during fall and spring semesters. Be sure to check with your campus on how to maintain your student enrollment status while interning.

Will I be provided housing for an internship?
Do not let the location of a company deter you. Some employers will provide housing, while others will help connect you with resources and fellow interns to find an apartment in the area.

I think that first year-students in engineering should do a co-op or internship. It was extremely valuable to my education. Now that I am back in the classroom, I know what to focus on and why what I am learning is important.

Before the experience I did not know what I wanted to do with my major, and I didn't fully understand word problems that were presented in a manner that applied to manufacturing or real life. Being in industry and working in a number of different departments, I figured out that I liked one area more than any other area, and that is where I am focusing my emphasis area studies during my senior year.

K. Glast, IE

WISE WORDS: DESCRIBE A PROJECT YOU WORKED ON DURING YOUR CO-OP OR INTERNSHIP

The large project I worked on was an upgrade of an insulin production facility. A small project I worked on for 2 weeks was the design of a pressure relief valve for a heat exchanger in the plant.

D. Jones, BE

I have been working on a series of projects, all designed to make the production of electric power meters more efficient. I am rewriting all the machine vision programs to make the process more efficient and to provide a more sophisticated graphical user interface for the operators. These projects have challenged me by requiring that I master a new "machine vision" programming language, as well as think in terms of efficiency rather than simply getting the job done.

R. Izard, ME

A transmission fluid additive was not working correctly and producing harmful emissions, so I conducted series of reactions adding different amounts of materials in a bioreactor. I determined the best fluid composition by assessing the activation energy and how clean it burned.

C. Darling, Biosystem

At Boeing, I worked with Liaison Engineering. Liaison engineers provide engineering solutions to discrepancies on the aircraft that have deviated from original engineering plans. As one example, I worked closely with other engineers to determine how grain properties in titanium provide a sound margin or safety in the seat tracks.

J. Compton, ME

The site I worked at designs and manufactures radar systems (among others). During my internship I wrote C code that tests the computer systems in a certain radar model. The code will eventually be run by an operator on the production floor before the new radars are sent out to customers.

D. Rollend, EE

One of my last projects I worked on during my first term was building a new encoder generator box used to test the generator encoder on a wind turbine. The goal was to make a sturdier box that was organized inside so that if something had broken, somebody who has no electrical skills could fix it. I enjoyed this project because it allowed me to use my skills I have learned both from school and my last internship.

C. Balon, EE

Study Abroad

*Contributed by: Mrs. J. P. Creel, International Programs
Coordinator, College of Engineering and Science, Clemson University*

In today's global economy, it is important for engineering students to recognize the importance of studying abroad. A few reasons to study abroad include the following:

- Taking undergraduate courses abroad is an exciting way to set your résumé apart from those of your peers. Prospective employers will generally inquire about your international experiences during the interview process, giving you the chance to make a lasting impression that could be beneficial.
- Studying abroad will give you a deeper, more meaningful understanding of a different culture. These types of learning experiences are not created in traditional classrooms in the United States, and cannot be duplicated by traveling abroad on vacation.
- Students who study abroad generally experience milestones in their personal development as a result of stepping outside of their comfort zone.
- There is no better time to study abroad than now! Students often think that they will have the opportunity to spend significant amounts of time traveling the world after graduation. In reality, entering the workforce typically becomes top priority.
- Large engineering companies tend to operate on a global scale. For instance, a company's headquarters may be in the United States, but that company may also control factories in Sweden, and have parts shipped to them from Taiwan. Having an international experience under your belt will give you a competitive edge in your career because you will have global knowledge that your co-workers may not possess.

While programs differ between universities, many offer a variety of choices:

- **Exchange Programs:** Several institutions are part of the Global Engineering Education Exchange, or GE^3. This consortium connects students from top engineering schools in the United States with foreign institutions in any of 18 countries. In an ideal situation, the number of international students on exchange in the United States would be equal to the number of U.S. students studying abroad during the same time frame.

  Generally, students participating in these types of exchange programs will continue to pay tuition at their home institution; however, you should check your university's website for more information. Also, by entering "Global E3" into a web browser, you will be able to access the website to determine if your institution currently participates, which foreign institutions offer courses taught in English, and which schools offer courses applicable to your particular major. If your institution is not a member of GE^3, consult with someone in the international, or study abroad, office at your institution to find out about other exchange opportunities.

- **Faculty-Led Programs:** It is not unusual for faculty members to connect with institutions or other professionals abroad to establish discipline-specific study-abroad programs. These programs typically allow students to enroll in summer classes at their home institution, then travel abroad to complete the coursework. Faculty-led programs offer organized travel, lodging, administration, and excursions, making the overall experience hassle-free. Consult with professors in your department to find out if they are aware of any programs that are already in place at your institution.

- **Third-Party Providers:** Many universities screen and recommend providers of programs for students. If there's a place you want to go for study and your university does not have an established program at that particular location, you will probably

I studied abroad twice. The first time I spent a semester in the Netherlands, experiencing a full immersion in the Dutch culture and exploring my own heritage, and the second time I spent a summer in Austria taking one of my core chemical engineering classes. Both countries are beautiful and unique places and they will always hold a special place in my heart.

Through studying abroad, I was able to expand my own comfort zone by encountering novel situations and become a more confident individual. Although the experiences were amazing and the memories are truly priceless, the biggest thing I gained from studying abroad was that I was able to abandon many perceptions about other cultures and embrace new perspectives.

R. Lanoie, studying abroad in the Netherlands and Austria

find a study program of interest to you by discussing these options with the study-abroad office at your university.

■ **Direct Enrollment:** If you are interested in a particular overseas institution with which your home university does not have an established program, there is always the option of direct enrollment. This process is basically the same as applying to a university in the United States. The school will likely require an admissions application, a purpose statement, transcripts, and a letter of reference. Be cautious if you choose this route for study abroad, as it can be difficult to get credits transferred back to your home institution. It is a good idea to get your international courses preapproved prior to your departure. On a positive note, it can be cheaper to directly enroll in a foreign institution than it is to attend at your home university. This option is best suited for students who want to go abroad for a semester or full year.

I think study abroad is a wonderful learning experience everyone should have. In going abroad you get to meet people you would never meet otherwise, and experience things you never thought you would see.

I learned about independently navigating a new country, as well as adjusting to foreign ways of doing things. It is interesting to take classes that you would have taken at home and experience them a completely different way.

Studying abroad teaches you how to easily adjust to new situations and allows you to learn about cultures you had no idea of before.

B. Brubaker, studying abroad in Scotland

Typically, students go abroad during their junior year, though recently there has been an increase in the number of second-year students participating. The timing of your experience should be agreed upon by you and your academic advisor. Each program offers different international incentives, and some are geared for better opportunities later in your academic career, whereas others may be better toward the beginning.

If you would like to go abroad for a semester or year but do not feel entirely comfortable with the idea, why not get your feet wet by enrolling in a summer program first? This will give you a better idea of what it is that you are looking for without overwhelming you. Then, you can plan for a semester or year abroad later in your academic career. In fact, more students are electing to spend a semester or year of study abroad, and increasingly more opt for an academic semester plus a semester internship combination.

There are three basic principles to follow when deciding on a location:

- **Personal Preference:** Some people are more interested in Asia than Europe, or maybe Australia instead of Latin America. There are excellent opportunities worldwide, regardless of the location.
- **Program Opportunities:** Certain countries may be stronger or have more options in certain fields. For example, Germany is well known for innovations in mechanical engineering, while the Japanese tend to be more widely recognized for their efforts in computer engineering. Listen to your professors and weigh your options.
- **Language:** You may not feel like you are ready to study in a foreign country, speaking and reading in a foreign language. There are numerous institutions that offer courses taught in English. However, if you have taken at least 2 years of the same foreign language, you should be knowledgeable enough to succeed in courses taught in that language. Do not let your fears restrict you!

India is quite an eye-opening country. All of your senses work on overload just so you can take everything in at once. My time studying in India was simply phenomenal. I had an excellent opportunity placed before me to travel to India to earn credit for Electrical Engineering courses and I would have been a fool to pass up that chance. I regret nothing, I would do it again, and I would urge everyone with the slightest inkling of studying abroad to put their worries aside and have the experience of a lifetime.

J. Sadie, studying abroad in India

Writing Assignments

For each question, write a response according to the directions given by your instructor. Each response should contain correct grammar, spelling, and punctuation. Be sure to answer the question completely, but choose your words carefully so as to not exceed the word limit if one is given. There is no right or wrong answer; your score will be based upon the strength of the argument you make to defend your position.

1. On a separate sheet, write a one-page résumé for one person from the following list. Include information such as education, job experience, primary accomplishments (inventions, publications, etc.), and references. If you want, you can include a photo or likeness no larger than 2 inches by 3 inches. You can add some "made up" material such as current address and references, but do not overdo this.

- Ammann, Othmar
- Ampere, Andre Marie
- Arafat, Yasser
- Archimedes
- Avogadro, Armedeo
- Bernoulli, Daniel
- Bessemer, Henry
- Bezos, Jeffrey P.
- Birdseye, William
- Bloomberg, Michael
- Bohr, Niels
- Boyle, Robert
- Brezhnev, Leonid
- Brown, Robert
- Brunel, Isambard
- Calder, Alexander
- Capra, Frank
- Carnot, Nicolas
- Carrier, Willis Haviland
- Carter, Jimmy
- Cauchy, Augustin Louis
- Cavendish, Henry
- Celsius, Anders
- Clausius, Rudolf
- Coulomb, Charles
- Cray, Seymour
- Crosby, Philip
- Curie, Marie
- Dalton, John
- Darcy, Henri
- de Coriolis, Gaspard
- de Mestral, George
- Deming, W. Edwards

- Diesel, Rudolf
- Dunbar, Bonnie
- Eiffel, Gustave
- Euler, Leonhard
- Fahrenheit, Gabriel
- Faraday, Michael
- Fleming, Sandford
- Ford, Henry
- Fourier, Joseph
- Fung, Yuan-Cheng
- Gantt, Henry
- Gauss, Carl Friedrich
- Gibbs, Josiah Willard
- Gilbert, William
- Gilbreth, Lillian
- Goizueta, Robert
- Grove, Andrew
- Hancock, Herbie
- Henry, Beulah
- Hertz, Heinrich Rudolf
- Hitchcock, Alfred
- Hooke, Robert
- Hoover, Herbert
- Hopper, Grace
- Iacocca, Lee
- Joule, James Prescott
- Juran, Joseph Moses
- Kelvin, Lord
- Kraft, Christopher, Jr.
- Kwolek, Stephanie
- Landry, Tom
- Laplace, Pierre-Simon
- Leibniz, Gottfried

- LeMessurier, William
- MacCready, Paul
- Mach, Ernst
- McDonald, Capers
- Midgley, Thomas Jr.
- Millikan, Robert
- Navier, Claude-Louis
- Newton, Isaac
- Nielsen, Arthur
- Ochoa, Ellen
- Ohm, Georg
- Pascal, Blaise
- Poiseuille, Jean Loius
- Porsche, Ferdinand
- Prandtl, Ludwig
- Rankine, William
- Rayleigh, Lord
- Resnik, Judith

- Reynolds, Osborne
- Rømer, Ole
- Sikorsky, Igor Ivanovich
- Stinson, Katherine
- Stokes, George
- Sununu, John
- Taguchi, Gen'ichi
- Taylor, Fredrick
- Teller, Edward
- van der Waals, Johannes
- Venturi, Giovanni
- Volta, Count Alessandro
- von Braun, Wernher
- von Kármán, Theodore
- Watt, James
- Welch, Jack
- Wyeth, Nathaniel
- Yeltsin, Boris

Personal Reflections

2. Please address the following questions in approximately one page. You may write, type, draw, sketch your answers . . . whatever form you would like to use.

(a) Where are you from?

(b) What type of engineering are you interested in? Why?

(c) Are there any engineers in the circle of your friends and family?

(d) How would you classify yourself as a student (new freshman, transfer, or upperclassman)?

(e) What are your activities and hobbies?

(f) What are you most proud of?

(g) What are you fantastic at?

(h) What are you passionate about?

(i) What has been the most difficult aspect of college so far?

(j) What do you expect to be your biggest challenge this term?

(k) Do you have any concerns about this class?

(l) Anything else you would like to add?

3. In 2008, experts convened by the National Academy of Engineering (NAE) met and proposed a set of Grand Challenges for Engineering, a list of the 14 key goals for engineers to work toward during the 21st century. For any aspiring engineer, reading this list should feel like reading a description of the challenges you will face throughout your career. To read a description of each Challenge, as well as a description of some connected areas within each, visit the NAE's Grand Challenges for Engineering website at http://www.engineeringchallenges.org.

   After reading through the list, write a job description of your dream job. The job description should include the standard components of a job description: title, responsibilities (overall and specific), a description of the work hierarchy (whom you report to, whom you work with), as well as any necessary qualifications. Do not include any salary requirements, as this is a completely fictional position.

   After writing the description of your dream job, identify which of the 14 Grand Challenges for Engineering this position is instrumental in working toward solving. Cite specific examples of the types of projects your fictional job would require you to do, and discuss the impact of those projects on the Grand Challenges you have identified.

4. Choose and explain your choice of major, and the type of job you envision yourself doing in 15 years. Consider the following:

   (a) What skills or talents do you possess that will help you succeed in your field of interest?
   (b) How passionate are you about pursuing a career in engineering? If you do not plan on being an engineer, what changed your mind?
   (c) How confident are you in your choice of major?
   (d) How long will it take you to complete your degree?
   (e) Will you obtain a minor?
   (f) Will you pursue study abroad, co-op, or internship?
   (g) Do you plan to pursue an advanced degree, or become a professional engineer (PE)?
   (h) What type of work (industry, research, academic, medical, etc.) will you pursue?

5. In 2008, the University of Memphis made national headlines when the Memphis Tigers played in the NCAA basketball national championship game. When interviewed by a local newspaper about how he helped his team "own" the tournament, Coach Calipari revealed that before each game, he had his star player write an essay about how the game would play out. This particular player was prone to nervousness, so to help him focus, the coach told him to mentally envision the type of plays and how he himself would react to them.

   In this assignment, write a short essay on how you will prepare for and take the final exam. Consider the following:

   (a) What will you do to study? What materials will you gather, and how will you use them? Where will you study? Will it be quiet? Will you play music?
   (b) What kinds of things could go wrong on the day of the exam, and how would you avoid them? (List at least three.)
   (c) What will the exam look like, and how will you work through it?

   Thank you to Dr. Lisa Benson for contributing this assignment.

6. Read the essay "Engineering is an ... itch!" in the Engineering Essentials introduction. Reflect on what it means to have performance-focused versus mastery-focused learning goals.

   (a) Describe in your own words what it means to be a performance-based learner compared with a mastery-based learner.
   (b) What learning goals do you have? Are these goals performance based or mastery based?
   (c) Is it important to you to become more mastery focused?
   (d) Do you have different kinds of learning goals than you had in the past, and do you think you will have different learning goals in the future?

7. An article in *Science News* addressed the topics of nature and technology. In our electronic world, we are in constant contact with others, through our cell phones, iPods, Facebook, e-mail, etc. Researchers at the University of Washington have determined that this effect may create long-term problems in our stress levels and in our creativity.

Scenes of Nature Trump Technology in Reducing Low-Level Stress

Technology can send a man to the moon, help unlock the secrets of DNA, and let people around the world easily communicate through the Internet. But can it substitute for nature?

"Technology is good and it can help our lives, but let's not be fooled into thinking we can live without nature," said Peter Kahn, a University of Washington associate professor of psychology who led the research team.

We are losing direct experiences with nature. Instead, more and more we're experiencing nature represented technologically through television and other media. Children grow up watching Discovery Channel and Animal Planet. That's probably better than nothing. But as a species we need interaction with actual nature for our physical and psychological well-being.

Part of this loss comes from what the researchers call environmental generational amnesia. This is the idea that across generations the amount of environmental degradation increases, but each generation views conditions it grew up with as largely non-degraded and normal. Children growing up today in the cities with the worst air pollution often, for example, don't believe that their communities are particularly polluted.

"This problem of environmental generational amnesia is particularly important for children coming of age with current technologies," said Rachel Severson, a co-author of the study and a University of Washington psychology doctoral student. "Children may not realize they are not getting the benefits of actual nature when interacting with what we're calling technological nature."

> *[University of Washington (2008, June 16). "Scenes of Nature Trump*
> *Technology in Reducing Low-level Stress."* Science Daily*]*

Go someplace quiet and spend at least 10 minutes clearing your head. Only after these 10 minutes, get out a piece of paper and sketch something you see; are thinking of; want to create or invent or imagine, etc. On the same sheet of paper, write a poem about your sketch; something you are thinking of; or quiet, wonder of the universe, lack of technology, etc.

Rules:

- Must be done all by hand.
- Must be original work (no copied poems or artwork).
- Ability does not count—draw like you are 5 years old!
- Draw in any medium you want: use pencil, pen, colored pencils, markers, crayons, watercolors.
- Poetic form does not matter: use rhyme, no rhyme, haiku, whatever you want it to be.
- It does not need to be elaborate; simple is fine.

CHAPTER 2
ETHICS

Every day, we make numerous ethical decisions, although most are so minor that we do not even view them as such.

- When you drive your car, do you knowingly violate the posted speed limit?
- When you unload the supermarket cart at your car, do you leave it in the middle of the parking lot, or spend the extra time to return it to the cart corral?
- You know that another student has plagiarized an assignment; do you rat him or her out?
- A person with a mental disability tries to converse with you while waiting in a public queue. Do you treat him or her with respect or pretend he or she does not exist?
- In the grocery, a teenager's mother tells her to put back the package of ice cream she brought to the cart. The teenager walks around the corner and places the ice cream on the shelf with the soft drinks and returns to the buggy. Do you ignore this or approach the teenager and politely explain that leaving a package of ice cream in that location will cause it to melt thus increasing the cost of groceries for everyone else, or do you replace it in the freezer yourself?
- When going through a public door, do you make a habit of looking back to see if releasing the door will cause it to slam in someone's face?
- You notice a highway patrolman lying in wait for speeders. Do you flash your lights at other cars to warn them?
- A cashier gives you too much change for a purchase. Do you correct the cashier?
- You are on the lake in your boat and notice a person on a JetSki chasing a great blue heron across the lake. The skier stops at a nearby pier. Do you pilot your craft over to the dock and reprimand him for harassing the wildlife?

> Good people do not need laws to tell them to act responsibly, while bad people will find a way around the laws.
>
> *Plato*

On a grand scale, none of these decisions is particularly important, although some might lead to undesirable consequences. However, as an aspiring engineer, you may face numerous decisions in your career that could affect the lives and well-being of thousands of people. Just like almost everything else, practice makes perfect, or at least better. The more you practice analyzing day-to-day decisions from an ethical standpoint, the easier it will be for you to make good decisions when the results of a poor choice may be catastrophic.

In very general terms, there are two reasons people try to make ethical decisions.

- They wish to make the world a better place for everyone—in a single word, altruism.
- They wish to avoid unpleasant consequences, such as fines, incarceration, or loss of job.

In an ideal society, the second reason would not exist. However, history is replete with examples of people, and even nations, who do not base their decisions solely on whether or not they are acting ethically. Because of the common occurrence of unethical behavior and the negative impact it has on others, almost all societies have developed rules, codes, and laws to specify what is and is not acceptable behavior, and the punishments that will be meted out when violations occur.

The major religions all have fairly brief codes summarizing how one should conduct their life. Some examples are given below; other examples exist as well.

- Judaism, Christianity, and derivatives thereof have the Decalogue, or Ten Commandments.
- Islam has the Five Pillars in addition to a slightly modified and reorganized form of the Decalogue.
- Buddhism has the Noble Eightfold Path.
- Bahá'í has 12 social principles.
- In Hinduism, Grihastha dharma has four goals.

Secular codes of conduct go back more than four millennia to the Code of Ur-Nammu. Although by today's standards, some of the punishments in the earliest codes seem harsh or even barbaric, it was one of the earliest known attempts to codify crimes and corresponding punishments.

Admittedly, although not specifically religious in nature, these codes are usually firmly rooted in the prevailing religious thought of the time and location. Through the centuries, such codes and laws have been expanded, modified, and refined so that most forms of serious antisocial behavior are addressed and consequences for violations specified. These codes exist from a local to a global level. Several examples are given below.

- Most countries purport to abide by the Geneva Conventions, which govern certain types of conduct on an international scale.
- Most countries have national laws concerning murder, rape, theft, etc.
- In the United States, it is illegal to purchase alcohol unless you are 21 years of age. In England, the legal age is 18.
- In North and South Dakota, you can obtain a driver's license at age $14\frac{1}{2}$. In most other states, the legal age is 16.
- It is illegal to say "Oh boy!" in Jonesboro, Georgia.
- Many cities, such as Santa Fe, New Mexico, have ordinances prohibiting use of cell phones while driving.

2.1 ETHICAL DECISION MAKING

Some ethical decisions are clear-cut. For example, essentially everyone (excluding psychopaths) would agree that it is unethical to kill someone because you do not like his or her hat. Unfortunately, many real-world decisions that we must make are far from "black and white" issues, instead having many subtle nuances that must be considered to arrive at what one believes is the "best" decision.

There is no proven algorithm or set of rules that one can follow to guarantee that the most ethical decision possible is being made in any particular situation. However, numerous people have developed procedures that can guide us in considering questions with ethical ramifications. A four-step procedure is discussed here, although there are various other approaches.

Step 1: Determine *What* the issues are and *Who* might be affected by the various alternative courses of action that might be implemented.

We will refer to the *Who* as **stakeholders**. Note that at this point, we are not trying to determine how the stakeholders will be affected by any particular plan of action.

- The issues (What) can refer to a wide variety of things, including, for example, personal freedom, national security, quality of life, economic issues, fairness, and equality.
- The term stakeholders (Who) does not necessarily refer to people, but might be an individual, a group of people, an institution, or a natural system, among other things.

● **EXAMPLE 2-1**

Consider the question of whether to allow further drilling for oil in the Alaska National Wildlife Refuge (ANWR). List several issues and stakeholders.

Issues:

- *Oil independence*
- *The price of gasoline*
- *Possible impacts on the ecosystem*

Stakeholders:

- *Oil companies*
- *The general population of the United States*
- *Other countries from whom we purchase oil*
- *The flora and fauna in ANWR*
- *The native people in Alaska*

Step 2: Consider the effects of alternative courses of action from different perspectives.

Here, we look at three perspectives: consequences, intent, and character.

Perspective 1: Consequences

When considering this perspective, ask how the various stakeholders will be affected by each alternative plan being contemplated. In addition, attempt to assign a relative level of importance (weight) to each effect on each stakeholder. For instance, an action that might affect millions of people adversely is almost always more important than an action that would cause an equivalent level of harm to a dozen people.

● **EXAMPLE 2-2**

Should all U.S. children be fingerprinted when entering kindergarten and again each third year of grade school (3, 6, 9, 12)? Identify the stakeholders and consequences.

Stakeholders:

- *All U.S. children*
- *All U.S. citizens*
- *Law enforcement*
- *The judicial system*
- *The U.S. Constitution*

Consequences:

■ *Provides a record to help identify or trace missing children (not common, but possibly very important in some cases)*
■ *Affords an opportunity for malicious use of the fingerprint records for false accusation of crime or for identity theft (probability unknown, but potentially devastating to affected individuals)*
■ *Could help identify perpetrators of crimes, thus improving the safety of law-abiding citizens (importance varies with type of crime)*
■ *Raises serious questions concerning personal freedoms, possibly unconstitutional (importance, as well as constitutionality, largely dependent on the philosophy of the person doing the analysis)*

This list could easily be continued.

Fingerprint technology has advanced in recent years with the implementation of computer recognition for identification. Originally in the United States, the Henry Classification System was used to manually match fingerprints based on three main patterns: arches, loops, and whorls (shown below from left to right).

Today, the Automated Fingerprint Identification System (AFIS) uses algorithmic matching to compare images. Future work of AFIS systems is in the adoption and creation of secure multitouch devices like mobile computers and tablets, which can identify different security levels for the operator of the device. For example, a multitouch computer owner might be able to issue permissions to an administrator that might not be available to a 5-year old, all without providing a single password!

Perspective 2: Intent

The intentions of the person doing the acting or deciding are considered in this perspective, sometimes called the "rights" perspective. Since actions based on good intentions can sometimes yield bad results, and vice versa, the intent perspective avoids this possible pitfall by not considering the outcome at all, only the intentions.

It may be helpful when considering this perspective to recall Immanuel Kant's Categorical Imperative: "Act only according to that maxim whereby you can at the same time will that it should become a universal law." To pull this out of the eighteenth century, ask yourself the following questions:

(a) Is the action I am taking something that I believe everyone should do?
(b) Do I believe that this sort of behavior should be codified in law?
(c) Would I like to be on the receiving end (the victim) of this action?

● **EXAMPLE 2-3**   Should you download music illegally over the Internet?

Rephrasing this question using the suggestions above yields:

(a) *Should everyone illegally download the music they want if it is there for the taking?*

(b) *Should the laws be changed so that anyone who obtains a song by any means can post it on the web for everyone to get for free?*

(c) *If you were a struggling musician trying to pay the bills, would you like your revenue stream to dry up because everyone who wanted your music got it for free?*

Perspective 3: Character

Character is the inherent complex of attributes that determines a person's moral and ethical actions and reactions. This perspective considers the character of a person who takes the action under consideration. There are different ways of thinking about this. One is to simply ask: Would a person of good character do this? Another is to ask: If I do this, does it enhance or degrade my character? Yet another way is to ask yourself if a person you revere as a person of unimpeachable character (whoever that might be) would take this action.

● **EXAMPLE 2-4**   Your friends are deriding another student behind her back because she comes from a poor family and does not have good clothes.
   Do you:

(a) *Join in the criticism?*

(b) *Ignore it, pretend it is not happening, or simply walk away?*

(c) *Tell your friends that they are behaving badly and insist that they desist?*

- *Which of these actions would a person of good character take?*
- *Which of these actions would enhance your character and which would damage it?*
- *What would the founder of your religion do? (Moses or Jesus or Buddha or Mohammed or Bahá'u'lláh or Vishnu or whoever.) If you are not religious, what would the person who, in your opinion, has the highest moral character do?*

Step 3: Correlate perspectives.

Now look back at the results of considering the issues from the three perspectives. In many cases, all three perspectives will lead to the same or a similar conclusion. When this occurs, you have a high level of confidence that the indicated action is the best choice from an ethical standpoint.

If the three perspectives do not agree, you may wish to reconsider the question. It may be helpful to discuss the issue with people whom you have not previously consulted in this matter. Did you omit any factors? For complicated issues, it is difficult to make sure you have included all possible stakeholders and consequences. Did you properly assign weights to the various aspects? Upon reconsideration, all three perspectives may converge.

If you cannot obtain convergence of all three perspectives, no matter how hard you try to make sure you left nothing out, then go with two out of three.

Step 4: Act.

This is often the hardest step of all to take, since ethical action often requires courage. The whistle-blower who risks losing his or her job, Harriet Tubman repeatedly risking her life to lead slaves to freedom via the Underground Railroad, the elected official standing up for what she knows to be right even though it will probably cost her the next election, or even something as mundane as risking the ridicule of your friends because you refuse to go along with whatever questionable activities they are engaging in for "fun." Ask yourself the question: "Do *I* have the courage to do what I know is right?"

● **EXAMPLE 2-5**

Your company has been granted a contract to develop the next generation of electronic cigarette, also known as a "nicotine delivery system," and you have been assigned to the design team. Can you in good conscience contribute your expertise to this project?

NOTE

In the interest of brevity, this is not an exhaustive analysis but shows the general procedure.

Step 1: *Identify the issues (What) and the stakeholders (Who).*

Issues:

- *Nicotine is poisonous and addictive*
- *These devices eliminate many of the harmful components of tobacco smoke*
- *Laws concerning these devices range from completely legal, to classification as a medical device, to banned, depending on country*
- *There are claims that such devices can help wean tobacco addicts off nicotine*
- *The World Health Organization does not consider this an effective means to stop smoking*
- *Whether an individual chooses to use nicotine should be a personal decision, since its use does not generally degrade a person's function in society*
- *The carrier of the nicotine (80–90% of the total inhaled product) is propylene glycol, which is relatively safe, but can cause skin and eye irritation, as well as other adverse effects in doses much larger than would be obtained from this device*
- *A profit can be made from nicotine products or anti-smoking devices*

Stakeholders:

- *You (your job and promotions)*
- *Your company and stockholders (profit)*
- *Cigarette manufacturers and their employees and stockholders (lost revenue)*
- *Tobacco farmers (less demand)*
- *The public (less second-hand smoke)*
- *The user (various health effects, possibly positive or negative)*

Step 2: *Analyze alternative courses of action from different perspectives.*

1. *Consequences*

    - *You may lose your job or promotion if you refuse*
    - *If you convince management to abandon the project, the company may lose money*
    - *If you succeed brilliantly, your company may make money hand over fist, and you receive a promotion*
    - *If the project goes ahead, the possibility of future lawsuits exists*
    - *Users' health may be damaged*
    - *Users' dependence on nicotine may either increase or decrease*

2. *Intent*

- *Should everyone use electronic cigarettes, or at least condone their use?*
- *Should use of electronic cigarettes be unrestricted by law?*
- *Would I like to risk nicotine addiction because of using these devices?*
- *Would I be able to kick my tobacco habit by using these devices?*

3. *Character*

- *Would a person of good character develop this device, use it, or condone its use?*
- *Would work on this project (thus implicitly condoning its use) or use of the device itself enhance or degrade my character?*
- *Would my personal spiritual leader, or other person I revere, condone development or use of this product?*

Step 3: *Correlate perspectives.*

Here we enter the realm of subjective judgment. The individual author responsible for this example has a definite personal answer, but it is in the nature of ethical decision making that different people will often arrive at different results in good conscience. You would have to weigh the various factors (including any that have been overlooked or knowingly omitted) to arrive at your own conclusion. We refuse to dictate a decision to you.

Step 4: *Act on your decision.*

If your decision was that working on this project poses no threat to your soul (if you happen to believe in such), probably little courage is required to follow through, since your career may blossom, or at least not be curtailed.

On the other hand, if you believe that the project is unethical, you need to have the intestinal fortitude to either attempt to change the minds of management or refuse to work on the project, both of which may put your career at risk.

2.2 PLAGIARISM

> **Did you know?** There are Internet services available that will accept a document and search the web for exact or similar content. Also, there are programs that will scan multiple documents and search for exact or similar content.
>
> ---
>
> **Did you know?** Prior to the romantic movement of the eighteenth century, European writers were encouraged not to be inventive without good reason and to carefully imitate the work of the great masters of previous centuries.

You probably know what plagiarism is—claiming someone else's work as your own. This is most often used in reference to written words, but may be extended to other media as well. From a legal standpoint, plagiarism per se is not illegal, although it is widely considered unethical. However, if the plagiarism also involves copyright infringement, then this would be a violation of the law. Certainly, in the context of your role as a student, plagiarism is almost universally regarded as academic dishonesty, and subject to whatever punitive actions your school deems appropriate.

In some cases, plagiarism is obvious, as when an essay submitted by a student is almost identical to one found on the Internet, or is the same as that submitted by another student. It is amazing how frequently students are caught cheating because they copied verbatim from another student's work, complete with strange mistakes and bizarre phrasing that grab the grader's attention like an 18-wheeler loaded with live pigs locking its brakes at 80 miles per hour. (*Thanks to Gilbert Shelton for that image.*)

In other cases, things are far less clear. For example, if you were writing a short story for your English class and used the simile "her lips were like faded tulips, dull and wrinkled," can you (or the professor) really be sure whether that was an original phrase or if you had read it at some time in the past, and your brain dragged it up from your subconscious memory as though it were your own?

We all hear or read things during our lives that hang around in our brains whether we are consciously aware of them or not. We cannot go through life in fear of being accused of plagiarism because our brain might drag up old data masquerading as our own original thought, or even worrying about whether our own original thoughts have ever been concocted by another person completely independently.

Any reasonable person (although admittedly, there is a surfeit of unreasonable people) will take the work as a whole into account. If there is simply a single phrase or a couple of instances of wordings that are similar to another source, this is most likely an innocent coincidence. On the other hand, if a work has many such occurrences, the probability that the infractions are innocent is quite low.

We arrive here at intent. Did you knowingly copy part of someone else's work and submit it as your own without giving proper credit? If you did not, stop worrying about it. If you did, Big Brother, also known as your professor, is watching, possibly with the assistance of high-tech plagiarism detection tools. (*A tip of the hat to George Orwell.*)

2.3 ENGINEERING CREED

Ethical decisions in engineering have, in general, a narrow focus specific to the problems that arise when designing and producing products or services of a technical nature. Engineers and scientists have, by the very nature of their profession, a body of specialized knowledge that is understood only vaguely, if at all, by most of the population. This knowledge can be used for tremendous good in society, but can also cause untold mischief when used by unscrupulous practitioners. Various engineering organizations have thus developed codes of conduct specific to the profession. Perhaps the most well known is the Code of Ethics for Engineers developed by the National Society of Professional Engineers (NSPE). The entire NSPE Code of Ethics is rather long, so we list only the Engineer's Creed and the Fundamental Canons of the Code here.

Engineer's Creed

As a Professional Engineer, I dedicate my professional knowledge and skill to the advancement and betterment of human welfare. I pledge:

- To give the utmost of performance
- To participate in none but honest enterprise
- To live and work according to the laws of man and the highest standards of professional conduct
- To place service before profit, the honor and standing of the profession before personal advantage, and the public welfare above all other considerations

In humility and with need for Divine Guidance, I make this pledge.

Fundamental Canons

Engineers, in the fulfillment of their professional duties, shall

- Hold paramount the safety, health, and welfare of the public
- Perform services only in areas of their competence
- Issue public statements only in an objective and truthful manner
- Act for each employer or client as faithful agents or trustees
- Avoid deceptive acts
- Conduct themselves honorably, responsibly, ethically, and lawfully so as to enhance the honor, reputation, and usefulness of the profession

The complete code can easily be found online at a variety of sites. When this book went to press, the URL for the Code of Ethics on the NSPE site was http://www.nspe.org/Ethics/CodeofEthics/index.html.

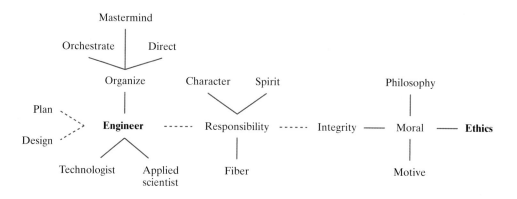

2.4 SOCIAL RESPONSIBILITY

Contributed by: Jason Huggins, P.E., Executive Councilor for Tau Beta Pi,
the National Engineering Honor Society, 2006–2014.

> **NOTE**
>
> Social responsibility is the ideology that an individual has an obligation to act to benefit society at large.

As a freshman engineering student, you are just beginning your journey to join the Engineering Profession. Have you thought about what it will mean to be a part of a profession? Being a professional means we hold the public's trust and confidence in our training, skills, and knowledge of engineering. As a profession, we recognize the importance of this trust in the Engineering Canons and the Engineering Creed that define our standards for ethics, integrity, and regard for public welfare. So, does adherence to the Engineering Canons and the Engineering Creed fulfill our social responsibilities as engineers?

Traditionally, professions have always been held in very high regard by society, largely due to the extensive amount of training, education, and dedication required for membership. With this come high expectations of how the members of a profession conduct themselves both in their professional and private lives: doctors save lives, lawyers protect people's rights, and engineers make people's lives better. I did not really make this connection or understand what it meant until I was initiated into Tau Beta Pi, the National Engineering Honor Society. The Tau Beta Pi initiation ceremony has remained largely unchanged for over 100 years, and emphasizes the obligation as engineers and members of Tau Beta Pi to society that extends beyond the services we offer to our employers and our clients.

Over the years, I have taken these obligations to mean that as a profession we are not elevated above anyone else in society. We are affected by the same problems as the general public and we must have an equal part in addressing them. In your lifetime, you will be impacted by issues such as the strength of the economy, the effectiveness of the public educational system, unemployment, the increasing national debt, national security, and environmental sustainability. You cannot focus your talents as an engineer on solving only technical issues and assume the rest of society will address the nontechnical issues. The same skill sets you are currently developing to solve technical issues can be applied to solve issues outside the field of engineering. Your ability as an engineer to effectively examine and organize facts and information in a logical manner and then present our conclusions in an unbiased fashion allows others to more fully understand complex issues and in turn help develop better solutions.

This does not mean that as an engineering profession, we are going to solve all of the world's problems. It simply means that it is our responsibility to use our skills and talents as engineers in helping to solve them. It is our obligation to actively use our skills and talents to act upon issues impacting our local, national, and global communities, not merely watching as passive observers.

I challenge you to pick one issue or problem facing society that you feel passionate about and get involved. Once you do, you will be surprised at the impact you can have, even if on a small scale. By adhering to the Engineering Canon and Creed in your professional life and getting actively involved trying to solve societal issues in your personal life, you will be fulfilling your social responsibility.

In-Class Activities

ICA 2-1

For each of the following situations, indicate whether you think the action is ethical or unethical or you are unsure. Do not read ahead; do not go back and change your answers.

Situation	Ethical	Unethical	Unsure
1. Not leaving a tip after a meal because your steak was not cooked to your liking			
2. Speeding 5 miles per hour over the limit			
3. Killing a roach			
4. Speeding 15 miles per hour over the limit			
5. Having plastic surgery after an accident			
6. Killing a mouse			
7. Driving 90 miles per hour			
8. Using Botox			
9. Not leaving a tip after a meal because the waiter was inattentive			
10. Killing a healthy cat			
11. Driving 90 miles per hour taking an injured child to the hospital			
12. Killing a healthy horse			
13. Dyeing your hair			
14. Killing a person			
15. Having liposuction			

ICA 2-2

For each of the following situations, indicate whether you think the action is ethical or unethical or you are unsure. Do not read ahead; do not go back and change your answers.

Situation	Ethical	Unethical	Unsure
1. Using time at work to IM your roommate			
2. Accepting a pen and pad of paper from a company trying to sell a new computer system to your company			
3. Obtaining a fake ID to purchase alcohol			
4. Using time at work to plan your friend's surprise party			
5. Accepting a wedge of cheese from a company trying to sell a new computer system to your company			
6. Taking a company pen home from work			
7. Taking extra time at lunch once a month to run a personal errand			
8. Accepting a set of golf clubs from a company trying to sell a new computer system to your company			
9. Drinking a beer while underage at a party in your dorm			
10. Using the company copier to copy your tax return			
11. Drinking a beer when underage at a party, knowing you will need to drive yourself and your roommate home			
12. Taking extra time at lunch once a week to run a personal errand			
13. Borrowing company tools			
14. Going to an NC-17 rated movie when underage			
15. Accepting a Hawaiian vacation from a company trying to sell a new computer system to your company			

ICA 2-3

For each of the following situations, indicate whether you think the action is ethical or unethical or you are unsure. Do not read ahead; do not go back and change your answers.

Situation	Ethical	Unethical	Unsure
1. Acting happy to see an acquaintance who is spreading rumors about you			
2. Letting a friend who has been sick copy your homework			
3. Shortcutting by walking across the grass on campus			
4. "Mooning" your friends as you drive by their apartment			
5. Registering as a Democrat even though you are a Republican			
6. Cheating on a test			
7. Shortcutting by walking across the grass behind a house			
8. Saying that you lunched with a coworker, rather than your high school sweetheart, when your spouse asks who you ate lunch with			
9. Helping people with their homework			
10. Shortcutting by walking through a building on campus			
11. Not telling your professor that you accidentally saw several of the final exam problems when you visited his or her office			
12. Suppressing derogatory comments about the college because the dean has asked you not to say anything negative when he or she invited you to meet with an external board evaluating the college			
13. Letting somebody copy your homework			
14. Shortcutting by walking through a house			
15. Not telling your professor that your score on a test was incorrectly totaled as 78 instead of the correct 58			

ICA 2-4

For each of the following situations, indicate how great you feel the need is in the world to solve the problem listed. Do not read ahead; do not go back and change your answers.

Situation	Urgent	Great	Somewhat	Little	None
1. Teaching those who cannot read or write					
2. Helping starving children in poor nations					
3. Helping people locked in prisons					
4. Helping to slow population growth					
5. Helping to reduce dependence on foreign oil					
6. Helping to reduce greenhouse gas emissions					
7. Helping people persecuted for sexual orientation					
8. Helping to reduce gun ownership					
9. Helping those who are mentally disabled					
10. Helping to supply laptops to poor children					
11. Helping prevent prosecution of "victimless" crimes					
12. Helping to end bigotry					
13. Helping to prevent development of WMD (weapons of mass destruction)					
14. Helping prosecute "hate" crimes					
15. Helping to eliminate violence in movies					
16. Helping homeless people in your community					
17. Helping people with AIDS					
18. Helping people in warring countries					
19. Helping endangered species					

ICA 2-5

Discuss the possible actions, if any, that you would take in each of the following situations. In each case, use the four-step analysis procedure presented in Section 2.1 to help determine an appropriate answer.

(a) Your roommate purchased a theme over the Internet and submitted it as his or her own work in English class.

(b) Your project team has been trying to get your design to work reliably for 2 weeks, but it still fails about 20% of the time. Your teammate notices another team's design that is much simpler, that is easy to build, and that works almost every time. Your teammate wants your group to build a replica of the other team's project at the last minute.

(c) You notice that your professor forgot to log off the computer in lab. You are the only person left in the room.

(d) The best student in the class, who consistently wrecks the "curve" by making 15–20 points higher than anyone else on every test, accidentally left her notes for the course in the classroom.

(e) You have already accepted and signed the paperwork for a position as an intern at ENGR-R-US. You then get an invitation to interview for an intern position (all expenses paid) at another company in a city you have always wanted to visit. What would you do? Would you behave differently if the agreement was verbal, but the papers had not been signed?

(f) One of your professors has posted a political cartoon with which both you and your friend vehemently disagree. The friend removes the cartoon from the bulletin board and tears it up.

ICA 2-6

Discuss the possible actions, if any, that you would take in each of the following situations. In each case, use the four-step analysis procedure presented in Section 2.1 to help determine an appropriate answer.

(a) You witness several students eating lunch on a bench on campus. When they finish, they leave their trash on the ground.

(b) You see a student carving his initials in one of the largest beech trees on campus.

(c) You see a student writing graffiti on a trash dumpster.

(d) There is a squirrel in the road ahead of a car you are driving. You know that a squirrel's instinct is to dart back and forth rather than run in a straight line away from a predator (in this case a vehicle) making it quite likely it will dart back into the road at the last instant.

(e) You find a wallet containing twenty-three $100 bills. The owner's contact information is quite clear. Does your answer change if the wallet contained three $1 bills?

ICA 2-7

Read the Engineer's Creed section of this chapter.

If you are planning to pursue a career in engineering: Type the creed word for word, then write a paragraph (100–200 words) on what the creed means to you, in your own words, and how the creed make you feel about your chosen profession (engineering).

If you are planning to pursue a career other than engineering, does your future discipline have such a creed? If so, look this up and type it, then write a paragraph (100–200 words) on what the creed means to you, in your own words, and how the creed makes you feel about your chosen profession. If not, write a paragraph (100–200 words) on what items should be included in a creed if your profession had one and how the lack of a creed makes you feel about your chosen profession.

ICA 2-8

Engineers often face workplace situations in which the ethical aspects of the job should be considered.

Table 2-1 lists a variety of types of organizations that hire engineers, and one or more possibly ethical issues that might arise.

Pick several of the organizations from the table that interest you (or those assigned by your professor) and answer the following:

(a) Can you think of other ethical problems that might arise at each of these organizations?

(b) Apply the four-step ethical decision-making procedure to gain insight into the nature of the decision to be made. In some cases, you may decide that an ethical issue is not really involved, but you should be able to justify why it is not.

(c) List 10 other types of organizations at which engineers would confront ethical problems, and explain the nature of the ethical decisions to be made.

(d) How does one find a balance between profit and environmental concerns?

(e) Under what circumstances should an engineer be held liable for personal injury or property damage caused by the products of his or her labor?

(f) Under what situations would you blow the whistle on your superior or your company?

(g) Should attorneys specializing in personal injury and property damage litigation be allowed to advertise, and if so, in what venues?

Table 2-1 Industries and issues for ICA 2-8

Organization/Occupation	Possible Issues
Alternative energy providers	Use of heavy metals in photovoltaic systems Effect of wind generators on bird populations Aesthetic considerations (e.g., NIMBY) Environmental concerns (e.g., Three Gorges project)
Environmental projects	Fertile floodplains inundated by dams/lakes Safety compromised for cost (e.g., New Orleans levees) Habitat destruction by projects Habitat renovation versus cost (e.g., Everglades)
Chemical engineers	Toxic effluents from manufacturing process Pesticide effect on ecosystem (e.g., artificial estrogens) Insufficient longitudinal studies of pharmaceuticals Non-biodegradable products (e.g., plastics)
Civil engineers	Runoff/erosion at large projects Disruption of migration routes (freeways) Quality of urban environments Failure modes of structures
Computer engineers	Vulnerability of software to malware Intellectual property rights (e.g., illegal downloads) Safety issues (e.g., programmed medical devices, computer-controlled transportation)
Electrical engineers	Toxic materials in batteries Cell phone safety concerns Power grid safety and quick restoration in crises Possible use to break the law (e.g., radar detectors)
Electric power industry	Shipment of high sulfur coal to China Disposal of nuclear waste Environmental issues (e.g., spraying power-line corridors)
Food processing industry	Health possibly compromised by high fat/sugar/salt products Use of genetically engineered organisms Sanitation (e.g., *Escherichia coli*, *Salmonella*) Use of artificial preservatives
Manufacturing companies	Manufacturing in countries with poor labor practices Lax safety standards in some countries Domestic jobs lost Environmental pollution due to shipping distances Trade imbalance
Industrial engineers	Quality/safety compromised by cost Efficiency versus quality of working environment Management of dangerous tools and materials
Mechanical engineers	Automotive safety versus cost Environmental impact of fossil fuels Robot failsafe mechanisms

CHAPTER 3
DESIGN AND TEAMWORK

Regardless of your selected engineering discipline or career path, communication is critical for survival as an engineer. Due to the complexity of many analysis and design projects, it is necessary for all engineers to operate effectively on a team. For a project to be completed successfully, there needs to be a synergy of all entities engaged, including design and teamwork. This chapter introduces the two together to emphasize the importance of teamwork in the design process.

3.1 THE DESIGN PROCESS

Design is a creative process that requires problem definition, idea searches, solutions development, and solutions sharing. Design is inherently multiobjective, so one expects that any problem addressed will have multiple solutions. While a particular solution might address some objectives very well, other objectives might not be well suited at all. The goal is to identify the design that meets the most important objectives. To evaluate ideas and communicate them to others, engineers commonly sketch possible solutions and even build models of their work.

There are many different versions of "the design process," but they have a lot of similarities. Recognizing that design is a creative process, it is more important to think about "a design process." The process we will use is the one used to design engineering education itself. ABET (pronounced with a long "A") was once the Accreditation Board for Engineering and Technology but is known now by just the acronym because it accredits computing and applied science programs as well. In 1932 ABET was established as the Engineers' Council for Professional Development (ECPD). ECPD was formed to fill the apparent need for a "joint program for upbuilding engineering as a profession." Currently, ABET accredits some 2,700 programs at more than 550 colleges and universities nationwide. Accreditation is a non-governmental, peer-review process. ABET accreditation is assurance that a college or university program meets the quality standards established by the profession for which it prepares its students.

The engineering profession is self-governing; the Engineering Accreditation Commission of ABET is made up of member engineering societies, and the Commission operates under a philosophy of outcomes assessment. This approach focuses on determining the outcomes we desire in engineering graduates and allowing the engineering degree program the flexibility of determining how to achieve those outcomes and demonstrating how they have been achieved.

The design process for engineering programs is made up of two iterative processes, shown in Figure 3-1. The iterative loop on the left comes first because a new program would begin there; it includes getting input from constituencies, determining educational objectives, and evaluating or assessing how well those objectives are being

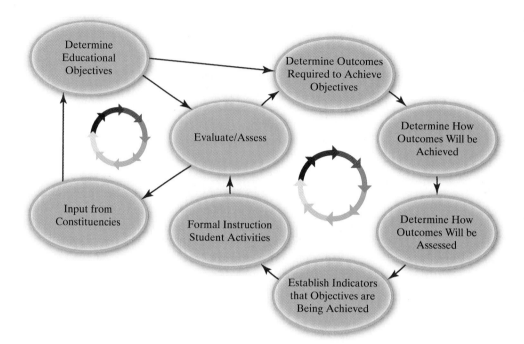

Figure 3-1 The ABET design approach.

achieved. In other engineering processes, these steps might be called something like "problem definition." "Constituencies" may also be called users, clients, stakeholders, or other terms. The process is iterative because it is important to confirm that the constituencies are pleased with the results, to adapt to changing needs, and to achieve the continuous improvement expected by the engineering profession.

With the problem identified (knowing our educational objectives), the iterative process on the right side of Figure 3-1 begins. This process occurs primarily in the designer's workspace, whatever that is. In the case of designing engineering curriculum, the process takes place within the walls of the university or college. Knowing the objectives, the design team determines the outcomes that will accomplish those objectives, how those outcomes will be achieved, how they will be assessed, and what indicators will demonstrate success before any students are actually taught. Once students have had learning experiences (including extracurricular experiences), evaluation and assessment guide both processes into another cycle.

The step "determine outcomes required to achieve objectives" is called "problem definition" or "specification" in many other design processes. This step is critical because it shapes all the others. "Determine how the outcomes will be achieved" is a particularly creative step in the process and is commonly referred to as "generating ideas," "innovating," "developing possible solutions," or something similar, and might include "research." "Determine how the outcomes will be assessed" is a step that might not be mentioned if most agree on how to measure the success of a design, but in designs with more complex objectives, designers must think carefully about what will be measured and how, finishing up with determining indicators of success. These steps are commonly called "analysis" in a more general design process, breaking down the design to examine its assumptions, benefits, and risks. The remaining part of the process is one of "prototyping," "implementation," and "testing."

What follows in this chapter describes the tools engineers use in this process.

3.2 BRAINSTORMING IN THE DESIGN PROCESS

Most high-school graduates have heard the term **brainstorming**, and many have participated in a process called by that name. This term refers to one particular process by which ideas are generated. Idea generation follows three common rules: encourage a lot of ideas, encourage a wide variety of ideas, and do not criticize. The third rule is important if the first two are to be achieved. It has been said: "The best way to get a good idea is to get lots of ideas!"

Where is quality in this process? We have often been encouraged on "quality, not quantity." While some idea-generation methods facilitate the identification of a small number of quality solutions, in brainstorming, "quality *is* quantity." The nature of generating and evaluating ideas is that the greater the number of ideas generated, the more likely that several high-quality ideas are found.

The following story is taken from Alex Osborn's *Applied Imagination* (out of print), illustrating the importance of generating ideas before evaluating them.

> *In November 1952, in Washington State, the local telephone company had to clear the frost from 700 miles of telephone lines to restore long distance service. The company believed strongly enough in the importance of variety in the process that ALL the company's employees were asked to participate in a brainstorming session, executives and secretaries, engineers and linemen.*
>
> *After some time of idea generation, it was clear that the participants needed a break. One of the sessions overheard one lineman say to another at the water fountain, "Ever had a bear chase you up one of the poles? That would sure shake the ice off the lines!" The facilitator encouraged the lineman to repeat himself when the session reconvened.*
>
> *The facilitator hoped that the unusual suggestion would encourage new ideas, and the lineman sheepishly offered his suggestion. "How should we encourage the bears to climb all the poles?" asked the facilitator. "We could put honey pots at the tops of all the poles!" shouted someone else in the room. "How should we put honey pots on the tops of all the poles?" asked the facilitator. "Use the company helicopters that the executives use!" piped in another participant. "Hmm," said one of the company secretaries calmly. "When I served in the Korean War, I was impressed by the force of the downdraft off helicopter blades. I wonder if that would be enough to shake the ice off the power lines."*

The idea was so intriguing that it was tested immediately—it provided a successful and economically viable solution. This story clearly illustrates the benefit of variety and the value of avoiding criticism in a brainstorming session. The rest of the story illustrates how essential quantity is to the process.

A problem-solving group composed of five veteran air force helicopter pilots was convened to address this same problem. Each was unfamiliar with the solution that had already been implemented. It was hoped that because of their background, they would eventually arrive at the same solution. In fact, they did, but it was the 36th idea on their list. If they had stopped after generating 35 ideas, they may very well have had an acceptable solution, but it might not have been as elegant or as successful as the proven solution.

NOTE

Brainstorming Rules:

- More is better.
- Variety is better.
- Do not criticize.

NOTE

Another great example of brainstorming is baking soda. Originally marketed for culinary use, today it is found in hundreds of products. How do you sell more product? Find more ways to use it!

3.3 EXPERIMENTAL DESIGN: PERIOD ANALYSIS

Experiments enable engineers to come up with a creative solution to a problem and test the validity of the proposed idea. An experiment is a test of a proposed explanation of a problem. A good design of an experiment is a critical part of the scientific method.

What Constitutes the Scientific Method?

1. Observation: Observe the problem and note items of interest.
2. Hypothesis: Search for a known explanation of the phenomenon or attempt to formulate a new explanation.
3. Prediction: Create a model or prediction of behavior based on that hypothesis.
4. Experiment: Test your predictions. If necessary, modify your hypothesis and retest.

Why Is Experimental Design Important?

As you move through your college career, you will be inundated with many equations and theories. These are useful in solving a wide variety of problems. However, as you will see, often the equations are only really useful in solving the most basic type of problems.

As an example, suppose you are interested in the speed of a ball as it rolls across the floor after rolling down a ramp. In physics, you will learn the equations of motion for bodies moving under the influence of gravity. If you are good, you can use these to examine rolling balls. What you will quickly find, however, is that numerous complicating factors make it difficult to apply the basic equations to obtain an adequate answer. Let us suppose you are interested in smooth balls (such as racquetballs), rough balls (tennis balls), heavy balls (bowling balls), and lightweight balls (ping-pong balls). The simplified equations of motion predict that all these will behave in essentially the same way. You will discover, however, that the drag of the air affects the ping-pong ball, the fuzz affects the tennis ball, and the flexible nature of the racquetball will allow it to bounce at steep ramp angles. It is difficult to predict the behavior analytically. Often, one of the quickest ways to learn about the performance of such complex situations is to conduct experiments.

What Are Experimental Measurements?

NOTE

Do you know what these unique measurement instruments do?

- Durometer
- Dynamometer
- Euidometer
- Galvanometer
- Gyroscope
- Manometer
- Opisometer
- Pycnometer
- Tachymeter
- Thiele tube

Most scientific experiments involve measuring the effect of variability of an attribute of an object. In an experiment, the **independent variable** is the variable that is controlled. The **dependent variable** is a variable that reacts to a change in the independent variable. A **control variable** is part of the experiment that can vary but is held constant to let the experimenter observe the influence of the independent variable on the dependent variable. Keeping control variables constant throughout an experiment eliminates any confounding effects resulting from excess variability.

Any measurement acquired in an experiment contains two important pieces of information. First, the measurement contains the actual value measured from the instrument. In general, a measurement is some physical dimension that is acquired with some man-made data-collection instrument. As with any man-made device, there may be some imperfection that can cause adverse effects during data collection. Thus, the second piece of information that goes along with any measurement is the level of uncertainty.

Any uncertainty in measurement is not strictly by instrumentation error. Systematic error is any error resulting from human or instrumentation malfunction. Random error is caused by the limits of the precision of the data-collection device. It is possible to minimize the systematic error in an experiment, but random error cannot be completely eliminated.

What Measurements Do You Need to Make?

You need to develop a coherent experimental program. You should make enough measurements to answer any anticipated questions, but you do not usually have the time or money to test every possible condition. Points to consider:

■ What are the parameters of interest?
■ What is the range of these parameters—minimum values, maximum values?
■ What increments are reasonable for testing (every 10 degrees, every 30 seconds, etc.)?
■ What order is best to vary the parameters? Which should be tested first, next, etc.?

Here is an acronym (PERIOD) that can help you remember these important steps. As an example, it is applied to the problem of the ramp and rolling balls, described above.

P – Parameters of interest determined

■ Parameter 1 is the ramp angle.
■ Parameter 2 is the distance up the ramp that we release the ball.
■ Parameter 3 is the type of ball.

E – Establish the range of parameters

■ Ramp angle can vary between 0 and 90 degrees in theory, but in reality can only vary between 10 degrees (if too shallow, ball would not move) and 45 degrees (if too steep, ball will bounce).
■ The distance we release the ball up the ramp can vary between 0 and 3 feet in theory, assuming that the ramp is 3 feet long. We cannot release the ball too close to the bottom of the ramp or it would not move. In reality, we can only vary between 0.5 and 3 feet.
■ We will test as many types of balls as we have interest in.

R – Repetition of each test specified

■ The ramp angle will be set according to the height of the ramp from the floor, so there is not much room for error in this measurement; only one measurement is needed for such geometry.
■ Each placement of the ball before release will vary slightly and may cause the ball to roll slightly differently down the ramp; this is probably the most important factor in determining the speed, so three measurements at each location are needed.
■ We will assume that every ball is the same, and the actual ball used will not change the outcome of the experiment; only one ball of each type is needed.

I – Increments of each parameter specified

■ We will test every 10 degrees of ramp angle, starting at 10 degrees and ending at 40 degrees.
■ We will release the balls at a height of 0.5, 1, 1.5, 2, 2.5, and 3 feet up the ramp.
■ We will test five types of balls: racquetball, baseball, tennis ball, ping-pong ball, and bowling ball.

O – Order to vary the parameters determined

- We will set the ramp angle and then test one ball type by releasing it at each of the four different distances up the ramp.
- We will repeat this process three times for each ball.
- We will then repeat this process for each type of ball.
- We will then change the ramp angle by 10 degrees and repeat the process.
- This process is repeated until all conditions have been tested.

D - Determine number of measurements needed and Do the experiment

It is always important to determine before you start how many measurements you need to make. Sometimes you can be too ambitious and end up developing an experimental program that will take too much effort or cost too much money. If this is the case, then you need to decide which increments can be relaxed, to reduce the number of overall measurements.

The number of measurements (N) you will need to make can be easily calculated by the following equation for a total of n parameters:

N = (# increments parameter 1 * number of repetitions for parameter 1) *

(# increments parameter 2 * number of repetitions for parameter 2) *. . .

(# increments parameter n * number of repetitions for parameter n) * . . .

Continuing the examples given above, the number of actual measurements that we need to make is calculated as

N = (4 angles)*(6 distances * 3 repetitions)*(5 types of balls) = 360 measurements

In this example, 360 measurements may be extreme. If we examine our plan, we can probably make the following changes without losing experimental information:

- We decide to test every 10 degrees of ramp angle, starting at 20 degrees and ending at 40 degrees. This will lower the angle testing from four to three angles.
- We will release the balls at a height of 1, 2, 2.5, and 3 ft up the ramp. This will lower the distances from six to four.
- We will test three types of balls: racquetball, ping-pong ball, and bowling ball. This will lower the type of balls from five to three.

The number of actual measurements that we now need to make is calculated as

N = (3 angles)*(4 distances * 3 repetitions)*(3 types of balls) = 108 measurements

This result seems much more manageable to complete than 360!

3.4   CRITERIA AND EVALUATION

We can generate and evaluate problem definitions, implementation strategies, and even the methods we use to evaluate our implementations. Brainstorming problem definitions stresses how the way we define a problem affects the solutions we consider. Is the problem that there is ice on the electrical lines, that there is water on the lines that can freeze, or that the lines and surrounding air are too cold? Looking at the problem these three different ways suggests different solution approaches. The second might suggest covering the lines to keep them dry or coating them with a material with special hydrophobic properties. The third might suggest heating the lines or moving them to somewhere warmer (maybe underground).

There are countless examples of how redefining a problem has been commercially profitable. When Dum Dum Pops® are made, the process is continuous, which causes the manufacture of some lollipops that combine two of the flavors. This might have been considered a problem, and some companies would have designed their manufacturing process so as to stop and clean the machinery between flavor runs. Instead, the Dum Dum company wraps these "combination" flavor lollipops and labels them "Mystery" flavor. By their defining the problem a different way, a creative solution emerged, saving money and providing a market attraction. In all designs, the way in which the problem is defined will affect the set of solutions explored. Once a particular solution approach is selected, it might be appropriate to generate and evaluate ideas for how to implement that solution.

Criteria: Deciding What Is Important

First, the problem definition must be clear so that we can generate an appropriate set of criteria. Then, when we stop generating solutions and begin evaluating the various solutions, we use those **criteria** to narrow our choices objectively. We might be particularly concerned about cost, so certain approaches become less attractive. Other proposed solutions might be hazardous, thus further narrowing our options. The criteria for evaluating potential solutions must be well defined before evaluation begins so that all ideas can be considered fairly—this helps avoid arguments based on hidden criteria.

Defining criteria before brainstorming solutions seems as if it might threaten the "do not criticize" objective—proposed solutions could be steered to match the expectations established by the criteria. Unfortunately, while defining criteria first might constrain idea generation, generating solutions first might cause participants to develop an opinion of a "best solution" and then force their criteria choices to fit their preferred idea. In this way, the process of generating solutions and the process of generating criteria to evaluate those solutions are linked. The second interaction is more damaging to the design process, so we will focus on identifying constraints and criteria first.

COMPREHENSION CHECK 3-1

We frequently express criteria in terms that are not clear and measurable. Write a clear criterion to replace each of these vague criteria for the products.

Product	Computer	Automobile	Bookshelf
Inexpensive	Less than $300		
Small			
Easy to assemble			Requires only a screwdriver
Aesthetically pleasing			
Lightweight			
Safe			
Durable			
Environmentally friendly		Has an estimated MPG of at least 50	

In general, using fewer criteria keeps things simple. A rule of thumb is that if you can think of 10 criteria that are meaningful, you should consider only the two most important criteria to compare solutions. In this way, you can be sure that the important criteria maintain their importance in your decision-making. Seek consensus on what the most important criteria are. Experts, specialists, managers, customers, research, etc., can help you focus on the most important criteria. In certain situations, many criteria are used—for example, when a magazine rates consumer products, extra criteria should be included to be sure that all the criteria important to the magazine's readers have been considered.

Another situation, in which a large number of criteria are included, occurs when very complex decisions are being evaluated. The U.S. Green Building Council has long lists of criteria used for certifying various types of projects according to the Leadership in Energy and Environmental Design (LEED) Green Building Rating System. A subset of items from the LEED for Homes project checklist is shown below.

LEED is an internationally recognized green building certification system, providing third-party verification that a building or community was designed and built using strategies aimed at improving performance across all the metrics that matter most: energy savings, water efficiency, CO_2 emissions reduction, improved indoor environmental quality, and stewardship of resources and sensitivity to their impacts.

From the U.S. Green Building Council website: www.usgbc.org

LEED
FOR HOMES

LEED for Homes Simplified Project Checklist
Addendum: Prescriptive Approach for Energy and Atmosphere (EA) Credits

				Max Points	Project Points	
					Preliminary	Final
Points cannot be earned in both the Prescriptive (below) and the Performance Approach (pg 2) of the EA section						
Energy and Atmosphere (EA)	(No minimum Points Required)	OR		Max	Y/Pts Maybe No	Y/Pts
2. Insulation	2.1 Basic Insulation			Prereq		
	2.2 Enhanced Insulation			2	0 0	0

These checklist items are neither clear nor measurable, however, as shown above. The full LEED for Homes rating system and the reference guide associated with the rating system provide the additional detail needed to make the insulation criteria both clear and measurable. Note that these criteria build from existing standards, meaning that even greater levels of detail can be found elsewhere (e.g., Chapter 4 of the 2004 International Energy Conservation Code and The National Home Energy Rating Standards).

Prerequisites

2.1 Basic Insulation. Meet all the following requirements:

(a) Install insulation that meets or exceeds the R-value requirements listed in Chapter 4 of the 2004 International Energy Conservation Code. Alternative wall and insulation systems, such as structural insulated panels (SIPs) and insulated concrete forms (ICFs), must demonstrate a comparable R-value, but thermal mass or infiltration effects cannot be included in the R-value calculation.

(b) Install insulation to meet the Grade II specifications set by the National Home Energy Rating Standards (Table 16). Installation must be verified by an energy rater or Green Rater conducting a predrywall thermal bypass inspection, as summarized in Figure 3.

Note: For any portion of the home constructed with SIPs or ICFs, the rater must conduct a modified visual inspection using the ENERGY STAR Structural Insulated Panel Visual Inspection Form.

Note that the LEED criteria specify only those characteristics that are important to the overall goal of green building. This leaves the choice of the type of insulation (fiberglass, cellulose) up to the designer and the consumer. Because homes, outdoor facilities, and factories face different challenges, different LEED criteria apply. It is important to recognize which choices are "options" rather than criteria—even the most energy efficient factory cannot substitute for a house. Further, the LEED criteria allow different thresholds for different size houses, so as not to penalize multifamily households—even though multifamily households will require more material to build and more energy to operate, the marginal cost is not as high as building another home.

When designing a product or a process, you must first listen to what a client wants and needs. The "client" might be from a department such as manufacturing, marketing, or accounting; a customer; your boss; the government; etc. Subsequently, you will need to translate those wants and needs into engineering specifications. Pay particular attention to separating the wants (criteria) from the needs (requirements). Requirements can be imposed legally (e.g., Corporate Average Fuel Economy regulations, building codes), they can be from within your company (e.g., the finished product must not cost more than $50 to manufacture), or they may come from your client (e.g., the product must be made from at least 90% recycled material).

COMPREHENSION CHECK 3-2

You have been asked to improve the fuel efficiency of an automobile by 20%. Convert this request into engineering criteria. What changes might be made to the automobile to achieve this objective?

Evaluation: Deciding on Options

Simply identifying criteria is not enough information for a decision—each proposed solution must be evaluated against the criteria. The first step will be to eliminate any solutions that do not meet the minimum requirements—those solutions are out-of-bounds and need not be considered further.

Once the minimum requirements have been satisfied, there are many ways of applying the remaining criteria to select a solution. One approach is to make **pairwise comparisons**. In this approach, you use a table for each criterion to summarize how each of the solutions compares with others. An example is shown in Table 3-1 for the criterion "safety." Among the table entries, 0 indicates that that option in that column is worse than the option in that row; 1 indicates that both options rank equally; and 2 indicates that that option in that column is better than the option in that row.

Table 3-1 Comparing options

Safety	Option 1	Option 2	Option 3
Option 1		0	1
Option 2	0		0
Option 3	1	2	
Total	1	2	1

Since a high level of safety is preferred, "better" means "safer." In the example, the first column indicates that option 1 is less safe than option 2, but the same as option 3. The resulting totals indicate that option 2 ranks best in terms of safety. To complete the process, generate similar tables for each of the criteria being considered and then sum the totals to identify the best solution.

A disadvantage of the pairwise comparisons approach is that all criteria have equal weight, whereas some criteria are likely to be more important than others. An alternative approach is to use **weighted benefit analysis**, shown in Table 3-2. In this approach, each option is scored against each of the criteria.

Table 3-2 Options with a weighted benefit analysis

	Weights	Option 1	Option 2	Option 3
Cost	2	2	6	10
Safety	8	10	4	6
Weight	10	7	7	2
Wow	5	2	4	6
Totals		4 + 80 + 70 + 10 = 164	12 + 32 + 70 + 20 = 134	20 + 48 + 20 + 30 = 118

Table 3-3 Sample scoring rubric

Score	Meaning
0	Not satisfactory
1	Barely applicable
2	Fairly good
3	Good
4	Very good; ideal

In Table 3-2, the weights are in the first column and each option has been assigned a score from 0 to 10, indicating how well that option meets each criterion. This approach may be inconsistent in that a "7" for one rater may be different from a "7" for another rater, so it can help to better define the scale. For example, the options may be scored on the scale shown in Table 3-3 as to how well the option fits each criterion.

Reality: Checking It Out

When a solution has been identified as the best fit to the criteria, it is best to stop for a reality check before moving forward. After the evaluation process, some ideas will be left on the cutting room floor. Do any of these merit further consideration? Are there important elements of those ideas that can be incorporated into the chosen design? If the reality check reveals that an idea really should not have been eliminated, then a change in the selection criteria may be appropriate.

After deciding, you will implement your chosen solution. Undoubtedly, both carrying out the design and using the design once it is complete will provide new information about how the design might be improved. In this way, design tends to be iterative—design, build, test, redesign, and so on. Even when a design performs as expected, it may be important to test a model extensively or even build multiple models to be sure that the design is reliable.

WISE WORDS: IN YOUR JOB, DO YOU WORK ALONE OR ON A TEAM?

I often work alone on projects and analyses, but I do need to interact with client company teams, sometimes leading the team as an outside expert, but still a "member" of the company I am trying to help. Some of the teams I will work with also have mechanical, chemical, industrial, and process engineers involved.

E. Basta, MSE

I work on a team made up of all electrical and computer engineers.

E. D'Avignon, CpE

Team: chemist, package engineer, process development engineer (chemical engineer), industrial designer, line engineer (mechanical engineer or chemical engineer), process engineer (chemical engineer), planner, quality engineer, marketing, and market research.

S. Forkner, ChE

I work mostly alone on my assignments. I am given tasks and I have to find solutions for my problems on my own. Only when needed I consult someone for questions and guidance on my task.

V. Gallas Cervo, ME

I worked the first 5 years of my career as a "sole contributor" in an engineering role. Since then, I have worked in a team setting managing technical employees.

L. Gascoigne, CE

I work with a chemical engineer, an instrumentation engineer, a mechanical engineer, and a piping engineer.

D. Jones, BE

Every project involves a team. My typical team includes surveyors, structural engineers, environmental scientists (wetlands, endangered species), permit specialists, electrical engineers, land appraisers, archaeologists, and architects.

J. Meena, CE

I work alone, but seek advice from my management team.

C. Pringle, IE

I work on a team including mechanical, electrical, aeronautical and systems hardware designers, software designers, integrators, power engineers, and other liaisons like myself.

E. Styles, EE

The Hubble Space Telescope servicing mission project is made up of a team of mechanical design engineers, human factors engineers, environmental test engineers, thermal engineers, underwater test engineers, safety engineers, mechanical technicians, documentation specialists, space scientists, and systems engineers.

R. Werneth, ME

3.5 SUSTAINABILITY

Contributed by: Dr. Leidy Klotz, Associate Professor,
Civil Engineering, Clemson University

The most common definition of sustainability is "meeting the needs of the present without compromising the ability of future generations to do the same."[1] Notice that this definition is fundamentally about people. There is no mention of hippies or saving trees just for the sake of saving trees (of course, we are dependent on the ecosystem services trees provide). Notice also that the definition includes future generations as well as present ones. Sustainability is not just an issue for our children and grandchildren, it is an issue that is affecting all of us right now.

"Environmental" and sustainability are often used interchangeably; however, sustainability also has social and economic dimensions. All three of these dimensions must be balanced for truly sustainable engineering solutions. The figures show the relationships between these dimensions. Our society would not exist if our environment did not support human life. Our economy would not exist if we did not have a stable society (most people do not want to start businesses in failed states). These relationships seem quite obvious, but can be overlooked if we just focus on one dimension of sustainability. As business leader Peter Senge points out: "the economy is the wholly owned subsidiary of nature, not the other way around."

You can apply a basic understanding of sustainability to your own engineering solutions. Sustainability is not a stand-alone topic. It cannot be bolted onto an engineering design at the end of the project. For example, in a new building project, one of the first sustainability considerations should be whether this project is even necessary. Perhaps similar goals could be achieved by more efficient use of existing facilities. This is quite an ethical dilemma! Imagine telling a potential client they do not need to hire and pay your engineering firm to design a new building. Assuming the building project is necessary, some of the best sustainability opportunities occur early on in the project, during project planning and design. This is where engineers play a key role. Teamwork and communication in the process are vital because sustainable solutions require consideration of multiple issues. We must be able to work with engineers from different disciplines and with non-engineers such as architects, contractors, and lawmakers. We must be able to communicate with the end users who will occupy and operate the building. After all, the end user is the recipient of your design. These basic ideas apply across disciplines, whether you are designing a building, an engine, or a new material.

You may be wondering how much humanity currently considers sustainability. Maybe we are already on a sustainable path? Unfortunately, this is not the case. Our use of critical resources, such as energy and water, and less critical resources, such as tequila and chocolate, cannot be sustained at current rates. Increasing population and affluence will stress these resources even more. Allocation of resources for the present and future is a huge ethical question engineers must consider. Should you build a reservoir that will provide water for an impoverished area, but restrict availability downstream? Do the risks associated with nuclear energy outweigh the fact that it is a carbon-free source of energy?

Creating solutions for sustainability issues requires expanding the boundaries of single-discipline thinking, being able to recognize relationships between systems and the associated problems and opportunities. For example, our fossil-fuel based energy system has increased standards of living all over the world but this same system also contributes to climate change, which is already having significant negative impacts, with more predicted for the future. In addition, the system contributes to inequalities between those who

---

[1] From the UN's Brundtland report.

have energy and those who do not, which is a major source of poverty and conflict. These complex relationships can make problems seem overwhelming; however, these relationships also offer opportunities. Engineers creating sustainable energy solutions will have positive impacts in multiple areas, such as helping to curb climate changing emissions, while reducing energy poverty and resource conflict. You can make a conscious effort to build your skills in the broad, systems-thinking needed to identify these opportunities.

PERSONAL REFLECTION ON SUSTAINABILITY

Dr. Leidy Klotz

I see sustainability issues as challenges, but also as incredible opportunities for engineers. Is your goal to save the world? Here is your chance. Is your goal to make as much money as possible? Engineering solutions that address sustainability issues offer huge opportunities for profit. Those who figure out ways to make solar energy more economical or provide greater access to clean water will be the Bill Gates and Steve Jobs of their time.

In particular, young engineers must play a key role. It is unlikely that an engineer who graduated before 2005 was exposed to sustainability topics during college. This is an area where older engineers need your help, where you can be a leader right away. History shows us that groundbreaking advances, like those needed in engineering for sustainability, are often made by young people. Albert Einstein had his most groundbreaking year at age 26, the same age at which Martin Luther King Jr. led the Montgomery bus boycott. Thomas Jefferson wrote the declaration of independence at 33, and Harriet Tubman started the Underground Railroad at 28. We need similar innovative ideas and bold actions in all areas of engineering for sustainability. I think young people are our best shot.

Please work hard to learn more about sustainability in engineering. In your area of engineering, learn as much as possible about the fundamentals and how they are related to sustainability. Develop your broad, systems-thinking skills. Take classes that provide additional information on engineering for sustainability. Pursue opportunities for hands-on practice with engineering for sustainability on your campus and beyond. I think you will truly enjoy the collaborative process and unique design challenges associated with creating sustainable solutions!

3.6 WORKING IN TEAMS

> The ability to work in a team is one of the most critical traits an engineer needs. Even if you're the greatest engineer in the world, you will not know all of the answers or have all of the right ideas. We can always learn something from our peers.
>
> *A. Thompson, EE*

Group: *A number of people who come together at the same place, at the same time.*

Team: *Individuals cooperating to accomplish a common goal.*

As a student and in the workplace, you will complete some assignments individually and complete some as part of a team. When you work independently, you are mostly free to choose when and how you will work. When you work as part of a team, make sure the team has ground rules for how it will operate. Any time several people are asked to work closely together, there is a potential for much good from a diversity of ideas and skills but there is also a potential for conflict. Because conflict can be both productive and unproductive, you need to manage it.

Team Behavior

The most critical task for a team, particularly a new team, is to establish its purpose, process (its way of doing things), and a means of measuring team progress. Here are several topics regarding team behavior that you may wish to consider.

- **Ground rules:** Each team needs to come to a consensus about acceptable and unacceptable individuals as well as team behavior.

We trained hard but it seemed that every time we were beginning to form teams we would be reorganized. I was to learn later in life that we tend to meet every situation by reorganizing, and a wonderful method it can be for creating the illusion of progress while producing confusion, inefficiency, and demoralization.

Petronius

- **Decision making:** Teams by necessity make decisions. Each team needs to decide how these decisions will be made. For example, will they be done through consensus, majority vote (either secret or show of hands), or by other methods?
- **Communication:** This is often one of the hardest parts of working effectively as a team. Team members need to recognize the value of real listening and constructive feedback. During the course of team meetings, *every* team member needs to participate *and* listen.
- **Roles:** You may adopt various roles on your team. In a long-term project, roles should rotate so that everyone has a chance to learn each role.
- **Participation:** Decide as a team how work will be distributed. Your team should also consider how to handle shifts in workload when a team member is sick or otherwise unavailable.
- **Values:** The team as a whole needs to acknowledge and accept the unique insights that each team member can contribute to their work.
- **Outcomes:** Discuss and agree on what types of measures will be used to determine that the team has reached its final goal.

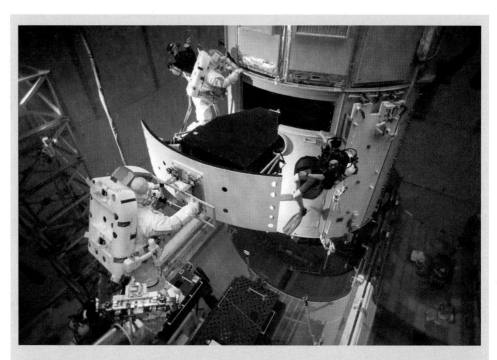

Underwater Training for a Hubble Space Telescope Servicing Mission. NASA engineers (on SCUBA) and astronauts (in modified space suits) take advantage of the effects of neutral buoyancy to practice replacing a Wide Field Camera in preparation for a Hubble Space Telescope (HST) servicing mission. Extensive teamwork is required in a 6.2-million gallon pool in Houston, TX, to develop, refine, and practice the procedures to be used on spacewalks in orbit. Engineers from Goddard Space Flight Center in Greenbelt, MD, and Johnson Space Center in Houston work together to perfect the procedures and hardware, including specialized astronaut tools. The team uses models representing the flight items for conducting end-to-end tasks in the neutral buoyancy facility. This unique example of engineering teamwork has resulted in five very successful HST servicing missions involving complex astronaut spacewalks. The challenge to the team is to develop the nominal and contingency procedures and tools on the ground (and in the water!) to be used for mission success with HST in orbit 300 miles up.

R. Werneth

Teammate Evaluation: Practicing Accountability

As previously discussed, engineering is a self-governing profession. ABET is charged with accrediting engineering programs in the United States through membership from the profession. Among a set of required outcomes of engineering graduates, ABET requires that engineering students graduate with an ability to function on multidisciplinary teams. To ensure that each student achieves this outcome, individual accountability is needed. Realizing that much of the activity of a team happens when the team is meeting privately (without a professor), an effective and increasingly common way of addressing this tenet is to have team members rate one another's performance.

It is important to learn how to be an effective team member now, because most engineering work is done in teams, and studies show that most engineering graduates will have supervisory responsibility (at least project management) within 5 years of graduation. You have worked in teams before, so you have probably noticed that some team members are more effective than others. Consider these three snapshots of interactions with engineering students. All three are true stories.

- Three team members approach the professor, concerned that they have not seen the fourth team member yet. The professor speaks with the student, who quickly becomes despondent, alerting the professor to a number of serious burdens the student is bearing. The professor alerts an advisor and the student gets needed help.
- A student, acting as the team spokesperson, tells the professor that one team member never comes to meetings. The professor speaks to the nonparticipating student, who expects to be contacted about meetings by cell phone, not email. The professor explains that the student's expectations are unrealistic.
- A student is insecure about being able to contribute during team activities. After the team's ratings of that student are in, the professor talks to the class about the importance of participating and the different ways students can contribute to a team. In the next evaluation, the student receives the highest rating on the team.

Peer evaluations are a useful way for team members to communicate to one another and to their professor about how the members of a team are performing. Reviewing and evaluating job performance is a marketable skill and is as useful to the employee seeking a job or a job advancement as it is to the supervisor.

Focus on What Your Teammates Do Rather Than What You Think of Them

It is challenging to give a team member a single rating on their effectiveness as a teammate because some team members will be helpful to the team in some ways, but engage in some behaviors that hinder the team. Another difficulty is that each team member is likely to consider some ways of contributing more valuable than others, so the evaluation of a particular teammate will be overly influenced by that teammate's performance in certain areas. The only way to be fair is to focus on behaviors—what your teammates do—rather than opinions such as how you feel about them. One way to focus on behaviors would be to ask you to take an inventory of what behaviors your teammates demonstrate and how often. The result would be that you might need to answer 50 or more questions about each member of the team. It is difficult to stay focused on answering accurately when completing such a long survey. A better way to focus on behaviors is by using sample behaviors to anchor each point of a rating scale. A peer evaluation instrument that is widely used in engineering education is the

Comprehensive Assessment of Team-Member Effectiveness (CATME, see www.catme.org). CATME measures five different types of contributions to a team using such a behaviorally anchored rating scale. Each scale includes representative behaviors describing exceptional, acceptable, and deficient performance in each area. Recognizing that an individual team member may exhibit a combination of behaviors, the CATME instrument also includes "in-between" ratings. The five types of contributions are described below the associate behaviors.

Contributing to the Team's Work describes a team member's commitment to the effort, quality, and timeliness of completing the team's assigned tasks.

- A student who is exceptional at contributing to the team's work
  - Does more or higher-quality work than expected
  - Makes important contributions that improve the team's work
  - Helps to complete the work of teammates who are having difficulty
- A student who does an acceptable job at contributing to the team's work
  - Completes a fair share of the team's work with acceptable quality
  - Keeps commitments and completes assignments on time
  - Fills in for teammates when it is easy or important
- A student who is deficient at contributing to the team's work
  - Does not do a fair share of the team's work. Delivers sloppy or incomplete work
  - Misses deadlines. Is late, unprepared, or absent for team meetings
  - Does not assist teammates. Quits if the work becomes difficult

Interacting with Teammates measures how a team member values and seeks contributions from other team members.

- A student who is exceptional at interacting with teammates
  - Asks for and shows an interest in teammates' ideas and contributions
  - Improves communication among teammates. Provides encouragement or enthusiasm to the team
  - Asks teammates for feedback and uses their suggestions to improve
- A student who does an acceptable job at interacting with teammates
  - Listens to teammates and respects their contributions
  - Communicates clearly. Shares information with teammates. Participates fully in team activities
  - Respects and responds to feedback from teammates
- A student who is deficient at interacting with teammates
  - Interrupts, ignores, bosses, or makes fun of teammates
  - Takes actions that affect teammates without their input. Does not share information
  - Complains, makes excuses, or does not interact with teammates. Accepts no help or advice

Keeping the Team on Track describes how a team member monitors conditions that affect the team's progress and acts on that information as needed.

- A student who is exceptional at keeping the team on track
  - Watches conditions affecting the team and monitors the team's progress
  - Makes sure teammates are making appropriate progress
  - Gives teammates specific, timely, and constructive feedback
- A student who does an acceptable job at keeping the team on track
  - Notices changes that influence the team's success
  - Knows what everyone on the team should be doing and notices problems
  - Alerts teammates or suggests solutions when the team's success is threatened
- A student who is deficient at keeping the team on track
  - Is unaware of whether the team is meeting its goals
  - Does not pay attention to teammates' progress
  - Avoids discussing team problems, even when they are obvious

Expecting Quality is about voicing expectations that the team can and should do high-quality work.

- A student who is exceptional at expecting quality
  - Motivates the team to do excellent work
  - Cares that the team does outstanding work, even if there is no additional reward
  - Believes that the team can do excellent work
- A student who does an acceptable job at expecting quality
  - Encourages the team to do good work that meets all requirements
  - Wants the team to perform well enough to earn all available rewards
  - Believes that the team can fully meet its responsibilities
- A student who is deficient at expecting quality
  - Is satisfied even if the team does not meet assigned standards
  - Wants the team to avoid work, even if it hurts the team
  - Doubts that the team can meet its requirements

Having Relevant Knowledge, Skills, and Abilities accounts for both the talents a member brings to the team and those talents a member develops for the team's benefit.

- A student who has exceptional knowledge, skills, and abilities
  - Demonstrates the knowledge, skills, and abilities to do excellent work
  - Acquires new knowledge or skills to improve the team's performance
  - Is able to perform the role of any team member if necessary.
- A student who has an acceptable level of knowledge, skills, and abilities
  - Has sufficient knowledge, skills, and abilities to contribute to the team's work
  - Acquires knowledge or skills needed to meet requirements
  - Is able to perform some of the tasks normally done by other team members
- A student who has deficient knowledge, skills, and abilities is
  - Missing basic qualifications needed to be a member of the team
  - Unable or unwilling to develop knowledge or skills to contribute to the team
  - Unable to perform any of the duties of other team members

Research shows that team performance can be enhanced if team members reflect on their own and their teammates' performance and give each other high-quality feedback. "High-quality" ratings are consistent with observed behavior, which may or may not be "high" ratings. We also know that rating quality (again, consistency with observed behavior) improves with practice. Guided practice in giving and receiving feedback and in practicing self- and peer-evaluations using behavioral criteria will help you improve. Please take your time in evaluating the members of the fictitious team below.

Pat   Friendly and very well-liked, makes working fun, and keeps everyone excited about working together. Relies on teammates to make sure everything is going okay. Pays attention to keeping the team upbeat but does not seem to notice if the team's work is getting done. Struggles to keep up with the rest of the team and often asks teammates for explanations. The team has to assign Pat the least difficult jobs because Pat does not have the skills to do more complex work. Offers ideas when able, but does not make suggestions that add anything unique or important to the final product. Always shows up for meetings, prepares beforehand, and does everything promised. Is confident that the team can do everything that is essential. Agrees that the team should meet all explicit task requirements.

Chris   Okay as a person and does not interfere with the contributions of others but rubs teammates the wrong way by frequently griping about the work and making excuses for not following through on promises to the team. Chris has the brains and experience to make a unique and valuable contribution, but does not try. The fact that Chris is so smart frustrates some teammates who have to try hard to accomplish tasks that would be easy for Chris. Ignores assigned tasks or does a sloppy job because "Robin will redo the work anyway." Misses meetings or shows up without assigned work. Contributes very little during meetings. Was late to one meeting because "no one told me the meeting time." Missed another meeting because "the alarm clock did not go off." After missing meetings, he asks lots of questions to make sure that everyone is making progress and the team's work is being accomplished. Spends more time checking that everyone else is doing their work than getting the job done. Chris always seems sure that the team will do fine and says that the team should do good work that fully meets the standards for acceptable performance. In response to a teammate's question about Chris' failure to deliver a promised piece of work, Chris said, "Why should I bother? Robin won't let the team fail."

Robin   Very bright. Has far greater knowledge of the subject than any of the other team members. Extremely skilled in problem solving. Robin has very high standards and wants the team's work to be impressive, but Robin worries whether the team's work will be good enough to stand out. Robin completes a big chunk of the team's work and takes on a lot of the really difficult work. Does the work that Chris leaves unfinished. The quality of Robin's work is consistently outstanding. Tends to just work out the solutions and discourages teammates' attempts to contribute. Reluctant to spend time explaining things to others. Does not like to explain "obvious" things. Is particularly impatient with Pat's questions and once told Pat "You are not smart enough to be on this team." Complains that Chris is a "lazy freeloader." Sometimes gets obsessed with grand plans and ignores new information that would call for changes. Does not pay attention to warning signs that the current plan might not be effective until the problems are obvious. Then handles the situation as a crisis and takes over without getting team input. Robin is reluctant to acknowledge or discuss problems in the team until they affect his work.

Terry   Not nearly as bright as Robin, but works to develop enough knowledge and skills to do the assigned tasks. Terry can usually fill in for other team members if given specific directions, but does not understand most of the tasks that other team members normally perform. Does more grunt work than any of the other team members, but does not do as good a job as Robin and does not take on difficult tasks. Sometimes makes mistakes on the more complex work. Super responsible, spends a lot of time giving one-on-one help to Pat. Always on time to meetings. Often calls to remind everyone (especially Chris) about meetings and usually makes some nice comment about one of the teammate's strengths or a valuable contribution that the teammate has recently made. Terry is outgoing and highly supportive of teammates when well-rested, but is sometimes too tired to get excited about teammates' ideas. Is not defensive when teammates' offer feedback, but does not ask for teammates' suggestions, even when teammates' input could help Terry to do better work. Terry thinks that the team can do great work and encourages teammates to do their best. When Robin expresses doubts if the team can do superior work, Terry reassures everyone that the team is capable of outstanding work. When the team is headed in the wrong direction, Terry is quick to notice and say something, but usually does not suggest a way to fix it. Terry reviews the team's objectives and alerts the team to anything that comes up that would affect the team. Terry was reluctant to press the issue when Robin's plan ignored one of the guidelines specified for the project.

To test your ability to focus on individual behaviors, go to https://www.catme.org/login/survey_instructions and rate each team member on each type of contribution to the team. On the Scenario Results page, a green arrow indicates that your rating matches the expected rating. If your rating does not match the expert rating, the blue arrow shows your rating and the red arrow indicates the rating experts would have assigned. If you count one point for every level separating your rating from the expert rating on the five different types of contribution, a low score is best, indicating the greatest agreement with the expert ratings. You can "mouse over" the red arrows to read the rationales underlying the expert ratings.

3.7 PROJECT TIMELINE

To complete a project successfully, on schedule and satisfying all constraints, careful planning is required. The following steps should help your team plan the completion of project work.

Step 1: Create a project timeline.

The first consideration is the project's due date. All team members need to note this on a calendar. Examine the due date within the context of other assignments and classes. For example, is there a calculus test in the fourth week? When is the first English paper due?

Next, look at the project itself and break into individual tasks and subtasks. Create a list, making it as specific, detailed and thorough as possible. Your list should include:

- All tasks needed to complete the project
- Decisions that need to be made at various times
- Any supplies/equipment that will need to be obtained

NOTE

Choose a weekly team meeting time and STICK TO IT!

Divide the workload and require updates at your weekly meeting.

If a team member is not completing assigned tasks as required, speak with your professor. Do not wait until the end of the project to raise concerns about a teammate who is AWOL.

Carefully consider the order in which the tasks should be completed. Does one task depend on the results of another? *Then, working backwards from the project due date, assign each task, decision, or purchase its own due date on the calendar.*

Finally, your team should consider meeting at least once a week at a consistent time and location for the duration of the semester. More meetings will be necessary, but there should be at least once per week when the entire team can get together and review the project status. A standing meeting time will prevent issues of "I did not know we were going to meet" or "I did not get the message."

Step 2: Create a responsibility matrix.

List the project's tasks and subtasks one by one down the left side of the paper. Then, create columns beneath each team member's name, written side by side across the top. Put a check mark in the column beneath the name of the member who agrees to perform each task. It then becomes the responsibility of that team member to successfully perform the task by the due date that was agreed upon in Step 1. An alternate grid is shown in Table 3-4.

Table 3-4 Sample responsibility matrix

Task	Completed by	Checked by
Purchase supplies	Pat and Chris by 9/15 team mtg	
Write initial proposal	Terry—Email to Robin by 9/22	Robin by 9/25 team mtg
Conduct preliminary calculations on height	Robin—Email to Pat by 9/22	Pat by 9/25 team mtg
Build prototype in lab	All—Lab: 7–9 p.m., 9/28	

In assigning the tasks, consider the complexity and time required for the job. One team member may have five small tasks while another may have one major task, with the goal being an equal distribution of effort. A second team member should be assigned to each task to assist or check the work completed by the first team member. Be sure all team members are comfortable with the assignments.

Step 3: **Consider team dynamics.**

Communication: The success of any project depends to a great extent on how well the team members communicate. Do not hesitate to share ideas and suggestions with the group and consider each member's input carefully. Do not be afraid to admit that you are having difficulties with a task or that the task is taking longer than expected. Be ready and willing to help one another.

Trust and respect: Remember the team is working toward a shared objective. Therefore, you must choose to trust and respect one another. Treat fellow team members with simple courtesy and consideration. Follow through with promises of completed tasks, remembering the team is counting on your individual contributions. Try to deal honestly and openly with disagreements. However, do not hesitate to ask for help from faculty if problems begin to escalate.

Nothing is carved in stone: It is important to plan the project as carefully as possible; however, unforeseen problems can still occur. Treat both the Project Timeline and Responsibility Matrix as working documents. Realize they were created to serve as guides, not as inflexible standards. Watch the project progress relative to the timeline, and do not hesitate to redesign, reallocate, or reschedule should the need arise. Review your matrix each week and adjust as needed.

Finally, do not forget to have fun!

In-Class Activities

ICA 3-1

With your team, compose a plan to build the longest bridge possible using the K'Nex™ pieces provided by your instructor. The longest part of your bridge will be defined as the longest stretch of K'Nex pieces that are not touching another surface (table, floor, chair, etc.).

During the planning phase (15 minutes), your team will only be allowed to connect as many pieces as your team has hands. For example, a team of four will only be allowed to have eight K'Nex pieces connected together at any one moment. In addition, every team member must have their hand on a K'Nex piece if it is connected together during planning. As soon as your team has finished the planning phase, disconnect all K'Nex pieces and place them in the provided container. The container will be shaken before you begin, so do not bother attempting to order the pieces in any way.

During the building phase (60 seconds), the restriction on the number of connected pieces goes away, but your team will not be allowed to talk. Your instructor will say go, and then stop after 60 seconds has elapsed. At the end of the building phase, your team must step away from the bridge and remove all hands and other body parts from the K'Nex bridge. If any pieces falls after time is called, you are not allowed to stabilize the structure in any way.

ICA 3-2

With your team, come up with a plan to build the tallest tower possible using the K'Nex™ pieces provided by your instructor. The tallest portion of your tower will be defined as the longest stretch of K'Nex pieces that are not touching another surface (table, floor, chair, etc.).

During the planning phase (15 minutes), your team will only be allowed to connect as many pieces as your team has hands. For example, a team of four will only be allowed to have eight K'Nex pieces connected together at any one moment. In addition, every team member must have their hand on a K'Nex piece if it is connected together during planning. As soon as your team has finished the planning phase, disconnect all K'Nex pieces and place them in the provided container. The container will be shaken before you begin, so do not bother attempting to order the pieces in any way.

During the building phase (60 seconds), the restriction on the number of connected pieces goes away, but your team will not be allowed to talk. Your instructor will say go, and then stop after 60 seconds has elapsed. At the end of the building phase, your team must step away from the tower and remove all hands and other body parts from the K'Nex tower. If any pieces fall after time is called, you are not allowed to stabilize the structure in any way.

ICA 3-3

Following the rules for brainstorming (encourage a lot of ideas, encourage a wide variety of ideas, and do not criticize), develop ideas for the following with your team.

(a) A better kitty litter box

(b) A new computer interface device

(c) A new kind of personal transportation device

(d) Reducing noise pollution

(e) Reducing light pollution

(f) A new board game

(g) A squirrel-proof bird feeder

(h) A no-kill mole trap

(i) A tub toy for children of 4 years or younger

(j) A jelly bean dispenser

(k) A new musical instrument

(l) A self-cleaning bird bath

(m) A new smart phone application

Chapter 3 MINI DESIGN PROJECTS

This section provides a wide range of design projects, varying in difficulty and time commitment. Your instructor may assign projects required for your specific course and provide more details.

Category I: Demonstrate a Physical Law or Measure a Material Property

1. Prove the law of the lever.
2. Demonstrate conservation of energy (PE + KE = constant).
3. Determine the coefficient of static and sliding friction for a piece of wood.
4. Prove that the angle of incidence is equal to the angle of reflection.
5. Demonstrate momentum conservation ($F = ma$).
6. Demonstrate the ideal gas law.
7. Obtain a series of data points from an experiment you conduct that, when plotted, exhibit a normal distribution.
8. Show that forces can be resolved into horizontal and vertical components.
9. Find the center of gravity of an irregular piece of plywood.
10. Show that for circular motion, $F = mv^2/r$.
11. Show that for circular motion, $v = \omega r$.
12. Measure the effective porosity of a sand sample.
13. Prove the law of the pendulum.
14. Prove Hooke's law for a spring.
15. Prove Hooke's law for a metal rod (in deflection).
16. Measure the coefficient of thermal expansion for a solid rod or bar.
17. Estimate the heat capacity for several objects; compare with published results.
18. Prove Archimedes' law of buoyancy.
19. Determine the value of pi experimentally.
20. Prove the hydrostatic pressure distribution.
21. Relate the magnetic strength to the radius.
22. Determine the density and specific gravity of a rock.

Category II: Solve a Problem

23. What is the volume of a straight pin?
24. Determine the thickness of a specified coin or a piece of paper.
25. How many pennies are needed to sink a paper cup in water?
26. Determine the specific gravity of your body.
27. What is the volumetric flow rate from your shower?
28. Use a coat hanger to make a direct reading scale for weight.

Category III: Design a Solution

29. THE GREAT EGG DROP

    You have no doubt seen the "Odyssey of the Mind" type of assignment in which you are to design protection for an egg that is to be dropped from some height without being broken. This assignment is to have the same end product (i.e., an unbroken egg) but in a different way.

    You cannot protect the egg in any way but are allowed to design "something" for it to land on. You will be allowed three drops per team, and will be assigned at random the heights from which you will drop the egg. The egg must free-fall after release. If the egg breaks open, the height will be taken as zero. If the egg shell cracks, the height from which you dropped it will be divided by 2. If you drop the egg and miss the catching apparatus, that is your tough luck (and a zero height will be used)—suggesting that you need to devise a way to always hit the "target."

Your grade will be determined by

   Ranking = (height from which egg is dropped)/(weight of catching apparatus)

After demonstrations, the average ranking number for each team will be calculated, and the value truncated to an integer. The teams will then be ranked from highest to lowest value. The heights and weights will be measured in class. The actual grade corresponding to your class ranking will be determined by your instructor.

30. TREE HEIGHT

We have been contacted by a power company to conduct a study of tree height and interference with high-voltage wires. Among other requirements, the company is looking for a quick, easy, and inexpensive method to measure the height of a tree.

   Your project is to develop different methods of measuring the height of a tree. As a test case, use a tree designated by your instructor. You may *not* climb the tree as one of the methods! In addition, you may not harm the trees or leave any trace of your project behind.

   To sell your methods to the customer, you must create a poster. The poster should showcase your measurement methods, including instructions on how to conduct the experiment, any important calculations, graphs or photos, and your resulting measurements. The poster will be graded on neatness, organization, spelling and grammar and mechanics, formatting, and strength of conclusions. You may use any piece of poster board commercially available, or a "science fair" board. The poster can be handwritten, or typed and attached, or ... here is a chance to use your creativity! If you present more than one method, you must indicate and justify your "best" choice.

Your grade will be determined as follows:

- Method 1:   30 points
- Method 2:   20 points
- Method 3:   10 points
- Presentation Board: 40 points

31. FIRE EXTINGUISHER

Make a fire extinguisher for a candle. The candle will be lit, and the extinguisher will put out the flame at a predetermined time after the candle is lit (say, 20 seconds). The only thing that the participant can do to start the time is to light the candle. The candle can be mounted anywhere you like. The results will be scored as follows:

   You will be allowed three trials. You will be allowed to use your best trial for grading.

   You must extinguish the candle between 19 and 21 seconds; for every second or fraction thereof outside this range you will lose five points.

32. CLEPSYDRA

Construct a clepsydra (water clock). When you bring it to class to demonstrate its performance, the following test will be used:

   You will have three times to measure: a short time, a medium time, and an extended period. The actual times to be measured will be given to you at the time of demonstration, and you will have 2 minutes to set up your apparatus.

- Short: Between 10 and 30 seconds (in 2-second increments)
- Medium: Between 1 and 4 minutes (in 30-second increments)
- Extended: Between 5 and 10 minutes (in 1-minute increments)

You will start the clock and tell the timekeeper to begin. You will then call out the times for each of the three intervals and the timekeeper will record the actual times. Your clock must "run" for the total time.

Your grade will be determined by the average percent error of your timings.

For example, if the specified "medium" time was 2 minutes and 30 seconds and you said "mark" at an actual time of 2 minutes and 50 seconds, the absolute value of the percent error would be $(20 \text{ s})/(150 \text{ s}) = 13\%$. The absolute values of the three errors will be summed, divided by 3, and subtracted from 100% to get a numerical grade.

33. ON TARGET

    Each team will design, build, and test a device that will allow you to successfully hit a target with a table tennis ball. The target will be a flat sheet of poster paper placed on the floor with a bull's-eye and two other rings around it for scores of 100, 90, and 80 with a score of 60 for hitting the paper. The target will be placed at a location of 15 feet from the point at which you release the ball. Once the ball is released, you cannot touch it again, and it must be airborne before it hits the target. When demonstrating your device, you will not be allowed any trial run.

    Your grade will be determined as follows:

    Average numerical score + bonus (10, 8, or 6 points for creativity and simplicity)

    The class (each person) will be given a slip of paper on which they will rank their top three teams with respect to creativity and also with respect to simplicity. These will be tallied and the top three teams in each category will receive 10, 8, or 6 bonus points.

34. KEEPING TIME

    Each team is to build a "clock." When you say "go," a stopwatch will be started, and you are to tell the timekeeper when 10, 30, and 60 seconds have elapsed. Differences between the actual times and the predicted times at the three checkpoints will be noted and the percent error calculated. Average the absolute values of these three errors and subtract from 100 to obtain your final score. You may not use any store-bought device that is designed to measure time. No electronic devices may be used.

    Your final grade will be determined as follows:

    Average numerical score + bonus points (10, 8, or 6 for creativity and simplicity)

    The class (each person) will be given a slip of paper on which they will rank their top three teams with respect to creativity and simplicity. These will be tallied and the top three teams in each category will receive 10, 8, or 6 bonus points.

Category IV: Additional Projects

35. Develop a device that can be placed into a container of water and used to measure the pressure as a function of depth. Take measurements and plot them against theory for a hydrostatic pressure distribution.
36. A 2-liter soft-drink container, nearly full of water and open to the atmosphere, is placed on the floor. Where could you locate an orifice in the side of the bottle so that the jet of water that squirts out will have the maximum range? Keep the bottle filled by continuously pouring water in the container as the tests are conducted. Justify your answer with theory by discussion rather than equations.
37. Design and build a device that will allow a ping-pong ball to hit a target (small circle) between 5 and 15 feet away from the point at which you release the ball with the device on the floor. Points will be given for accuracy (distance from the target center). From a hat, you will draw two slips of paper: a short distance (3–7 feet) and a long distance (8–15 feet). The slips of paper will have values in 1-foot increments on them (3, 4, 5, 6, or 7 feet for the short distances, and similar for the long distances). When you set up your device, you will be given 2 minutes to set the device for the first test and 2 minutes to set the device for the second (long distance) test.

38. Build a vehicle that will travel over a flat surface (hallway in the building). Points will be awarded for the distance traveled divided by the total (initial) weight of the vehicle. The vehicle must move under its own internal "engine"—the team can only release the stationary vehicle when the test begins. You will have two attempts, and the best value will be recorded. Your instructor may impose an allowable maximum weight. No batteries or electricity can be used.

39. Build a thermometer. You are to be able to measure the temperature of cold water in a bucket, room temperature, and the temperature of hot water in a container (degrees Celsius). The total percent errors will be summed (absolute values), averaged, and subtracted from 100.

40. Without moving more than 10 feet from your initial location, position a person a distance of 50 feet (or 100 feet) away from your initial location; bonus points for doing it several ways; the person must initially start beside you. You cannot be connected to the other person in any way (e.g., string, rope). Points are given for accuracy.

41. Roll an object of your design down an inclined plane (provided and the same for all participants). The object is to knock over a small piece of wood placed at a distance of 5 feet from the base of the incline and 5 feet to one side.

42. We are interested in rolling plastic drink bottles across the floor. Rather than hold a race to see which bottle will roll the fastest, we want to determine which will roll the farthest. It is important that each bottle be given a fair chance, so the starting conditions must be the same for each. Each team will use the same ramp (18 inches wide by 24 inches long) and supported by the 4-inch dimension of a 2×4 inches board. The bottles to be used will be clear-plastic soft-drink bottles. The 2-liter size is probably the best, but smaller bottles could also be used. The objective is to determine the answer to questions. You must develop a defendable test program, carry out the tests, present your results in an easy-to-understand manner, and defend your conclusions.

  (a) How much water should the bottle contain in order to roll the farthest distance (until it stops)?

  (b) How much water should the bottle contain in order to roll the shortest distance (until it stops)?

  (c) As a part of your test program, you will release two bottles simultaneously on the ramp (with differing amounts of water in them, including one empty). Do not let them roll all the way until they stop, but catch them about 1 foot after they leave the bottom of the ramp. Which moves the fastest, which the slowest, and why?

CHAPTER 4
ENGINEERING COMMUNICATION

It is a common joke that most engineers cannot construct a grammatically correct sentence, and there is all too much truth in this anecdote. In reality, the most successful engineers have developed good communication skills, not only oral and written, but also those involving multimedia formats. You might have the best idea in the known universe, but at some point you are going to have to convince someone to supply the $200 million needed to develop it. You *must* be able to communicate effectively not only that you have this great idea for a practical antigravity device, but also that it will actually work and that you are the person to lead the team developing it.

Our intent here is not to make you expert communicators, but to at least make you aware of the importance of good communication skills in engineering, as well as to give you a bit of guidance and practice developing these skills.

WISE WORDS: HOW IMPORTANT ARE COMMUNICATION SKILLS AT YOUR JOB?

When working with clients, we deliver our approach, analysis structure, status, findings, and final deliverable by presentation.

M. Ciuca, ME

Absolutely essential. While I thankfully have a job that allows me to dig into the math and analysis, I still deal with a lot of people. Being able to communicate effectively, where two people (or more) really understand what each is saying is very important on a complex project, and is often more difficult than one would think; or at least it involves more active participation than many are inclined to put into it.

J. Dabling, ME

I have to write design documentation, edit customer specs, produce design review presentations, and sometimes present to our customers.

E. D'Avignon, CpE

Being able to connect with customers, and internal team members, is the means to develop relationships and win new business.

B. Dieringer, ME

Absolutely critical. I can think of people that I have worked with who do not have these skills, and they are not easy or fun to work with.

S. Forkner, ChE

Social skills are crucial to success when working at a company. Many times the person who is most successful is not necessarily the one who has the best or brightest ideas, but rather the individual who has solid ideas and is able to communicate in a manner that allows others to easily understand the vision and path to goal achievement.

L. Gascoigne, CE

Good social skills are very important because they improve productivity, teamwork, and goodwill. I have also found socials skills can be a deciding factor in job advancement.

R. Holcomb, IE

4.1 BASIC PRESENTATION SKILLS

Since most students consider giving an oral presentation a more daunting task than submitting written documents, we focus on live presentations first, although many of the suggestions apply to all forms of engineering communication.

Many years ago, one of us was responsible for a program to recruit high school students into engineering. Each engineering department made a short presentation to the visiting students, extolling the glories of its particular discipline. One department sent its most personable and able communicator about half the time, and Professor X came the other half. Both used the same set of PowerPoint™ slides, but when Professor X showed up, every single student seemed to be completely brain-dead within 3 minutes. It was *awful*! The other professor maintained their rapt attention for the entire 15 minutes, with supposedly the same presentation.

With this in mind, you need to focus on several factors when planning a presentation. Note that the first item in our list is *who* the audience is, although the other factors mentioned are equally important.

Preplanning

NOTE

5 Ws and 1 H

- Who
- What
- Where
- When
- Why
- How

- **Who is my audience?** Know the age group, demographics, prior knowledge about the topic, and what positions or opinions they may hold.
- **What is my purpose?** What do I hope to accomplish? What response do I expect? What will the audience get out of my speech?
- **Where is all the equipment I need?** Where will the talk be held?
- **When am I on the program agenda?** Will I be the first presenter (when audience is most alert) or the last one before lunch (when they are becoming restless) or after lunch (when they are sleepy)? What will I need to do to keep my listeners attentive?
- **Why am I giving this talk?** Why is the audience here?
- **How long should I talk?** Remember that only few people can focus for more than 20 minutes. Trim your talk so that people will ask for more information rather than thinking "When will he sit down?"

Preparing the Verbal Elements

The preceding list focused primarily on logistics. In addition to these considerations, the structure of your presentation is vital. As a simple example, which of the following two sentences is easier to understand?

> Sentence A: While perambulating in the antithesis of the metropolis to evade the intemperate brouhaha thereof, my visual cortex perceived an ophidian.
>
> Sentence B: I saw a snake while taking a relaxing walk in the woods.

Although sentence A may be phrased in a more intriguing manner, it tends to obscure the underlying meaning. This is perhaps desirable in poetry or fancy fiction but generally detrimental to a professional engineering presentation.

To help you avoid such pitfalls, we offer the 4-S formula for structuring presentations.

NOTE

The 4-S Formula

- Short
- Simple
- Strong
- Sincere

- **Shortness:** Use short sentences, avoid too many details, and do not talk too long.
- **Simplicity:** Avoid wordy, lengthy phrases. Remember the old saying: "Eschew obfuscation!"
- **Strength:** Use active voice and action verbs, not passive voice and "to be" verbs.
- **Sincerity:** Convey empathy, understanding, and respect for the audience.

Three Structural Parts

Keep in mind the purpose of discrete elements of a speech.

- **Introduction:** Purpose: to capture the interest of the audience. Your first task is to *hook your audience*. What is it about your subject that *they* (and not necessarily you) would find most interesting and relevant?
- **Body:** Purpose: to keep your audience interested. They will continue to pay attention if you keep the material interesting and relevant to them.
  - Divide the presentation into two or three main points, and organize your thoughts and ideas around them.
  - Use one or more simple examples to illustrate each major point.
  - Include slides, transparencies, or other visual aids.
- **Conclusion:** Purpose: to pull it all together.
  - Summarize major points.
  - Show appreciation for your audience's attention.
  - Allow for a few questions, but be sensitive to your audience and the other speakers.

Preparing Visual Aids

Most of you have seen presentations that used slides with unreadable text, incomprehensible graphics, or annoying special effects. Well-designed graphics can greatly enhance your presentation, not only making it easier for the audience to understand, but also keeping their attention focused. A picture really is worth about 2^{10} words! The guidelines below will help you design visual aids that are easily understood without being boring. Although our focus is on PowerPoint presentations, these suggestions apply to other media as well.

Helpful Hints

- Keep each slide simple, with one concept per slide. As a rule, use no more than *six lines per slide*. Each slide should correspond to an average of *60 seconds* of speech.
- If possible, make slides in *landscape* format.
- Present data in simple graphs rather than in lists or tables. Avoid excessively complex graphs with extensive data. If you must present tables, divide them among several slides.
- Pictures, diagrams, and video simulations all may enhance your presentation. Be sure that all are large enough to be seen by the audience, and have color schemes that do not appear washed out when projected. Often, such items are designed for viewing on a small screen and do not project well. Be sure to test them prior to your presentation.
- Use bullet points with important phrases to convey ideas. Avoid complete sentences.
- Large size text is best. A font size of at least 18 points and preferably no less than 24 points should be used. This includes *all* objects, such as axis and legend captions, table headings, figure symbols, and subscripts.
- Use high-contrast colors. Avoid fancy fonts, such as cursive, or light colors, such as yellow or other pastels. Avoid using all capitals.
- Use a light background and dark print to keep the room brighter. Avoid black slides with white text.
- Keep background styles simple and minimize animation to avoid distracting from the presentation. Keep all the slide backgrounds the same throughout a single presentation.

4.2 SAMPLE PRESENTATIONS

To illustrate the visual aids caveats, we critiqued three student presentations.

Sample Student Presentation 1

Original Presentation: Critique

- Slide 1 font is difficult to read, poor choice of abbreviation for *approximately*.
- Slide 2 dates and text appear disjointed due to text size and graphic; graphic is too large.
- Slide 3 graph is difficult to read; too many gridlines.
- Slide 4 too many words; graphic is too large.

Improved Presentation:

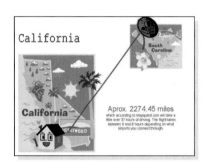

Sample Student Presentation 2

Original Presentation: Critique

- Different graphics on every page is distracting.
- White color is hard to project unless on black background.
- Slide 4 graphic is difficult to read; yellow highlights make it worse.

(1)

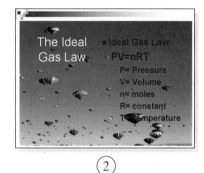

(2)

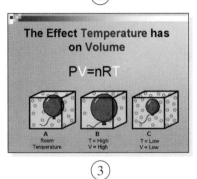

(3)

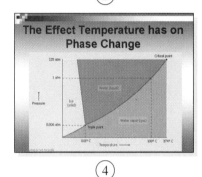

(4)

Improved Presentation:

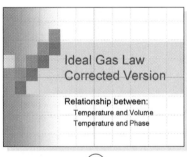

(1)

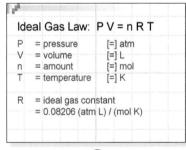

(2)

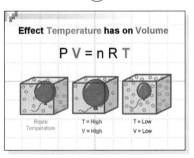

(3)

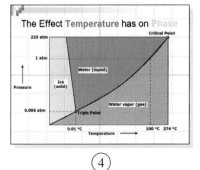

(4)

Sample Student Presentation 3

Original Presentation: Critique

- Green backgrounds with white text do not project well; blue text is especially hard to read.
- Slides 2 and 4 have too many words; should use bullets, not sentences.
- No graphics; some pictures of acid rain damage would be nice.

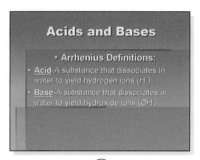

(1)

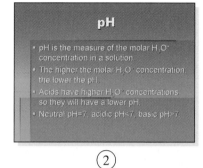

(2)

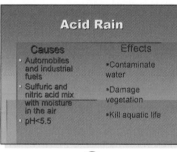

(3)

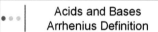

(4)

Improved Presentation:

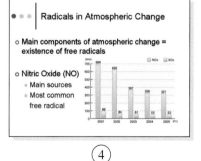

(1)

(2)

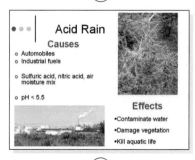

(3)

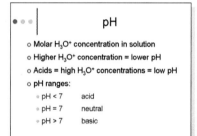

(4)

Making the Presentation

Oral presentations present several challenges for effective communication. How many of you have had an instructor who simply reads the contents of the slides with no embellishment? How many have had an instructor who seems to be terrified of the audience, cowering in fear and trying to disappear into the wall? How many have tried desperately to read the contents of a slide containing hundreds of words in a minuscule font? If you have not yet suffered through one or more such painful presentations — you will! Do not inflict such things on your own audiences.

Presentation Dos and Don'ts

NOTE

The key to improving presentation skills is practice, practice, practice!

When delivering a presentation, do:

- Relax!
- Speak slowly and clearly, making good eye contact.
- When your hands are not busy, drop them to your sides.
- Rehearse your presentation *out loud* multiple times. If possible, have a friend critique.
- Arrive early enough to make sure that all technology is present and working, and resolve any problems you may discover.

When delivering a presentation, do not:

- Lean on your surroundings, turn your back to the audience, or cover your mouth while speaking.
- Read your presentation from a prepared text.
- Tell inappropriate jokes.
- Stammer, overuse the pronoun "I," or repeatedly say "um" or "uh." Do not be afraid of a little silence if you need to glance at notes or collect your thoughts.
- Chew gum, remove coins from pockets, crack your knuckles, etc.
- Shuffle your feet or slouch; move repetitively, for example, pace back and forth or sway.
- Play with your notes.

4.3 BASIC TECHNICAL WRITING SKILLS

Although most of you probably consider written communications much easier than oral presentations (the fear factor is largely absent), technical documents you produce will often be far more important to your company and your career than a live presentation of the same information.

General Guidelines

In addition to many of the points made earlier, effective technical writing requires its own set of guidelines.

- **Be clear**; use precise language. Keep wording efficient without losing meaning. Do not exaggerate.
- Ensure that the finished copy logically and smoothly **flows** toward a conclusion. Beware of "choppiness" or discontinuity. Avoid extremely long sentences because they may confuse the reader.
- If possible, use **10-point font size and 1.5 line spacing**.
- **Generally, prefer past tense verbs.** Keep verb tenses in agreement within a paragraph.

■ **Define any terms** that might be unfamiliar to the reader, including acronyms and symbols within equations.
■ **Present facts or inferences** rather than personal feelings.
■ **Maintain a professional tone.** Do not be emotional or facetious.
■ **Number and caption all tables, figures, and appendices.** Refer to each from within the body of the text, numbering them in order of appearance within the text.
  • Tables are numbered and captioned *above* the table.
  • Figures are numbered and captioned *below* the figure.

Table 4-1 Example of a properly formatted table

Current (I) [A]	2	6	10	14	16
Energy of Inductor #1 (E1) [J]	0.002	0.016	0.050	0.095	0.125
Energy of Inductor #2 (E2) [J]	0.010	0.085	0.250	0.510	0.675
Energy of Inductor #3 (E3) [J]	0.005	0.045	0.125	0.250	0.310

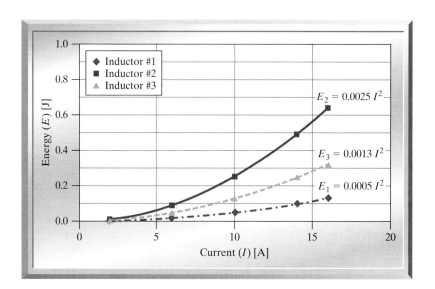

Figure 4-1 Example of a properly formatted figure.

■ **Proofread and edit several times.**
  • **Remember to include** headings, figures, tables, captions, and references.
  • Do not assume that the spell check on the computer will catch everything! It will not distinguish between such words as *whether* and *weather*, or *was* and *as*.

■ **Read it twice:** once for technical content and once for flow.
  • As you proofread, look for and *remove* the following: unnecessary words; sentences that do not add to the message; superfluous paragraphs.
  • Do one proofreading aloud. When you encounter commas, semicolons, colons, or periods, pause. Read a comma as a brief pause. Read a colon or semicolon as a longer pause. Read a period as a complete stop before the next sentence. Read what is actually written, not what you "think" it should say. If the text sounds stilted or blurred when read, you probably need to reconsider your use of these punctuation marks.
  • If possible, have someone not associated with the project (a roommate, a friend, or a mentor) read it, and ask that person for suggestions.

- **Spell out a number that starts a sentence.** If the number is large (e.g., a date), reword the sentence.
  - 23 points were outliers. (Unacceptable)
  - Twenty-three points were outliers. (OK)

- **Keep the leading zero with a decimal.**
  - The bridge cost .23 dollars per gram. (Unacceptable)
  - The bridge cost $0.23 per gram. (OK)

- **For long numbers, do not spell out.**
  - The average was one thousand, two hundred fifty-five grams. (Unacceptable)
  - The average was 1,255 grams. (OK)

- **Use the dollar symbol.**
  - The bridge cost four thousand dollars. (Unacceptable)
  - The bridge cost $4,000. (OK)

- **Watch for significant figures. Keep it reasonable!**
  - The bolt is 2.5029 inches long. (Unacceptable)
  - The bolt is 2.5 inches long. (OK)

WISE WORDS: HOW IMPORTANT ARE COMMUNICATION SKILLS AT YOUR JOB?

Projects are successful when the people who work together on them are able to communicate clearly with each other and work together to achieve a common goal. Misunderstanding and miscommunication leads to delays, poor quality, and frustration.

A. Hu, EE

As an engineer you communicate at all levels from the least senior production employee on the factory floor to the president of the company. Tailoring the message to the audience is the difference in acceptance and rejection.

J. Huggins, ME

Writing clear and concise specifications for construction can make the difference between an under-budget, on-time project and an over-budget, late, and unsafe final product.

L. Johnson, CE

As a consultant, extremely. If clients can't get along with you, they won't hire you. Every job requires a proposal and an interview.

J. Meena, CE

Much of my work in my current job involves researching what is going on in the world, and then putting that information into a format that makes sense to people and helps other people draw conclusions from it. Good written communication skills are essential for what I do every day.

M. Peterson, EE

All the social skills are *extremely* important because of the different functions and technical levels that I interface with.

E. Styles, EE

Communication—along with teamwork—really separates bad engineers from good ones. Someone could have the best idea in the world, but if he isn't able to describe the invention or provide reasons as to why it should be developed, the idea is useless. Plus, engineers are trained to be rational and thus perfect for managerial positions. If you have good communication skills, one can easily expect you to climb quickly up the corporate ladder.

A. Thompson, EE

Proper Use of References

Contributed by Ms. Jan Comfort—Engineering Reference Librarian, Clemson University Libraries

In today's wired age, most students immediately go to the Internet to find information. Although this can be an excellent source, particularly for preliminary research, there are definite risks associated with using online sources, since essentially anyone can put anything they want on the web. For example, type "flat earth society" into your favorite search engine and check out the "truth" concerning our home planet, or explore how you can save the endangered Pacific Northwest Tree Octopus.

When making presentations or writing reports, it is important to verify the veracity of any sources you consult. These guidelines will help you avoid egregious errors in your own technical communications.

The **ABCs of evaluating information** offer a useful start.

- **Authority:** Is it clear who is responsible for the site? What are the author's credentials? Is the author an expert in the field? Is it a .com or .gov or .edu site?
- **Bias:** What is the purpose of the article? Is it free of obvious bias? Is the author presenting an objective view of the subject matter?
- **Currency:** When was the information created or last updated?

But there is more to evaluating resources than that. Good students take it to the next level. Here is how you can, too.

- **Use sources that have been reviewed by experts.** Instead of searching for hours trying to find websites that meet stringent requirements, try using library sources to identify good quality sources that have already been through a review process.
- **Secure a peer review:** An expert in the appropriate field evaluates something proposed (as for research or publication).[1] Academic Search Premier and Expanded Academic ASAP are the names of two very good multisubject databases that contain scholarly (peer-reviewed) as well as popular articles. One or both of them should be available at your library.
- **Compare the information found in your article or website with content from other websites, or from reviewed sources.** Comparing sources can also alert you to controversial information or bias that will need further study. Are facts from one website the same as those of another? How about depth of coverage? Maybe one site has better-quality information. Does the site have photos or other unique features that make it a good choice? Or perhaps a journal article from a library database is a better source. Until you compare several sources, you will not know what you are missing!
- **Corroborate the information.** Verify the facts from your source—regardless of where you found it—against one or more different sources. Do not take the word of one person or organization. A simple rule might be: "Do not use information unless you have corroborated it. Corroboration with varied and reviewed sources increases the probability of success."[2]

---

[1]Peer review. (2009). In *Merriam-Webster Online Dictionary*. Retrieved May 13, 2009, from http://www.merriam-webster.com/dictionary/ peer review

[2]Meola, Marc. (2004). "Chucking the checklist: A contextual approach to teaching undergraduates web-site evaluation." Portal: *Libraries and the Academy*, 4(3), 331–344.

4.4 COMMON TECHNICAL COMMUNICATION FORMATS

Technical communications can take on a variety of formats. Here, we will specifically address e-mail, memos, and short technical reports. Other, usually longer formats will probably be addressed later in your engineering career, but the same general guidelines apply regardless of form or length of content.

E-mail

Many students believe that the rules they use for instant messaging (IM), Twitter, etc. apply to e-mail also. When sending informal messages to friends, this is acceptable. However, when using e-mail in a professional context (including e-mail to professors!), more formal rules should be followed. The suggestions below will help you write e-mail that is clear, concise, and appropriate for the recipient.

After you have composed your e-mail, ask yourself if you would mind the president of the university, the CEO of your company, or your parents reading it. If the answer to any of these is no, then you probably should reword it.

E-mail Etiquette

- Be sure to correctly address the recipient. If you are unsure of a person's proper title (Dr., Mrs., Prof.), look it up!
- Use an appropriate subject line. Avoid silly subjects (Hey—Read this!) or omitting the subject line—this may cause the e-mail to end up in the Junk Mail folder.
- Sign your *full* name and include contact information for e-mail, phone, fax, or mailing address if appropriate. When sending e-mail about a class, including your course number and course day and time is often helpful.
- Change your sending name to your full name (such as Elizabeth Stephan) or an appropriate nickname (Beth Stephan). Do not leave your account as Student or the computer default setting (such as Noname Stephan). The easier it is for your reader to identify who you are, the better your chance for a response.
- Keep it brief. Do not use one continuous paragraph—make it easy to read.
- If you expect a response from your reader, be sure that action items are clearly defined.
- Use correct capitalization and punctuation. Spelling does count—even in e-mail! Avoid IM speak (e.g., LOL, IMHO, IIRC).
- Avoid putting anything in e-mail you would not say in person. Do not use e-mail to "vent" or write anything that can be easily misinterpreted by the reader.
- To avoid sending an e-mail before you have a chance to check over your work, fill in the To: and CC: lines last.
- When waiting for a reply, allow a grace period of 48 to 72 hours. If you have not received a reply after 48 hours and a deadline is approaching, you can resend your message, inquiring politely if it was received. Items do sometimes get lost in cyberspace! If the matter is critical, try the phone or request a face-to-face meeting if the first contact does not elicit a response.

Sample E-mail

To...	R. Swarthmore, Ph.D. [swart@reactorsealsrus.com]
cc...	C. Ohland [carson@reactorsealsrus.com];
	K. Stephan [katie@reactorsealsrus.com]
Subject:	Leaky gel reactor seal

Dr. Swarthmore:

The gel reactor seals in B4L3 are leaking and causing production losses (over 200K for FY 2001). The Materials Engineering Lab was asked to test other seal materials. Laboratory tests identified six material couples that produced better wear resistance than the current seal. A prototype seal was made with a new material, self-mated cemented carbide, but the carbide on the seal cracked during fabrication.

The purpose of this e-mail is to request an additional $40,000 and four months' project time to fabricate and test another new seal configuration.

Your approval of this program before Friday noon will allow us to proceed with the project as quickly as possible without any delay. If you have further questions or would like more information, please contact me.

Sincerely,
J. Brock
Design Team Manager, Reactor Seals R Us
(123) 456-7890 x 1234
jbrock@reactorsealsrus.com

● **EXAMPLE 4-1**   **Samples of poorly written (actual student) e-mail**

Subject: Carrier Fair
I hoping to be balle to show up to the Friday class, or take the quiz early on Wed. to allow more time for myself at the Carrier Fair on Wed.

Subject: FW: John Doe fri 8
Im sorry for missing class this morning, I stayed up really lait working on an english paper. Im not really doing that well in any of my classes, but im confident that im going to pass the other ones. Thank god for that. Anyway last night i ate a bunch of fish too, and i think fish dosn't go to well on my stomach because i feel really sick. Could i meet with you sometime this weekend?

Subject: excel
I don't know why it sent 10 other e-mails but sorry. I'm writing to you about the excell test today. I've been sick for the last three days and i wasn't feeling great today either. Also i had a hard time getting started on the test because of downlad problems. Anyway by the time I had figured out the format of the test there were only 15 minutes left so if i could come and talk to you about that then that and one other thing then that would be great.

Subject: Hey, I need help

Hey,

I can't figue out how to do a complex interpolation problem. I understand how to do the simple one, and i used to no how to do the complex one but now everytime I do it i get it wrong. Do you have any office hours I can come in a lean how to do it? If you can't could you send my a worked example of how to do one?

Thank You

Subject: classes for next semester

from what I have heard the math classes we pick at our time pick classes are only for a time. and that we dont pick the teacher we want. is there any way where we can pick the teacher we want for next semester?

Memo (One Page Limit)

The following is a template for a brief memo. Your professor may ask you to adhere to this format, or may suggest a different one.

NOTE

You should use a 10-point font such as Times New Roman or Verdana, with 1 to 1.5 line spacing.

Margins should be set to 1 inch all around.

Be sure to use correct spelling and grammar.

Include the headings given here, in bold.

To:	Dr. Engineering
From:	Ima Tiger, Section 000 (IMT@school.edu)
Subject:	Memo Guidelines
Date:	May 21, 2011

Introduction: The first three or four sentences should explain the purpose behind the memo. You should attempt to explain what you were asked to do, what questions you are trying to solve, what process you are attempting to determine, etc.

Results: Place any experimental results, in tabular and/or graphical format, here. As space is limited, this normally only includes two items: two tables, two figures, or one table and one figure. Be sure that each is clear enough to stand alone, with one to two sentences of explanation. Include a table caption at the top of each table, and a figure caption at the bottom of each figure. The caption should include a number and a word description. When a figure is used within a document, a title is not necessary on the graph and is replaced by the caption. The two items should be pasted side by side from EXCEL into WORD using the **Paste Special > JPEG** command or similar picture format command (PNG, Bitmap) and then sized appropriately.

Discussion: In this section, discuss how you obtained your data, the meaning of any trends observed, and significance of your results. Refer to the tables or figures shown in results by name (Table 1 or Figure 1). Explain any errors in your data (if possible) and how your data differs from theory. If you are deciding among several alternatives, be sure to explain why you did NOT choose the other options.

References: List any sources you use here. You do not need to reference your textbook (Thinking like an Engineer). You may use a new page for references if necessary. Any reference format is acceptable; Modern Language Association (MLA) citation style is preferred.

● EXAMPLE 4-2　Sample of a poorly written student memo

To:	Dr. Engineering
From:	Ima Student
Subject:	Memo
Date:	April 1, 2009

Introduction: We are given the job to analyze the cost of upgrade a machine line, which produces widgets. We were given three companies to choose from, to figure out witch would be the best for the cost and its production. Just by graphing the variables would allow us to find our answer.

Results:

	Klein Teil	
Varible Cost	0.95	
Fixed Cost	5.00E+06	
Material Cost	0.75	
Energy Cost	0.15	
Labor Cost	0.05	
Selling Price	3	
apacity per da	6500	
antity Produc	*ein Teil Total Co*	*Revenue*
0.00E+00	*5.00E+06*	*0.00E+00*

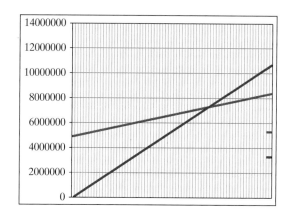

Discussion: We got the data by taking all variables from the information provided, then graphing the results together. This allowed us to see which machine line would provide the better outcome for the situation at hand. Considering the cost of the machine, material, labor, and energy into consideration with what would produce the quantity and quality product we're striving for. Figure 1 display all three solutions': total cost, revenue for us, and our breakeven point to ensure us of our choice. From observation of the graph we see that the Klein Teil machine is better. Its breakeven is at $2,400,000 and the profit is twice as much.

Summary: So to answer the question, Klein Teil would be our best option. The results yielded that the Klein Teil machine would give us the most quality for its price, a better production rate, and more money in return. From this research I hope you choose to take the Klein Teil machine.

Comments on this memo:

There are so many problems with this submission that we address only the major problems.

■ *The subject line simply informs us that this is a memo.*
■ *The introduction tells us very little about what the memo will address.*
■ *The same data is presented in both the graph and the table. The formatting of the table is very poor; the formatting of the graph is worse.*
■ *The Discussion does not explain how the data was analyzed, and the justification of the final recommendation is essentially nonexistent.*
■ *Similarly, the Summary says almost nothing. Although not an appropriate summary, the three words "Buy Klein Teil" would probably have been more effective.*

How many more problems can you find in this sample memo?

Short Report (Two to Four Pages)

The following template is for a short report. Again, your professor may ask you to follow these guidelines or provide a somewhat modified version.

Introduction Type the introduction here. This should be four or five sentences. What is the problem that will be addressed in this memo? What can the reader expect to learn by reading your report?

Procedure Type the procedure here. This should be at most ¾ page. It must be in bulleted format. You should generalize the procedure used to include the basic steps, but you do not need to include every detail. The reader should gain an understanding of how you collected your data and performed your analysis.

Results Insert the results here, but do not discuss them or draw any conclusions. This may include a maximum of three illustrations, in a combination of figures and tables. Be sure that each is clear enough to stand alone, with one or two sentences of explanation. Include a table caption at the top of the table, and a figure caption at the bottom of the figure. The caption should include a number and a word description.

Be aware that your tables and figures should illustrate *different* ideas; they should not contain the same data. Do *not* include large tables of raw data or *every* graph generated. This section should be a *sample* of those items, used to illustrate the points of your discussion below.

Table 4-2 Example of a properly formatted table

Section	Instructor	E-mail	Time
−030	Dr. Stephan	beths	M 8:00–9:55
−031	Dr. Park	wpark	M 12:20–2:15
−032	Dr. Sill	sillb	W 8:00–9:55

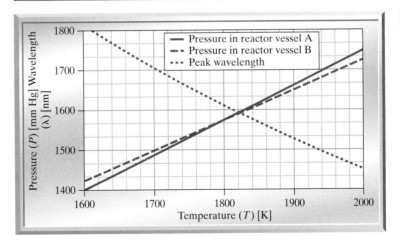

Figure 4-2 Example of a properly formatted figure.

Discussion Talk about your results here. This can be up to a maximum of one page. Refer to the table and figure shown in results by name. Be sure to include the items specifically requested in the original project description.

Summary What is the final conclusion? As your boss, I may only have time to read the introduction (what the problem is) and the summary (what the answer is) and glance at the results (how you got the answer) before I head to a big meeting. This should be four to five sentences long, and answer the initial questions asked in the introduction and summarize any important findings.

Poster Presentation

The following template is for posters. Again, your professor may ask you to follow these guidelines or provide a somewhat modified version. The template provided online is a PowerPoint format, but is meant to be printed. The default is set to $8\frac{1}{2} \times 11$ printing, which will allow you to submit this to your instructor without the need for a plotter. This could easily be changed, however, and this template be used to create a large poster.

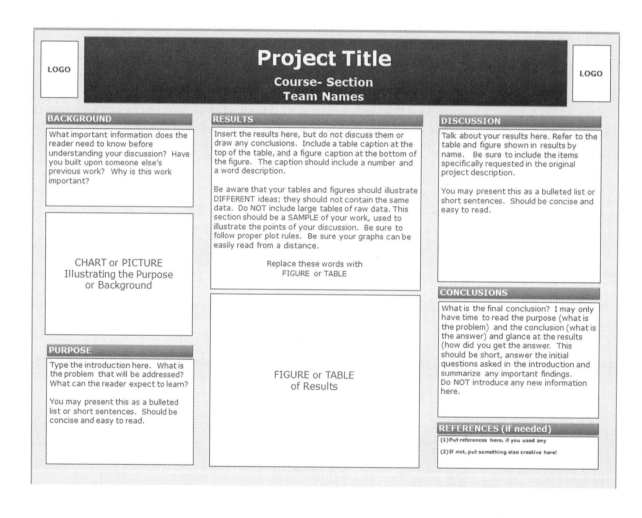

In-Class Activities

POWER POINT

ICA 4-1

Critique the following student presentation, discussing improvement strategies.

Purpose

In this Business and Engineering Collaborative Project we were given to task to design a product that would make dorm life safer, more enjoyable, or more productive. Our design was driven by our desire to meet customers needs, which would correspond to the ability to make a profit. We will show you all the steps it took to make our product marketable. Finally, we would like to thank everyone for coming and we hope that you agree with us that our product is marketable and of course a masterpiece.

Idea Generation

Here are some of our best ideas....

Screening

In our next step we took all the ideas that we had come up with and narrowed them down to a select three. The reason all the others were thrown out is because some of them already existed on the market, some of them would be too complex to the manufacturer, and some of them we didn't think that we had technical expertise needed to produce them. The three products we chose are....

Marketing Programs

Production Processes

1st: The wood comes off of the trucks and goes strait to cutting.

2nd: After the wood is cut, it is sent to be dulled.

3rd: Then all of the wood is sanded down, to make a nice surface.

4th: The loft is then put together by 5 employees.

5th: While the loft is still together it is tested for any flaws.

6th: The loft is then taken back apart by another 5 employees.

7th: The loft is then packaged and sent to be shipped.

8th: The last and final step, the loft is loaded into trucks.

Conclusion

Developing and introducing new products is frequently time-consuming, expensive, and risky. Thousands of new products are introduced annually but the failure rate is between 60 and 75 percent. If our product was put into production, we are confident that we would not be a part of that percentage. We think we have a feasible and well thought out marketing plan. We would like to thank you for coming and listening to our marketing pitch!

ICA 4-2

Critique the following student presentation, discussing improvement strategies.

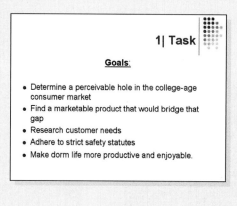

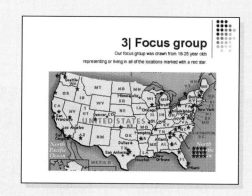

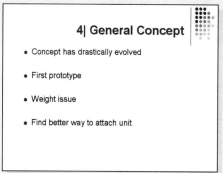

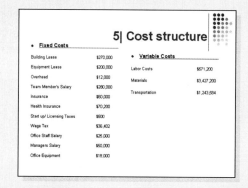

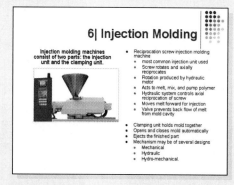

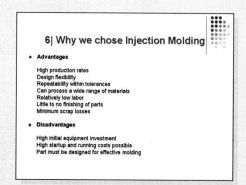

The following three ICAs are possible presentations you may be asked to make in this course. For your presentation, you may be required to choose one of the topics from the list provided, your instructor may specify different topics, or you may even be allowed to choose one of your own. Your instructor will make clear which options you have. Remember, you are giving a *technical* presentation. *Any discussion of the topic that is considered inappropriate by the instructor will be subject to a grade reduction.*

ICA 4-3

This ICA is meant to give you practice getting up in front of a group. The topics are informal, as the purpose is simply to help you gain confidence in public speaking. The presentation topic will be assigned to you at random shortly before your scheduled presentation; your instructor will determine how "shortly" before. The presentation should last *one minute*. If your speech is shorter or longer by more than 15 seconds, 10 points will be deducted from your presentation grade. For your presentation, here are some possible approaches you may want to take.

Make it informative.

Suppose your word is "WHALE." To give an informative talk, make a mental note of important things you know about whales. They are big. Some species are almost extinct. They are mammals. The blue whale is the largest animal that has ever lived on Earth. They were hunted for their blubber and oil that was used in lamps before electric lights. They hold their breath while under the water.

Tell a story.

If you know *Moby Dick*, you could give a brief outline of the most important points or of a particular scene in the book. You could tell of a visit to Sea World where you saw the trained killer whales perform. You could make up a short bedtime story that you relate as one you would tell your child or that was told to you as a child.

Be innovative.

You could do something like this: Whale is spelled W-H-A-L-E. The W stands for "water" where the whales live (talk about the ocean for a while); H stands for "huge," which is what we think of when we say whale; A stands for "animal" and the whale is the largest; L stands for "large," which is a lot like huge; and E stands for "eating" since the whale has to eat a lot, or E could stand for "enormous" like huge and large.

Or

You could do something like this: I dreamed about a whale that was in the parking lot at K-Mart. It kept smiling, and let out an enormous burp every few minutes. Being an animal lover, I went up next to the whale and stroked its sides. I was amazed at how smooth it was.

Or

You could do something like this: I have always wanted to scuba dive and in particular to ride on the back of a killer whale. I know that it would be dangerous, but it would be something I could remember the rest of my life.

ICA 4-3: Presentation Topic Suggestions

- Person who impressed you
- A hobby
- Favorite sport
- Favorite team mascot
- Favorite course
- Importance of cars
- Importance of space exploration
- Why engineers are neat
- Why calculus is important
- My favorite food
- My favorite dessert
- The worst insect
- How I would improve my school
- Coolest animal
- Favorite day of the week

- A childhood memory
- My favorite relative
- Someone I admire and why
- The best car
- Favorite month of the year
- Why I came to this college
- My dream job (after college)
- Why I want to be an engineer
- Place I would like to visit
- If I had a million dollars
- The ultimate dorm room
- Favorite phone app
- Favorite vacation
- Why I would like to be a professor
- The best thing about being in college

- The best thing about computers
- What kind of boss I would make
- Why I would like to be the president
- My opinion on Napster
- Something I learned this week
- The best thing about my school
- Advice for incoming freshman
- How to improve this classroom
- A feature to add to the car
- A good book I have read
- My favorite year in grade school
- The farthest I have ever been from my hometown

ICA 4-4

In this ICA, the speeches will discuss the graphical representation of various phenomena, such as "A total eclipse" or "Using a toaster." The presentations should last 1 *minute*, 30 *seconds*. If your speech is more than 15 seconds longer or shorter than this, 10 points will be deducted from your presentation grade.

Later in this text, much time is spent on graphing mechanics and interpretation. You are to choose a topic from the following list and represent it graphically. A variety of presentations of the available data that tell different things are possible. Do not forget to use all your senses and imagination.

You must prepare a proper graph, incorporated in PowerPoint, and use it during your presentation. You are not allowed to copy a premade graph from the Internet! In your speech, you should discuss the process, explain how information is shown on the graph, and be prepared to answer questions about your process. *You are limited to three slides*: (1) a title slide, (2) an introductory slide, and (3) a graph. If needed, you will be allowed a single 3×5 inch index card with notes on one side only.

ICA 4-4: Presentation Topic Suggestions

- Moving a desk down a set of stairs
- Letting go of a helium balloon
- Feedback from an audio system
- A cow being picked up by a tornado
- A kangaroo hopping along
- A glass of water in a moving vehicle
- People in Florida
- People in Michigan
- The flight of a hot air balloon
- Using a toaster
- A train passing
- Dropping ice in a tub of warm water
- Slipping on a banana peel
- Sound echoing in a canyon
- Detecting a submarine using sonar
- Pouring water out of a bottle
- Boiling water in a whistling teapot
- Hammering nails
- Spinning a Hula-Hoop™
- Playing with a yo-yo
- Snow blowing over a roof
- Climbing a mountain
- Pumping air into a bicycle tire
- The plume from a smokestack
- Formation of an icicle
- A rabbit family
- Firing a bullet from a rifle
- Hiking the Appalachian Trail
- Eating a stack of pancakes
- A solar eclipse

- Taking a bath
- Throwing a ball
- Driving from Clemson to Greenville
- Diving into a swimming pool
- A person growing up
- Burning a pile of leaves
- Driving home from work
- An oak tree over the years
- Airplane from airport to airport
- Football game crowd
- Train passing through town
- A coastal river
- The moon
- Daily electric power consumption
- A typical day
- A thunderstorm
- A day in the life of a chicken
- Popping corn
- Feeding birds at a bird feeder
- Skipping a stone on water
- A burning candle
- Pony Express
- Talking on a cell phone
- Using Instant Messenger
- An engineer's salary
- Baseball in play
- Exercising
- Running of the bulls
- Drag racing
- D.O.R. possum ("dead on the road")

- Political affiliations
- Traffic at intersections
- River in a rainstorm
- A tiger hunting
- Hair
- Grades in calculus class
- Forest fire
- Cooking a Thanksgiving turkey
- Brushing your teeth
- Baking bread
- Eating at a fast-food restaurant
- Studying for an exam
- Power consumption of your laptop
- Power usage on campus
- Oil supply
- Air temperature
- Student attention span during class
- Baseball
- A diet
- Sleeping
- Bird migration
- Strength of concrete
- Human population
- A mosquito
- A video game
- Taking a test
- A trip to Mars
- The North Pole
- Space elevators

ICA 4-5

This presentation covers topics related to your future major. The presentations should last *three minutes*. If your speech is more than 15 seconds longer or shorter than this, 10 points will be deducted from your presentation grade.

In 2000, the National Academy of Engineering compiled a list of "Greatest Engineering Achievements of the 20th Century." The following list contains some of the achievements listed, as well as some of the greatest failures of the past century. The full list can be found at www.greatachievements.org.

It is your responsibility to research the topic you choose so that you can intelligently discuss it in reference to the engineering involved and how it affects everyday life, both then and now.

You must prepare a PowerPoint presentation and use it in this speech. You should have between five and eight slides, of which the first must be a title slide, and the last must be a slide listing references. Other visual aids may be helpful and are encouraged. If needed, you will be allowed a single 3×5 inch index card with notes on one side only.

ICA 4-5: Presentation Topic Suggestions

Electricity
- Grand Coulee Dam (1942)
- Electric light bulbs

Automobiles
- Mass production (1901: Olds; 1908: Ford)
- Octane ratings (1929)
- Goodrich "tubeless" tires (1947)

Airplanes
- Hindenburg (1937)
- Concord

Water Supply and Distribution
- Purification of water supply (1915: Wolman)
- Cuyahoga River pollution (1970s)
- Hinkley, California

Electronics
- Integrated circuits (1958: Kilby)
- Handheld calculator (1967: TI)

Radio and TV
- Color television (1928; 1954)
- Phonograph (Edison)

Agricultural Mechanization
- Mechanical cotton picker (1949: Rust)
- Irrigation equipment (1948: Zybach)

Computers
- Language compiler (1952: Hopper)
- Transistor (1947)

Telephones
- Touch-tone dialing (1961)

Air Conditioning and Refrigeration
- Air conditioning (1902: Carrier)
- Freezing cycle for food (1914: Birdseye)

Highways
- PA turnpike
- Tacoma Narrows Bridge
- Golden Gate Bridge
- Chunnel

Spacecraft
- Space shuttle *Challenger* explosion
- Hubble telescope
- *Apollo 1* and *Apollo 13* failures
- *Gemini* spacecraft

Internet
- E-mail (1972: Tomlinson)
- WWW (1992)

Imaging
- Sonar (1915: Langevin)
- Radar (1940)
- Ultrasound (1958)
- MRI (1980)

Household Appliances
- Dishwasher (1932)
- Clothes dryer (1935: Moore)
- Microwave (1946)

Health
- Defibrillator (1932: Kauwenhoven)
- Artificial heart (1980: Jarvik)

Petroleum
- Distillation (1920s: Fischer and Tropsch)
- Ethyl gas (1921)
- Alaskan pipeline

Laser and Fiber-optics
- Chemical vapor deposition (1974)
- Optical fibers (1970: Maurer, Corning)

Nuclear Technologies
- Van de Graff generator (1937)
- Manhattan Project (1939–1945)
- Three Mile Island power plant (1979)

High-Performance Materials
- Vulcanization of rubber (1926)
- Polyvinyl chloride—"PVC" (1927)
- Teflon (1938: Plunkett)
- Kevlar (1971: Kwolek)

ICA 4-6 and 4-7 were written to address the question of what obligations humans may have to protect extraterrestrial life if it is discovered.

ICA 4-6

Critique the following writing assignment, discussing improvement strategies.

If life were to be discovered elsewhere in the universe, it would be our duty to protect that life only to the extent in which we do not compromise our own stability. The nation as a whole currently already has debt issues and is going through a depression. America could not afford to spend money on a life when they are struggling to take care of themselves. According to Hodges, "America has become more a debt 'junkie'—than ever before with total debt of $53 Trillion—and the highest debt ratio in history." (2) Globally there are people who are dying from starvation each day. "On Tuesday September 11, 2001, at least 35,615 of our brother and sisters died from the worst possible death, starvation. Somewhere around 85% of these starvation deaths occur in children 5 years of age or younger." (1) What would make this new life form a higher priority than the neighbors right around the corner? If they have managed to survive this long we shouldn't worry about helping them until we have taken care of ourselves. Even if they are extremely advanced with many discoveries we have not made, what good would it do to spend billions on reverse engineering but leave our own world in desperation. Now if we were to reach a point in the future in which we could help them; we would naturally have to approach with caution. Politically one of the worst first impressions is when new people meet and one group starts a biological plague. In 1493 right after Columbus came to America he went back to Hispaniola, bringing livestock in order to start a colony there. "Influenza, probably from germs carried by the livestock, swept through the native people, killing many of them. Modern researchers believe that American Indian traders carried the disease to Florida and throughout the Caribbean." (3) We would not know what would hurt them as well as what could hurt us. In order to avoid this terrible introduction, we would need to do research with probes and satellites. These observations from afar would make a seamless introduction possible. After the introduction it would be top priority to keep diplomatic relations intact. Finding another life out in space would be great, but not if we made them our enemies.

ICA 4-7

Critique the following writing assignment, discussing improvement strategies.

For centuries, people have wondered if we were all alone in our universe. Still today, even with our advancements in science we still have not been able to find life anywhere else. Recently the Phoenix Lander discovered what seems to be incontrovertible evidence that water exists on Mars. Since life is dependent on water eye brows have again been raised about the probability that life may exist, or has existed on Mars. So what if life is discovered somewhere other than on earth? Are we obligated to protect that life? Does that obligation depend on how advanced the life is? How do we stop ourselves from not destroying the non-terrestrial life? If life is discovered somewhere other than earth, some people will feel threaten, others will feel that maybe somehow human life could be supported by the non-terrestrial's planet and that life is elsewhere out there, maybe life like ours. As far as our obligation to protect that life, it is recommended that we do not interfere. We are having problems with our own planet, people pollute, litter, kill wildlife, start wildfires, cut down forests, and destroy whole ecosystems. Imagine what we will do to life if it doesn't matter to us if it lives or not. No matter how advance the life is we should leave the life to itself, unless the life was able to communicate with us somehow. The only way to keep ourselves from destroying the non-terrestrial life is to just to leave it alone. If we leave it, it wouldn't be our fault if it survives or eventually dies off. Yet still we haven't been able to find life anywhere besides earth so until we do we should try finding ways to turn our planet in the right direction before we lose what supports us.

ICA 4-8 and 4-9 were written to address the question of whether we should pursue manned exploration of space or restrict such activities to unmanned robotic devices.

ICA 4-8

Critique the following writing assignment, discussing improvement strategies.

Manned, unmanned space travel has been a big controversy for many years; it's too risky or space bots gets just as much done as humans. I agree 100% that robots are far better off into space than humans; manned space travel is too dangerous. Almost all the money for such a mission would be spent simply to keep the people alive. (John Tierney) Space bots are more reliable, don't get tired, and do what it's programed to do, but more proficient. Things humans need for survival includes: air, food, water, things we must take with us for space travel, which adds to the cost for each individual we fly out into space. While space bots don't need any of these sources for survive, which makes space travel less expensive and more productive. Also the bots can retrieve more data because there aren't any stopping periods for breaks or any other reasons for that matter. It cost about 1.3 billion dollars per shuttle launched out into orbit, while it cost far more per individual we send out into space. Manned space travel is not all bad; things humans can do like, making quick and on the spot decision or human senses to evaluate our surroundings are things robots can't imitate. It's nothing like being somewhere physically and knowing what's there instead of watching it through another pair of eyes. By using unmanned space travel, in reality we can only see what the robot sees, but is a human life worth the risk just for data. A human life is priceless, it like time when it's gone; it's gone, so why not send robotic material into space to collect data. It's only scrap metal, but the life of any human is far more valuable just for the expense of useless data.

ICA 4-9

Critique the following writing assignment, discussing improvement strategies.

Since the early 1960s, humans have been venturing beyond the Earth's atmosphere, a few times as far as the moon, a quarter of a million miles away. NASA currently plans to return humans to the moon by 2020, and plans Mars missions after that. Much criticism has been leveled at the entire idea of manned spaceflight, claiming that unmanned craft can do essentially all of the jobs people can with less cost and less risk. There are of course two sides to the situation. The idea for space exploration and those against it.

Space flight is a very integral part of exploration and adventure. Discovering the worlds around us I a very key part to discovering space. NASA's mission is to pioneer the future in space exploration, scientific discovery and aeronautics research. To do that, thousands of people have been working around the world—and off of it—for almost 50 years, trying to answer some basic questions. What's out there in space? How do we get there? What will we find? What can we learn there, or learn just by trying to get there, that will make life better here on Earth? (NASA). The ideal of many working together to the common goal of space exploration makes it a very important part. Manned spaceflight can be very dangerous. Many deaths and tragedies have occurred because of the exploration of space.

I think that space should still be explored by mankind. With technology advances mentioned in In the first 25 years of its existence, NASA conducted five manned spaceflight programs: Mercury, Gemini, Apollo, Skylab, and Shuttle.

The latter four programs produced spacecraft that had on-board digital computers. The Gemini computer was a single unit dedicated to guidance and navigation functions. Apollo used computers in the command module and lunar excursion module, again primarily for guidance and navigation. Skylab had a dual computer system for attitude control of the laboratory and pointing of the solar telescope. NASA's Space Shuttle is the most computerized spacecraft built to date, with five general-purpose computers as the heart of the avionics system and twin computers on each of the main engines. The Shuttle computers dominate all checkout, guidance, navigation, systems management, payload, and powered flight functions. The computers helped and as they advance manned spaceflight gets better. In the long run the risk is worth the gain for knowledge.

Chapter 4 REVIEW QUESTIONS

Writing Assignments

For each question, write a response according to the directions given by your instructor. Each response should contain correct grammar, spelling, and punctuation. Be sure to answer the question completely, but choose your words carefully so as to not exceed the word limit if one is given. There is no right or wrong answer; your score will be based upon the strength of the argument you make to defend your position.

1. In August 2007, the space shuttle *Endeavor* suffered minor damage to the heat shield tiles during liftoff. After the *Columbia* burned up on reentry in 2003 due to damaged tiles, NASA developed a protocol for repairing damaged tiles while in orbit. After much consideration, NASA decided *not* to attempt to repair the tiles on *Endeavor* while in orbit on this mission.

    Find a minimum of three different references discussing this incident and NASA's decision. Summarize the reasons both for and against repairing the tile while in orbit. If you had been on the team making this decision, would you have argued for or against the repair and why?

2. Research the failure of the Teton Dam. Describe what you think are the fundamental ethical issues involved in the failure. In terms of ethics, compare the Teton Dam failure in 1976 to the failure of levees in New Orleans in 2005.

3. In recent years, many studies have been conducted on the use of cell phones while driving. University of Utah psychologists have published a study showing that motorists who talk on handheld or hands-free cellular phones are as impaired as drunken drivers.

    > *"We found that people are as impaired when they drive and talk on a cell phone as they are when they drive intoxicated at the legal blood-alcohol limit of 0.08 percent, which is the minimum level that defines illegal drunken driving in most U.S. states," says study co-author Frank Drews, an assistant professor of psychology. "If legislators really want to address driver distraction, then they should consider outlawing cell phone use while driving."*

    > (*Strayer and Drews, "Human Factors."* The Journal of the Human Factors and Ergonomics Society, *Summer 2006.*)

    Another way to approach this problem is to force manufacturers of cell phones to create devices on the phones that would restrict usage if the user is driving, much like a safety on a gun trigger. Research and report current statistics on cell phone or other texting device usage during driving. Discuss the hazards (or lack of hazards) of driving while using a cell phone or other texting device. Discuss and justify whether cell phone manufacturers should be required to provide "safety" features on their devices to prevent usage while driving.

4. According to an article in the *Christian Science Monitor* (January 16, 2008, issue), environmentalists claim that a development around a remote lake 140 miles north of Augusta, Maine, would emit 500,000 tons of carbon dioxide over 50 years, including estimated emissions from cars traveling to and from the development. An environmental group has presented this "carbon footprint" to the state, and is requesting that the impact on the environment become part of the process for granting development permits.

    As many as 35 states have adopted climate-action plans, but there are few cases like this in which environmental impact factors into government approval of land development. This could have a significant effect on engineers involved with land development, structures, land-use planning, or environmental impact assessments in the future. The original article can be found at: http://www.csmonitor.com/2008/0116/p01s04-wogi.html.

    Discuss both sides of this issue, and take a stance for or against mandating carbon footprint assessment for new developments. Justify your position, including information from at least three sources. Ideas to include in your essay are cases in other states, climate-action plans, calculation of a "carbon footprint," or the land-development approval process. Most importantly, consider how an engineer would view this issue.

    Thank you to Dr. Lisa Benson for contributing this assignment.

5. From 2005 to 2009, much of the southeast experienced a serious rainfall deficit. Lake levels were at record lows, water restrictions were debated and enacted, and the ecology of the area was changed. Amid all of this are nonessential services that are extremely heavy users of water. Perhaps the most egregious examples are sports fields, and in particular, golf courses. On average, an 18-hole golf course in the United States uses about 100 million gallons per year.

Summarize the arguments both in favor and against the heavy use of water for nonessential services, particularly in areas where water supply is limited. Consider the following:

- If you were in a position to recommend legislation for such water use, what policies would you recommend and why? Consider this from an engineer's point of view.
- Consider the impact on local environments for building dams for water supplies and hydropower. As an engineer, what would you consider to be the benefits and drawbacks of this type of project, and why?

6. Since the early 1960s, humans have been venturing beyond Earth's atmosphere, a few times as far as the moon, a quarter of a million miles away. NASA currently plans to return humans to the moon by 2020, and plans Mars missions after that.

Much criticism has been leveled at the entire idea of manned spaceflight, claiming that unmanned craft can do essentially all the jobs people can with less cost and less risk.

Summarize the arguments on both sides of this issue.

7. In summer 2008, the *Phoenix Lander* discovered what seems to be incontrovertible evidence that water exists on Mars. Since life as we know depends on water (for a variety of reasons), this discovery raises the probability that life may exist, or did exist in the past, on Mars.

If life is discovered somewhere other than on Earth, what are our obligations to protect that life? Does that obligation depend on how "advanced" we perceive that life to be? What steps, if any, should be taken to ensure that we do not destroy the non-terrestrial life?

8. Advances in genetic engineering may make it possible to bring extinct species back to life.

(a) Should we attempt to restore populations of recently extinct animals, such as the Passenger pigeon, Carolina parakeet, ivory-billed woodpecker, Tasmanian tiger, or Formosan clouded leopard, for which there are many preserved specimens from which DNA could be acquired? There is currently an effort to obtain enough DNA to reestablish a population of Woolly mammoths. What are the ethical issues involved with bringing back a mammal extinct for 3,500 years?

(b) It was announced in February 2009 that a first draft of the genome of *Homo sapiens neanderthalensis* (Neanderthal), comprising about two-thirds of the base pairs, has been completed. If future advances make it possible, should we attempt to bring Neanderthals back to life?

9. If you could go back in time and be part of any engineering achievement in the past, what would you choose?

Your choice must involve something that was accomplished before 1970, although it might still be in use today. You may go as far back in time as you wish, but not more recently than 1970.

Your essay must include the following.

- Why did you choose this specific thing?
- Discuss the actual development of the item in question. Include issues such as why the item was desired, what problems the designers confronted, and specific design decisions that were made.
- What effects has the development of this device (or process) had on society and the planet, including positive, negative, and neutral effects.
- A bibliography with at least three distinct entries. All may be online references, but no more than one may be Wikipedia or similar sites.

10. Choose a bridge innovation to research from a list of potential topics given below, and write a memo or presentation that includes the following:

   - The name and location of the structure, when it was constructed, and other significant attributes.
   - If possible, a photo or sketch.
   - A summary table of the design features. Potential features to consider including, if appropriate:
     - Type of bridge design (arch, beam, suspension, etc.)
     - Dimensions (span, height)
     - Unique design features
     - Cost
     - Awards or superlatives (longest, highest, most cost-effective, costliest)
   - A discussion of the aspects of the bridge design that contributed to its success (materials, structural design, geographical, topographical, or climatic challenges, aesthetic qualities, etc.).

   (a) Akashi Kaikyo Bridge, linking the islands of Honshu and Shikoku in Japan
   (b) Cooper River Bridge in Charleston, South Carolina
   (c) Forth Bridge, Scotland
   (d) Gateshead Millennium Bridge, spanning the River Tyne in England
   (e) Hanzhou Bay Bridge, China
   (f) Humber Bridge, England
   (g) Lake Pontchartrain Causeway, Louisiana
   (h) Mackinac Bridge, Michigan
   (i) Millau Viaduct, Millau, France
   (j) Natchez Trace Bridge in Franklin, Tennessee
   (k) Penobscot Narrows Bridge, Maine
   (l) Rio-Antirio Bridge, Greece
   (m) Sundial Bridge at Turtle Bay in Redding, California
   (n) Sunshine Skyway Bridge in Tampa, Florida
   (o) Sydney Harbor Bridge, Australia
   (p) Tower Bridge, London
   (q) Woodrow Wilson Bridge (proposed), Washington, DC
   (r) Zakim Bunker Hill Bridge in Boston in Massachusetts.

11. Choose a major transportation structure failure to research from the list of potential topics that follows, and write a memo or presentation that includes the following:

   - The name and location of the structure, when it was constructed, and other significant attributes.
   - If possible, a photo or sketch.
   - A summary of the event(s) surrounding the structure's failure, including dates. Discuss aspects of the design that contributed to its failure (materials, structural flaws, misuse, improper maintenance, climatic conditions, etc.). If it was rebuilt, repaired, or replaced, how was the design modified or improved?

   (a) Angers Bridge over the Maine River in Angers, France
   (b) Autoroute 19 Overpass, Quebec
   (c) Arroyo Pasajero Twin Bridges in Coalinga, California
   (d) Banqiao Dam, China
   (e) Charles de Gaulle Airport, France
   (f) Hartford Civic Center, Connecticut
   (g) Hyatt Regency hotel walkway in Kansas City, Missouri
   (h) Kemper Arena, Missouri
   (i) L'Ambiance Plaza in Bridgeport, Connecticut

(j) Loncomilla Bridge, Chile

(k) Mianus River Bridge in Greenwich, Connecticut

(l) Millennium Footbridge, London

(m) Sampoong Department Store, South Korea

(n) Seongsu Bridge, spanning the Han River in Seoul, South Korea

(o) Sgt. Aubrey Cosens VC Memorial Bridge in Latchford, Ontario

(p) Silver Bridge, between Point Pleasant, WV, and Kanauga, Ohio

(q) The Big Dig in Boston, Massachusetts

(r) West Gate Bridge in Melbourne, Victoria, Australia

12. Choose an environmental issue to research from a list of potential topics given below, and write a memo or presentation that includes:

■ A clear definition of the problem.

■ If possible, a photo or sketch.

■ Discussion of the issue; potential features to consider include, if appropriate:

- Affected areas (local, region, continent, global)
- Causes, sources
- Effects (human, animal, vegetation, climate)
- Prevention or reversal options (including feasibility and cost)
- Important legislation (or lack thereof)

The topic should be discussed from an *engineering* viewpoint.

(a) Acid rain

(b) Air purification

(c) Bioremediation

(d) Clean Air Act

(e) Cuyahoga River pollution/Clean Water Act

(f) Deforestation

(g) Erosion control

(h) Groundwater pollution

(i) Kyoto Protocol

(j) Lake eutrophication/hypoxia

(k) Loss of wetlands/ecosystems/pollinators

(l) Ocean acidification/pollution

(m) Purification/safe drinking water

(n) RCRA (Resource Conservation and Recovery Act)

(o) Smog/particle reduction

(p) Soil contamination

(q) Solar energy

(r) Superfund/CERCLA

(s) Water Pollution Control Act

(t) Wind energy

13. Technology promises to revolutionize many aspects of life and society. In a memo or presentation, discuss the advantages and disadvantages of one or more of the following. You may need to research some of these topics first.

(a) Nano-robots designed to destroy cancer cells

(b) More powerful particle accelerators for probing the structure of matter

(c) The James Webb Space Telescope

(d) Cloaking technology at visible wavelengths

(e) Practical antigravity devices

(f) Drugs to inhibit soldiers' fears

(g) Nano-devices to rewrite an organism's DNA

(h) Practical (time and cost) interplanetary travel

(i) Artificial chlorophyll to harness energy from the sun

(j) Genetically engineered microbes to attack invasive species
(k) Direct neural interface to computers
(l) Replacement human limbs with full functionality
(m) Terraforming Mars, Venus, or the moons of Jupiter or Saturn
(n) Quantum computers
(o) Room-temperature superconductors
(p) Materials harder than diamonds
(q) Artificial spider silk
(r) Full-color night-vision goggles
(s) Home-scale fusion reactors
(t) Fully automated cars
(u) Space elevators
(v) Bionic implants for enhancing sight, sound, and/or smell

CHAPTER 5
ESTIMATION

Enrico Fermi was a Nobel laureate and one of many brilliant scientists and engineers involved in the Manhattan Project, which developed the first nuclear weapons during World War II. He wrote the following after witnessing the first test of an atomic bomb, called the Trinity Test (see Figure 5-1). Of particular note is the final paragraph.

> On the morning of the 16th of July, I was stationed at the Base Camp at Trinity in a position about ten miles from the site of the explosion.
>
> The explosion took place at about 5:30 A.M. I had my face protected by a large board in which a piece of dark welding glass had been inserted. My first impression of the explosion was the very intense flash of light, and a sensation of heat on the parts of my body that were exposed. Although I did not look directly towards the object, I had the impression that suddenly the countryside became brighter than in full daylight.
>
> I subsequently looked in the direction of the explosion through the dark glass and could see something that looked like a conglomeration of flames that promptly started rising. After a few seconds the rising flames lost their brightness and appeared as a huge pillar of smoke with an expanded head like a gigantic mushroom that rose rapidly beyond the clouds probably to a height of the order of 30,000 feet. After reaching its full height, the smoke stayed stationary for a while before the wind started dispersing it.
>
> About 40 seconds after the explosion the air blast reached me. I tried to estimate its strength by dropping from about six feet small pieces of paper before, during and after the passage of the blast wave. Since at the time there was no wind I could observe very distinctly and actually measure the displacement of the pieces of paper that were in the process of falling while the blast was passing. The shift was about 2½ meters, which, at the time, I estimated to correspond to the blast that would be produced by ten thousand tons of T.N.T.
>
> *Citation: U.S. National Archives, Record Group 227, OSRD-S1 Committee, Box 82 folder 6, "Trinity."

Before the test, no one knew what would happen. Speculation among the many people involved concerning the results of the test ranged from nothing (no explosion at all) to setting the planetary atmosphere on fire and destroying all life on Earth. When all the data from the blast were analyzed, the true strength of the blast was calculated to be 19 kilotons. By simply observing the behavior of falling bits of paper 10 miles from ground zero, Fermi's estimation of 10 kilotons was in error by less than a factor of 2.

After the war, Fermi taught at the University of Chicago, where he was noted for giving his students problems in which so much information was missing that a solution seemed impossible. Such problems have

Figure 5-1 The first atomic bomb test, Alamogordo, New Mexico.

Courtesy of: Truman Presidential Museum & Library.

been named **Fermi problems**, and in general they require the person considering them to determine an answer with far less information than would really be necessary to calculate an accurate value. Engineers are often faced with solving problems for which they do not have all the information. They must be adept at making initial estimates. This skill helps them identify critical information that is missing, develop their reasoning skills to solve problems, and recognize what a "reasonable" solution will look like.

Most practical engineering problems are better defined than Fermi problems and can be estimated more easily and accurately in general. The following are just a few examples of real engineering problems. See if you can estimate answers for these problems. These should be done without reference to the web or any other source of information.

- How many cubic yards of concrete are needed to pave 1 mile of interstate highway (two lanes each direction)?
- How many feet of wire are needed to connect the lighting systems in an automobile?
- What is the average flow rate in gallons per minute of gasoline moving from the fuel tank to the fuel injectors in an automobile cruising at highway speed?

An accomplished engineer knows the answer to most problems before doing any calculations. This does not mean an answer to three significant figures, but a general idea of the range of values that would be reasonable. For example, if you throw a baseball as high as possible, how long will it take for the ball to hit the ground? Obviously, an answer of a few milliseconds is unreasonable, as is several months. Several seconds seems more realistic. When you actually do a calculation to determine this time, you should ask yourself, "Is my answer reasonable?"

Sample Fermi Problems

- Estimate the total number of hairs on your head.
- Estimate the number of drops of water in all of the Great Lakes.
- Estimate the number of piano tuners in New York City.

● EXAMPLE 5-1

Every year, numerous people run out of fuel while driving their vehicle on the road. Determine how many gallons of gasoline are carried to vehicles with empty fuel tanks each year in the United States so that the vehicle can be driven to the nearest gas station. Do this without reference to any other material, such as the Internet or reference books.

Estimations

In almost all cases, the first step is to estimate unknown pieces of information that are not available. In general, there are numerous paths to a solution, and different people may arrive at different answers. Often, the answer arrived at is only accurate to within an order of magnitude (a factor of 10) or less. Nonetheless, such problems can provide valuable insight not only into the problem itself but also into the nature of problem solving in general.

Remember, these are estimates, not accurate values. Someone else making these estimates might make different assumptions.

- *Number of people in the United States: 500,000,000 persons.*
- *Fraction of people in the United States that drive: drivers per 10 persons.*
- *Average times a person runs out of gas per year: one "out of gas" per 4 years per driver.*
- *Fraction of "out of gas" incidents in which gas is brought to the car (rather than pushing or towing the car to a station): 24 "bring gas to car" per 25 "out of gas".*

■ *Average amount of gas carried to car: 1.5 gallons per "bring gas to car".*

Calculation

This is where you combine your estimates to arrive at a solution. In the process, you may realize that you need further information to complete the computation.

$$\text{Drivers in United States} = (5 \times 10^8 \text{ people})\left(\frac{7 \text{ drivers}}{10 \text{ people}}\right) = 3.5 \times 10^8 \text{ drivers}$$

$$\text{Number out of gas per year} = (3.5 \times 10^8 \text{ drivers})\left(\frac{1 \text{ out of gas}}{(4 \text{ years})(1 \text{ driver})}\right)$$

$$= 8.75 \times 10^7 \frac{\text{out of gas}}{\text{year}}$$

$$\text{Number bring gas to car per year} = \left(\frac{8.75 \times 10^7 \text{ out of gas}}{\text{year}}\right)\left(\frac{24 \text{ bring to car}}{25 \text{ out of gas}}\right)$$

$$= 8.4 \times 10^7 \frac{\text{bring gas to car}}{\text{year}}$$

$$\text{Amount of gas to cars} = \left(\frac{8.4 \times 10^7 \text{ bring gas to car}}{\text{year}}\right)\left(\frac{1.5 \text{ gallons of gas}}{\text{bring gas to car}}\right)$$

$$= 1.26 \times 10^8 \frac{\text{gallons}}{\text{year}}$$

Thus, we have estimated that about 125 million gallons of gas are taken to "out of fuel" vehicles each year in the United States. It would be perfectly valid to give the answer as "about 100 million" gallons, since we probably have only about one significant digit worth of confidence in our results, if that.

A few things to note about this solution:

■ *"Units" were used on all numerical values, although some of these units were somewhat contrived (e.g., "out of gas") to meet the needs of the problem. Keeping track of the units is critical to obtaining correct answers and will be highly emphasized, not only in this text but also throughout your engineering education and career.*

■ *The units combine and cancel according to the regular algebraic rules. For example, in the first computation, the unit "persons" appeared in both the numerator and the denominator, and thus canceled, leaving "drivers."*

■ *Rather than the computation being combined into one huge string of computations, it was broken into smaller pieces, with the results from one step used to compute the next step. This is certainly not an immutable rule, but for long computations it reduces careless errors and makes it easier to understand the overall flow of the problem.*

Many cars today use a plastic fuel tank. It is more durable and lighter than a conventional steel tank. The picture below shows a blow-molded fuel tank with six layers of plastics. Six layers are required to eliminate any permeation of the fuel through the tank.

The picture on the right shows a model of how the melted plastic flows through the extruder head of a blow-molding machine. The six layers are combined at the bottom to form a parison, which is then blow molded to form the fuel tank.

Photos courtesy of Kautex Machines.

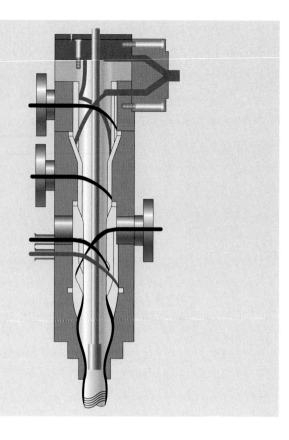

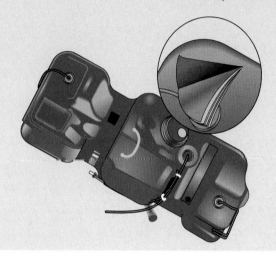

5.1   GENERAL HINTS FOR ESTIMATION

As you gain more knowledge and experience, the types of problems you can estimate will become more complicated. Here we give you a few hints about making estimates.

- Try to determine the accuracy required. Is order of magnitude enough? How about ±25%?
  - What level of accuracy is needed to calculate a satellite trajectory?
  - What level of accuracy is needed to determine the amount of paint needed to paint a specified classroom?

The term "**orders of magnitude**" is often used when comparing things of very different scales, such as a small rock and a planet. By far the most common usage refers to factors of 10; for example, three orders of magnitude refer to a difference in scale of $10^3 = 1,000$. If we wanted to consider the order of magnitude between 10,000,000 and 1,000, we would calculate the logarithm of each value ($\log(10,000,000) = 7$ and $\log(1,000) = 3$), thus there are $7 - 3 = 4$ orders of magnitude difference between 10,000,000 and 1,000.

- Remember that a "ballpark" value for an input parameter is often good enough.
  - What is the square footage of a typical house?
  - What is the maximum high temperature to expect in Dallas, Texas, in July?
  - What is the typical velocity of a car on the highway?

■ Always ask yourself if it is better to err on the high side or the low side. There are two primary reasons for asking this question.

- Safety and practical considerations. Will a higher or lower estimate result in a safer or more reliable result?
    - ■ If estimating the weight a bridge can support, it is better to err on the low side, so that the actual load it can safely carry is greater than the estimate.
    - ■ For the bridge mentioned above, if estimating the load a single beam needs to support, it is better to err on the high side, thus giving a stronger beam than necessary. Be sure you understand the difference between these two points.

■ Estimate improvement. Can the errors cancel each other?

- If estimating the product of two numbers, if one of the terms is rounded low, the other should be rounded high to counteract for the lower term.
- If estimating a quotient, if you round the numerator term on the low side, should the denominator term be rounded low or high?

■ Do not get bogged down with second-order or minor effects.

- If estimating the mass of air in the classroom, do you need to correct for the presence of furniture?
- In most instances, can the effect of temperature on the density of water be neglected?

The best way to develop your ability to estimate is through experience. An experienced painter can more easily estimate how much paint is needed to repaint a room because experience will have taught the painter such things as how many coats of one paint color it will take to paint over another, how different paint brands differ in their coverage, and how to estimate surface area quickly. In *Outliers: The Story of Success*, Malcolm Gladwell provides examples from diverse career pathways that demonstrate 10,000 hours of practice are required to develop world-class expertise in any area. Fortunately for aspiring engineers, much of this expertise can be developed starting at a young age and outside of formal schooling. For example, how many hours have you spent observing the effects of gravity? Of course, some important engineering concepts stem from phenomena that are not so easily observed, and some lend themselves to misinterpretation. As a result, it helps to have a systematic approach to developing estimates—particularly where we have less experience to guide us. Estimating an approximate answer of a calculation including known quantities is a valuable skill—such as approximating the square root of 50 as about 7, approximating the value of pi as 3 for quick estimates, etc. These mathematical approximations, however, assume that you have all the numbers to begin with, and that you can use shortcuts to estimate the precisely calculated answer to save time or as a check against your more carefully calculated answer. Estimation is discussed here in a broader sense—estimating quantities that cannot be known, are complicated to measure, or are otherwise inconvenient to obtain. It is in these cases that the following strategies are recommended.

The Windows interface estimates the time needed to copy files. The estimate is dynamic and appears to be based on the total number of files and the assumption that each file will take the same amount of time to copy. As a result, when large files are copied, the estimate will increase—sometimes significantly. Similarly, as a large number of small files are copied, the estimate will decrease rapidly. A better estimation algorithm might be based on the percentage of the total file size.

5.2 ESTIMATION BY ANALOGY

One useful strategy for estimating a quantity is by comparison to something else we have measured previously or otherwise know the dimension of. The best way to prepare for this approach to estimation is to learn a large number of comparison measures for each type of quantity you might wish to estimate. Each of these comparison measures becomes an anchor point on that scale of measurement. This book provides some scale anchors for various measurable quantities—particularly in the case of power and energy, concepts with which many people struggle.

● EXAMPLE 5-2

Estimate the size of a laptop computer using analogy.

Laptop computers come in different sizes, but it is not difficult to estimate the size of a particular laptop. Laptops were first called "notebook" computers—a good starting estimate would be to compare the particular laptop to notebook paper, which is 8.5 inches by 11 inches in the United States.

● EXAMPLE 5-3

Estimate the size of an acre and a hectare of land using analogy.

American football field—playing area is 300 feet by 160 feet = approximately 50,000 square feet. An acre is 43,560 square feet. Using this data, we have a better sense of how much land an acre is—about 90% of the size of the playing area of an American football field. Soccer fields are larger, but vary in size. The largest soccer field that satisfies international guidelines would be about 2 acres.

   A hectare, or 10,000 square meter, is equivalent to 108,000 square feet and is much larger than an acre—about the maximum size of the pitch in international rugby competition.

5.3 ESTIMATION BY AGGREGATION

Another useful strategy is to estimate the quantity of something by adding up an estimate of its parts. This can involve multiplication in the case of a number of similarly sized parts, such as estimating the size of a tile by comparing it to your foot (estimation by analogy), counting the number of floor tiles across a room, and multiplying to estimate the total length of the room. In other cases, aggregation may involve adding together parts that are estimated by separate methods. For example, to estimate the volume of a two-scoop ice cream cone, you might estimate the volume of the cone and then separately estimate the volume of each scoop assuming they are each spheres.

● EXAMPLE 5-4

Estimate by aggregation the amount of money students at your school spend on pizza each year.

Ask students around you how often they purchase a pizza and how much it costs;
Convert this estimate into a cost per week;
Multiply your estimate by the number of weeks in an academic year;
Multiple that result by the number of students at your school.

5.4 ESTIMATION BY UPPER AND LOWER BOUNDS

An important part of estimating is keeping track of whether your estimate is high or low. In the earlier example of estimating the volume of a two-scoop ice cream cone, we would have over-estimated, because one of the scoops of ice cream is pressed inside the cone. The effect of pressing the scoops together is not that important, because the same amount of ice cream is still there, but if the scoop is pressed into an ellipsoid, it may be difficult to estimate the original radius of the scoop.

Engineers frequently make "conservative" estimates, which consider the "worst-case" scenario. Depending on the situation, the worst case may be a lower limit (such as estimating the strength of a structure) or an upper limit (such as estimating how much material is needed for a project).

● **EXAMPLE 5-5**

If you are to estimate how many gallons of paint are needed to paint the room you are in, what assumptions will you need to make? Where will you need to make assumptions to ensure that you have enough paint without running out?

In estimating the wall area, you should round up the length and height.

Noting that paint (for large jobs) is sold in 5-gallon pails, you will want to round your final estimate to the next whole 5-gallon pail.

Close estimates allow for subtracting 21 square feet per doorway (if the doors are not being painted the same color). In making a rough estimate, if there are not a lot of doorways, it would be conservative to leave in the door area.

5.5 ESTIMATION USING MODELING

In cases that are more complicated or where a more precise estimate is required, mathematical models and statistics might be used. Sometimes dimensionless quantities are useful for characterizing systems, sometimes modeling the relationship of a small number of variables is needed, and at other times, extrapolating even a single variable from available data is all that is needed to make an estimate.

● **EXAMPLE 5-6**

You would like to enjoy a bowl of peas, but they are too hot to eat. Spreading them out on a plate allows them to cool faster. Describe why this happens and devise a model of how much faster the peas on the plate will cool. Mice have a harder time keeping warm compared to elephants. Explain how this is related to the bowl of peas. How does this relate to the fact that smaller animals have higher heart rates? Canaries and humming-birds can have heart rates of 1,200 beats per minute, whereas human heart rates should not exceed 150 beats per minute even during exercise.

The peas cool faster when spread out because of the increase in surface area. The ratio of surface area (proportional to cooling) to volume (proportional to the heat capacity for a particular substance) is therefore important. Similarly, smaller animals have a harder time staying warm because they have a higher ratio of surface area to volume. The higher heart rate is needed to keep their bodies warm. This also relates to why smaller animals consume a much larger amount of food compared to their body mass.

● EXAMPLE 5-7

A large sample of sunflower seeds is collected and their lengths are measured. Using that information, estimate the length of the longest sunflower seed you are likely to find if you measure one billion seeds.

Given a large sample, its average and standard deviation can be calculated. Assuming that the length of sunflower seeds is normally distributed, the one-in-a-billion largest sunflower seed would be expected to be six standard deviations greater than the sample average.

5.6 SIGNIFICANT FIGURES

Significant figures or "sig figs" are the digits considered reliable as a result of measurement or calculation. This is not to be confused with the number of digits or decimal places. The number of **decimal places** is simply the number of digits to the right of the decimal point. Example 5-8 illustrates these two concepts.

● EXAMPLE 5-8

NOTE

Decimal places is the number of digits to the right of the decimal point.

Significant figures are the digits considered reliable.

Mantissa: gives the numerical value of significant figures.

Exponent: specifies the location of the decimal point.

Number	Decimal Places	Significant Figures
376	0	3
376.0	1	4
376,908	0	6
3,760,000	0	3
3,760,000.	0	7
0.376	3	3
0.37600	5	5
0.0037600	7	5
376×10^{-6}	0*	3

* There is no universal agreement concerning whether numbers in scientific or engineering notation should be considered to have the number of decimal places indicated in the mantissa (as shown), or the number that would be present if the number were written out in standard decimal notation (6 in the last example above).

For those who did not run across the term mantissa in high school (or have forgotten it)—the two parts of a number expressed in either scientific or engineering notation are the mantissa and the exponent. The **mantissa** is the part that gives the numerical values of the significant figures; the **exponent** specifies the location of the decimal point, thus the magnitude of the overall number. In the last row of the table above, the mantissa is 376, the exponent is −6. The mantissa can also contain a decimal point, for example, 3.76×10^{-4}: in this case the mantissa is 3.76 and the exponent is −4.

The Meaning of "Significant"

All digits other than zero are automatically considered significant. Zero is significant when:

■ It appears between two nonzero numbers
- 306 has three significant figures
- 5.006 has four significant figures
■ It is a "terminal" zero in a number with a decimal point
- 2.790 has four significant figures
- 2000.0 has five significant figures

Zero is not significant when:

■ It is used to fix a decimal place
- 0.0456 has three significant figures
■ It is used in integers without a decimal point that could be expressed in scientific notation without including that zero
- 2000 has one significant figure (2×10^3)
- 35,100 has three significant figures (3.51×10^4)

COMPREHENSION CHECK 5-1

Determine the number of significant figures and decimal places for each value.

(a) 0.0050
(c) 3.00
(d) 447×10^9
(e) 75×10^{-3}
(f) 7,790,200
(h) 20.000

Calculation Rules

As an engineer, you will likely find that your job involves the design and creation of a product. It is imperative that your calculations lead to the design being "reasonable." It is also important that you remember that others will use much of your work, including people with no technical training.

Engineers must not imply more accuracy in their calculations than is reasonable. To assist in this, there are many rules that pertain to using the proper number of "significant figures" in computations. These rules, however, are cumbersome and tedious. In your daily life as an engineer, you might use these rules only occasionally. The rules given below provide a reference if you ever need them; in this text, however, you are simply expected to be *reasonable,* the concept of which is discussed in Section 5.8. In general, asking yourself if the number of significant figures in your answer is reasonable is usually sufficient. However, it is a good idea to be familiar with the rules, or at least know how to find them and use them if the need ever arises.

Multiplication and Division

A quotient or product should contain the same number of significant figures as the number with the fewest significant figures. Exact conversions do not affect this rule.

● EXAMPLE 5-9

$(2.43)(17.675) = 42.95025 \cong 43.0$

- *2.43 has three significant figures.*
- *17.675 has five significant figures.*
- *The answer (43.0) has three significant figures.*

● EXAMPLE 5-10

$(2.479\,\text{h})(60\,\text{min}/\text{h}) = 148.74 \cong 148.7\,\text{min}$

- *2.479 hours has four significant figures.*
- *60 minutes/hour is an exact conversion.*
- *The answer (148.7) has four significant figures.*

Addition and Subtraction

The answer resulting from an addition or subtraction should show significant figures only as far to the right as the least precise number in the calculation. For addition and subtraction operations, the "least precise" number should be that containing the lowest number of decimal places.

● EXAMPLE 5-11

$1725.463 + 489.2 + 16.73 = 1931.393$

- *489.2 is the least precise.*
- *The answer should contain one decimal place: 1931.4.*

● EXAMPLE 5-12

$903,000 + 59,600 + 104,470 = 1,067,070$

- *903,000 is the least precise.*
- *The answer should be carried to the thousands place: 1,067,000.*

Rounding

If the most significant figure dropped is 5 or greater, then increase the last digit retained by 1.

● EXAMPLE 5-13

Quantity	Rounded to	Appears as
43.48	3 significant figures	43.5
43.48	2 significant figures	43
0.0143	2 significant figures	0.014
0.0143	1 significant figures	0.01
1.555	3 significant figures	1.56
1.555	2 significant figures	0.6
1.555	1 significant figures	2

At what point in a calculation should I round my values?

Calculators are quite adept at keeping track of lots of digits—let them do what they are good at. In general, it is neither necessary nor desirable to round intermediate values in a calculation, and if you do, maintain at least two more significant figures for all intermediate values than the number you plan to use in the final result. The following example illustrates the risk of excessively rounding intermediate results.

$$\text{Evaluate } C = 10{,}000\left[0.6 - (5/9 + 0.044)\right]$$

■ Using calculator with no intermediate rounding:

$$C = \textbf{4.4}$$

■ Rounding value in inner parenthesis to two significant figures:
$$C = 10{,}000(0.6 - (0.599555)) = 10{,}000(0.6 - (0.60)) = \textbf{0}$$

■ Rounding 9/5 to two significant figures:
$$C = 10{,}000(0.6 - (0.55555 + 0.044)) = 10{,}000(0.6 - (0.56 + 0.044))$$
$$= 10{,}000(0.6 - (0.604)) = \textbf{-40}$$

As you can see, rounding to two significant figures at different points in the calculation gives dramatically different results: 4.4, 0, and −40. **Be very sure you know what effect rounding of intermediate values will have if you choose to do so!**

Some numbers, such as certain unit conversions, are considered "exact" by definition. Do not consider them in the determination of significant figures. In calculations with a *known constant* (such as pi (π), which is defined to an infinite number of significant figures), *include at least two more significant figures* in the constant than are contained in the other values in the calculation.

**COMPREHENSION
CHECK 5-2**

Express the answer to the following, using the correct number of significant digits.

(a) $102.345 + 7.8 - 169.05 =$
(b) $20. * 3.567 + 175.6 =$
(c) $(9.78 - 4.352)/2.20 =$
(d) $(783 + 8.98)/(2{,}980 - 1{,}387.2) =$

5.7 REASONABLENESS

In the preceding discussion of estimation, the word *reasonable* was mentioned in several places. We consider two types of reasonableness in answers to problems in this section.

■ **Physically reasonable.** Does the answer make sense in light of our understanding of the physical situation being explored or the estimates that we can make?
■ **Reasonable precision.** Is the number of digits in the answer commensurate with the level of accuracy and precision available to us in the parameters of the problem?

When Is Something Physically Reasonable?

Here are a few hints to help you determine if a solution to a problem is physically reasonable.

- First, ask yourself if the answer makes sense in the physical world.
  - You determine that the wingspan of a new airplane to carry 200 passengers should be 4 feet. This is obvious rubbish.
  - You determine that a sewage treatment plant for a community of 10,000 people must handle 100,000 pound-mass of sewage effluent per day. Since a gallon of water weighs about 8 pound-mass, this is about 1.25 gallons per person per day, which is far too low to be reasonable.
  - In an upper-level engineering course, you have to calculate the acceleration of a 1982 Volkswagen® Rabbit with a diesel engine. After performing your calculations, you find that the time required to accelerate from 0 to 60 miles per hour is 38 seconds. Although for a similarly sized gasoline engine, this is a rather low acceleration; for a small diesel engine, it is quite reasonable.
  - You are designing playground equipment, including a swing set. The top support (pivot point) for the swings is 10 feet above the ground. You calculate that a child using one of the swings will make one full swing (forward, then backward) in 3.6 seconds. This seems reasonable.

- If the final answer is in units for which you do not have an intuitive feel, convert to units for which you do have an intuitive feel.
  - You calculate the speed of a pitched baseball to be 2×10^{13} millimeters per year. Converting this to miles per hour gives over 1,400 miles per hour, obviously too fast for a pitched baseball.
  - You are interested in what angle a smooth steel ramp must have before a wooden block will begin to slide down it. Your calculations show that the value is 0.55 radians. Is this reasonable? If you have a better "feel" for degrees, you should convert the value in radians to degrees, which gives 32 degrees, this value seems reasonable.
  - You have measured the force of a hammer hitting a nail by using a brand-new sensor. The result is a value of 110 million dynes. Do you believe this value? Since few engineers work in dynes, it seems reasonable to convert this to units that are more familiar, such as pound-force. This conversion gives a value of 240 pound-force, which seems reasonable.
  - You are told by a colleague that a ¾-inch pipe supplying water to a chemical process delivers 10 cubic meters of water per day. Converting to gallons per minute gives 1.8 gallons per minute, which seems completely reasonable.

- If your solution is a mathematical model, consider the behavior of the model at very large and very small values.
  - You have determined that the temperature (T in degrees Fahrenheit) of a freshly forged steel ingot can be described as a function of time (t in minutes) by this expression: $T = 2,500 - 10t$. Using this equation to calculate the temperature of the ingot, you discover that after less than 300 minutes (6 hours), the temperature of the ingot will be less than absolute zero!
  - You have determined the temperature (T in degrees Celsius) of a small steel rod placed over a Bunsen burner with a flame temperature of 1,000 degrees Celsius is given by $T = 960 - 939e^{-0.002t}$, where t is the time in seconds from the first application of the flame to the rod. At time $t = 0$, $T = 21$ degree Celsius (since the exponential term will become $e^0 = 1$). This seems reasonable, since it implies that

the temperature of the rod at the beginning of the experiment is 21 degrees Celsius, which is about room temperature. As the value of time increases, the temperature approaches 960 degrees Celsius (since $e^{-\infty} = 0$). This also seems reasonable since the ultimate temperature of the rod will probably be a bit less than the temperature of the flame heating, because of inefficiencies in the heat transfer process.

- A large tank is filled with water to a depth of 10 meters, and a drain in the bottom is opened so that the water begins to flow out. You determine that the depth (D) of water in the tank is given by $D = 5t^{-0.1}$, where t is the time in minutes after the drain was opened at $t = 0$. As t increases, D approaches zero, as we would expect since all of the water will eventually drain from the tank. As t approaches zero, however, D approaches infinity, an obviously ridiculous situation; thus, the model is probably incorrect.

When Is an Answer Reasonably Precise?

> **NOTE**
>
> Most common measuring devices can reliably measure the parameter in question to only three or four significant figures at best, and sometimes less than that. To obtain a really repeatable measurement (e.g., to seven significant figures) in general requires sophisticated (and expensive) equipment.

First, we need to differentiate between the two terms: *accurate* and *precise*. Although laypersons tend to use these words interchangeably, an aspiring engineer should understand the difference in meaning as applied to measured or calculated values.

Accuracy is a measure of how close a calculation or measurement (or an average of a group of measurements) is to the actual value. For measured data, if the average of all measurements of a specific parameter is close to the actual value, then the measurement is accurate, whether or not the individual measurements are close to each other. The difference between the measured value and the actual value is the error in the measurement. Errors come about due to lack of accuracy of measuring equipment, poor measurement techniques, misuse of equipment, and factors in the environment (e.g., temperature or vibration).

Repeatability is a measure of how close together multiple measurements of the same parameter are, whether or not they are close to the actual value.

Precision is a combination of accuracy and repeatability, and is reflected in the number of significant figures used to report a value. The more significant figures, the more precise the value is, assuming it is also accurate.

To illustrate these concepts, consider the distribution of hits on a standard "bulls-eye" target. The figure shows all four combinations of accuracy and repeatable.

- Neither repeatable nor accurate.
- Repeatable, but not accurate.
- Accurate, but not repeatable.
- Both repeatable and accurate. This is called precise.

When considering if the precision of a numeric value is reasonable, always ask yourself the following questions:

- How many significant figures do I need in my design parameters? The more precision you specify in a design, the more it will cost and the less competitive it will be unless the extra precision is really needed.
  - You can buy a really nice 16-ounce hammer for about $20. If you wanted a 16 ± 0.0001-ounce hammer, it would probably cost well over a hundred dollars, possibly thousands.

Precise

- What are the inherent limitations of my measuring equipment? How much is the measurement affected by environmental factors, user error, etc.?
  - The plastic ruler you buy at the discount store for considerably less than a dollar will measure lengths up to 12 inches with a precision of better than 0.1 inch, but not as good as 0.01 inch. On the other hand, you can spend a few hundred dollars for a high-quality micrometer and measure lengths up to perhaps 6 inches to a precision of 0.0001 inch.
  - Pumps at gas stations all over the United States often display their gas price and the amount of gas pumped to three decimal places. When gas prices are high, it is extremely important to consumers that the pumps are correctly calibrated. The National Institute of Standards and Technology (NIST) requires that for every 5 gallons pumped, the amount must not be off by more than 6 cubic inches. To determine if a gas pump is calibrated correctly, you need to be able to see to three decimal places the amount pumped since 6 cubic inches is approximately 0.026 gallons.

  You should report values in engineering calculations in a way that does not imply a higher level of accuracy than is known. Use the fewest number of decimal places without reducing the usefulness of the answer. Several examples, given below, illustrate this concept.

- We want to compute the area (A) of a circle. We measure the diameter (D) as 2.63 centimeters. We calculate

$$A = \frac{1}{4}\pi D^2 = \frac{1}{4}\pi (2.63)^2 = 5.432521 \text{ cm}^2$$

  The value of π is known to as many places as we desire, and ¼ is an exact number. It seems reasonable to give our answer as 5.4 or 5.43 square centimeters since the original data of diameter is given to two decimal places. Most of the time, reporting answers with two to four significant digits is acceptable and reasonable.

- We want to compute the area (A) of a circle. We measure the diameter (D) as 0.0024 centimeters. We calculate

$$A = \frac{1}{4}\pi D^2 = \frac{1}{4}\pi (0.0024)^2 = 0.0004446 \text{ cm}^2$$

  If we keep only two decimal places, our answer would be 0.00 square centimeters, which has no meaning. Consequently, when reporting numerical results, particularly those with a magnitude much smaller than 1, we use significant figures, not decimal places. It would be reasonable to report our answer as 0.00044 or 4.4×10^{-4} square centimeters.

- We want to determine a linear relationship for a set of data, using a standard software package such as Excel®. The program will automatically generate a linear relationship based on the data set. Suppose that the result of this exercise is

$$y = 0.50236x + 2.0378$$

  While we do not necessarily have proof that the coefficients in this equation are nice simple numbers or even integers, a look at the equation above suggests that the linear relationship should probably be taken as

$$y = 0.5x + 2$$

- If calculations and design procedures require a high level of precision, pay close attention to the established rules regarding significant digits. If the values you generate are small, you may need more significant digits. For example, if all the values are between 0.02 and 0.04 and you select one significant figure, *all* your values will read 0.02, 0.03, or 0.04; going to two significant figures gives values such as 0.026 or 0.021 or 0.034.

- Calculators are often set to show eight or more decimal places.

  - If you measure the size of a rectangle as $2\frac{1}{16}$ inches by $5\frac{1}{8}$ inches, then the area is calculated to be 6.4453125 square inches since the calculator does not care about how many significant digits result. It is unreasonable that we can determine the area of a rectangle to seven decimal places when we made two measurements, the most accurate of which was 0.0625 inches, or four decimal places.

  - If a car has a mass of 1.5 tons, should we say it has a mass of 3010.29 pound-mass?

- Worksheets in Excel often have a default of six to eight decimal places. Two important reasons to use fewer are that: (1) long decimal places are often unreasonable and (2) columns of numbers to this many decimal places make a worksheet difficult to read and unnecessarily cluttered.

NOTE

In general, it is a good idea to set your calculator to show answers in engineering format (or generally less desirable, scientific format) with two to four decimal places.

COMPREHENSION CHECK 5-3

In each of the cases below, a value of the desired quantity has been determined in some way, resulting in a number displayed on a calculator or computer screen. Your task is to round each number to a *reasonable* number of significant digits— *up* if a higher value is conservative, *down* if a lower value is conservative, and to the *nearest* value if it does not make a difference. Specify why your assumption is conservative.

(a)	The mass of an adult human riding on an elevator	178.8 pounds
(b)	The amount of milk needed to fill a cereal bowl	1.25 cups
(c)	The time it takes to sing *Happy Birthday*	32.67 seconds

Increasingly, engineers are working at smaller and smaller scales. On the left, a vascular clamp is compared to the tip of a match. The clamp is made from a bio-absorbable plastic through the process of injection molding.

Photo courtesy of E. Stephan

5.8 NOTATION

When discussing numerical values, there are several different ways to represent the values. To read, interpret, and discuss values between scientists and engineers, it is important to understand the different styles of notation. For example, in the United States a period is used as the decimal separator and a comma is used as a digit group separator, indicating groups of a thousand (such as 5,245.25). In some countries, how-ever, this notation is reversed (5.245,25) and in other countries a space is used as the digit group separator (5 245.25). It is important to always consider the country of origin when interpreting written values. Several other types of notations are discussed below.

Engineering Notation Versus Scientific Notation

NOTE

Scientific Notation:

$$\#.\#\#\# \times 10^N$$

N = integer

_____

Engineering Notation:

$$\#\#\#.\#\#\# \times 10^M$$

M = integer multiple of 3

In high school, you probably learned to represent numbers in scientific notation, par-ticularly when the numbers were very large or very small. Although this is indeed a useful means of representing numeric values, in engineering, a slight modification of this notation, called engineering notation, is often more useful. This is particularly true when the value of a parameter can vary over many orders of magnitude. For example, electrical engineers routinely deal with currents ranging from 10^{-15} amperes or less to 10^2 amperes or more.

Scientific notation is typically expressed in the form $\#.\#\#\# \times 10^N$, where the digit to the left of the decimal point is the most significant nonzero digit of the value being represented. Sometimes, the digit to the right of the decimal point is the most signifi-cant digit instead. The number of decimal places can vary, but is usually two to four. N is an integer, and multiplying by 10^N serves to locate the true position of the decimal point.

Engineering notation is expressed in the form $\#\#\#.\#\#\# \times 10^M$, where M is an inte-ger multiple of 3, and the number of digits to the left of the decimal point is either 1, 2, or 3 as needed to yield a power of 10 that is indeed a multiple of 3. The number of digits to the right of the decimal point is typically between two and four.

● **EXAMPLE 5-14**

Standard	Scientific	Engineering
43,480,000	4.348×10^7	43.48×10^6
0.0000003060	3.060×10^{-7}	306.0×10^{-9}
9,860,000,000	9.86×10^9	9.86×10^9
0.0351	3.51×10^{-2}	35.1×10^{-3}
0.0000000522	5.22×10^{-14}	52.2×10^{-15}
456200	4.562×10^8	456.2×10^6

COMPREHENSION CHECK 5-4

Express each of the following values in scientific and engineering notation.

(a) 58,093,099
(b) 0.00458097
(c) 42,677,000.99

Calculator E-notation

NOTE

Some computer programs like MATLAB give you an option to display numbers in scientific notation with an uppercase or lowercase "E."

Most scientific calculators use the uppercase "E" as shorthand for both scientific and engineering notation when representing numbers. To state the meaning of the letter E in English, it is read as "times 10 raised to the __." For example, 3.707 E –5 would be read as "3.707 times 10 to the negative 5." When transcribing numbers from your calculator, in general it is best *not* to use the E notation, showing the actual power of 10 instead. Thus 3.707 E –5 should be written as 3.707×10^{-5}.

Never use a lowercase "e" for transcribing these values from the calculator, such as $3.707 \, e^{-5}$, since this looks like you are multiplying by the number e ($\cong 2.717$) raised to the negative 5. But if you do use a capital E (which is occasionally, though rarely, justifiable), *do not* superscript the number following the E (e.g., $3.707 \, E^{-5}$) since this looks like you are raising some value E (whatever it may be) to a power.

Situations for Use of an Exponential Notation

NOTE

Use exponential notation when the magnitude greater than 10,000 or less than 0.0001.

In general, if the magnitude of a number is difficult to almost instantly determine when written in standard notation, use an exponential notation like scientific or engineering notation. Although there are no definite rules for this, if the magnitude is greater than 10,000 or less than 0.0001, you probably should consider using exponential notation. For larger numbers, using the comma notation can extend this range somewhat, for example, 85,048,900 is quickly seen to be 85 million plus a bit. However, there is no similar notation for very small numbers: 0.0000000483 is difficult to simply glance at and realize that it is about 48 billionths.

Note that it is never actually incorrect to use either exponential or standard notation; it is merely a matter of readability. To write 5×10^{20} as 500000000000000000000 is not wrong, but it may leave the readers' eyes vibrating trying to keep track of all the zeros. On the other hand, it is usually silly to write a number like 7 as 7×10^0.

Representation of Fractions and Use of Constants

Many of you have been previously taught that representing a numeric result exactly as a fraction is preferable to giving an approximate answer. This is seldom the preferred method of reporting values in engineering, for two reasons:

- Fractions are often difficult to glance at with instant comprehension of the actual value.
  - Quick! What does 727/41 equal? Did you immediately recognize that it is a little less than 18?
  - Is it easier to know the magnitude of 37/523 or 0.071?
- Seldom do engineers need a precision of more than three or four digits; thus, there is no need to try to represent values exactly by using the harder-to-read fractions.

Similarly, you may have learned earlier that when calculating with constants such as pi, it is better to leave answers in terms of that constant. For the same reasons cited above, it is generally better to express such values as a decimal number, for example, 27 instead of 8.6π.

There are, of course, exceptions to these rules, but in general, a simple decimal number is more useful to engineers.

In-Class Activities

ICA 5-1

With your team, you are to determine common, readily available or understood quantities to help you estimate a variety of parameters. For example, a 2-liter bottle of soda weighs about 4 pounds (this is an understood quantity since almost everyone has picked up one of these many times), or the end joint of your middle finger is about an inch long (this is a readily available quantity since it goes everywhere you do). To determine the benchmarks or "helpers," you may use whatever measuring tools are appropriate (rulers, scales, watches, etc.) to determine the values of the common objects or phenomena you choose. Try to determine at least two different estimation "helpers" for the following units:

- **Lengths:** millimeter, centimeter, inch (other than the example given above), foot, meter, kilometer
- **Areas:** square centimeter, square inch, square foot, square meter, acre
- **Volumes:** cubic centimeter, cubic inch, cubic foot, cubic meter, gallon
- **Weights and masses:** gram, newton, pound, kilogram, ton
- **Time:** second, minute, hour (the "helpers" you choose cannot be any form of device designed for measuring time).

ICA 5-2

Materials
    Ruler
    Tape measure
    Calipers

Procedure
The following measurements and estimations are to be completed individually, not using one set of measurements per team. This activity is designed to help you learn the size of common items to help you with future estimates. Be sure give both the value and the unit for all measurements.

Measure the following:

(a) Your height in meters
(b) Your arm-span (left fingertip to right fingertip) in meters
(c) The length of your index finger, in centimeters
(d) The width of your thumb, in centimeters
(e) The width of the palm of your hand, in centimeters
(f) The length of your shoe, in feet
(g) The length of your "pace," in feet. A pace is considered as the distance between the toe of the rear shoe and the toe of the lead shoe during a normal step. Think about how to make this measurement before doing it.

Determine the following relationships:

(h) How does your arm span measurement compare to your height?
(i) How does your knee height compare to your overall height?
(j) How does the length of your index finger compare to the width of your thumb?

Determine through estimation:

(k) The height of a door in units of feet
(l) The length of a car in units of yards
(m) The area of the floor of the classroom in units of square meters
(n) The volume of your body in units of gallons

ICA 5-3

(a) Estimate by aggregation how many gallons of gasoline are used by cars each year in the United States.

(b) Estimate by aggregation the volume of a person. A rough approximation of the volume of a person would be a cylinder approximately 1.75 m tall with a radius of 0.25 m. How different was your estimate from the cylindrical approximation?

(c) At the time of publication, the website logging the progress of the Eagle Empowerment Youth Tour 2005 (http://www.eagle-empowerment.org/youthtour2005updates2.html) reported the height of the Empire State Building as 12,500 feet. Use estimation by analogy and estimation by aggregation to prove this is incorrect.

(d) Allow S to be the number of atoms along one edge of a cube of N total atoms. During X-ray diffraction, the scattering from each of those N atoms interacts with the scattering of every other atom. If a simulation of the diffraction from a cube with $S = 10$ takes 1 second to calculate, how long will a simulation a cube with $S = 100$ take to calculate?

(e) If you were leaving on a trip of 1,000 miles (1,600 kilometers), and you could not stop for money along the way, how much cash would you need to carry to be able to buy gas along the way? Estimate using upper and lower bounds.

ICA 5-4

(a) If estimating the amount of time to design a new product, should you err on the high side or the low side?

(b) If estimating the switching speed of the transistors for a faster computer, is it better to err on the low side (slower switching) or high side?

(c) How many square yards of fabric are needed to cover the seats in a typical minivan? Is it better to err on the high side or the low side?

(d) Estimate the dimensions of the classroom, in feet. Using these values and ignoring the fact that you would not paint over the windows and doors, estimate the gallons of paint required to paint the classroom. A gallon of paint covers 400 square feet. Would it be better to round the final answer up or down to the nearest gallon?

ICA 5-5

In each of the cases below, a value of some desired quantity has been determined in some way, resulting in a number displayed on a calculator or computer screen. Your task is to

- Round each number to a *reasonable* number of significant digits—*up* if a higher value is conservative, *down* if a lower value is conservative, and to the *nearest* value if it does not make a difference.
- Specify why your assumption is conservative.

(a) Paint needed to cover a single room	4.36 gallons
(b) Paint needed to paint all rooms in a given building	1,123.05 gallons
(c) The distance from Tampa to New York	1,484.2 miles
(d) The flight time from Charlotte to Atlanta	3 hours, 56 minutes
(e) The brightness of a light bulb	1,690 lumens
(f) The size of a bolt head	0.500 inch
(g) The mass of a typical paperclip	1.125 grams

ICA 5-6

(a) How many toothpicks can be made from an 8-foot long 2-inch by 4-inch board?

(b) Who travels farther: a star player during a college basketball game or a golfer playing 18 holes? By how much farther? State all estimations that determine your answer.

(c) Noah's ark has been described as having the following dimensions: 300 cubits long × 50 cubits wide × 30 cubits high. If a cubit is 18 inches, how many people could fit into the ark?

(d) Due to drought, the water level of a 10-acre pond is five feet below normal. If you wanted to fill the pond to normal capacity by using a hose connected to your kitchen faucet, how long would it take to fill the pond? Select an appropriate unit of time to report your answer.

(e) A cubic meter of air has a mass of about 1.2 kilograms. What is the total mass of air in your home or in a designated building?

(f) How many basketballs can be placed in the classroom?

(g) A portable device uses two D cells for power and can be operated for 50 hours before the batteries must be replaced. How long would the same device be able to operate if two AA cells were used for power instead of D cells?

(h) How many leaves are on the tree pointed out by the professor? What is the mass of that tree (above ground portion only)?

(i) How many carrots are used to make all of the canned soup consumed in the United States in 1 year? (Alternatively, how many acres are used to grow these carrots?)

(j) Estimate the mass of planet Earth.

Chapter 5 REVIEW QUESTIONS

1. How many cubic yards of concrete is required to construct 1 mile of interstate highway?
2. How many gallons of gasoline are burned when the students in this class leave for school break, assuming only one-way travel?
3. If all the land (both currently habituated and all the inhabitable land) were divided equally among all the people now living, how much land would each one have?
4. How many times do my rear tires rotate if I drive around the perimeter of campus?
5. A cubic meter of air has a mass of about 1.2 kilograms. What is the total mass of air within 1,000 feet of sea level on planet Earth?
6. How much water would be saved if everyone in the United States who does not turn off the faucet while they brush their teeth, did so?
7. In 1978, cars that got about 40 miles per gallon were readily available. If the average fuel economy of all cars sold since then was 40 miles per gallon (instead of the lower average mileage of the cars that were actually sold), how many billions of gallons of gas would have been saved in the United States since 1978?
8. If it is raining steadily at a rate of 1 inch per hour (a container with perpendicular sides and a flat bottom would accumulate water to a depth of 1 inch in 1 hour), how many raindrops are in each cubic yard of air approximately 10 feet above the ground in an open field at any given instant?

CHAPTER 6
SOLVEM

Everyone solves problems in different ways. There are, however, some procedures and techniques that can help in developing a generally successful problem-solving approach. Although it is not possible to write down a specific recipe that will always work, some broad approaches will help.

6.1 DEFINING SOLVEM

One problem-solving approach has been given the acronym **SOLVEM:**

Sketch
Observations or **O**bjectives
List
Variables and
Equations
Manipulation

Note that this approach is equally useful for problems involving estimation and more precise calculations. Each step is described below.

Sketch

Figure 6-1 illustrates how a drawing can help you visualize a problem. In sketching a problem, you are subconsciously thinking about it. Be sure to draw the diagram large enough so that everything is clear, and label the things that you know about the problem in the diagram. For some problems, a before-and-after set of diagrams may be helpful. In very complex problems, you can use intermediate diagrams or subdiagrams as well.

**COMPREHENSION
CHECK 6-1**

We use SOLVEM to complete this problem in the Comprehension Checks in this chapter. Create a sketch for the following problem.

Calculate the mass in kilograms of gravel stored in a rectangular bin 18.5 feet by 25.0 feet. The depth of the gravel bin is 15 feet, and the density of the gravel is 97 pound-mass per cubic foot.

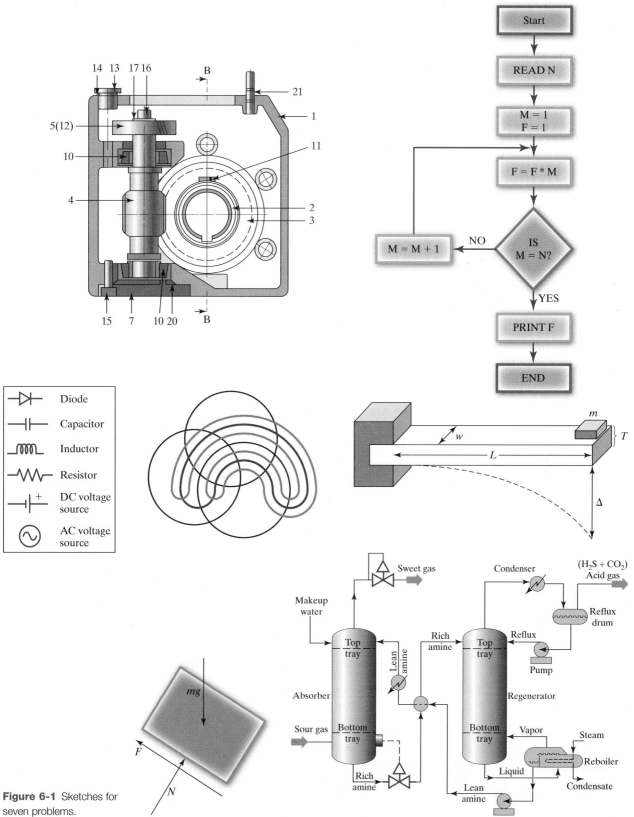

Figure 6-1 Sketches for seven problems.

Observations, Objectives

These can be in the form of simple statements, questions, or anything else that might acquaint you with the problem at hand. It often helps to divide your observations and objectives into several categories. Some of the easiest to remember are:

■ Objective to be achieved
■ Observations about the problem geometry (size, shape, etc.)
■ Observations about materials and material properties (density, hardness, etc.)
■ Observations about parameters not easily sketched (temperature, velocity, etc.)
■ Other miscellaneous observations that might be pertinent

You will almost always find after writing down some observations that you actually know more about the problem than you originally thought.

Here are some typical examples:

Objectives

■ Find the velocity, force, flow rate, time, pressure, etc., for a given situation.
■ Profitably market the device for less than $25.
■ Fit the device into a 12-cubic-inch box.

Observations

PROBLEM GEOMETRY

■ The liquid has a free surface.
■ The submerged plate is rectangular.
■ The support is vertical.
■ The tank is cylindrical.
■ The cross-sectional area is octagonal.
■ The orbit is elliptical.

MATERIALS AND MATERIAL PROPERTIES

■ The gate is steel.
■ The coefficient of static friction is 0.6.
■ The specific gravity is 0.65.
■ Ice will float in water.
■ The alloy superconducts at 97 kelvin.
■ The alloy melts at 543 degrees Fahrenheit.

OTHER PARAMETERS

■ If depth increases, pressure increases.
■ If temperature increases, resistance increases.
■ The flow is steady.
■ The fluid is a gas and is compressible.
■ The pulley is frictionless.
■ The magnetic field is decreasing.
■ Temperature may not fall below 34 degrees Fahrenheit.

MISCELLANEOUS

■ The force will act to the right.
■ Gravity causes the ball to accelerate.
■ The sphere is buoyant.
■ Drag increases as the speed increases.

Remember to include those quantities whose value is zero! Often such quantities are hidden with terms such as:

■ Constant (implies derivative $= 0$)
■ Initially (at time $= 0$)
■ At rest (no motion)
■ Dropped (no initial velocity)
■ At the origin (at zero position)
■ Constant velocity (force $= 0$)
■ Melts (changes phase, temperature is constant)

**COMPREHENSION
CHECK 6-2**   State the objective and any relevant observations for the following problem.
Calculate the mass in kilograms of gravel stored in a rectangular bin 18.5 feet by 25.0 feet. The depth of the gravel bin is 15 feet, and the density of the gravel is 97 pound-mass per cubic foot.

Finally, one of the most important reasons to make *many* observations is that you often will observe the "wrong" thing. For example, write down things as you read this:

A bus contains 13 passengers.
At the first stop, four get off and two get on.
At the next stop, six get off and one gets on.
At the next stop, nobody gets off and five get on.
At the next stop, eight get off and three get on.
At the next stop, one gets off.
At the last stop, four get off and four get on.

After putting your pencil down and without looking again at the list, answer the question given below textbox discussing "The Importance of Observations."

THE IMPORTANCE OF OBSERVATIONS

An excerpt adapted from The Crooked Man *by Sir Arthur Conan Doyle*
Dr. Watson writes: I looked at the clock. It was a quarter to twelve. This could not be a visitor at so late an hour. A patient, evidently, and possibly an all-night sitting. With a wry face I went out into the hall and opened the door. To my astonishment, it was Sherlock Holmes who stood upon my step.

"Ah, Watson, I hoped that I might not be too late to catch you."

"My dear fellow, pray come in."

"You look surprised, and no wonder! Relieved, too, I fancy! Hum! You still smoke the Arcadia mixture of your bachelor days, then! There's no mistaking that fluffy ash upon your coat. It's easy to tell that you've been accustomed to wear a uniform, Watson; you'll never pass as a pure-bred civilian as long as you keep that habit of carrying your handkerchief in your sleeve. Could you put me up for the night?"

"With pleasure."

"You told me that you had bachelor quarters for one, and I see that you have no gentleman visitor at present. Your hat-stand proclaims as much."

"I shall be delighted if you will stay."

"Thank you. I'll find a vacant peg, then. Sorry to see that you've had the British workman in the house. He's a token of evil. Not the drains, I hope?"

"No, the gas."

"Ah! He has left two nail marks from his boot upon your linoleum just where the light strikes it. No, thank you, I had some supper at Waterloo, but I'll smoke a pipe with you with pleasure."

I handed him my pouch, and he seated himself opposite to me, and smoked for some time in silence. I was well aware that nothing but business of importance could have brought him to me at such an hour, so I waited patiently until he should come round to it.

"I see that you are professionally rather busy just now."

"Yes, I've had a busy day. It may seem very foolish in your eyes, but I really don't know how you deduced it."

"I have the advantage of knowing your habits, my dear Watson. When your round is a short one you walk, and when it is a long one you use a hansom (a carriage). As I perceive that your boots, although used, are by no means dirty, I cannot doubt that you are at present busy enough to justify the hansom."

"Excellent!"

"Elementary. It is one of those instances where the reason can produce an effect which seems remarkable to his neighbor, because the latter has missed the one little point which is the basis of the deduction."

While we cannot all be as observant as Sherlock Holmes, we can improve our powers of observation through practice. This will pay dividends as we seek to be engineers with high levels of analytical skills.

QUESTION: For the bus problem, how many stops did the bus make?
The lesson here is that often we may be observing the wrong thing.

List of Variables and Constants

Go over the observations previously determined and list the variables that are important. It may help to divide the list into several broad categories—those related to the geometry of the problem, those related to the materials, and a properties category—although for some types of problems those categories may not be appropriate. Include in your list the written name of the variable, the symbol used to represent the quantity, and, if the value of the variable is known, list the numeric value, including units. If a value is a constant you had to look up, record where you found the information.

INITIAL AND FINAL CONDITIONS

- Initial temperature (T_0) 60 [°F]
- Initial radius (r_i) 5 [cm]
- Mass of the object (m) 23 [kg]

CONSTANTS

- Acceleration of gravity (g) 32.2 [ft/s^2]
- Ideal gas constant (R) 8,314 [(Pa L)/(mol K)]

GEOMETRY

- Length of beam (L) [m]
- Cross-sectional area of a pipe (A) [cm^2]
- Volume of a reactor vessel (V) [gal]

MATERIALS

- Steel
- Polyvinyl chloride (PVC)
- Plasma
- Gallium arsenide
- Medium-density balsa wood

PROPERTIES

- Dynamic viscosity of honey (μ) 2,500 [cP]
- Density of PVC (ρ) 1,380 [kg/m³]
- Spring constant (k) 0.05 [N/m]
- Specific gravity   (SG)   1.34

COMPREHENSION CHECK 6-3

Create a list of variables and constants for the following problem.

     Calculate the mass in kilograms of gravel stored in a rectangular bin 18.5 feet by 25.0 feet. The depth of the gravel bin is 15 feet, and the density of the gravel is 97 pound-mass per cubic foot.

Equations

Only after completing the steps above (S-O-L-V) should you begin to think about the equations that might govern the problem. It is useful to make a list of the pertinent equations in a broad sense before listing specific expressions. For example:

- Conservation of energy
- Conservation of mass
- Conservation of momentum
- Frequency equations
- Ideal gas law

- Newton's laws of motion
- Stress–strain relations
- Surface areas of geometric solids
- Volumes of geometric solids
- Work, energy relations

You may need "subequations" such as:

- Distance = (velocity)(time)
- Energy = (power)(time)
- Force = (pressure)(area)

- Mass = (density)(volume)
- Voltage = (current)(resistance)
- Weight = (mass)(gravity)

     For an equation, list the broad category of the equation (Hooke's law) and then the actual expression $(F = kx)$ to help with problem recognition.

     Do not substitute numerical values of the parameters into the equation right away. Instead, manipulate the equation algebraically to the desired form.

COMPREHENSION CHECK 6-4

Create a list of equations for the following problem.

     Calculate the mass in kilograms of gravel stored in a rectangular bin 18.5 feet by 25.0 feet. The depth of the gravel bin is 15 feet, and the density of the gravel is 97 pound-mass per cubic foot.

Manipulation

Most of the time, you need to manipulate pertinent equations before you can obtain a final solution. ***Do not substitute numerical values of the parameters into the equation right away.*** Instead, manipulate the equation algebraically to the desired form. Often you will discover terms that will cancel, giving you a simpler expression to deal with. By doing this, you will:

- Obtain general expressions useful for solving other problems of this type
- Be less likely to make math errors
- Be able to judge whether your final equation is dimensionally consistent
- Better understand the final result

The SOLVEM acronym does not contain a word or step for "numerical solution." In fact, this process helps you analyze the problem and obtain an expression or procedure so that you can find a numerical answer. The thought here is that engineers need training to be able to analyze and solve problems. If you can do everything except "substitute numbers" you are essentially finished—as an engineer, you will "be paid the big bucks" for analysis, not for punching a calculator.

6.2 REPRESENTING FINAL RESULTS

When you have completed all the steps to SOLVEM, plug in values for the variables and constants and solve for a final answer. Be sure to use reasonableness. The final answer should include both a numeric value and its unit. In addition, it is often useful to write a sentence describing how the answer meets the objectives. Box your final answer for easy identification.

Repeated use of SOLVEM can help you develop a better "gut-level understanding" about the analysis of problems by forcing you to talk and think about the generalities of the problem before jumping in and searching for an equation into which you can immediately substitute numbers.

NOTE
Don't plug values into the equation until the final step.

COMPREHENSION CHECK 6-5

Manipulate and solve for the following problem, using the information from Comprehension Checks 1–4.

Calculate the mass in kilograms of gravel stored in a rectangular bin 18.5 feet by 25.0 feet. The depth of the gravel bin is 15 feet, and the density of the gravel is 97 pound-mass per cubic foot.

6.3 AVOIDING COMMON MISTAKES

Erroneous or argumentative thinking can lead to problem-solving errors. For example,

NOTE
Be sure your equation is the right one for the problem you are solving.

- **I can probably find a good equation in the next few pages.** Perhaps you read a problem and rifle through the chapter to find the proper equation so that you can start substituting numbers. You find one that looks good. You do not worry about whether the equation is the right one or whether the assumptions you made in committing to the equation apply to the present problem. You whip out your calculator and produce an answer. *Don't do this!*

- **I hate algebra, or I cannot do algebra, or I have got the numbers, so let us substitute the values right in.** Many problems become much simpler if you are willing to do a little algebra before trying to find a numerical solution. Also, by doing some manipulation first, you often obtain a general expression that is easy to apply to another problem when a variable is given a new value. By doing a little algebra, you can also often circumvent problems with different sets of units. *Do some algebra!*
- **It is a simple problem, so why do I need a sketch?** Even if you have a photographic memory, you will need to communicate with people who do not. It is usually much simpler to sort out the various parts of a problem if a picture is staring you right in the face. *Draw pictures!*
- **I do not have time to think about the problem, I need to get this stuff finished.** Well, most often, if you take a deep breath and jot down several important aspects of the problem, you will find the problem much easier to solve and will solve it correctly. *Take your time!*

● EXAMPLE 6-1

Estimate how many miles of wire stock are needed to make 1 million standard paper clips.

Sketch:

See the adjacent diagram.

Objective:

Determine the amount of wire needed to manufacture a million paper clips.

Observations:

- *Paper clips come in a variety of sizes*
- *There are four straight segments and three semicircular sections in one clip*
- *The three semicircular sections have slightly different diameters*
- *The four straight sections have slightly different lengths*

List of Variables and Constants:

- *L — Overall length of clip*
- *W — Overall width of clip*
- *L_1, L_2, L_3, L_4 — Lengths of four straight sections*
- *D_1, D_2, D_3 — Diameters of three semicircular sections*
- *P_1, P_2, P_3 — Lengths of three semicircular sections*
- *A — Total amount (length) of wire per clip*

Estimations and Assumptions:

- *Length of clip: $L = 1.5$ in*
- *Width of clip: $W = \frac{3}{8}$ in*
- *Diameters from largest to smallest*
  - *$D_1 = W = \frac{3}{8}$ in*
  - *$D_2 = \frac{5}{6}$ in*
  - *$D_3 = \frac{1}{4}$ in*

- *Lengths from left to right in sketch*
  - L_1 = To be calculated
  - L_2 = 0.8 in
  - $L_3 \approx L4 = 1$ in

Equations:

- *Perimeter of semicircle:* $P = \pi D / 2$ *(half of circumference of circle)*
- $L_1 = L - D_1/2 - D_2/2$
- *Total length of wire in clip:* $A = L_1 + L_2 + L_3 + L_4 + P_1 + P_2 + P_3$

Manipulation:

In this case, none of the equations need to be manipulated into another form.

Length of longest straight side:	$L_1 = 1.5 - {}^3/_{16} - {}^5/_{32} \approx 1.2$ in
Lengths of semicircular sections:	$P_1 = \pi\, {}^3/_{16} \approx 0.6$ in
	$P_2 = \pi\, {}^5/_{32} \approx 0.5$ in
	$P_3 = \pi\, {}^1/_{8} \approx 0.4$ in
Overall length for one clip:	$A = 1.2 + 0.8 + 1 + 1 + 0.6$
	$\quad\quad + 0.5 + 0.4 = 5.5$ in/clip
Length of wire for 1 million clips:	$(5.5 \text{ in/clip}) \ (1 \times 10^6 \text{clips}) = 5.5 \times 10^6$ in
Convert from inches to miles:	$(5.5 \times 10^6 \text{ in}) \ (1 \text{ ft}/12 \text{ in})$
	$(1 \text{ mile}/5{,}280 \text{ ft}) \approx 86.8$ miles

One million, 1.5-inch paper clips require about 87 miles of wire stock.

● **EXAMPLE 6-2**

A spherical balloon has an initial radius of 5 inches. Air is pumped in at a rate of 10 cubic inches per second, and the balloon expands. Assuming that the pressure and temperature of the air in the balloon remain constant, how long will it take for the surface area to reach 1,000 square inches?

Sketch:

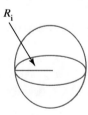 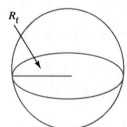

Objective:

Determine how long it will take for the surface area of the balloon to reach 1,000 in^2

Observations:

- *The balloon is spherical*
- *The balloon, thus its volume and surface area, gets larger as more air is pumped in*
- *The faster air is pumped in, the more rapidly the balloon expands*

List of Variables and Constants:

- *Initial radius:* $R_i = 5$ in
- *Final radius:* R_f [in]
- *Initial surface area:* A_i [in²]
- *Final surface area:* A_f [in²]
- *Change in volume:* ΔV [in³]
- *Initial volume:* V_i [in³]
- *Final volume:* V_f [in³]
- *Fill rate:* $Q = 10$ [in³/s]
- *Time since initial size:* t [s]

Equations:

- *Surface area of sphere:* $A = 4\pi R^2$
- *Volume of sphere:* $V = 4\pi R^3/3$
- *Change in volume:* $\Delta V = Qt$

Manipulation:

There are a few different ways to proceed. The plan used here is to determine how much the balloon volume changes as air is blown into the balloon and to equate this to an expression for the volume change in terms of the balloon geometry (actually the radius of the balloon).

Radius of balloon in terms of surface area:
$$R = \left(\frac{A}{4\pi}\right)^{1/2}$$

Final balloon radius in terms of surface area:
$$R_f = \left(\frac{A_f}{4\pi}\right)^{1/2}$$

Final volume of balloon in terms of surface area:
$$V_f = \left(\frac{4\pi}{3}\right)\left(\frac{A_f}{4\pi}\right)^{3/2}$$

Volume change in terms of air blown in:
$$\Delta V = V_f - V_i = Qt$$

Volume change in terms of geometry:
$$V_f - V_i = \left(\frac{4\pi}{3}\right)\left(\frac{A_f}{4\pi}\right)^{3/2} - \left(\frac{4\pi}{3}\right)R_i^3$$

Solve for time to blow up balloon:
$$t = \left(\frac{4\pi}{3Q}\right)\left(\frac{A_f}{4\pi}\right)^{3/2} - \left(\frac{4\pi}{3Q}\right)R_i^3$$

And simplifying:
$$t = \left(\frac{4\pi}{3Q}\right)\left\{\left(\frac{A_f}{4\pi}\right)^{3/2} - R_i^3\right\}$$

It takes just over 4 minutes to increase the volume to 1,000 cubic inches.

NOTE

Whenever you obtain a result in equation form, you should check to see if the dimensions match in each term.

IMPORTANT

Recall that you should manipulate the equations *before* inserting known values. Note that the final expression for elapsed time is given in terms of initial radius (R_i), flow rate (Q), and final surface area (A_f). So you now have a general equation that can be solved for any values of these three parameters. If you had begun substituting numbers into equations at the beginning and then wanted to obtain the same result for different starting values, you would have to resolve the entire problem.

In-Class Activities

ICA 6-1

Each of these items should be addressed by a team. List as many things about the specified items as you can determine by observation. One way to do this is to let each team member make one observation, write it down, and iteratively canvass the team until nobody can think of any more additions (it is fine to pass). Remember that you have five senses. Also note that observations are things you can actually detect during the experiment, *not* things you already know or deduce that cannot be observed.

To help, Examples A and B are given here before you do one activity (or more) on your own. Remember that not all observations will be important for a particular problem, but write them down anyway—they may trigger an observation that is important.

Example A: A loudspeaker reproducing music

- Electrical signal is sent to speaker.
- Speaker vibrates from electrical signal.
- Speaker motion imparts kinetic energy to the air.
- Diaphragm accelerates.
- Speaker is gray.
- A magnet is involved.
- Speaker is circular.
- Speaker is about 10 inches in diameter.
- Speaker diaphragm is made of paper; vibrating air creates sound.
- Gravity acts on speaker.
- Speaker is in a "box."

Example B: Drinking a soft drink in a can through a straw

- Liquid is cold.
- Straw is cylindrical.
- Can is cylindrical.
- Low pressure in your mouth draws the liquid up.
- Liquid assumes the shape of the container.
- Gravity opposes the rise of the liquid.
- Moving liquid has kinetic energy.
- As the liquid rises, it gains potential energy.
- Table supports the weight of the can, liquid, and straw.
- Can is opaque.
- Liquid is brown.
- Liquid is carbonated (carbonic acid).
- Liquid contains caffeine.
- Friction between straw and lips allows you to hold it.
- Plan view of the can is a circle.
- Silhouette of the can is the same from any direction.
- Silhouette of the can is a rectangle.
- Can is painted.
- Can is metal.
- Liquid surface is horizontal.
- Liquid surface is circular.

From the list below or others as selected by the instructor, list as many observations as you can about the following topic:

(a) An object provided by the professor (placed on the desk)
(b) Candle placed on the desk and lit
(c) Ball dropped from several feet, allowed to bounce and come to rest
(d) A weight on a string pulled to one side and released; watched until it comes to rest
(e) Coin spinning on the desk
(f) Cup of hot coffee placed on desk
(g) A glass of cold water placed on desk
(h) Book pushed across the desk

(i) Large container guided smoothly up a ramp
(j) Ruler hanging over the desk
(k) Your computer (when turned off)
(l) Your chair
(m) Your classroom
(n) A weight tied on a string and twirled

Final Assignment of this ICA: You have done several observation exercises. In these, you thought of observation as just a "stream of consciousness" with no regard to organization of your efforts. With your previous observation as a basis, generalize the search for observations into several (three to six or so) categories. The use of these categories should make the construction of a list of observations easier in the future.

ICA 6-2

What diameter will produce a maximum discharge velocity of a liquid through an orifice on the side at the bottom of the cylindrical container? Consider diameters ranging from 0.2 to 2 meters.
Use the SOLVEM method to determine the answer.

ICA 6-3

A hungry bookworm bores through a complete set of encyclopedias consisting of n volumes stacked in numerical order on a library shelf. The bookworm starts inside the front cover of volume 1, bores from page 1 of volume 1 to the last page of the last volume, and stops inside the back cover of the last volume. Note that the book worm starts inside the front cover of volume 1 and ends inside the back cover of volume n.

Assume that each volume has the same number of pages. For each book, assume that you know how thick the cover is, and that the thickness of a front cover is equal to the thickness of a back cover; assume also that you know the total thickness of all the pages in the book. How far does the bookworm travel? How far will it travel if there are 13 volumes in the set and each book has 2 inches of pages and a $\frac{1}{8}$-inch thick cover?
Use the SOLVEM method to determine the answer.

ICA 6-4

Two cargo trains each leave their respective stations at 1:00 p.m. and approach each other, one traveling west at 10 miles per hour and the other on separate tracks traveling east at 15 miles per hour. The stations are 100 miles apart. Find the time when the trains meet and determine how far the eastbound train has traveled.
Use the SOLVEM method to determine the answer.

ICA 6-5

Water drips from a faucet at the rate of three drops per second. What distance separates one drop from the following drop 0.65 seconds after the leading drop leaves the faucet? How much time elapses between impacts of the two drops if they fall onto a surface that is 6 feet below the lip of the faucet?

Your sketch should include the faucet, the two water drops of interest, and the impact surface. Annotate the sketch, labeling the each item shown and denote the relevant distances in symbolic form, for example, you might use d_1 to represent the distance from the faucet to the first drop.
Use the SOLVEM method to determine the answer.

ICA 6-6

During rush hour, cars back up when the traffic signal turns red. When cars line up at a traffic signal, assume that they are equally spaced (Δx) and that all the cars are the same length (L). You do not begin to move until the car in front of you begins to move, creating a reaction time (Δt) between the time the car in front begins to move and the time you start moving. To keep things simple, assume that when you start to move, you immediately move at a constant speed (v). If the traffic signal stays green for some time (t_g), how many cars (N) will make it through the light?

After you have analyzed this problem with SOLVEM, answer the following:

(a) If the light remains green for twice the time, how many more cars will get through the light?

(b) If the speed of each car is doubled when it begins to move, will twice as many cars get through the light? If not, what variable would have to go to zero for this to be true?

(c) For a reaction time of zero and no space between cars, find an expression for the number of cars that will pass through the light. Does this make sense?

ICA 6-7

Suppose that the earth were a smooth sphere and you could wrap a 25,000-mile-long band snugly around it. Now let us say that you lengthen the band by 10 feet, loosening it just a little. What would be the largest thing that could slither under the new band (assume that it is now raised above the earth's surface equally all the way around so that it doesn't touch anywhere): an amoeba, a snake, or an alligator?

Use the SOLVEM method to determine the answer.

Chapter 6 REVIEW QUESTIONS

1. A motorcycle weighing 500 pounds-mass plus a rider weighing 300 pounds-mass produces the following chart. Predict a similar table if a 50-pound-mass dog is added as a passenger. Use SOLVEM.

Velocity (v) [mi/h]	Time (t) [s]
0	0.0
10	2.3
20	4.6
30	6.9
40	9.2

2. A circus performer jumps from a platform onto one end of a seesaw, while his or her partner, a child of age 12, stands on the other end. How high will the child "fly"? Analyze this problem with SOLVEM.

3. Your college quadrangle is 85 meters long and 66 meters wide. When you are late for class, you can walk (well, run) at 7 miles per hour. You are at one corner of the quad and your class is at the directly opposite corner. How much time can you save by cutting across the quad rather than walking around the edge? Analyze this problem with SOLVEM.

4. I am standing on the upper deck of the football stadium. I have an egg in my hand. I am going to drop it and you are going to try to catch it. You are standing on the ground. Apparently, you do not want to stand directly under me; in fact, you would like to stand as far to one side as you can so that if I accidentally release it, it won't hit you on the head. If you can run at 20 feet per second and I am at a height of 100 feet, how far away can you stand and still catch the egg if you start running when I let go? Analyze this problem with SOLVEM.

5. A 1-kilogram mass has just been dropped from the roof of a building. I need to catch it after it has fallen exactly 100 meters. If I weigh 80 kilograms and start running at 7 meters per second as soon as the object is released, how far away can I stand and still catch the object? Analyze this problem using SOLVEM.

6. Neglect the weight of the drum in the following problem. A sealed cylindrical drum has a diameter of 6 feet and a length of 12 feet. The drum is filled exactly half-full of a liquid having a density of 90 pound-mass per cubic foot. It is resting on its side at the bottom of a 10-foot deep drainage channel that is empty. Suppose a flash flood suddenly raises the water level in the channel to a depth of 10 feet. Determine if the drum will float. The density of water is 62 pound-mass per cubic foot. Analyze this problem with SOLVEM.

Part 2

UBIQUITOUS UNITS

You may not be sure what the word "ubiquitous" means ... we suggest you look it up!

Ubiquitous: **yoo·bik·we·teous** ~ adjective;
    definition _____

LEARNING OBJECTIVES

The overall learning objectives for this unit include:

- Apply basic principles from mathematical and physical sciences, such as trigonometry, Hooke's law, and the ideal gas law, to solve engineering problems.
- Convert units for physical and chemical parameters such as density, energy, pressure, and power as required for different systems of units.
- Identify basic and derived dimensions and units; accurately express relevant observations and parameters such as distance, force, and temperature in appropriate units.
- Use "practical" skills, such as visualizing common units and conducting simple measurements, calculations, and comparisons to make estimations.

Imagine you are in a small boat with a large stone in the bottom of the boat. The boat is floating in the swimming pool in the campus recreation center. What happens to the level of water in the pool if you throw the stone overboard? Assume no water splashes out of the pool or into the boat.

    Archimedes was a Greek scientist and mathematician. Most people know Archimedes for his discovery of buoyancy. According to legend, the king asked Archimedes to determine if his new crown was made of pure gold. Before this, no method had been developed for measuring the density of irregularly shaped objects. While taking a bath, Archimedes noted that the water rose in proportion to the amount of his body in the tub. He shouted "Eureka (I have found it)!" and ran though the streets naked because he was so excited he forgot to get dressed. While Archimedes never recounts this tale himself, he does outline Archimedes' principle in his treatise *On Floating Bodies: A body immersed in a fluid is buoyed up by a force*

143

NOTE

"Give me a place to stand on, and I will move the Earth."
—*Archimedes*

In addition to buoyancy, Archimedes made many contributions to science, including the explanation of the lever, and is considered one of the greatest mathematicians.

equal to the weight of the displaced fluid. Before we can begin to answer the question of the boat and the stone (the answer is found on the final page of Chapter 8), we need to understand the principles of dimensions and units.

HOW I STUDY

The Myers–Briggs Type Indicator, often referred to as the MBTI, is a method for assessing individuals—their attitudes and modes of actions, learning, and interacting with others. By answering a series of questions, an individual is categorized according to four dichotomies. You can think of these as preferences for particular ways of conducting your life. This brief introduction is by no means intended to acquaint you with the intricacies of the MBTI, but simply to help you gain insight that will help you study, learn, and interact with others more effectively. An individual's preference relative to each of the four dichotomies is indicated by a single letter. When combined, these create a four-letter code that describes your preferences. The four dichotomies are as follows:

Engraving from Mechanics Magazine published in London in 1824

Cognitive Attitude:	Extrovert or Introvert	E vs. I
Mode of Data Gathering:	Sensing or Intuition	S vs. N
Mode of Decision Making:	Thinking or Feeling	T vs. F
Lifestyle Preference:	Judging or Perceiving	J vs. P

Cognitive Attitude (or how one directs one's energy)
- The Extrovert (E) prefers action, and focuses on people and objects.
- The Introvert (I) prefers reflection, and focuses on concepts and ideas.

Mode of Data Gathering (or how one perceives things)
- The Sensing (S) individual tends to trust data gathered by the five senses and is most interested in details and facts.
- The Intuitive (N) individual tends to trust information that is theoretical or abstract and is most interested in how data fit a theory or a pattern.

Mode of Decision Making (or how one judges things)
- The Thinking (T) person makes decisions with an emotionally detached attitude, being concerned with what is most logical and what established rules dictate.
- The Feeling (F) person makes decisions by empathizing with the situation, being concerned with what will promote the most harmonious outcome.

Lifestyle Preference (or how one relates to the outside world)
- The Judging (J) individual prefers making decisions to obtaining information, whether gathering that information by thinking or feeling.
- The Perceiving (P) individual prefers gathering data to making decisions, whether gathering that information by sensing or intuition.

The Myers Briggs Type Index is based on Dr. Carl Jung's original work on Personality types and results in the individual being identified as a single type as represented by the 4 letters. However, we aren't the same in all situations, and the MBTI fails to show any information as to degree of preference. Dr. Lois Breur Krause did further research on applying Dr. Jung's work specifically to learning and studying. The resulting material on learning and studying specific to the individual learner's profile is found in the book "*How We Learn and Why We Don't.*" Before proceeding further, it would be helpful for you to take the inventory yourself. It may be done online at http://www. CognitiveProfile.com.

HOW ATTITUDES ARE ASSESSED: YOUR FULL PROFILE

Dr. Krause's Cognitive Profile Inventory uses two of Jung's four bipolar descriptors and results in a Profile that describes your proportion of strength in each of 4 quadrants, ST, SF, NT and NF. It does not include the Judger/Perceiver or Introvert/ Extrovert bipolar descriptors as they were found to be less critical to learning and studying. This allows the model to be reduced to a more manageable size, and is actually more complete in its application. It shows where you are strongest or most comfortable, and identifies where you can build skills to strengthen proportionally weaker areas. By using study techniques that complement the way your brain wants to learn, you can make your study time more productive and more efficient, and probably improve your grades. By identifying and building skills in your weaker areas, you will be more able to tackle previously difficult types of tasks.

Knowing your profile and being able to recognize characteristics in others will help you in your studies, your interactions with our professors and fellow students and eventually with your coworkers on the job.

Typical Characteristics

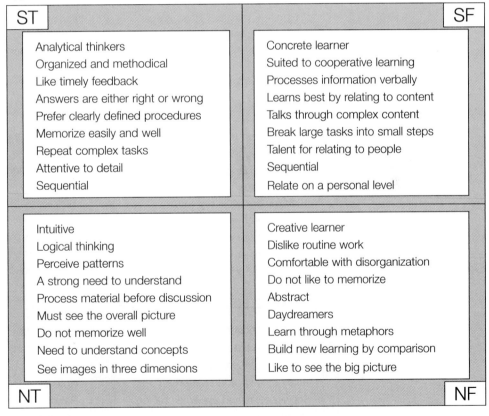

ST	SF
Analytical thinkers	Concrete learner
Organized and methodical	Suited to cooperative learning
Like timely feedback	Processes information verbally
Answers are either right or wrong	Learns best by relating to content
Prefer clearly defined procedures	Talks through complex content
Memorize easily and well	Break large tasks into small steps
Repeat complex tasks	Talent for relating to people
Attentive to detail	Sequential
Sequential	Relate on a personal level
Intuitive	Creative learner
Logical thinking	Dislike routine work
Perceive patterns	Comfortable with disorganization
A strong need to understand	Do not like to memorize
Process material before discussion	Abstract
Must see the overall picture	Daydreamers
Do not memorize well	Learn through metaphors
Need to understand concepts	Build new learning by comparison
See images in three dimensions	Like to see the big picture
NT	NF

Compare your profile from the inventory with the table above. You will see that you have greater or lesser areas in each quadrant. The quadrant in which you have the greatest area is your dominant quadrant, where you are most comfortable, most fluent, and most efficient. You may or may not have a second quadrant that is nearly as strong, or runs a close second. You will also see that you probably have a quadrant or two that have considerably less area. That identifies what skills you may need to build in order to accomplish some types of tasks, or where you may need to find ways to adapt material to better fit your dominant quadrant. If you have what is called a "kite shaped" profile, where your two strongest quadrants are opposite one another, it is likely that you have already developed strong coping skills to balance your natural abilities. This is frequently found in strong creative, NF dominant individuals who find that ST skills of organization work well to bring success. It doesn't mean your ST is native to you, but that you have learned that concrete, methodical organization works when you need it and have come to trust and rely on those skills.

To simplify things a little, we focus here on only two of the dichotomies: how you gather information (sensing or intuition) and how you make decisions (thinking or feeling). This limitation reduces the number of possible dichotomy combinations to 4 instead of 16, which we will call profiles: ST, SF, NT, and NF.

Remember that people who are dominant in a particular quadrant will certainly have talents that draw from the other quadrants as well, since essentially everyone has some of the characteristics of the other types. It is critical that you recognize that you do not act, think, and learn solely within your dominant quadrant. The dominant quadrant is likely, however, to describe the areas in which you are most comfortable and talented.

Study Strategies for Each of the Four Profiles

By now, you must have realized that some teachers present information in ways that you grasp easily, while other instructors are hard for you to follow. In college, you can get lost much faster because the pace and expectations are considerably higher than in high school. One way to prevent this is to know how you learn.

Study Strategies

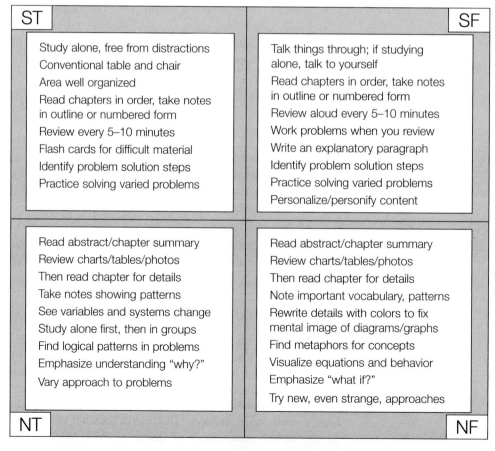

ST

Study alone, free from distractions
Conventional table and chair
Area well organized
Read chapters in order, take notes in outline or numbered form
Review every 5–10 minutes
Flash cards for difficult material
Identify problem solution steps
Practice solving varied problems

SF

Talk things through; if studying alone, talk to yourself
Read chapters in order, take notes in outline or numbered form
Review aloud every 5–10 minutes
Work problems when you review
Write an explanatory paragraph
Identify problem solution steps
Practice solving varied problems
Personalize/personify content

NT

Read abstract/chapter summary
Review charts/tables/photos
Then read chapter for details
Take notes showing patterns
See variables and systems change
Study alone first, then in groups
Find logical patterns in problems
Emphasize understanding "why?"
Vary approach to problems

NF

Read abstract/chapter summary
Review charts/tables/photos
Then read chapter for details
Note important vocabulary, patterns
Rewrite details with colors to fix mental image of diagrams/graphs
Find metaphors for concepts
Visualize equations and behavior
Emphasize "what if?"
Try new, even strange, approaches

The Cognitive Profile Inventory is specifically designed help you determine the combination of study strategies that will work best for you personally. It determines your complete profile, showing both your strengths and areas where you are less strong and emphasizes that you have some area in each quadrant, a profile, not just a single type. The book and associated website specifically discuss the following topics:

- Successful study strategies by quadrant.
- Working with faculty whose teaching style does not match your learning style.
- How to deal with and benefit from students whose profiles are different from yours.
- How to use knowledge of your profile to inform your career direction decisions.

It is critical to recognize that you do not act, think, and learn solely within your dominant quadrant. The dominant quadrant is likely, however, to describe the areas in which you are most comfortable and talented. The table below summarizes the learning strategies for each type.

Contributions to Engineering and Science from Each of the Four Profiles

This section lists typical contributions to engineering and science made by workers in each of the four quadrants. Engineers and scientists, like other professionals, need to develop skills in all four quadrants. You may find that you are most strongly drawn to work that is in your dominant quadrant.

Team Contributions and Careers

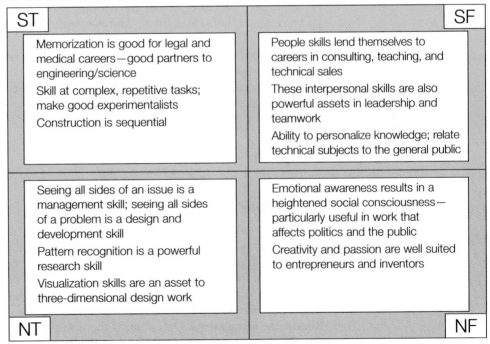

ST	SF
Memorization is good for legal and medical careers—good partners to engineering/science Skill at complex, repetitive tasks; make good experimentalists Construction is sequential	People skills lend themselves to careers in consulting, teaching, and technical sales These interpersonal skills are also powerful assets in leadership and teamwork Ability to personalize knowledge; relate technical subjects to the general public
Seeing all sides of an issue is a management skill; seeing all sides of a problem is a design and development skill Pattern recognition is a powerful research skill Visualization skills are an asset to three-dimensional design work	Emotional awareness results in a heightened social consciousness—particularly useful in work that affects politics and the public Creativity and passion are well suited to entrepreneurs and inventors
NT	NF

HOW KNOWING YOUR MBTI TYPE OR COGNITIVE PROFILE HELPS INDIVIDUAL AND TEAM INTERACTIONS

When you are working with a team or group or even with another individual student it helps to know their profile, to see their strengths, and then to take advantage of the strengths in the group. An ST will keep track of materials, details and tasks to be accomplished. An SF will see that everyone is participating and being listened to, the NT will determine the best way to accomplish the assigned task without getting bogged down or off track, and the NF will find a creative approach to the results, pulling all the details from different directions together to make a cohesive presentation in the required format.

In real life, this is how it worked out for two of the authors.

> *We—Dr. Stephan and Dr. Park—are complete MBTI opposites. I, Dr. Stephan, am ESTJ and he, Dr. Park, is INFP—notice no letters in common. Our first few years together at Clemson University, we very nearly came to blows. We each thought the other was "so smart they were stupid," mentally ill, or probably both. However, we have learned to work together quite effectively, and we each value what the other brings to the team. Admittedly, we frequently have difficulty understanding what the other is saying, occasionally even having to ask someone else to translate English to English for us. (And no, this is not an exaggeration!) At this point, we laugh about it a fair amount.*

The moral of our little tale is this: just because a colleague makes little sense to you does not necessarily means that he or she is not firing on all thrusters—it may simply be a difference in personality types. You should try to understand how your fellow students or coworkers think and act, not simply immediately condemn them. As always, there are exceptions to every rule and you will indeed sometimes stumble on others that really are not an asset to the team in any way, but you will have to deal with that by other methods. In general, everyone on a team brings their own unique skills to the table. The best teams, in fact, have a variety of personality types.

As a point of interest, each of the four two-letter profiles mentioned earlier is represented among the authors of this text. Hopefully, our melding of the various ways the five of us perceive the world and communicate our knowledge into one unified presentation will make sense to most of our readers.

The information below may help to illustrate how each of us uniquely developed different strategies for success in college.

How I Study

by Dr. Beth Stephan

Two weeks before the exam:

- **Review all book and lecture materials.** Take notes on important formulas and concepts as you reread the text (assuming you have already read it once). This will become the foundation of a note sheet. Work the example questions on your own, and compare the answers. Determine how the lectures were different from the book (different examples, different equation derivations). Determine if particular topics were emphasized more than others.
- **Reexamine assignments.** Determine if there was a recurring type of question. Does this match the emphasis areas given in the lecture materials? Decide if there are concepts you did not understand. Work these questions again, and determine if you now understand the concept or need to ask for more help.
- **Divide the studying over several days.** Create a list of materials to review each day. Study in small chunks, not hours at a time on one subject. If you become tired—stop! Get something to eat, take a walk, or work on a different subject. You will not absorb the material well if you cannot focus.

One week before the exam:

- **Figure out the exam structure.** If you do not already know what will be covered on the exam, ask the professor. Determine the material covered on the exam, the type and number of questions, and review problems if available. Note any special things to bring (e.g., calculator, extra batteries, laptop, note sheet, or textbook).
- **Attend any review session.** This is a good place to cover the text examples or assignment questions you still do not understand. Be specific—come prepared with a list of things to ask. Do not worry about your class image. If it is "uncool" to ask questions, imagine how "uncool" it will be when you fail an exam because you failed to ask.

Three days before:

- **Create a "brain dump" sheet.** The basic idea is you, your memory, alone, pencil, paper; fill it up with everything you can remember. Compare with book review notes sheet. Review any concepts you incorrectly listed or omitted on the brain dump sheet.
- **Make a notes sheet (see below for more information).**
- **Take a sample exam.** Create a sample exam and put it away, and then take it a few days later under actual exam conditions (time, materials allowed—not in front of the television). This is a great place to use a study group to generate questions and check your answers.

The night before and day of:

- **Pack all materials needed for exam before going to sleep.**
- **Get a good night's sleep.**
- **Get up early.** You do not want to be in a rush or panic to get to class.
- **Take a shower and get dressed up.** Confidence comes from knowing the material and feeling good about yourself. Do not wear a suit, but wear something you would wear to dinner with friends. Jeans and a tucked-in shirt improve your mood compared to sleep pants and shower shoes.
- **Eat some good food.** It is impossible to ignore a growling stomach during an exam.

Notice here it does not say "Study," "Cram," or "Pull an All Nighter." At this point, you will do better if you get a good night's sleep rather than studying all night. A rested brain can pull up information far better and faster than a tired one.

On the exam:

- **Read the directions and questions carefully.** Look over the whole exam before beginning to answer any questions. Be sure you are answering the right question. Watch for details of how the answer should be presented and what to include in your solution. But do not overanalyze—go with your instincts.
- **Answer the easy questions first.** Skip what you do not know—maximize your points! Exams are not always ordered easy to hard, so there may be some simple questions on page 5 buried behind several pages of derivations.
- **Try to answer all questions, even if you are unsure of what you are doing.** Sometimes in trying to work out a problem, the solution will become clear to you. Start by writing down what you know or equations you think may apply. If you get stuck, move on. You may answer a few more questions and the solution to the "stuck" question will come to you subconsciously.
- **Ask the instructor to explain unclear points.**
- **Use all of the time allotted for the test.** Go back and review that your answers match the question asked. Be sure you have used consistent unit systems. Double-check your calculations. Take the exam twice if time permits. It is difficult to find a mistake

when checking over your own work, especially under stress conditions. Think of it as proofreading: it is always better to have someone else check it for you or to read it out loud. By working the problem all over again, you may write something differently and be able to compare it to find your mistake.

- **Relax!** Have confidence to know you are prepared, and do your best. At this point, that is all you can do, so do not fret about the outcome!

What goes on a note sheet?

- **Important formulas and definitions.** Determine if any will be given to you during the exam. If they will be provided (such as a conversion table), do not waste the time and effort to put them on your note sheet and risk writing them incorrectly. Check your note sheet carefully against the text and lecture materials. Make sure the formulas are written correctly.
- **Procedures.** Write general steps that can be followed for any problem. Do *not* include example problems. You will not have time on an exam to figure them out if you have not already done so.
- **Reminders.** Include tips to yourself on things you have made mistakes on during the assignments, such as "Do not forget to convert 1 atm = 101,325 Pa" or "Check to make sure input power is always greater than output power." Be sure to write down important facts, such as "The density of water = 1 g/cc" even if you have this memorized. Under pressure, it becomes harder to recall small details.

Tips and Techniques

- **Be sure you can read it!!** Often, symbols have two or three different meanings. Be sure you can decipher what you write. Write neatly. If you cannot read it on the exam, it will not help! Do not copy old exams. Most professors are not lazy enough to give you the exact same question from the text all over again on the exam, but will alter it to stretch your mind.
- **Be sure it is your own work.** Each student understands the material in a unique way. Things you struggle with may be easy for your neighbor, and chances are your neighbor organizes material differently than you do. Do *not* copy anyone else's note sheet. Sometimes, students find it helpful to write encouraging meditative thoughts, such as "Remain CALM" or "You can do this!" or an inspirational quote. Write something that will ground you, and if you begin to panic, you can read your notes to yourself and try to relax.
- **Create a note sheet even if you cannot use it on the exam.** A note sheet is a great organizational tool to make sure you have the important stuff all in one place. It makes a great last-minute review paper to look over those last 10 minutes before the exam. Since you composed it, you should (in theory) understand everything on it. Flipping through the text at the last minute often causes panic if you come across a topic you are unsure of.

How I Study

by Dr. Bill Park

Engineering is about solving problems. I did not learn to solve the problems in my field (electrical engineering, EE) by reading books, or highlighting stuff, or listening to professors rattle on interminably; *I slogged through lots and lots of problems*! If you look at my old textbooks, you will find no handwritten notes, no highlighting, no dog-eared corners. By the way, I kept *all* of my texts, and I am very glad I did. Even today, decades later, I not infrequently look something up in them.

This is not to say that I did not go to class every day (unless I was deathly ill) and take detailed notes—I absolutely did. The professor's explanations or examples frequently proved invaluable in helping me solve the problems later. My first rule for success in college was *never* cut class unless absolutely necessary. You never know when the professor will say the one thing that will make everything fall into place in your brain. This happened to me many times during my formal education.

As an EE, I never had to do much unit conversion to a large extent; EEs just have to move the decimal place around efficiently and correctly using appropriate metric prefixes; because everything is in SI units. However, I would have studied the units material in this book the same way I studied electronics, communication theory, or digital signal processing.

When learning new material, I would look at the first problem at the end of the chapter being studied. Note that I did *not* say the first problem the professor assigned. It would be obvious that I knew how to work the problem, probably because I had worked a few similar exercise problems, or I thought I could work the problem but was not sure about all of the details, or I was totally clueless.

If the solution and all of the steps necessary was obvious, I would move on to the next problem. There is little to be gained from working problems with which you are already proficient, except possibly self-esteem. However, I promise you that there is more self-esteem to be had by conquering a new type of problem than by being bored to tears working the same old problems a zillion times.

As I read through the problems, eventually I would find one I was not certain about, so I would have a go at it. If I figured it out, I would move on to the next. If I got stuck, was uncertain, or came upon a problem that I had no idea how to approach, I would try to find an example problem that was similar, whether in the text or in my class notes. Working through a solved example or two (or ten) would usually give me the insight I needed to solve the problem at hand.

Some of you are thinking at this point, "This guy is insane. I do not have time to work that many problems!" Well, sorry folks. Engineering school is *far* more than a full-time job, at least if you wish to live up to your potential. Unless you have an IQ of about 175 or higher, you will need to spend far more than 40 hours per week to *excel* in a curriculum as difficult as engineering. There is nothing anyone can do to change that.

If working through the available examples failed, then, *and only then*, would I resort to reading the book, and even then I would try to isolate the sections that had bearing on the problem I was trying to work. Many engineering books are deadly dull (in my opinion) and are excellent cures for insomnia!

Sometimes, particularly if I had to resort to the book, things were still not clear. Time to go see the professor!

You will find that if you ask professors for help, most of them will quickly know whether or not you have put an appropriate amount of effort into learning on your own, or whether you tried for a few minutes and gave up. Professors' time is a limited commodity, and if you have obviously done your part, they are generally far more willing to help you. I *never* went to a professor about a problem unless I had beaten my head against the wall to the point where a few more blows would gelatinize my brain. All of this effort almost always paid off, because invariably I could articulate exactly where my comprehension failed and more often than not the professor could clear up my misunderstanding within 1–5 minutes—my brain was primed for the answer.

If you say to a professor, "I do not understand this problem about power," then the professor will have to start asking questions to try to find exactly where your problem is. If, on the other hand, you say "I understand how to get the total amount of energy required to lift the load, and I understand how to get the power since I know how long it takes, but I am confused about the relationship between input power, output power, and efficiency," then the professor immediately knows which part of the problem to explain, using less professorial time, and keeping her or him happy. Do you want a happy professor or an irritable professor? (No-brainer!)

Another issue: note sheets. I do not believe I ever made out a note sheet for a test, even when it was allowed. I recall in first semester calculus, we were allowed to have a note sheet on tests. For the first test, I showed up with no notes. The student next to me leaned over, almost aghast, just before the test was handed out: "Where is your note sheet?!?" My reply was, "What in the world would I put on the note sheet? That the derivative of x with respect to x is 1? If you had done enough problems, you'd know all this stuff!"

Another time, in second semester electronics, we had open-book tests. One of the problems required an equation I did not recall. Rather than waste my time flipping frantically through the book, I derived the necessary formula from fundamental principles—much quicker than finding it in the book. This is the level of understanding for which I strove.

More germane to the topic of units, when I teach this material, it is sometimes necessary to express a complex unit, like Pascal, in terms of base SI units. I *do not* have this stuff memorized, but I can easily derive it from my understanding of what pressure and force are. Pressure is force per area, and by Newton's second law, force is mass times acceleration. Stuffing in the base units gives:

$$(\text{kg m/s}^2)/\text{m}^2 = \text{kg}/(\text{m s}^2)$$

I do not use this frequently enough to have it memorized, but I can derive it in just a few seconds. Things I use frequently I memorize without trying; things I only use occasionally are not worth cluttering up my brain with.

To summarize my approach to studying engineering:

- The focus is on solving problems, not on reading books and notes. I work problems until the solutions to similar problems are almost blindingly obvious.
- I strive to understand concepts and the overall picture, not memorize formulae and details—the details I can fill in later. To me, memorization (at least in engineering) is double-plus unclever.
- I do my work before seeking help. Not only is this more effective for me, it uses less time on the part of the person from whom I seek help.

With all of this said, you need to tailor your study methods to the subject. I have taken courses (e.g., plant taxonomy and Chinese) that require rather prodigious feats of memorization. However, for me, engineering is not something to be memorized, but mastered by understanding concepts.

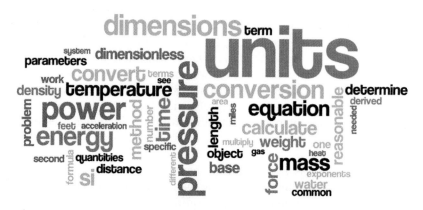

CHAPTER 7
FUNDAMENTAL DIMENSIONS AND BASE UNITS

As aspiring engineers you must learn to distinguish among many terms that laymen tend to use interchangeably. You must also understand the technical meaning of terms that are misunderstood by those untrained in science or engineering. One term that is often misunderstood is "dimension." To most people, a dimension refers to a straight line (length, one dimension), a flat surface (area, two dimensions), or a solid object (volume, three dimensions). Some slightly more educated folks might include time as a fourth dimension. The number of concepts classified as dimensions is far broader.

A **dimension** is a measurable physical idea; it generally consists solely of a word description with no numbers. A **unit** allows us to quantify a dimension—to state a number describing how much of that dimension exists in a specific situation. Units are defined by convention and related to an accepted standard.

- Length is a dimension. There are many units for length, such as mile, foot, meter, light-year, and fathom.
- Time is a dimension. There are many units for time, such as second, minute, hour, day, fortnight, year, and century.
- Temperature is a dimension. There are many units for temperature, such as Celsius, Fahrenheit, and kelvin.

The dimensions of length, time, and temperature are familiar to us, but in reality, we do not often use these words since they are fairly vague.

We do not say . . .	We do say . . .
It is really a long **length** to Lumberton.	Lumberton is about 175 **miles** away.
Bake the cake for a **time**.	Bake the cake for 35 **minutes**.
Set the oven to a high **temperature**.	Set the oven to 450 **degrees** **Fahrenheit**.

The difference between the left and the right columns is that the statements on the left refer to dimensions and those on the right refer to established standards or units.

NOTE

One of Leonardo da Vinci's most famous drawings is the Vitruvian Man. He drew this figure to depict the proportions of the human body, based on the work of the Roman architect Vitruvian. Shown on the drawing are nine historical units in their proper ratios.

7.1   THE METRIC SYSTEM

The SI system (Le Système International d'Unités), commonly known as the metric system, is the standard set of units for most of the world. Originally developed by French scientists under King Louis XVI, the SI system was finalized by the international scientific community as the standard unit system in 1971. This system defines seven base units, from which all others are derived. Table 7-1 shows the seven base units and their corresponding fundamental dimensions.

Table 7-1 Fundamental dimensions and base units

Dimension	Symbol	Unit	Symbol
Length	L	meter	m
Mass	M	kilogram	kg
Time	T	second	s
Temperature	Θ	kelvin	K
Amount of substance	N	mole	mol
Light intensity	J	candela	cd
Electric current	I	ampere	A

SI Prefixes

The SI system is based upon multiples of 10. By using an **SI prefix** when reporting numbers, we avoid scientific notation or long strings of zeros. For example, instead of saying, "The distance to Atlanta is 198,000 meters," we would say, "The distance to Atlanta is 198 kilometers."

For a list of SI prefixes, refer to the inside cover of this book or to Table 7-2. Note that the abbreviations for all SI prefixes from kilo- down to yocto- are lower case, whereas from Mega- up to Yotta- are upper case.

Table 7-2 SI prefixes (example: 1 millimeter [mm] = 1 × 10^{-3} meters [m])

Numbers Less than One			Numbers Greater than One		
Power of 10	Prefix	Abbreviation	Power of 10	Prefix	Abbreviation
10^{-1}	deci-	d	10^{1}	deca-	da
10^{-2}	centi-	c	10^{2}	hecto-	h
10^{-3}	milli-	m	10^{3}	kilo-	k
10^{-6}	micro-	μ	10^{6}	Mega-	M
10^{-9}	nano-	n	10^{9}	Giga-	G
10^{-12}	pico-	p	10^{12}	Tera-	T
10^{-15}	femto-	f	10^{15}	Peta-	P
10^{-18}	atto-	a	10^{18}	Exa-	E
10^{-21}	zepto-	z	10^{21}	Zetta-	Z
10^{-24}	yocto-	y	10^{24}	Yotta-	Y

● **EXAMPLE 7-1**

Express the following values using the correct SI prefix.

Quantity	SI Prefix	SI Abbreviation
(a) 3,100 joules	3.1 kilojoules	3.1 kJ
(b) 2.51×10^6 watts	2.51 megawatts	2.51 MW
(c) 4×10^{-18} seconds	4 attoseconds	4 as
(d) 3×10^{-6} grams	3 micrograms	3 μg

NOTE

In engineering contexts using SI prefixes, it is common practice to adjust the decimal point such that there are one, two, or three digits to the left of the decimal point coupled with the appropriate SI prefix.

Determining the appropriate SI prefix to use becomes simple when the number is placed in engineering notation: just examine the exponent.

As a reminder, scientific and engineering notation are defined as follows:

Scientific notation is typically expressed in the form $\#.\#\#\# \times 10^N$, where the digit to the left of the decimal point is the most significant nonzero digit of the value being represented. Sometimes, the digit to the right of the decimal point is the most significant digit instead. The number of decimal places can vary, but is usually two to four. N is an integer, and multiplying by 10^N serves to locate the true position of the decimal point.

Engineering notation is expressed in the form $\#\#\#.\#\#\# \times 10^M$, where M is an integer multiple of 3, and the number of digits to the left of the decimal point is 1, 2, or 3 as needed to yield a power of 10 that is indeed a multiple of 3. The number of digits to the right of the decimal point is typically between two and four.

● **EXAMPLE 7-2**

Express the following values using scientific notation, engineering notation, or with the correct SI prefix.

Standard	Scientific	Engineering	With Prefix
(a) 43,480,000 m	4.348×10^7 m	43.48×10^6 m	43.48 Mm
(b) 0.0000003060 V	3.060×10^{-7} V	306.0×10^{-9} V	306.0 nV
(c) 9,860,000,000 J	9.86×10^9 J	9.86×10^9 J	9.86 GJ
(d) 0.0351 s	3.51×10^{-2} s	35.1×10^{-3} s	35.1 ms
(e) 0.0000000522 μA	5.22×10^{-14} A	52.2×10^{-15} A	52.2 fA
(f) 456200 kW	4.562×10^8 W	456.2×10^6 W	456.2 MW

Note that the numeric values of the mantissa are the same in the last two columns, and the exponent in engineering notation specifies the metric prefix.

COMPREHENSION CHECK 7-1

In January 2008, *Scientific American* reported that physicists Peter Sutter and Eli Sutter of Brookhaven National Laboratories made a pipette to measure droplets in units of a zeptoliter. Previously, the smallest unit of measure in a pipette was an attoliter. Express the answers to the following questions in scientific notation.

(a) Convert the measurement of 5 zeptoliters into units of picoliters.
(b) Convert the measurement of 80 zeptoliters into units of microliters.
(c) Convert the measurement of 300 zeptoliters into units of liters.

Official SI Rules

When reporting units using the SI system, follow these official rules.

- **If a unit abbreviation appears as a capital letter, it has been named after a person; all other abbreviations appear as lowercase letters.** For example, the abbreviation "N" stands for "newton," the SI unit of force named after Isaac Newton.

  Correct: The book weighs 5 N.   Incorrect: The book weighs 5 n.
  Correct: The rod is 5 m long.   Incorrect: The rod is 5 M long.

  The one exception to this rule is the volumetric unit of liter. The abbreviation is shown as L, since a lowercase l can be confused with both the number 1 and the uppercase letter I.

- **Symbols of units are not shown as plural.**

  Correct: 10 centimeters = 10 cm   Incorrect: 10 centimeters ≠ 10 cms

- **Symbols are not shown with a period unless they appear at the end of a sentence.**

  Correct: The rod is 5 mm long.   Incorrect: The rod is 5 mm. long.

- **Symbols are written in upright Roman type (m, k, L) to distinguish them from mathematical variables (m, k, l), which are indicated by italics.**

- **One space separates the number and symbol, except with the degree symbol referring to an angle.**

  Correct: 5 mm or 5°   Incorrect: 5mm or 5 °

- **Spaces or commas may be used to group digits by threes.**

  Correct: 1 000 000 or 1,000,000

- **Symbols for derived units formed by multiple units are joined by a space or center dot. Care must be taken to avoid confusing SI prefixes with units.**

  Correct: kg m or kg · m   Incorrect: kgm or mkg

  This is particularly important when confusion might arise. For example, "ms" stands for millisecond, but "m s" stands for meter second. In cases like this, using a center dot is preferable since it is less likely to be misunderstood.

- **Symbols for derived units formed by dividing units are joined by a virgule (the "slash" /) or shown with a negative exponent. Care must be taken to appropriately display the entire denominator.**

  Correct: $N/(m\ s^2)$ or $N\ m^{-1}\ s^{-2}$ Incorrect: $N/m\ s^2$

- **Do not combine prefixes to form compound prefixes. Use the single correct prefix.**

  Correct: picojoules (pJ)   Incorrect: millinanojoules (mnJ)
  Correct: Gigaseconds (Gs)   Incorrect: kiloMegaseconds (kMs)

COMPREHENSION CHECK 7-2

Indicate if the following units are correctly expressed according to the official SI rules. If the unit is incorrectly displayed, show the correction.

(a) Reading this sentence took 5 Secs.
(b) The average person's pupils are 60mms. apart.
(c) One gallon is the same as 380 microkiloliters.

7.2 OTHER UNIT SYSTEMS

Prior to the adoption of the SI unit system by the scientific community, several other systems of units were used and are still used today, particularly in the United States. The other countries that use non-SI units are Liberia and Myanmar. Great Britain officially converted to metric in 1965, but it is still common there to see nonmetric units used in communications for the general public.

It is important to know how to convert between all unit systems. Table 7-3 compares several systems. The system listed as AES (American Engineering System) is in common use by the general public in the United States. The USCS (United States Customary System) is commonly called "English" units.

Table 7-3 Comparison of unit system, with corresponding abbreviations

Dimension	SI (MKS)	AES	USCS
Length {L}	meter [m]	foot [ft]	foot [ft]
Mass {M}	kilogram [kg]	pound-mass [lb_m]	slug
Time {T}	second [s]	second [s]	second [s]
Relative temperature {Θ}	Celsius [°C]	Fahrenheit [°F]	Fahrenheit [°F]
Absolute temperature {Θ}	kelvin [K]	Rankine [°R]	Rankine [°R]

Accepted Non-SI Units

The units in Table 7-4 are not technically in the SI system, but due to their common usage, are acceptable for use in combination with the base SI units.

Table 7-4 Acceptable non-SI units

Unit	Equivalent SI	Unit	Equivalent SI
Astronomical unit [AU]	1 AU = 1.4959787×10^{11} m	day [d]	1 d = 86,400 s
Atomic mass unit [amu]	1 amu = $1.6605402 \times 10^{-24}$ g	hour [h]	1 h = 3,600 s
Electronvolt [eV]	1 eV = $1.6021773 \times 10^{-19}$ J	minute [min]	1 min = 60 s
Liter [L]	1 L = 0.001 m^3	year [yr]	1 yr = 3.16×10^7 s
		degree [°]	1° = 0.0175 rad or 1 rad = 57.3°

NOTE

1 liter does not equal 1 cubic meter!

7.3 CONVERSION PROCEDURE FOR UNITS

We use conversion factors to translate from one set of units to another. This must be done correctly and consistently to obtain the right answers. Some common conversion factors can be found inside the cover of this book, categorized by dimension. Although many more conversions are available, all the work for a typical engineering class can be accomplished using the conversions found in this table.

LENGTH

1 m = 3.28 ft

1 km = 0.621 mi

1 in = 2.54 cm

1 mi = 5,280 ft

1 yd = 3 ft

Let us examine the conversions found for the dimension of length, as shown in the box, beginning with the conversion: 1 meter [m] = 3.28 feet [ft]. By dividing both sides of this equation by 3.28 feet, we obtain

$$\frac{1 \text{ m}}{3.28 \text{ ft}} = 1$$

or in other words, "There is 1 meter per 3.28 feet." If we divide both sides of the original expression by 1 meter, we obtain

$$1 = \frac{3.28 \text{ ft}}{1 \text{ m}}$$

or in other words, "In every 3.28 feet there is 1 meter."

The number 1 is dimensionless, a pure number. *We can multiply any expression by 1 without changing the expression.* We do this so as to change the units to the standard we desire.

For example, on a trip we note that the distance to Atlanta is 123 miles [mi]. How many kilometers [km] is it to Atlanta? From the conversion table, we can find that 1 kilometer [km] = 0.621 miles [mi], or

$$1 = \frac{1 \text{ km}}{0.621 \text{ mi}}$$

By multiplying the original quantity of 123 miles by 1, we can say

$$(123 \text{ mi})(1) = (123 \text{ mi})\left(\frac{1 \text{ km}}{0.621 \text{ mi}}\right) = 198 \text{ km}$$

Note that we could have multiplied by the following relationship:

$$1 = \frac{0.621 \text{ mi}}{1 \text{ km}}$$

We would still have multiplied the original answer by 1, but the units would not cancel and we would be left with an awkward, meaningless answer.

$$(123 \text{ mi})(1) = (123 \text{ mi})\left(\frac{0.621 \text{ mi}}{1 \text{ km}}\right) = 76 \frac{\text{mi}^2}{\text{km}}$$

As a second example, we are designing a reactor system using 2-inch [in] diameter plastic pipe. The design office in Germany would like the pipe specifications in units of centimeters [cm]. From the conversion table, we find that 1 inch [in] = 2.54 centimeters [cm], or

$$1 = \frac{1 \text{ in}}{2.54 \text{ cm}}$$

By multiplying the original quantity of 2 inches by 1, we can say

$$(2 \text{ in})(1) = (2 \text{ in})\left(\frac{2.54 \text{ cm}}{1 \text{ in}}\right) = 5 \text{ cm}$$

In a final example, suppose a car travels at 40 miles per hour (abbreviated mph). Stated in words, "a car traveling at a rate of 40 mph will take 1 hour to travel 40 miles if the velocity remains constant." By simple arithmetic this means that the car will travel 80 miles in 2 hours or 120 miles in 3 hours. In general,

$$\text{Distance} = (\text{velocity})(\text{time elapsed at that velocity})$$

NOTE

Place the units you are converting TO in the NUMERATOR; place the units you are converting FROM in the DENOMINATOR.

Suppose the car is traveling at 40 mph for 6 minutes. How far does the car travel? Simple calculation shows

$$\text{Distance} = (40)(6) = 240$$

Without considering units, the preceding example implies that if we drive our car at 40 mph, we can cover the distance from Charlotte, North Carolina, to Atlanta, Georgia, in 6 minutes! What is wrong? Note that the velocity is given in miles per hour, and the time is given in minutes. If the equation is written with consistent units attached, we get

$$\text{Distance} = \left(\frac{40\text{ mi}}{\text{h}}\right)\left(\frac{6\text{ min}}{}\middle|\frac{1\text{ h}}{60\text{ min}}\right) = 4\text{ mi}$$

It seems more reasonable to say "traveling at a rate of 40 miles per hour for a time period of 6 minutes will allow us to go 4 miles."

To convert between any set of units, the following method demonstrated in Examples 7-3 to 7-6 is very helpful. This procedure is easy to use, but take care to avoid mistakes. If you use one of the conversion factors incorrectly, say, with 3 in the numerator instead of the denominator, your answer will be in error by a factor of 9.

IMPORTANT CONCEPT

Be sure to *always* include units in your calculations *and* your final answer!

● **EXAMPLE 7-3**

Convert the length 40 yards [yd] into units of feet [ft].

Method	Steps
(1) Term to be converted	40 yd
(2) Conversion formula	1 yd = 3 ft
(3) Make a fraction (equal to one)	$\dfrac{3\text{ ft}}{1\text{ yd}}$
(4) Multiply	$\dfrac{40\text{ yd}}{}\left\|\dfrac{3\text{ ft}}{1\text{ yd}}\right\|$
(5) Cancel, calculate, be reasonable	120 ft

COMPREHENSION CHECK 7-3

The highest mountain in the world is Mount Everest in Nepal. The peak of Mount Everest is 29,029 feet above sea level. Convert the height from feet to miles.

COMPREHENSION CHECK 7-4

In North America, the greatest depth of snow on the ground was recorded at 1,145.5 centimeters at Tamarack, California. Convert the depth from centimeters to inches.

COMPREHENSION CHECK 7-5

The shortest race in the Olympics is the 100-meter dash. Convert the length from meters to feet.

7.4 CONVERSIONS INVOLVING MULTIPLE STEPS

Sometimes, more than one conversion factor is needed. We can multiply by several conversion factors, each one of which is the same as multiplying by 1, as many times as needed to reach the desired result. For example, suppose we determined that the distance to Atlanta is 123 miles [mi]. How many yards [yd] is it to Atlanta? From the conversion table, we do not have a direct conversion between miles and yards, but we see that both can be related to feet. We can find that 1 mile [mi] = 5,280 feet [ft], or

$$1 = \frac{5{,}280 \text{ ft}}{1 \text{ mi}}$$

We can also find that 1 yard [yd] = 3 feet [ft], or

$$1 = \frac{1 \text{ yd}}{3 \text{ ft}}$$

By multiplying the original quantity of 123 miles by 1 using the first set of conversion factors, we can say:

$$(123 \text{ mi})(1) = (123 \text{ mi})\left(\frac{5{,}280 \text{ ft}}{1 \text{ mi}}\right) = 649{,}440 \text{ ft}$$

If we multiply by 1 again, using the second set of conversion factors and applying reasonableness:

$$(649{,}440 \text{ ft})(1) = (649{,}440 \text{ ft})\left(\frac{1 \text{ yd}}{3 \text{ ft}}\right) = 216{,}000 \text{ yd}$$

This is usually shown as a single step:

$$(123 \text{ mi})\left(\frac{5{,}280 \text{ ft}}{1 \text{ mi}}\right)\left(\frac{1 \text{ yd}}{3 \text{ ft}}\right) = 216{,}000 \text{ yd}$$

Unit Conversion Procedure

1. Write the value and unit to be converted.
2. Write the conversion formula between the given unit and the desired unit.
3. Make a fraction, equal to 1, of the conversion formula in Step 2, such that the original unit in Step 1 is located either in the denominator or in the numerator, depending on where it must reside so that the original unit will cancel.
4. Multiply the term from Step 1 by the fractions developed in Step 3.
5. Cancel units, perform mathematical calculations, and express the answer in "reasonable" terms (i.e., not too many decimal places).

● **EXAMPLE 7-4**

Convert the length 40 yards [yd] into units of millimeters [mm].

Method	Steps			
(1) Term to be converted	40 yd			
(2) Conversion formula	1 yd = 3 ft 1 ft = 12 in 1 in = 2.54 cm 1 cm = 10 mm			
(3) Make fractions (equal to one)	$\frac{3 \text{ ft}}{1 \text{ yd}}$	$\frac{12 \text{ in}}{1 \text{ ft}}$	$\frac{2.54 \text{ cm}}{1 \text{ in}}$	$\frac{10 \text{ mm}}{1 \text{ cm}}$
(4) Multiply	$\frac{40 \text{ yd}}{} \left\vert \frac{3 \text{ ft}}{1 \text{ yd}} \right\vert \frac{12 \text{ in}}{1 \text{ ft}} \left\vert \frac{2.54 \text{ cm}}{1 \text{ in}} \right\vert \frac{10 \text{ mm}}{1 \text{ cm}}$			
(5) Cancel, calculate, be reasonable	37,000 mm			

● **EXAMPLE 7-5**

Convert 55 miles per hour [mph] to units of meters per second [m/s].

Note that we have two units to convert here, miles to meters, and hours to seconds.

TIME
1 d = 24 h
1 h = 60 min
1 min = 60 s
1 yr = 365 d

Method	Steps	
(1) Term to be converted	55 mph	
(2) Conversion formula	1 km = 0.621 mi 1 km = 1,000 m	1 h = 60 min 1 min = 60 s
(3) Make fractions (equal to one) (4) Multiply	$\frac{55 \text{ mi}}{\text{h}} \left\vert \frac{1 \text{ km}}{0.621 \text{ mi}} \right\vert \frac{1,000 \text{ m}}{1 \text{ km}} \left\vert \frac{1 \text{ h}}{60 \text{ min}} \right\vert \frac{1 \text{ min}}{60 \text{ s}}$	
(5) Cancel, calculate, be reasonable	24.6 m/s	

● **EXAMPLE 7-6**

Convert the volume of 40 gallons [gal] into units of cubic feet [ft³].

By examining the "Volume" box in the conversion table, we see that the following facts are available for use:

$$1 \text{ L} = 0.264 \text{ gal} \quad \text{and} \quad 1 \text{ L} = 0.0353 \text{ ft}^3$$

By the transitive property, if a = b and a = c, then b = c. Therefore, we can directly write

$$0.264 \text{ gal} = 0.0353 \text{ ft}^3$$

VOLUME
1 L = 0.264 gal
1 L = 0.0353 ft³
1 L = 33.8 fl oz
1 mL = 1 cm³

Method	Steps
(1) Term to be converted	40 gal
(2) Conversion formula	$0.264 \text{ gal} = 0.0353 \text{ ft}^3$
(3) Make a fraction (equal to one)	$\frac{0.0353 \text{ ft}^3}{0.264 \text{ gal}}$
(4) Multiply	$\frac{40 \text{ gal}}{} \left\vert \frac{0.0353 \text{ ft}^3}{0.264 \text{ gal}} \right.$
(5) Cancel, calculate, be reasonable	5.3 ft^3

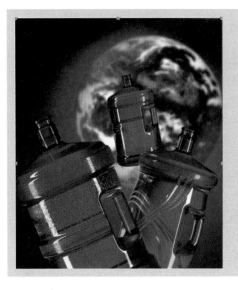

This picture shows 5-gallon water bottles made from polycarbonate. Millions of these bottles are made each year around the world to transport clean water to remote locations.

The use of polycarbonate to contain products for consumption has raised safety concerns because bisphenol A is leached from the plastic into the stored liquid. As of March 2009, six U.S. manufacturers have stopped using polycarbonate in bottles and cups used by infants and small children.

Photo courtesy of Kautex Machines

One frequently needs to convert a value that has some unit or units raised to a power, for example, converting a volume given in cubic feet to cubic meters. It is critical in this case that the power involved be applied to the *entire* conversion factor, both the numerical values and the units.

● EXAMPLE 7-7

Convert 35 cubic inches [in^3] to cubic centimeters [cm^3 or cc].

NOTE

When raising a quantity to a power, be sure to apply the power to both the value and the units.

Method	Steps
(1) Term to be converted	35 in^3
(2) Conversion formula	1 in = 2.54 cm
(3) Make fractions (equal to one)	$\dfrac{35 \text{ in}^3}{} \left\| \dfrac{(2.54 \text{ cm})^3}{(1 \text{ in})^3} \right.$
(4) Multiply	$\dfrac{35 \text{ in}^3}{} \left\| \dfrac{(2.54)^3 \text{ cm}^3}{1 \text{ in}^3} \right.$
(5) Cancel, calculate, be reasonable	574 cm^3

Note that in some cases, a unit that is raised to a power is being converted to another unit that has been defined to have the same dimension as the one raised to a power. This is difficult to say in words, but a couple of examples should clarify it.

If one is converting square meters [m^2] to acres, the conversion factor is *not* squared, since the conversion provided is already in terms of length squared: 1 acre = 4,047 m^2.

If one is converting cubic feet [ft^3] to liters [L], the conversion factor is *not* cubed, since the conversion provided is already in terms of length cubed: 1 L = 0.0353 ft^3.

● **EXAMPLE 7-8**

Convert a velocity of 250 kilometers per second [km/s] to units of millimeters per picosecond [mm/ps].

Method	Steps
(1) Term to be converted	250 km/s
(2) Conversion formula	$1 \text{ km} = 10^3 \text{ m}$ $1 \text{ mm} = 10^{-3} \text{ m}$ $1 \text{ ps} = 10^{-12} \text{ s}$
(3) Make fractions (equal to one) (4) Multiply	$\dfrac{250 \text{ km}}{\text{s}} \left\| \dfrac{10^3 \text{ m}}{1 \text{ km}} \right\| \dfrac{1 \text{ mm}}{10^{-3} \text{ m}} \left\| \dfrac{10^{-12} \text{ s}}{1 \text{ ps}} \right.$
(5) Cancel, calculate, be reasonable	250×10^{-6} mm/ps

- *Comment: Following the rules for use of prefixes given earlier would indicate that the term in the denominator should not have a prefix at all, and it should be transferred to the numerator, giving 250×10^6 mm/s. Beyond that, it was earlier stated that in general, the prefix should be adjusted to give one, two, or three digits to the left of the decimal place (with no power of 10). This would yield 250 km/s, right back where we started.*

- *Moral: **specific instructions** (such as "convert to mm/ps") **usually override the general rules**. In this case, perhaps we are studying the velocity of protons in a particle accelerator, and for comparison with other experiments, we need to know how many millimeters the particles go in one picosecond, rather than ending up with the units the general rules would dictate.*

COMPREHENSION CHECK 7-6

Officially, a hurricane is a tropical storm with sustained winds of at least 74 miles per hour. Convert this speed into units of kilometers per minute.

COMPREHENSION CHECK 7-7

Many toilets in commercial establishments have a value printed on them stating the amount of water consumed per flush. For example, a label of 2 Lpf indicates the consumption of 2 liters per flush. If a toilet is rated at 3 Lpf, how many flushes are required to consume 20 gallons of water?

RULES OF THUMB

1 quart ≈ 1 liter	1 cubic foot ≈ 7.5 gallons
1 cubic meter ≈ 250 gallons	1 cubic meter ≈ 5, 55-gallon drums
1 cup ≈ 250 milliliters	1 golf ball ≈ 1 cubic inch

LESSONS OF THE MARS CLIMATE ORBITER

Some of you may have heard that the loss of the Mars Climate Orbiter (MCO) spacecraft in 1999 was due to a unit conversion error. The complete story is rather more complicated and illustrates a valuable lesson in engineering design. Most engineering failures are not due to a single mistake, since built-in redundancies and anticipation of failure modes make this unlikely. Three primary factors (plus bad luck) conspired to send the MCO off course.

First, the spacecraft was asymmetrical, with the body of the spacecraft on one side and a large solar panel on the other. You might think shape is not an issue in the vacuum of space, but in fact it is, and the NASA engineers were well aware of the potential problems. The panel acted like a sail, causing the craft to slowly change its orientation and requiring the MCO to make occasional small corrections by firing thrusters onboard the craft. This was a perfectly manageable "problem."

Second, the software on the spacecraft expected thruster data in SI units, requiring the force expressed in newtons. On the Earth, a separate system calculated and sent instructions to the MCO concerning when and how long to fire its thrusters. The Earth-based system relied on software from an earlier Mars mission, and the thruster equations had to be modified to correct for the thrusters used on the new spacecraft. The original software had been written correctly, with the conversion factor from pound-force to newtons included. However, this conversion was neither documented nor obvious from the code, being buried in the equations. When the equations were rewritten, the programmers were unaware of the conversion factor and it was left out of the new code. This sent incorrect thruster-firing data to the MCO, specifically being too small by a factor of 4.45. This problem alone was manageable by comparing the calculated trajectory with tracking data.

Finally, after the third trajectory correction, the MCO entered "safe mode" while adjusting the solar panel, indicating a fault on the craft. At about the same time, the preliminary indications that the spacecraft trajectory was flawed began to come in. Unfortunately, the engineers spent the next several weeks trying to determine what caused the craft to enter safe mode, falsely assuming the preliminary trajectory data was in error and waiting for longer-term tracking to give a better estimate. In the end, the spacecraft arrived at Mars about 100 kilometers off course.

Here is where the bad luck comes in. Other configurations of the craft or trajectory might have caused the 100 kilometer error to be away from Mars or parallel to the surface, in which case the trajectory could have been corrected later. Unfortunately, the trajectory was 100 kilometers lower than expected, and the MCO was probably destroyed by heating and stresses as it plunged through the Martian atmosphere. Cost: well over $100 million.

7.5 CONVERSIONS INVOLVING "NEW" UNITS

In the past, many units were derived from common physical objects. The "inch" was the width of man's thumb, and the "foot" was the heel-to-toe length of a king's shoe. Obviously, when one king died or was deposed and another took over, the unit of "foot" changed, too. Over time, these units were standardized and have become common terminology.

New units are added as technology evolves; for example, in 1999 the unit of katal was added as an SI derived unit of catalytic activity used in biochemistry. As you proceed in your engineering field, you will be introduced to many "new" units. The procedures discussed here apply to *any* unit in *any* engineering field.

● EXAMPLE 7-9

According to the U.S. Food and Drug Administration (21CFR101.9), the following definition applies for nutritional labeling:

1 fluid ounce means 30 milliliters

Using this definition, how many fluid ounces [fl oz] are in a "U.S. standard" beverage can of 355 milliliters [mL]?

Method	Steps
(1) Term to be converted	355 mL
(2) Conversion formula	1 fl oz = 30 mL
(3) Make a fraction (equal to one) (4) Multiply	$\dfrac{355 \text{ mL}}{} \left\| \dfrac{1 \text{ fl oz}}{30 \text{ mL}}\right.$
(5) Cancel, calculate, be reasonable	11.8 fl oz

● EXAMPLE 7-10

The volume of water in a reservoir or aquifer is often expressed using the unit of acre-foot. A volume of 1 acre-foot is the amount of water covering an area of 1 acre to a depth of 1 foot.

Lake Mead, located 30 miles southeast of Las Vegas, Nevada, is the largest manmade lake in the United States. It holds approximately 28.5 million acre-feet of water behind the Hoover Dam. Convert this volume to units of gallons.

Method	Steps
(1) Term to be converted	28.5×10^6 acre feet
(2) Conversion formula	1 acre = 4,047 m^2 1 m = 3.28 ft 1 m^3 = $100^3 cm^3$ 1,000 cm^3 = 0.264 gal
(3) Make a fraction (4) Multiply	$28.5 \times 10^6 \text{ acre ft} \left\| \dfrac{4,047 \text{ m}^2}{1 \text{ acre}}\right\| \dfrac{1 \text{ m}}{3.28 \text{ ft}} \left\| \dfrac{100^3 \text{ cm}^3}{1 \text{ m}^3}\right\| \dfrac{0.264 \text{ gal}}{1,000 \text{ cm}^3}$
(5) Cancel, calculate, be reasonable	9.3×10^{12} gal

COMPREHENSION CHECK 7-8

A hogshead is a unit of volume describing a large barrel of liquid. Convert 10 hogsheads into units of cubic feet. Conversion factor: 1 hogshead = 63 gallons.

COMPREHENSION CHECK 7-9	A barrel in petroleum engineering is a unit of volume describing a measure of crude oil. Convert 1 barrel into units of cubic meters. Conversion factor: 1 barrel = 42 gallons.

COMPREHENSION CHECK 7-10	A drum is a shipping container used to transport liquids. Convert 5 drums into units of liters. Conversion factor: 1 drum = 55 gallons.

7.6 DERIVED DIMENSIONS AND UNITS

With only the seven base dimensions in the metric system, all measurable things in the known universe can be expressed by various combinations of these concepts. These are called **derived dimensions**. As simple examples, area is length squared, volume is length cubed, and velocity is length divided by time.

As we explore more complex parameters, the dimensions become more complex. For example, the concept of force is derived from Newton's second law, which states that force is equal to mass times acceleration. Force is then used to define more complex dimensions such as pressure, which is force acting over an area, or work, which is force acting over a distance. As we introduce new concepts, we introduce the dimensions and units for each parameter.

Sometimes, the derived dimensions become quite complicated. For example, electrical resistance is mass times length squared divided by both time cubed and current squared. Particularly in the more complicated cases like this, a **derived unit** is defined to avoid having to say things like "The resistance is 15 kilogram-meters squared divided by second cubed ampere squared." It is much easier to say "The resistance is 15 ohms," where the derived unit "ohm" equals one $(kg\ m^2)/(s^3\ A^2)$.

Within this text, dimensions are presented in exponential notation rather than fractional notation. If a dimension is not present in the quantity, it is noted by a zero exponent.

Quantity	Fractional Notation	Exponential Notation
Velocity	$\dfrac{L}{T}$	$M^0\ L^1\ T^{-1}\ \Theta^0$
Acceleration	$\dfrac{L}{T^2}$	$M^0\ L^1\ T^{-2}\ \Theta^0$

Currently, there are officially 22 named derived units in the SI system. All are named after famous scientists or engineers who are deceased. Five of the most common derived units can be found in Table 7.5. It is worth noting that numerous common derived dimensions do not have a corresponding named derived SI unit. For example, there is no named derived SI unit for the derived dimension velocity as there is for force (newton) or electrical resistance (ohm).

Table 7.5 Common derived units in the SI system

Dimension	SI Unit	Base SI Units	Derived from
Force (F)	newton [N]	$1\,N = 1\dfrac{kg\,m}{s^2}$	$F = ma$ Force = mass times acceleration
Energy (E)	joule [J]	$1\,J = 1\,N\,m = 1\dfrac{kg\,m^2}{s^2}$	$E = Fd$ Energy = force times distance
Power (P)	watt [W]	$1\,W = 1\dfrac{J}{s} = 1\dfrac{kg\,m^2}{s^3}$	$P = E/t$ Power = energy per time
Pressure (P)	pascal [Pa]	$1\,Pa = 1\dfrac{N}{m^2} = 1\dfrac{kg}{m\,s^2}$	$P = F/A$ Pressure = force per area
Voltage (V)	volt [V]	$1\,V = 1\dfrac{W}{A} = 1\dfrac{kg\,m^2}{s^3\,A}$	$V = P/I$ Voltage = power per current

- A note of caution: One letter can represent several quantities in various engineering disciplines. For example, the letter "P" can indicate pressure, power, or vertical load on a beam. It is important to examine and determine the nomenclature in terms of the context of the problem presented.
- Always remember to use the units and the symbol in calculations.

COMPREHENSION CHECK 7-11

Complete Table 7-6 by finding the exponents to express the parameters in the form $\{=\}\ M^a\ L^b\ T^c\ \Theta^d$.

Table 7-6 Dimensions of some common parameters

Quantity	Common Units	Exponents M	L	T	Θ
Fundamental Quantities					
Mass	kg	1	0	0	0
Length	m	0	1	0	0
Time	s	0	0	1	0
Temperature	K	0	0	0	1
Geometric Quantities					
Area	ft^2				
	acre	0	2	0	0
Volume	gal				
	L				
Rate Quantities					
Velocity	mi/h	0	1	−1	0
Acceleration	ft/s^2	0	1	−2	0
Flowrate	gal/min				
Evaporation	kg/h				

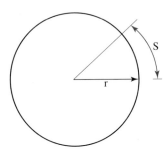

Special Unit: Radian

The derived unit of **radian** is defined as the angle at the center of a circle formed by an arc (S) equal in length to the radius (r) of that circle. In a complete circle there are 2π radians. Since by definition a radian is a length (S) divided by a length (r), it is a dimensionless ratio.

$$1 \text{ radian } [\text{rad}] = S/r$$

Thus, an angle has units, but is dimensionless! In addition to radians, another common unit used for angle is the degree [°]. There are 360° in a complete circle.

360° = 2π radians

7.7 EQUATION LAWS

Equations are mathematical "sentences" composed of "words" (terms) that are strung together with "punctuation marks" (mathematical symbols, such as $+$, $-$, $\times$, $\div$, and $=$). Just as there are rules in the English language that govern how a sentence is structured, there exists a set of "rules" for equations.

Addition and Subtraction

Suppose we are interested in the manufacture and use of sandpaper for furniture construction. We think for a while and then develop a list of the important quantities that affect the final product, along with their respective units and dimensions:

W	Wood removed	[in]	L
R	Roughness diameter	[mm]	L
D	Density of grains	[kg/m³]	$\dfrac{M}{L^3}$
A	Adhesive thickness	[mm]	L
H	How heavy the paper is	[N]	$\dfrac{M\,L}{T^2}$
O	Operation stroke length	[cm]	L
K	Kernel (grain) spacing	[mm]	L

Let us propose a simple equation with only plus and minus signs that could possibly relate several of these parameters. If we are interested in how heavy the product would be, we might assume this would depend on the thickness of the adhesive, the diameter of the roughness, and the grain density. We will try

$$H = A + R + D$$

Each of these terms represents something "real," and consequently we expect that each term can be expressed in terms of fundamental dimensions. Writing the equation in terms of dimensions given:

$$\frac{M\,L}{T^2} = L + L + \frac{M}{L^3}$$

IMPORTANT CONCEPT:
PLUS LAW

Every term being added or subtracted in an equation must have the same dimension.

It is obvious that this is just terrible! We cannot add length and mass or time; as the adage goes, "You can't add apples and oranges!" The same holds true for dimensions. As a result of this observation, we see that this cannot possibly be a valid equation. This gives one important "law" governing equations, the **Plus law**.

Let us try this again with another equation to see if we can determine how effective the sandpaper will be, or how much wood will be removed after each stroke. We might assume this depends on the operation stroke length, the roughness diameter, and the spacing of the grains.

$$W = O + R + K$$

Substituting dimensions,

$$L = L + L + L$$

We see that at least dimensionally, this can be a valid equation, based on the Plus law. Next, units can be inserted to give

$$\text{inches} = \text{centimeters} + \text{millimeters} + \text{millimeters}$$

IMPORTANT CONCEPT:
UNIT LAW

Every term in an equation must have the same units so that the arithmetic operations of addition and subtraction can be carried out.
_____

COROLLARY TO
UNIT LAW

A dimensionally consistent and unit consistent equation is not necessarily a valid equation in terms of physical meaning.

Dimensionally, this equation is fine, but from the perspective of units, we cannot carry out the arithmetic above without first converting all the length dimensions into the same units, such as millimeters. We can state an important result from this observation as well, forming the **Unit law**.

It is important to state a corollary to this observation. If two parameters have the same dimensions and units, it is not always meaningful to add or subtract them.

Two examples show this.

1. If Student A has a mass m_A [kilograms] and Student B a mass m_B [kilograms], then the total mass of both students [kilograms] is the sum of the two masses. This is correct and meaningful in both dimensions and units.
2. Suppose we assume that an equation to predict the mass of a car is this: mass of the car in kilograms = mass of an oak tree in kilograms + mass of an opossum in kilograms. This equation has three terms; all with the dimension of mass and units of kilograms; thus, the terms can be added, although the equation itself is nonsense.

Consequently, the requirement that each term must have the same dimensions and units is a necessary, but not a sufficient, condition for a satisfactory equation.

Multiplication and Division

There are many ways to express the rate at which things are done. Much of our daily life is conducted on a "*per*" or rate basis. We eat 3 meals *per* day, have 5 fingers *per* hand, there are 11 players *per* team in football, 3 feet *per* yard, 4 tires *per* car, 12 fluid ounces *per* canned drink, and 4 people *per* quartet.

Although it is incorrect to add or subtract parameters with different dimensions, it is perfectly permissible to divide or multiply two or more parameters with different dimensions. This is another law of dimensions, the **Per law**.

IMPORTANT CONCEPT:
PER LAW

When parameters are multiplied or divided, the dimensions and units are treated with the same operation rules as numerical values.

When we say 65 miles *per* hour, we mean that we travel 65 miles in 1 hour. We could say we travel at 130 miles per 2 hours, and it would mean the same thing. Either way, this rate is expressed by the "*per*" ratio, distance per time.

One of the most useful applications of your knowledge of dimensions is in helping to determine if an equation is dimensionally correct. This is easy to do and only involves the substitution of the dimensions of every parameter into the equation and simplifying the resulting expressions. A simple application will demonstrate this process.

● **EXAMPLE 7-11**

Is the following equation dimensionally correct?

$$t = \sqrt{\frac{d_{final} - d_{initial}}{0.5a}}$$

where t is time
d is distance
a is acceleration
0.5 is unitless

Determine the dimensions of each parameter:

Acceleration	(a)	$\{=\} L^1 T^{-2}$
Distance	(d)	$\{=\} L$
Time	(t)	$\{=\} T$

Substitute into the equation: $T = \sqrt{[L - L]\left|\frac{T^2}{L}\right.}$

Simplifying $T = \sqrt{L\left|\frac{T^2}{L}\right.} = \sqrt{T^2} = T$

Yes, the equation is dimensionally correct. Both sides of the equation have the same dimensions.

● **EXAMPLE 7-12**

We can use dimensional arguments to help remember formulas. We are interested in the acceleration of a body swung in a circle of radius (r), at a constant velocity (v). We remember that acceleration depends on r and v, and one is divided by the other, but cannot quite remember how. Is the acceleration (a) given by one of the following?

$$a = \frac{v}{r} \quad \text{or} \quad a = \frac{v}{r^2} \quad \text{or} \quad a = \frac{v^2}{r} \quad \text{or} \quad a = \frac{r}{v} \quad \text{or} \quad a = \frac{r}{v^2} \quad \text{or} \quad a = \frac{r^2}{v}$$

Determine the dimensions of each parameter:

Acceleration	(a)	$\{=\} L^1 T^{-2}$
Radius	(r)	$\{=\} L$
Velocity	(v)	$\{=\} L^1 T^{-1}$

Original Equation	Substituting into the Equation	Simplify	Correct?
$a = v/r$	$LT^{-2} = (LT^{-1})\, L^{-1}$	$LT^{-2} = T^{-1}$	No
$a = v/r^2$	$LT^{-2} = (LT^{-1})\, L^{-2}$	$LT^{-2} = L^{-1} T^{-1}$	No
$a = v^2/r$	$LT^{-2} = (LT^{-1})^2\, L^{-1}$	$LT^{-2} = LT^{-2}$	Yes
$a = r/v$	$LT^{-2} = L\, (LT^{-1})^{-1}$	$LT^{-2} = T$	No
$a = r/v^2$	$LT^{-2} = L\, (LT^{-1})^{-2}$	$LT^{-2} = L^{-1} T^2$	No
$a = r^2/v$	$LT^{-2} = L^2\, (LT^{-1})^{-1}$	$LT^{-2} = LT$	No

COMPREHENSION CHECK 7-12

The power absorbed by a resistor can be given by $P = I^2 R$, where P is power in units of watts [W], I is electric current in amperes [A], and R is resistance in ohms [Ω]. Express the unit of ohms in terms of fundamental dimensions.

COMPREHENSION
CHECK 7-13

Weirs are typically installed in open channels such as streams to determine discharge (flow rate). A Cipoletti weir is described by the following equation, where Q is flow rate [gallons per minute, or gpm], L is length [cm], and H is height [cm]. For the following equation to be dimensionally correct, what are the dimensions on the constant value of 3.37?

$$Q = 3.37 \, LH^{3/2}$$

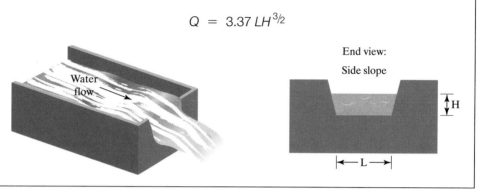

7.8 CONVERSION INVOLVING EQUATIONS

Engineering problems are rarely as simple as converting from one set of units to another. Normally, an equation is involved in the problem solution. To minimize the likelihood of mistakes, we adopt the following procedure for all problems. While this procedure may seem to overanalyze simple problems, it is relatively foolproof and will become more and more useful as the material progresses in difficulty.

Equation Procedure

1. Given a problem, first convert all parameters into base SI units, combinations of these units, or accepted non-SI units. Use the five-step conversion procedure previously described.
2. Perform all necessary calculations, as follows:
   (a) Determine the appropriate equation.
   (b) Insert the known quantities and units. Be sure to carry the units through until the end!
   (c) Calculate the desired quantity and express the answer in "reasonable" terms. This gives the answer in SI units.
3. Convert the final answer to the required units and express the answer in "reasonable" terms.

● EXAMPLE 7-13

On a trip from Alphaville to Betaville, you can take two main routes. Route 1, which goes through Gammatown, is 50 kilometers [km] long; however, you can only drive an average speed of 36 miles per hour [mph]. Route 2 travels along the freeway, at an average speed of 50 mph, but it is 65 km long. How long does it take to complete each route? State the time for each route in minutes [min].

Step One: Convert to Base SI Units		
Method	**Route 1**	**Route 2**
(1) Term to be converted	36 mph	50 mph
(2) Conversion formula (3) Make a fraction (equal to one) (4) Multiply	$\dfrac{36\ mi}{h}\left\vert\dfrac{1\ km}{0.621\ mi}\right.$	$\dfrac{50\ mi}{h}\left\vert\dfrac{1\ km}{0.621\ mi}\right.$
(5) Cancel, calculate, be reasonable	58 km/h	81 km/h
Step Two: Calculate		
Method	**Route 1**	**Route 2**
(1) Determine appropriate equation	Distance = (velocity) (time) which can be rewritten as . . . Time = distance/velocity	
(2) Insert known quantities	$Time = \dfrac{50\ km}{58\ \frac{km}{h}}$	$Time = \dfrac{65\ km}{81\ \frac{km}{h}}$
(3) Calculate, be reasonable	Time = 0.86 h	Time = 0.8 h
Step Three: Convert from Base SI Units to Desired Units		
Method	**Route 1**	**Route 2**
(1) Term to be converted	0.86 h	0.8 h
(2) Conversion formula (3) Make a fraction (equal to one) (4) Multiply	$\dfrac{0.86\ h}{}\left\vert\dfrac{60\ min}{1\ h}\right.$	$\dfrac{0.8\ h}{}\left\vert\dfrac{60\ min}{1\ h}\right.$
(5) Cancel, calculate, be reasonable	52 min	48 min

COMPREHENSION CHECK 7-14

Eclipses, both solar and lunar, follow a cycle of just over 18 years, specifically 6585.32 days. This is called the Saros Cycle. One Saros Cycle after any given eclipse an almost identical eclipse will occur due to fact that the Earth, the Moon, and the Sun are in essentially the same positions relative to each other. The Sun, and the entire solar system, is moving relative to the Cosmic Microwave Background Radiation (the largest detectable frame of reference) at roughly 370 kilometers per second. How far does our solar system travel through the universe in one Saros Cycle? Express your answer in the following units:

(a) meters, with an appropriately chosen prefix;
(b) miles;
(c) astronomical units. Check Table 7-4 for the definition of an AU;
(d) light-years. One light year = 9.46×10^{15} meters.

COMPREHENSION CHECK 7-15

A basketball has a diameter of approximately 27 centimeters. Find the volume of the basketball in units of gallons.

In-Class Activities

ICA 7-1

Complete the following table:

	Meters	Centimeters	Millimeters	Micrometers	Nanometers
Abbreviation	[m]				
Example	9E-08	9E-06	9E-05	0.09	90
(a)		50			
(b)				5	
(c)			300		

ICA 7-2

Complete the following table:

	Grams	Milligrams	Micrograms	Nanograms	Kilograms
Abbreviation	[g]				
(a)					0.4
(b)		20			
(c)				800	
(d)			3		

ICA 7-3

Complete the following table:

	Inches	Feet	Yards	Meters	Miles
Abbreviation	[in]				
(a)		90			
(b)					2
(c)				52	
(d)			53.5		

ICA 7-4

Complete the following table:

	Kilograms	Grams	Pound-Mass	Slugs	Ton
Abbreviation	[kg]				
(a)		700			
(b)					1.5
(c)				75	
(d)			150		

ICA 7-5

Complete the following table:

	Cubic Inch	Fluid Ounces	Gallon	Liter	Cubic Foot
Abbreviation	[in³]				
Example	716	400	3.12	11.8	0.414
(a)					3
(b)				5	
(c)			42		

ICA 7-6

Complete the following table:

	Square Feet	Square Meter	Acre	Square Inch	Square Mile
Abbreviation	[ft²]				
(a)		0.06			
(b)					3
(c)				500	
(d)			1.5		

ICA 7-7

Complete the following table:

	Miles per Hour	Kilometers per Hour	Yards per Minute	Feet per Second
Abbreviation	[mph] or [mi/h]			
(a)		100		
(b)	55			
(c)				110
(d)			2.25	

ICA 7-8

Complete the following table:

	Gallons per Minute	Cubic Feet per Hour	Liters per Second	Fluid Ounces per Day
Abbreviation	[gpm] or [gal/min]			
(a)		15		
(b)	20			
(c)				64
(d)			8	

ICA 7-9

In China, one "bu" is 1.66 meters. A "league" is defined as 3 miles. Convert 2 leagues to units of bu.

ICA 7-10

In China, one "bu" is 1.66 meters. The average height of a human is 5 feet, 7 inches. Convert this height to units of bu.

ICA 7-11

In China, one "cun" is 3.5 centimeters. A "cubit" is defined as 18 inches. Convert 50 cubits to units of cun.

ICA 7-12

In China, one "fen" is defined as 3.3 millimeters. Ten nanometers is the thickness of a cell membrane. Convert 10 nanometers to units of fen.

ICA 7-13

A blink of a human eye takes approximately 300–400 milliseconds. Convert 350 milliseconds to units of shake. One "shake" is equal to 10 nanoseconds.

ICA 7-14

As we all know, everyone gets their fifteen minutes of fame. Convert your 15 minutes of fame to units of shake. One "shake" is equal to 10 nanoseconds.

ICA 7-15

If the SI prefix system was expanded to other units, there would be such definitions as a "millihour," meaning 1/1,000 of an hour. Convert 1 millihour to units of shake. One "shake" is equal to 10 nanoseconds.

ICA 7-16

A "jiffy" is defined as 1/60 of a second. Convert 20 jiffys to units of shake. One "shake" is equal to 10 nanoseconds.

ICA 7-17

A football field is 120 yards long by 53.5 yards wide. Rather than using the traditional MKS (meter–kilogram–second) unit system, an unusual unit system is the FFF system: furlong–firkin–fortnight. One furlong is equal to 201 meters. Calculate the area of the football field, and convert this area into units of furlong squared.

ICA 7-18

A "knot" is a unit of speed in marine travel. One knot is 1.852 kilometers per hour. Rather than using the traditional MKS (meter–kilogram–second) unit system, an unusual unit system is the FFF system: furlong–firkin–fortnight. One furlong is equal to 201 meters and one fortnight is 14 days. Convert the speed of 20 knots to units of furlong per fortnight.

ICA 7-19

The Earth's escape velocity is 7 miles per second. Rather than using the traditional MKS (meter–kilogram–second) unit system, an unusual unit system is the FFF system: furlong–firkin–fortnight. One furlong is equal to 201 meters and one fortnight is 14 days. Convert this to units of furlong per fortnight.

ICA 7-20

A manufacturing process uses 10 pound-mass of plastic resin per hour. Rather than using the traditional MKS (meter–kilogram–second) unit system, an unusual unit system is the FFF system: furlong–firkin–fortnight. One firkin is equal to 40 kilograms and one fortnight is 14 days. Convert this rate to units of firkin per fortnight.

ICA 7-21

A "pinch" is defined as ⅛ of a teaspoon, or 0.616 milliliters [mL]. Convert 1 pinch to units of drops. One "metric drop" is equal to 5 microliters [µL].

ICA 7-22

One metric teaspoon is defined as 5 milliliters [mL]. Convert 2 teaspoons to units of drops. One "metric drop" is equal to 5 microliters [µL].

ICA 7-23

One cup is 250 milliliters [mL]. Convert ½ cup to units of drops. One "metric drop" is equal to 5 microliters [µL].

ICA 7-24

A standard can of soda is 355 milliliters [mL]. Convert 355 milliliters to units of drops. One "metric drop" is equal to 5 microliters [µL].

ICA 7-25

A category F5 tornado can have wind speeds of 300 miles per hour [mph]. What is this velocity in units of meters per second?

ICA 7-26

A new hybrid automobile with regenerative braking has a fuel economy of 55 miles per gallon [mpg] in city driving. What is this fuel economy expressed in units of feet per milliliter?

ICA 7-27

The AbioCor™ artificial heart pumps at a rate of 10 liters per minute. Express this rate in units of gallons per second.

ICA 7-28

If a pump moves water at 2 cubic feet per hour, what is this rate in units of cubic centimeters per second?

ICA 7-29

A circular window has a 10-inch radius. What is the surface area of one side of the window in units of square centimeters?

ICA 7-30

When shipping freight around the world, most companies use a standardized set of containers to make transportation and handling easier. The 40-foot container is the most popular container worldwide. If the container is 2.4 meters wide and has an enclosed volume of 2,385 cubic feet, what is the height of the container in units of inches?

ICA 7-31

A body traveling in a circle experiences an acceleration (a) of $a = v^2/r$, where v is the speed of the body and r is the radius of the circle. We are tasked with designing a large centrifuge to allow astronauts to experience a high "g" forces similar to those encountered on takeoff. One "g" is defined as 9.8 meters per second squared. Design specifications indicate that our design must create at least 5 "g"s. If we use a radius of 30 feet, what is the required speed of the rotating capsule at the end of the arm?

ICA 7-32

We have constructed a model car that can accelerate (a) over time (t) and travel a distance (d). A friend who is knowledgeable in physics tells us that we can compute the distance traveled with the equation $d = \frac{1}{2}at^2$. Given the following values, compute the distance traveled in units of miles: $a = 6.5$ feet per second squared, and $t = 0.25$ hour.

ICA 7-33

Complete the following table, expressing each unit in terms of fundamental dimensions:

	Quantity and/or Unit	Symbol	Exponents			
			M	L	T	Θ
	acre	acre	0	2	0	0
	atmospheres	atm	1	−1	−2	0
(a)	calorie	cal				
(b)	calories per minute	cal/min				
(c)	cubic feet	ft^3				
(d)	day	d				
(e)	degrees Fahrenheit	°F				
(f)	foot pound-force	ft lb$_f$				
(g)	gallon	gal				
(h)	horsepower	hp				
(i)	inches of mercury	in Hg				
(j)	joule	J				
(k)	kelvin	K				
(l)	kilogram	kg				
(m)	liter	L				
(n)	mile	mi				
(o)	newton	N				
(p)	pascal	Pa				
(q)	pound-force	lb$_f$				
(r)	pound-force per square inch	psi				
(s)	pound-mass	lb$_m$				
(t)	slug					
(u)	square feet	ft^2				
(v)	watt	W				

ICA 7-34

For each equation listed, indicate if the equation is a correct mathematical expression based on dimensional considerations. You must show you work to receive credit.

(a) Volume = (mass)/(density)

(b) Area = (acceleration)^2 (time)^3

(c) Speed = (time)^2 (acceleration)

(d) Speed = (mass)$^{1/3}$ ((time) (density)$^{1/3}$)

(e) Acceleration = (velocity)^2/(area)$^{1/2}$

(f) Energy = (mass) (speed) (area)$^{1/2}$

(g) Energy = (pressure) (volume)

(h) Power = (mass) (velocity)/(time)

(i) Time = (area)$^{1/2}$/(velocity)

ICA 7–35

Identify the following quantities through the use of fundamental dimensions. Choose from the multiple list shown below:

(a) Acceleration	**(d)** Power
(b) Energy	**(e)** Pressure
(c) Force	**(f)** Velocity

Quantity
(a) *mgH* where: m = mass [kg] g = gravity [m/s^2] H = height [m]
(b) *P/(mg)* where: m = mass [kg] g = gravity [m/s^2] P = power [W]
(c) *ρQgH* where: ρ = density [kg/m^3] Q = volumetric flowrate [m^3/s] g = gravity [m/s^2] H = height [m]
(d) $\sqrt{\dfrac{E}{H_\rho}}$ where: ρ = density [kg/m^3] E = energy [J] H = height [m]
(e) *mgv* where: m = mass [kg] g = gravity [m/s^2] v = velocity [m/s]
(f) *nRT* where: n = amount [mol] R = ideal gas constant [atm L/(mol K)] T = temperature [K]
(g) *PV* where: P = pressure [Pa] V = volume [m^3]

CHAPTER 7 REVIEW QUESTIONS

1. The Whos Horton heard down in Who-ville love basketball. A regulation Who-ville basketball court is 6 picometers by 10 picometers. What is the area of a regulation Who-ville basketball court in units of square nanometers?

2. The largest hailstone is the United States was 44.5 centimeters in circumference in Coffeyville, Kansas. What is the diameter of the hailstone in units of inches?

3. The largest hailstone is the United States was 44.5 centimeters in circumference in Coffeyville, Kansas. What is the volume of the hailstone in units of liters?

4. How large a surface area in units of square feet will 1 gallon of paint cover if we apply a coat of paint that is 0.1 centimeter thick?

5. How large a surface area in units of square feet will 1 gallon of paint cover if we apply a coat of paint that is 0.1 inches thick?

6. An unmanned X-43A scramjet test vehicle has achieved a maximum speed of Mach number = 9.68 in a test flight over the Pacific Ocean. Mach number is defined as the speed of an object divided by the speed of sound. Assuming the speed of sound is 343 meters per second, what is this record speed expressed in units of miles per hour?

7. One of the National Academy of Engineering Grand Challenges for Engineering is **Provide Access to Clean Water.** Only 5% of water is used for households—the majority is used for agriculture and industry. It takes 240 gallons of water to produce one pound of rice. How many liters of water are needed to produce one kilogram of rice?

8. One of the National Academy of Engineering Grand Challenges for Engineering is **Provide Access to Clean Water.** Only 5% of water is used for households—the majority is used for agriculture and industry. It takes 1,680 gallons of water to produce one pound of grain-fed beef. How many cubic feet of water are needed to produce one kilogram of beef?

9. One of the National Academy of Engineering Grand Challenges for Engineering is **Provide Access to Clean Water.** Only 5% of water is used for households—the majority is used for agriculture and industry. It is estimated that 528 gallons of water are required to produce food for one person for one day. How many liters per year are required to feed one person?

10. A common unit of volume for large water bodies is acre-foot, which is the product of the area of 1 acre multiplied by a depth of 1 foot. If a lake contains 43 million cubic feet of water, what is the volume of the lake expressed in units of acre-foot?

11. New plastic fuel tanks for cars can be molded to many shapes, an advantage over the current metal tanks, allowing manufacturers to increase the tank capacity from 77 liters to 82 liters. What is this increase in gallons?

12. If a liquid evaporates at a rate of 50 kilograms per minute, what is this evaporation rate in units of pounds-mass per second?

13. If a liquid evaporates at a rate of 50 kilograms per minute, what is this evaporation rate in units of slugs per hour?

14. If a liquid evaporates at a rate of 50 kilograms per minute, what is this evaporation rate in units of grams per second?

15. If a pump moves water at 70 gallons per minute, what is the volumetric flow rate in units of cubic inches per second?

16. If a pump moves water at 70 gallons per minute, what is the volumetric flow rate in units of cubic feet per hour?

17. If a pump moves water at 70 gallons per minute, what is the volumetric flow rate in units of liters per day?

18. In an attempt to convert the United States to the metric system, a proposal is introduced to force all car manufacturers to report engine power in units of kilowatts. Convert a 250 horsepower motor rating into kilowatts.

19. The oxgang is unit of area equal to 20 acres. Express an area of 12 oxgangs in units of square meters.

20. The distance travelled in an experimental spacecraft depends on several parameters:
    - The initial position from the starting point: x_0
    - The initial velocity: v_i
    - The rate of acceleration: a
    - The time of travel: t

    The equation that governs the location of the spacecraft is $x = x_0 + v_i t + 1/2\ at^2$.

    If an experimental spacecraft has a head start of 0.3 miles from the starting point, has an initial speed of 50 miles per hour, and accelerates at a constant rate of 10 feet per second squared, how far is it from the starting point after 1.5 hours? Express your answer in units of feet.

21. We know a speed boat can travel at 30 knots. How long (in minutes) will it take to cross the Chesapeake Bay at a place where the bay is 24 miles across? 1 knot = 1 nautical mile per hour; 1 nautical mile = 6,076 feet.

22. In an effort to modernize the United States interstate system, the Department of Transportation proposes to change speed limits from miles per hour to "flashes." A flash is equal to 10 feet per second. On a car speedometer, what will the new range be in units of "flashes" if the old scale was set to a maximum of 120 miles per hour?

23. In many engineering uses, the value of "g," the acceleration due to gravity, is taken as a constant. However, g is actually dependent upon the distance from the center of the Earth. A more accurate expression for g is:

$$g = g_0 \left(\frac{R_e}{R_e + A} \right)^2$$

Here, g_0 is the acceleration of gravity at the surface of the Earth, A is the altitude, and R_e is the radius of the Earth, approximately 6,380 kilometers. What is the value of g at an altitude of 20 miles in units of meters per second squared?

24. In many engineering uses, the value of "g," the acceleration due to gravity, is taken as a constant. However, g is actually dependent upon the distance from the center of the Earth. A more accurate expression for g is:

$$g = g_0 \left(\frac{R_e}{R_e + A} \right)^2$$

Here, g_0 is the acceleration of gravity at the surface of the Earth, A is the altitude, and R_e is the radius of the Earth, approximately 6,380 kilometers. If the value of g is 9 meters per second squared, what is the altitude in units of miles?

25. The Units Society Empire (USE) had defined the following set of "new" units: 1 foot = 10 toes. Convert 45 toes to units of meters.

26. The Units Society Empire (USE) had defined the following set of "new" units: 1 mile = 50 yonders. Convert 500 yards to units of yonders.

27. The Units Society Empire (USE) had defined the following set of "new" units: 1 leap = 4 years. Convert 64 leaps to units of months.

28. The Units Society Empire (USE) had defined the following set of "new" units:

    Length   1 car = 20 feet
    Time   1 class = 50 minutes

    Determine X in the following expression: speed limit, 60 miles per hour = X cars per class.

29. The Units Society Empire (USE) had defined the following set of "new" units:

    Length   1 stride = 1.5 meters
    Time   1 blink = 0.3 second

    Determine X in the following expression: Boeing 747 cruising speed, 550 miles per hour = X strides per blink.

30. The Units Society Empire (USE) had defined the following set of "new" units:

Length   1 stride = 1.5 meters
Time   1 blink = 0.3 second
Mass   1 heavy = 5 kilograms

Determine X in the following expression: force, 1 newton = X heavy stride per blink squared.

31. The Units Society Empire (USE) had defined the following set of "new" units:

Length   1 car = 20 feet
Time   1 class = 50 minutes
Mass   1 light = 2 pound-mass

Determine X in the following expression: force, 1 pound-force = X light car per class squared.

32. A box has a volume of 10 gallons. If two sides of the box measure 2.4 meters × 2.4 feet, what is the length of the third side of the box in units of inches?

33. We turn on our garden hose and point it straight up. It seems reasonable to assume that the height (H) to which the jet of water rises depends on the initial velocity of the water (v_0) and the acceleration due to gravity (g) as expressed by the relationship

$$H = K\frac{v_0^2}{g}$$

The constant (K) is unitless. If the value of K is 25, what initial velocity, in units of meters per second, will give a water height of 0.5 meters?

CHAPTER 8
UNIVERSAL UNITS

In Chapter 7, the concepts of derived dimensions and units were introduced. Five of the most common named units were given in Table 7-5, which is so critical it is repeated here as Table 8-1. Recall that numerous common derived dimensions do not have a corresponding derived SI unit. For example, there is no named SI unit for the derived dimension velocity as there is for force (newton) or electrical resistance (ohm).

The conversion table inside the front cover of this book groups unit conversion factors by dimension. Several fundamental dimensions, such as length and time, were previously discussed in Chapter 7. The remaining derived dimensions shown in the table are discussed below, such as force, along with several named derived dimensions that do not have corresponding simple unit conversions, such as density.

Table 8-1 Common derived units in the SI system

Dimension	SI Unit	Base SI Units	Derived from
Force (F)	newton [N]	$1\,\text{N} = 1\,\dfrac{\text{kg m}}{\text{s}^2}$	$F = ma$ Force = mass times acceleration
Energy (E)	joule [J]	$1\,\text{J} = 1\,\text{N m} = 1\,\dfrac{\text{kg m}^2}{\text{s}^2}$	$E = Fd$ Energy = force times distance
Power (P)	watt [W]	$1\,\text{W} = 1\,\dfrac{\text{J}}{\text{s}} = 1\,\dfrac{\text{kg m}^2}{\text{s}^3}$	$P = E/t$ Power = energy per time
Pressure (P)	pascal [Pa]	$1\,\text{Pa} = 1\,\dfrac{\text{N}}{\text{m}^2} = 1\,\dfrac{\text{kg}}{\text{m s}^2}$	$P = F/A$ Pressure = force per area
Voltage (V)	volt [V]	$1\,\text{V} = 1\,\dfrac{\text{W}}{\text{A}} = 1\,\dfrac{\text{kg m}^2}{\text{s}^3\,\text{A}}$	$V = P/I$ Voltage = power per current

8.1 FORCE

When you push a grocery cart, it moves. If you keep pushing, it keeps moving. The longer you push, the faster it goes; the velocity increases over time, meaning that it accelerates. If you push a full grocery cart that has a high mass, it does not speed up as much, meaning it accelerates less than a cart with low mass. Simply put, the acceleration (a) of a body depends on the force (F) exerted on it and its mass (m). This is a simple form of "Newton's second law of motion" and is usually written as $F = ma$.

Table 8-2 Dimensions of force

Quantity	Common Units	Exponents			
		M	**L**	**T**	**Θ**
Force	N	1	1	–2	0

The SI unit of force, the **newton** [N], is defined as the force required to accelerate a mass of one kilogram at a rate of one meter per second squared (see Table 8-2). It is named for Sir Isaac Newton (1643–1727). Newton's *Principia* is considered one of the world's greatest scientific writings, explaining the law of universal gravitation and the three laws of motion. Newton also developed the law of conservation of momentum, the law of cooling, and the reflecting telescope. He shares credit for the development of calculus with Gottfried Leibniz.

In the SI system, mass, length, and time are base units and force is a derived unit; force is found from combining mass, length, and time using Newton's second law. The SI system is called "coherent," because the derived unit is set at one by combing base units. The AES system is considered non-coherent as it uses units that do not work together in the same fashion as the SI units do. There are two uses of the term "pound" in the AES system, which occurred in common usage long before Newton discovered gravity. To distinguish mass in pounds and force in pounds, the unit of mass is given as pound-mass (lb_m) and the unit of force is given as pound-force (lb_f). One pound-force is the amount of force needed to accelerate one pound-mass at a rate of 32.2 feet per second squared. Since this relationship is not easy to remember or use in conversions, we will stick with SI units for problem solving, following the procedure discussed in Chapter 7.

● EXAMPLE 8-1

A professional archer is designing a new longbow with a full draw weight of 63 pounds-force [lb_f]. The draw weight is the amount of force needed to hold the bowstring at a given amount of draw, or the distance the string has been pulled back from the rest position. What is the full draw weight of this bow in units of newtons [N]?

Method	Steps
(1) Convert term	63 lb_f
(2) Apply conversion formula	1 N = 0.225 lb_f
(3) Make a fraction	$\dfrac{63\ lb_f}{} \bigg\vert \dfrac{1\ N}{0.225\ lb_f}$
(4) Multiply	
(5) Cancel, calculate, be reasonable	280 N

8.2 WEIGHT

The **mass** of an object is a fundamental dimension. Mass is a quantitative measure of how much of an object there is, or in other words, how much matter it contains. The **weight** (w) of an object is a force equal to the mass of the object (m) times the acceleration of **gravity** (g).

While mass is independent of location in the universe, weight is dependent upon both mass and gravity (Table 8-3).

On the Earth, the pull of gravity is approximately 9.8 meters per second squared [m/s²]. On the moon, gravity is approximately one-sixth this value, or 1.6 m/s². A one kilogram [kg] object acted on by Earth's gravity would have a weight of 9.8 N, but on the moon it would have a weight of 1.6 N. Unless otherwise stated, assume all examples take place on the Earth.

Table 8-3 Dimensions of weight

Quantity	Common Units	Exponents			
		M	**L**	**T**	**Θ**
Weight	N	1	1	−2	0

DEVILISH DERIVATION

- **Mass of an object:** A quantitative measure of how much of an object there is.
- **Weight of an object:** A quantitative measure of the force exerted on the object due to gravity; we are concerned about the force between the Earth (subscript e) and a body on the Earth (or another planet).

  Newton's law of universal gravitation

$$F = G \frac{mm_e}{R^2}$$

where:
  G is universal gravitational constant

$$G = 6.673 \times 10^{-11} \, (\mathrm{N\,m^2})/\mathrm{kg^2}$$

  m is the mass
  R is the distance between the centers of mass of the two bodies

On the Earth, the distance between the center of a body and the center of the earth is approximately the radius of the Earth, R_e.

The mass of one of the bodies (Earth) is the mass of the Earth (with an "e" subscript). Rewrite the equation:

$$F = m\left[G \frac{m_e}{R_e^2}\right]$$

The quantity in square brackets is a constant (call it "g"). We call the force "the weight (w) of the body." So,

$$w = mg$$

This is the common equation that relates weight and mass. The value for g is calculated to be 9.8 meters per second squared [m/s²], or 32.2 feet per second squared [ft/s²]. Note that g has the units of acceleration.

● **EXAMPLE 8-2**   What is the weight of a 225-kilogram [kg] bag of birdseed in units of newtons [N]?

Step One: Convert to Base SI Units
No conversion necessary

Step Two: Calculate	
Method	**Steps**
(1) Determine appropriate equation	$w = mg$
(2) Insert known quantities	$w = \dfrac{225 \text{ kg}}{} \left\| \dfrac{9.8 \text{ m}}{s^2}\right.$
(3) Calculate, be reasonable	$w = 2{,}205 \dfrac{\text{kg m}}{s^2}$

This is apparently our final answer, but the units are puzzling. If the unit of force is the newton, and if this is a valid equation, then our final result for force should be newtons. If we consider the dimensions of force:

		Exponents			
Quantity	**Common Units**	**M**	**L**	**T**	**Θ**
Force	N	1	1	−2	0

A unit of force has dimensions $F\{=\}ML/T^2$, *which in terms of base SI units would be* $F[=]\text{kg m/s}^2$. *As this term occurs so frequently it is given the special name "newton" (see Table 8-1). Anytime we see the term* $\left[\text{kg m/s}^2\right]$, *we know we are dealing with a force equal to a newton.*

(3) Calculate, be reasonable	$w = 2{,}205 \dfrac{\text{kg m}}{s^2} \left\| \dfrac{1 \text{ N}}{\frac{1 \text{ kg m}}{s^2}}\right. = 2{,}205 \text{ N}$

Step Three: Convert from Base SI Units to Desired Units
No conversion necessary

COMPREHENSION CHECK 8-1

The mass of the human brain is 1,360 grams. State the weight of the human brain in units of newtons on the Earth.

COMPREHENSION CHECK 8-2

The mass of the human brain is 1,360 grams. State the weight of the human brain in units of newtons on the moon. The gravity on the moon is 1.6 meters per second squared.

COMPREHENSION CHECK 8-3   What is the weight of a 5-kilogram bowling ball in units of pounds-force?

8.3 DENSITY

Density (ρ, Greek letter rho) is the mass of an object (m) divided by the volume the object occupies (V). Density should not be confused with weight—think of the old riddle: which weighs more, a pound of feathers or a pound of bricks? The answer is they both weigh the same amount, one pound, but the density of each is different. The bricks have a higher density than the feathers, since the same mass takes up less space.

Specific weight (γ, Greek letter gamma) is the weight of an object (w) divided by the volume the object occupies (V) (Table 8-4).

Table 8-4 Dimensions of density and specific weight

	Common	Exponents			
Quantity	Units	M	L	T	Θ
Density	kg/m^3	1	−3	0	0
Specific weight	N/m^3	1	2	2	0

● **EXAMPLE 8-3**   The density of sugar is 1.61 grams per cubic centimeter [g/cm^3]. What is the density of sugar in units of pound-mass per cubic foot [lb_m/ft^3]?

NOTE

Upon conversion from units of grams per cubic centimeter to pound-mass per cubic foot, the answer should be ≈ 60 times larger.

Method	Steps
(1) Term to be converted	1.61 g/cm^3
(2) Conversion formula	
(3) Make fractions	$1.61\dfrac{g}{cm^3}\left|\dfrac{2.205\ lb_m}{1,000\ g}\right|\dfrac{1,000\ cm^3}{0.0353\ ft^3}$
(4) Multiply	
(5) Cancel, calculate, be reasonable	101 lb_m/ft^3

● **EXAMPLE 8-4**   The density of a biofuel blend is 0.72 grams per cubic centimeter [g/cm^3]. What is the density of the biofuel in units of kilograms per cubic meter [kg/m^3]?

NOTE

Upon conversion from units of grams per cubic centimeter to kilograms per cubic meter, the answer should be 1,000 times larger.

Method	Steps
(1) Term to be converted	0.72 g/cm^3
(2) Conversion formula	
(3) Make fractions	$0.72\dfrac{g}{cm^3}\left|\dfrac{1\ kg}{1,000\ g}\right|\dfrac{(100\ cm)^3}{1\ m^3}$
(4) Multiply	
(5) Cancel, calculate, be reasonable	720 kg/m^3

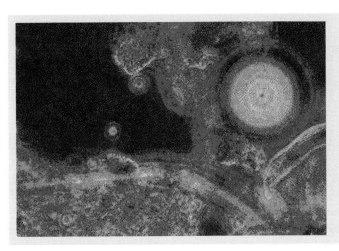

A vast array of valuable compounds can be formed by microbial cultures. Oil produced by the fungi *Pythium irregulare* can be extracted and used for biofuels or pharmaceutical compounds. Biosystems engineers culture the microorganism, design the bioreactor, and extract the valuable compounds using sustainable, ecoprocessing techniques.

Photo courtesy of C. Drapcho

● EXAMPLE 8-5

What is the weight of water, in units of pounds-force [lb_f], in a 55-gallon drum completely full? Assume the density of water to be 1 gram per cubic centimeter. Ignore the weight of the drum.

Step One: Convert to Base SI Units		
Method	**Steps**	
(1) Term to be converted	55 gal	1 g/cm^3
(2) Conversion formula (3) Make fractions (4) Multiply	$\dfrac{55 \text{ gal}}{} \left\lvert \dfrac{1{,}000 \text{ cm}^3}{0.264 \text{ gal}} \right\rvert \dfrac{1 \text{ m}^3}{100^3 \text{ cm}^3}$	$\dfrac{1 \text{ g}}{\text{cm}^3} \left\lvert \dfrac{1 \text{ kg}}{1{,}000 \text{ g}} \right\rvert \dfrac{100^3 \text{ cm}^3}{1 \text{ m}^3}$
(5) Cancel, calculate, be reasonable	0.208 m^3	1,000 kg/m^3

Step Two: Calculate	
Method	**Steps**
(1) Determine appropriate equation	$w = mg$
(2) Insert known quantities	$w = \dfrac{m}{} \left\lvert \dfrac{9.8 \text{ m}}{\text{s}^2} \right.$

For Unknown Quantities, Repeat the Process	
Method	**Steps**
(1) Determine appropriate equation	$m = \rho V$
(2) Insert known quantities	$m = \dfrac{1{,}000 \text{ kg}}{\text{m}^3} \left\lvert 0.208 \text{ m}^3 \right.$
(3) Calculate, be reasonable	$m = 208 \text{ kg}$
(2) Insert known quantities	$w = \dfrac{208 \text{ kg}}{} \left\lvert \dfrac{9.8 \text{ m}}{\text{s}^2} \right.$
(3) Calculate, be reasonable	$w = 2{,}038 \dfrac{\text{kg m}}{\text{s}^2} \left\lvert \dfrac{1 \text{ N}}{\frac{1 \text{ kg m}}{\text{s}^2}} \right. = 2{,}038 \text{ N}$

Step Three: Convert from Base SI Units to Desired Units	
Method	**Steps**
(1) Term to be converted	2,038 N
(2) Conversion formula (3) Make a fraction (4) Multiply	$\dfrac{2{,}038 \text{ N} \mid 0.225 \text{ lb}_f}{\mid 1 \text{ N}}$
(5) Cancel, calculate, be reasonable	460 lb$_f$

Specific Gravity

IMPORTANT CONCEPT

$$SG = \frac{\rho_{object}}{\rho_{water}}$$

In technical literature, density is rarely given; instead, the **specific gravity** is reported. The specific gravity (SG) of an object is a dimensionless ratio of the density of the object to the density of water (see Table 8-5). It is convenient to list density in this fashion so *any* unit system may be applied by our choice of the units of the density of water. The specific gravities of several common substances are listed in Table 8-6.

Table 8-5 Dimensions of specific gravity

		Exponents			
Quantity	**Common Units**	**M**	**L**	**T**	**(·)**
Specific gravity	–	0	0	0	0

IMPORTANT CONCEPT

Density of water

$= 1 \text{ g/cm}^3$

$= 1 \text{ kg/L}$

$= 1{,}000 \text{ kg/m}^3$

$= 62.4 \text{ lb}_m/\text{ft}^3$

$= 1.94 \text{ slug/ft}^3$

Table 8-6 Specific gravity values for common substances

Liquids	SG	Solids	SG
Acetone	0.785	Aluminum	2.70
Benzene	0.876	Baking soda	0.689
Citric acid	1.67	Brass	8.40–8.75
Gasoline	0.739	Concrete	2.30
Glycerin	1.26	Copper	8.96
Iodine	4.93	Gallium	5.91
Mercury	13.6	Gold	19.3
Mineral oil	0.900	Graphite	2.20
Olive oil	0.703	Iron	7.87
Propane	0.806	Lead	11.4
Sea water	1.03	Polyvinyl chloride (PVC)	1.38
Toluene	0.865	Silicon	2.33
Water	1.00	Zinc oxide	5.60

When calculating or considering specific gravities, it is helpful to keep in mind the range of values that you are likely to have.

The densest naturally occurring elements at normal temperature and pressure are osmium and iridium, both with a specific gravity close to 22.6. The *densest substances that a normal person is likely to encounter are platinum (SG = 21.5) and gold (SG = 19.3).* Thus, if you calculate a specific gravity to be higher than about 23, you have almost certainly made an error.

Most liquids are similar to water, with a specific gravity around 1. One notable exception is mercury, with a specific gravity of 13.

On the lower end of the scale, the *specific gravity of air is about 0.001,* whereas hydrogen has a specific gravity of slightly less than 0.0001.

Therefore, if you get a specific gravity value less than about 10^{-4}, you need to check your work very carefully.

● **EXAMPLE 8-6**

The specific gravity of butane is 0.599. What is the density of butane in units of kilograms per cubic meter?

Step One: Convert to Base SI Units	
No conversion needed	

Step Two: Calculate	
Method	**Steps**
(1) Determine appropriate equation	$\rho_{object} = (SG)(\rho_{water})$
(2) Insert known quantities	$\rho_{object} = (0.599)\left(1{,}000\ \dfrac{kg}{m^3}\right)$
(3) Calculate, be reasonable	$\rho_{object} = 599\ \dfrac{kg}{m^3}$

Step Three: Convert from Base SI Units to Desired Units	
No conversion needed	

● **EXAMPLE 8-7**

Mercury has a specific gravity of 13.6. What is the density of mercury in units of slugs per cubic foot?

Step One: Convert to Base SI Units	
No conversion needed	

Step Two: Calculate	
Method	**Steps**
(1) Determine appropriate equation	$\rho_{object} = (SG)(\rho_{water})$
(2) Insert known quantities	$\rho_{object} = (13.6)\left(62.4\ \dfrac{lb_m}{ft^3}\right)$
(3) Calculate, be reasonable	$\rho_{object} = 848.64\ \dfrac{lb_m}{ft^3}$

Step Three: Convert from Base SI Units to Desired Units	
Method	**Steps**
(1) Term to be converted	848.64 lb_m/ft^3
(2) Conversion formula	
(3) Make a fraction	$\dfrac{848.64\ lb_m}{ft^3} \left\| \dfrac{1\ slug}{32.2\ lb_m}\right.$
(4) Multiply	
(5) Cancel, calculate, be reasonable	26.4 $slug/ft^3$

COMPREHENSION CHECK 8-4

Convert 50 grams per cubic centimeter into units of pounds-mass per cubic foot.

COMPREHENSION CHECK 8-5

A 75-gram cylindrical rod is measured to be 10 centimeters long and 2.5 centimeters in diameter. What is the specific gravity of the material?

8.4 AMOUNT

Some things are really very large and some are very small. Stellar distances are so large that it becomes inconvenient to report values such as 235 trillion miles, or 6.4×10^{21} feet when we are interested in the distance between two stars or two galaxies. To make things better, we use a new unit of length that itself is large—the distance that light goes in a year; this is a very long way, 3.1×10^{16} feet. As a result, we do not have to say that the distance between two stars is 620,000,000,000,000,000 feet, we can just say that they are 2 light-years apart.

This same logic holds when we want to discuss very small things such as molecules or atoms. Most often we use a constant that has been named after Amedeo Avogadro, an Italian scientist (1777–1856) who first proposed the idea of a fixed ratio between the amount of substance and the number of elementary particles. The Avogadro constant has a value of 6.022×10^{23} particles per mole. If we have 12 of something, we call it a dozen. If we have 20, it is a score. If we have 6.022×10^{23} of anything, we have a mole. If we have 6.022×10^{23} baseballs, we have a mole of baseballs. If we have 6.022×10^{23} elephants, we have a mole of elephants, and if we have 6.022×10^{23} molecules, we have a mole of molecules. Of course, the mole is never used to define amounts of macroscopic things like elephants or baseballs, being relegated to the realm of the extremely tiny. In the paragraphs below we will see how this rather odd value originated and how this concept simplifies our calculations.

The mass of a nucleon (neutron or proton) is about 1.66×10^{-24} grams. To avoid having to use such tiny numeric values when dealing with nucleons, physicists defined the **atomic mass unit** [amu] to be approximately the mass of one nucleon. Technically, it is defined as one-twelfth of the mass of a carbon twelve atom. In other words, 1 amu = 1.66×10^{-24} g. The symbol "u" is often used for amu, which is also known as a **Dalton** [Da].

If there is $(1.66 \times 10^{-24}$ g)/(1 amu), then there is (1 amu)/$(1.66 \times 10^{-24}$ g). Dividing this out gives 6.022×10^{23} amu/g. This numeric value is used to define the **mole** [mol]. One mole of a substance (usually an element or compound) contains exactly 6.022×10^{23} fundamental units (atoms or molecules) of that substance. In other words, there are 6.022×10^{23} fundamental units per mole. This is often written as

$$N_A = 6.022 \times 10^{23} \text{ mol}^{-1}$$

As mentioned above, this is called Avogadro's constant or **Avogadro's number**, symbolized by N_A. So why is this important? Consider combining hydrogen and oxygen to get water (H_2O). We need twice as many atoms of hydrogen as atoms of oxygen for this reaction; thus, for every mole of oxygen, we need two moles of hydrogen, since one mole of anything contains the same number of fundamental units, atoms in this case.

The problem is that it is difficult to measure a substance directly in moles, but it is easy to measure its mass. *Avogadro's number affords a conversion path between moles and mass.* Consider hydrogen and oxygen in the above. The atomic mass of an atom in **atomic mass units** [amu] is approximately equal to the number of nucleons it contains. Hydrogen contains one proton, and thus has an atomic mass of 1 amu. We can also say that there is 1 amu per hydrogen atom. Oxygen has an atomic mass of 16; thus, there are 16 amu per oxygen atom. Since atomic mass refers to an individual specific atom, the term **atomic weight** is used, representing the average value of all isotopes of the element. This is the value commonly listed on periodic tables.

Let us use this information, along with Avogadro's number, to determine the mass of one mole of each of these two elements.

> **NOTE**
>
> If Element Z has an atomic mass of X amu, there are X grams per mole of Element Z.

$$\text{Hydrogen:} \quad \frac{1 \text{ amu}}{\text{H atom}} \left| \frac{1 \text{ g}}{6.022 \times 10^{23} \text{ amu}} \right| \frac{6.022 \times 10^{23} \text{atom}}{1 \text{ mol}} = \frac{1 \text{ g}}{1 \text{ mol H}}$$

$$\text{Oxygen:} \quad \frac{16 \text{ amu}}{\text{O atom}} \left| \frac{1 \text{ g}}{6.022 \times 10^{23} \text{ amu}} \right| \frac{6.022 \times 10^{23} \text{ atom}}{1 \text{ mol}} = \frac{16 \text{ g}}{1 \text{ mol O}}$$

The numerical value for the atomic mass of a substance is the same as the number of grams in one mole of that substance, often called the **molar mass**.

$$\text{Atomic weight} = \text{molar mass}$$

> **NOTE**
>
> If Molecule AB has a molecular weight of X amu, there are X grams per mole of Molecule AB.

Avogadro's number is the link between the two. Hydrogen has a molar mass of 1 gram per mole; oxygen has a molar mass of 16 grams per mole.

When groups of atoms react together, they form molecules. Consider combining hydrogen and oxygen to get water (H_2O). Two atoms of hydrogen combine with one atom of oxygen, so 2 * 1 amu H + 16 amu O = 18 amu H_2O. The **molecular mass** of water is 18 amu. By an extension of the example above, we can also state that one mole of water has a mass of 18 grams, called the **formula weight**.

$$\text{molecular weight} = \text{formula weight}$$

The difference between these ideas is summarized in Table 8-7.

This text assumes that you have been exposed to these ideas in an introductory chemistry class and so does not cover them in any detail. In all problems presented, you will be given the atomic weight of the elements or the formula weight of the molecule, depending on the question asked. This topic is briefly introduced because Avogadro's number (N_A) is important in the relationship between several constants, including the following:

- The gas constant (R [=] J/(mol K)) and the Boltzmann constant (k [=] J/K), which relates energy to temperature: $R = kN_A$.

- The elementary charge (e [=] C) and the Faraday constant (F [=] C/mol), which is the electric charge contained in one mole of electrons: $F = eN_A$.
- An electron volt [eV] is a unit of energy describing the amount of energy gained by one electron accelerating through an electrostatic potential difference of one volt: $1\ eV = 1.602 \times 10^{-19}$ J.

Table 8-7 Definitions of "amount" of substance

The quantity . . .	measures the . . .	in units of . . .	and is found by . . .
Atomic mass	Mass of one atom of an individual isotope of an element	[amu]	Direct laboratory measurement
Atomic weight	Average mass of all isotopes of an element	[amu]	Listed on Periodic Table
Molar mass	Mass of one mole of the atom	[g/mol]	Listed on Periodic Table
Molecular mass or molecular weight	Sum of average weight of isotopes in molecule	[amu]	Combining atomic weights of individual atoms represented in the molecule
Formula weight	Mass of one mole of the molecule	[g/mol]	Combining molar mass of individual atoms represented in the molecule

● **EXAMPLE 8-8**

Let us return to the problem of combining hydrogen and oxygen to get water. Assume you have 50 grams of oxygen with which you want to combine the proper mass of hydrogen to convert it completely to water. The atomic weight of hydrogen is 1 and the atomic weight of oxygen is 16.

First determine how many moles of oxygen are present.

$$\frac{50\ g\ O}{} \left| \frac{1\ mol\ O}{16\ g\ O} \right. = 3.125\ mol\ O$$

We need twice as many moles of hydrogen as oxygen (H_2O), so we need 6.25 moles of hydrogen. Converting to mass gives

$$\frac{6.25\ mol\ H}{} \left| \frac{1\ g\ H}{1\ mol\ H} \right. = 6.25\ g\ H$$

● **EXAMPLE 8-9**

Acetylsalicylic acid (aspirin) has the chemical formula $C_9H_8O_4$. How many moles of aspirin are in a 1-gram dose? Use the following facts:

- Atomic weight of carbon = 12
- Atomic weight of hydrogen = 1
- Atomic weight of oxygen = 16

First, determine how many grams are in 1 mole of aspirin (determine formula weight).

$$\text{FW of aspirin} = \left[\frac{12\frac{g}{mole}}{1 \text{ molecule C}} \middle| \frac{9 \text{ C molecules}}{} \right] + \left[\frac{1\frac{g}{mole}}{1 \text{ molecule H}} \middle| \frac{8 \text{ H molecules}}{} \right]$$

$$+ \left[\frac{16\frac{g}{mole}}{1 \text{ molecule O}} \middle| \frac{4 \text{ O molecules}}{} \right] = 180 \frac{g}{mole}$$

Finally, convert to moles per dose.

$$\frac{1 \text{ g aspirin}}{\text{dose}} \middle| \frac{1 \text{ mol aspirin}}{180 \text{ g aspirin}} = 5.56 \times 10^{-3} \frac{\text{mol aspirin}}{\text{dose}}$$

● EXAMPLE 8-10

Many gases exist as diatomic compounds in nature, meaning two of the atoms are attached to form a molecule. Hydrogen, oxygen, and nitrogen all exist in a gaseous diatomic state under standard conditions.

Assume there are 100 grams of nitrogen gas in a container. How many moles of nitrogen (N_2) are in the container? Atomic weight of nitrogen = 14.

First, determine how many grams are in 1 mole of nitrogen (determine the formula weight).

$$\text{FW of } N_2 = \frac{14\frac{g}{mol}}{1 \text{ mol N}} \middle| \frac{2 \text{ mol N}}{} = 28 \frac{g}{mol}$$

Next, convert mass to moles.

$$\frac{100 \text{ grams of } N_2}{} \middle| \frac{\text{mole}}{28 \text{ gram}} = 3.57 \text{ moles } N_2$$

8.5 **TEMPERATURE**

Temperature was originally conceived as a description of energy: heat (thermal energy) flows spontaneously from "hot" to "cold." But how hot is "hot"? The thermometer was devised as a way to measure the "hotness" of an object. As an object gets warmer, it usually expands. In a thermometer, a temperature is a level of hotness that corresponds to the length of the liquid in the tube. As the liquid gets warmer, it expands and moves up the tube. To give temperature a quantitative meaning, numerous temperature scales have been developed.

Many scientists, including Isaac Newton, have proposed temperature scales. Two scales were originally developed about the same time—**Fahrenheit** [°F] and **Celsius** [°C]—and have become widely accepted in laymen use. These are the most frequently used temperature scales by the general public. Gabriel Fahrenheit (1686–1736), a German physicist and engineer, developed the Fahrenheit scale in 1708. Anders Celsius (1701–1744), a Swedish astronomer, developed the Celsius scale in 1742. The properties of each scale are in Table 8-8.

You may wonder why the Celsius scale seems so reasonable, and the Fahrenheit scale so random. Actually, Mr. Fahrenheit was just as reasonable as Mr. Celsius. Mr. Celsius set the freezing point of water to be 0 and the boiling point to be 100. Mr. Fahrenheit took as 0 a freezing mixture of salt and ice, and as 100 body tempera-ture. With this scale, it just so happens that the freezing and boiling points of water work out to be odd numbers.

Table 8-8 Properties of water

Scale	Freezing Point	Boiling Point	Divisions Between Freezing and Boiling
Fahrenheit [°F]	32	212	180
Celsius [°C]	0	100	100
Kelvin [K]	273	373	100
Rankine [°R]	492	672	180

Some units can cause confusion in conversion. One of those is temperature. One reason for this is that we use temperature in two different ways: (1) reporting an actual temperature value and (2) discussing the way a change in temperature affects a material property. To clarify, we resort to examples.

Conversion Between Temperature Values

IMPORTANT CONCEPT

When actual temperature readings are converted:

$$\frac{T[°F] - 32}{180} = \frac{T[°C] - 0}{100}$$

$$T[K] = T[°C] + 273$$

$$T[°R] = T[°F] + 460$$

When an actual temperature reading is reported, such as "the temperature in this room is 70°F," how do we convert from one temperature unit to another? The scales have different zero points, so they cannot be converted with a single conversion factor as done previously but require a conversion formula. Most of you are familiar with the conversion between Fahrenheit and Celsius, but this equation is cumbersome to remember.

$$T[°F] = \frac{9}{5} T[°C] + 32$$

Let us imagine we have two thermometers, one with the Fahrenheit scale and the other with the Celsius scale. We set two thermometers side by side so that the freezing point and the boiling point of water are at the same location on both thermometers. We are interested in the relationship between these two scales. From this figure we see that the fraction of the distance from the freezing point to the boiling point in both scales is the same. This means that we can write

$$\frac{T[°F] - 32}{212 - 32} = \frac{T[°C] - 0}{100 - 0}$$

This relationship is really all we need to know to relate a temperature in degrees Fahrenheit to one in degrees Celsius. You can easily do the algebra to convert from Fahrenheit to Celsius, or vice versa. By remembering this form, you do not have to remember if the conversion is 9/5 or 5/9, or to add or subtract 32. This formula is determined by the method of interpolation.

There are numerous other temperature scales, but two are worth mentioning: **kelvin** [K] and **degrees rankine** [°R]. The kelvin scale is named for First Baron William Thomson Kelvin (1824–1907), an English mathematician and physicist. Kelvin first proposed the idea of "infinite cold," or absolute zero, in 1848, using the Celsius scale for comparison. The Rankine scale is named for William J. M. Rankine (1820–1872), a Scottish engineer and physicist, who proposed

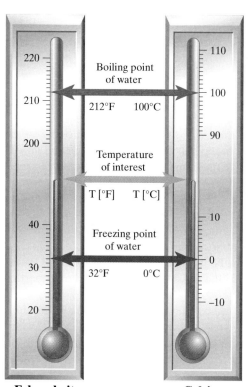

Fahrenheit   **Celsius**

an analogy to the kelvin scale, using the Fahrenheit scale. Both men made significant contributions to the field of thermodynamics.

The kelvin and Rankine scales are "absolute," which means that at absolute zero, the temperature at which molecules have minimum possible motion, the temperature is zero. Absolute temperature scales therefore have no negative values. In the kelvin scale, the degree sign is not used; it is simply referred to as "kelvin," not "degrees kelvin." It is the base SI unit for temperature and the most frequently used temperature unit in the scientific community.

IMPORTANT CONCEPT

Absolute temperature scales are never negative.

● **EXAMPLE 8-11**

The hottest temperature in the United States ever recorded by the National Weather Service, 56.7 degrees Celsius [°C], occurred in Death Valley, California, on July 10, 1913. State this value in units of degrees Fahrenheit [°F].

Step One: Convert to Base SI Units	
No conversion needed	
Step Two: Calculate	
Method	**Steps**
(1) Determine appropriate equation	$\dfrac{T[°F] - 32}{212 - 32} = \dfrac{T[°C] - 0}{100 - 0}$
(2) Insert known quantities	$\dfrac{T[°F] - 32}{180} = \dfrac{56.7}{100}$
(3) Calculate, be reasonable	$T = 134°F$
Step Three: Convert from Base SI Units to Desired Units	
No conversion needed	

Conversions Involving Temperature Within a Material Property

IMPORTANT CONCEPT

When properties that contain temperature are converted:

$$\frac{1°C}{1.8°F} \quad \frac{1K}{1°C} \quad \frac{1°R}{1°F}$$

For this type of conversion, we read the units under consideration as "*per* degree Fahrenheit," with the clue being the word "*per*."

When considering how a change in temperature affects a material property, we use a scalar conversion factor. In general, we encounter this in sets of units relating to the property of the material; for example, the units of the thermal conductivity are given by W/m K, which is read as "watts per meter kelvin." When this is the case, we are referring to the size of the degree, not the actual temperature.

To find this relationship, remember that between the freezing point and the boiling point of pure water, the Celsius scale contains 100 divisions, whereas the Fahrenheit scale contains 180 divisions. The conversion factor between Celsius and Fahrenheit is $100°C \equiv 180°F$, or $1°C \equiv 1.8°F$.

● **EXAMPLE 8-12**

The specific heat (C_p) is the ability of an object to store heat. Specific heat is a material property, and values are available in technical literature. The specific heat of copper is 0.385 J/(g °C), which is read as "joules per gram degree Celsius." Convert this to units of J/(lb$_m$°F), which reads "joules per pound-mass degree Fahrenheit."

Method	Steps
(1) Term to be converted	$0.385 \dfrac{J}{g\,°C}$
(2) Conversion formula (3) Make a fraction (4) Multiply	$\dfrac{0.385\ J}{g\,°C} \left\vert \dfrac{1{,}000\ g}{2.205\ lb_m} \right\vert \dfrac{1°C}{1.8°F}$
(5) Cancel, calculate, be reasonable	$97 \dfrac{J}{lb_m\,°F}$

A note of clarification about the term "PER": When reading the sentence: "Gravity on earth is commonly assumed to be 9.8 meters per second squared," there is often little confusion in translating the words to symbols: $g = 9.8\ m/s^2$. For a more complex unit however, this can present a challenge. For example, the sentence "The thermal conductivity of aluminum is 237 calories per hour meter degree Celsius," can be confusing because it can be interpreted as:

$$k = 237 \frac{cal}{h\,m\,°C} \quad \text{or} \quad k = 237 \frac{cal}{h}(m\,°C) \quad \text{or} \quad k = 237 \frac{cal}{h\,m}°C$$

Officially, according to SI rules, when writing out unit names anything following the word "per" appears in the denominator of the expression. This implies the first example listed is correct.

COMPREHENSION CHECK 8-6

The temperature of dry ice is –109.3 degrees Fahrenheit [°F]. Convert this temperature into units of kelvin [K].

COMPREHENSION CHECK 8-7

The specific heat capacity of copper is 0.09 British thermal units per pound-mass degree Fahrenheit [BTU/(lb_m °F)]. Convert into units of British thermal units per gram kelvin $[BTU/(g\ K)]$.

8.6 PRESSURE

Pressure is defined as force acting over an area, where the force is perpendicular to the area. In SI units, a **pascal** [Pa] is the unit of pressure, defined as one newton of force acting on an area of one square meter (Table 8-9). The unit pascal is named after Blaise Pascal (1623–1662), a French mathematician and physicist who made great contributions to the study of fluids, pressure, and vacuums. His contributions with Pierre de Fermat on the theory of probability were the groundwork for calculus.

Table 8-9 Dimensions of pressure

Quantity	Common Units	Exponents			
		M	L	T	Θ
Pressure	Pa	1	−1	−2	0

● **EXAMPLE 8-13**

An automobile tire is pressurized to a 40 pound-force per square inch [psi or lb_f/in^2]. State this pressure in units of atmospheres [atm].

By examining the "Pressure" box in the conversion table on the inside front cover, we see that the following facts are available for use: 1 atm = 14.7 psi.

Method	Steps
(1) Term to be converted	40 psi
(2) Conversion formula	1 atm = 14.7 psi
(3) Make a fraction	$\dfrac{40\ psi}{}$ $\dfrac{1\ atm}{14.7\ psi}$
(4) Multiply	
(5) Cancel, calculate, be reasonable	2.7 atm

PRESSURE

1 atm = 1.01325 bar

= 33.9 ft H_2O

= 29.92 in Hg

= 760 mm Hg

= 101,325 Pa

= 14.7 psi

COMPREHENSION CHECK 8-8

If the pressure is 250 feet of water [ft H_2O], what is the pressure in units of inches of mercury [in Hg]?

In this chapter, we consider four forms of pressure, all involving fluids. The general term **fluid** applies to a gas, such as helium or air, or a liquid, such as water or honey.

- **Atmospheric pressure**—the pressure created by the weight of air above us.
- **Hydrostatic pressure**—the pressure exerted on a submerged object by the fluid in which it is immersed.
- **Total pressure**—the combination of atmospheric and hydrostatic pressure.
- **Gas pressure**—the pressure created by a gas inside a closed container.

Atmospheric Pressure

PRESSURE

1 atm ~ 14.7 psi ~ 101 kPa

Atmospheric pressure results from the weight of the air above us, which varies with both altitude and weather patterns. Standard atmospheric pressure is an average air pressure at sea level, defined as one atmosphere [atm], and is approximately equal to 14.7 pound-force per square inch [psi].

Pressure Measurement

When referring to the measurement of pressure, two types of reference points are commonly used.

Absolute pressure uses a perfect vacuum as a reference point. Most meteorological readings are given as absolute pressure, using units of atmospheres or bars.

Gauge pressure uses the local atmospheric pressure as a reference point. Note that local atmospheric pressure is generally *not* standard atmospheric pressure at sea level. Measurements such as tire pressure and blood pressure are given as gauge pressure.

Absolute pressures are distinguished by an "a" after the pressure unit, such as "psia" to signify "pound-force per square inch absolute." Gauge pressure readings are distinguished by a "g" after the pressure unit, such as "psig" to signify "pound-force per square inch gauge." When using instrumentation to determine the pressure, be sure to note whether the device reads absolute or gauge pressure.

Gauge pressure, absolute pressure, and atmospheric pressure are related by

$$P_{absolute} = P_{gauge} + P_{atmospheric}$$

A few notes on absolute and gauge pressure. Except as otherwise noted, assume that local atmospheric pressure is 14.7 psi.

- 35 psig = 49.7 psia (35 psig + 14.7 psi)
- Using gauge pressure, local atmospheric pressure would be 0 psig, although this would seldom be used.
- If a gauge pressure being measured is less than the local atmospheric pressure, this is usually referred to as **vacuum pressure**, and the negative sign is dropped. Thus, a perfect vacuum created at sea level on the Earth would read about 14.7 psig vacuum pressure. (A perfect vacuum is 0 psia, and thus is about 14.7 psi less than atmospheric pressure.)
- A vacuum pressure of 10 psig is an absolute pressure of 4.7 psia (14.7 psi − 10 psi).
- To illustrate the effect of local atmospheric pressure, consider the following scenario. You fill your automobile's tires to 35 psig on the shore of the Pacific Ocean in Peru, and then drive to Lake Titicaca on the Bolivian border at about 12,500 feet above sea level. Your tire pressure now reads about 40 psig due to the decreased atmospheric pressure (9.5 psia at 12,500 feet altitude versus 14.7 psia at sea level).

Occasionally in industry, it may be helpful to use a point of reference other than atmospheric pressure. For these specific applications, pressure may be discussed in terms of **differential pressure**, distinguished by a "d" after the pressure unit, such as "psid."

Hydrostatic Pressure

Hydrostatic pressure (P_{hydro}) results from the weight of a liquid or gas pushing on an object. *Remember, weight is a force!* A simple way to determine this is to consider a cylinder with a cross-sectional area (A) filled with a liquid of density ρ.

The pressure (P) at the bottom of the container can be found by **Pascal's law**, named after (once again) Blaise Pascal. Pascal's law states the hydrostatic pressure of a fluid is equal to the force of the fluid acting over an area.

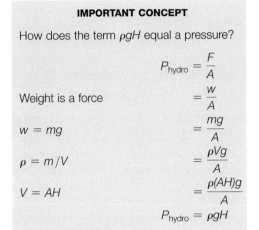

IMPORTANT CONCEPT

How does the term ρgH equal a pressure?

$$P_{hydro} = \frac{F}{A}$$

Weight is a force
$$= \frac{w}{A}$$

$w = mg$
$$= \frac{mg}{A}$$

$\rho = m/V$
$$= \frac{\rho V g}{A}$$

$V = AH$
$$= \frac{\rho(AH)g}{A}$$

$$P_{hydro} = \rho gH$$

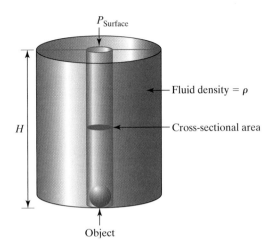

Recreational scuba diving takes place at depths between 0 and 20 meters. At deeper depths, additional training is usually required because of the increased risk of narcosis, a state similar to alcohol intoxication. The relationship between the depth and the level of narcosis is called the "Martini Effect," as it is said divers feel like they have drunk one martini for every 20 meters they descend.

Photo courtesy of E. Stephan

● **EXAMPLE 8-14**

We want to know the hydrostatic pressure in a lake at a depth of 20 feet in units of pascals.

For hydrostatic pressure, we need to know the density of the fluid in the lake. Since a density is not specified, we assume the density to be the standard density of water. We want all quantities in units of kilograms, meters, and seconds, so we use a density of 1,000 kilograms per cubic meter for water.

Step One: Convert to Base SI Units	
Method	**Steps**
(1) Term to be converted	20 ft
(2) Conversion formula	
(3) Make a fraction	$\dfrac{20\ \text{ft}}{} \Big| \dfrac{1\ \text{m}}{3.28\ \text{ft}}$
(4) Multiply	
(5) Cancel, calculate, be reasonable	6.1 m

Step Two: Calculate	
Method	Steps
(1) Determine appropriate equation	$P_{hydro} = \rho g H$
(2) Insert known quantities	$P_{hydro} = \dfrac{1{,}000\ kg}{m^3}\left\|\dfrac{9.8\ m}{s^2}\right\|\dfrac{6.1\ m}{}$
(3) Calculate, be reasonable	$P_{hydro} = 59{,}760\ \dfrac{kg}{m\ s^2}$

This is apparently our final answer, but the units are puzzling. If the units of pressure are pascals and if this is a valid equation, then our final result for pressure should be pascals. If we consider the dimensions of pressure:

Quantity	Common Units	Exponents			
		M	L	T	Θ
Pressure	Pa	1	−1	−2	0

A unit of pressure has dimensions, $P\{=\}M/(LT^2)$, which in terms of base SI units would be $P[=]\ kg/m\ s^2$. As this term occurs so frequently it is given the special name "Pascal." When we see this term, we know we are dealing with a pressure equal to a pascal.

(3) Calculate, be reasonable	$P_{hydro} = 59{,}760\ \dfrac{kg}{m\ s^2}\left\|\dfrac{1\ Pa}{1\ \frac{kg}{m\ s^2}}\right. = 59{,}760\ Pa$

Step Three: Convert from Base SI Units to Desired Units
No conversion needed

Total Pressure

We need to realize that Pascal's law is only a part of the story. Suppose we dive to a depth of 5 feet in a swimming pool and measure the pressure. Now we construct an enclosure over the pool and pressurize the air above the water surface to 3 atmospheres. When we dive back to the 5-foot depth, the pressure will have increased by 2 atmospheres.

Consequently, we conclude that total pressure at any depth in a fluid is the sum of hydrostatic pressure and pressure in the air above the fluid.

IMPORTANT CONCEPT

$P_{total} = P_{surface} + P_{hydro}$

$P_{total} = P_{surface} + \rho g H$

● **EXAMPLE 8-15**

When you dive to the bottom of a pool, at 12 feet under water, how much total pressure do you feel in units of atmospheres?

NOTE

A diver at 15 feet will feel a ½ atmosphere pressure increase.

A diver at 10 meters will feel a 1 atmosphere pressure increase.

Step One: Convert to Base SI Units	
Method	Steps
(1) Term to be converted	12 ft
(2) Conversion formula	
(3) Make a fraction	$\dfrac{12\ ft}{}\left\|\dfrac{1\ m}{3.28\ ft}\right.$
(4) Multiply	
(5) Cancel, calculate, be reasonable	3.66 m

For hydrostatic pressure, we need to know the density of the fluid in the pool. Since a density is not specified, we assume the density to be the standard density of water. We want all quantities in units of kilograms, meters, and seconds, so we use a density of 1,000 kilograms per cubic meter for water.

For total pressure, we need to know the surface pressure on top of the pool. Since a surface pressure is not specified, we assume the pressure to be 1 atmosphere. We want all quantities in units of kilograms, meters, and seconds, so we use a pressure of 101,325 pascals, or 101,325 kilograms per meter second squared.

Step Two: Calculate	
Method	**Steps**
(1) Determine appropriate equation	$P_{total} = P_{surface} + \rho g H$
(2) Insert known quantities	$P_{total} = \dfrac{101{,}325 \text{ kg}}{\text{m s}^2} + \dfrac{1{,}000 \text{ kg}}{\text{m}^3}\left\vert\dfrac{9.8 \text{ m}}{\text{s}^2}\right\vert 3.66 \text{ m}$
(3) Calculate, be reasonable	$P_{total} = 137{,}193\dfrac{\text{kg}}{\text{m s}^2}\left\vert\dfrac{\text{Pa}}{\frac{\text{kg}}{\text{m s}^2}}\right. = 137{,}193 \text{ Pa}$

Step Three: Convert from Base SI Units to Desired Units	
Method	**Steps**
(1) Term to be converted	137,193 Pa
(2) Conversion formula	
(3) Make a fraction	$\dfrac{137{,}193 \text{ Pa}}{}\left\vert\dfrac{1 \text{ atm}}{101{,}325 \text{ Pa}}\right.$
(4) Multiply	
(5) Cancel, calculate, be reasonable	1.35 atm

COMPREHENSION CHECK 8-9

An object is completely submerged in a liquid of density 0.75 grams per cubic centimeter at a depth of 3 meters. What is the total pressure on the object? State your answer in atmospheres.

NOTE

At an elevation of 3 miles, the pressure will be about half that at sea level. At an elevation of 10 miles, the pressure is about $\frac{1}{10}$ that at sea level.

Cakes and meats may require as much as ¼ more cooking time at elevations of 5,000 feet because of reduced pressure and lower boiling points.

8.7 GAS PRESSURE

IMPORTANT CONCEPT

Ideal Gas Law

$$PV = nRT$$

Only absolute temperature units (K or °R) can be used in the ideal gas equation.

Gas pressure results when gas molecules impact the inner walls of a sealed container. The **ideal gas law** relates the quantities of pressure (P), volume (V), temperature (T), and amount (n) of gas in a closed container:

$$PV = nRT$$

In this equation, R is a fundamental constant called the **gas constant**. It can have many different numerical values, depending on the units chosen for pressure, volume, temperature, and amount, just as a length has different numerical values, depending on whether feet or meters or miles is the unit being used. Scientists have defined an "ideal" gas as one where one mole [mol] of gas at a temperature of 273 kelvin [K] and a pressure of one atmosphere [atm] will occupy a volume of 22.4 liters [L]. Using these values to solve for the constant R yields

NOTE

$R = 8,314 \dfrac{\text{Pa L}}{\text{mol K}}$

$\quad = 0.08206 \dfrac{\text{atm L}}{\text{mol K}}$

$$R = \frac{PV}{nT} = \frac{1\,[\text{atm}]\ 22.4\,[\text{L}]}{1\,[\text{mol}]\ 273\,[\text{K}]} = 0.08206\ \frac{\text{atm L}}{\text{mol K}}$$

Note that we must *use* absolute *temperature units in the ideal gas equation.* We cannot begin with relative temperature units and then convert the final answer. Also, all pressure readings must be in absolute, not gauge, units.

In previous chapters, we have suggested a procedure for solving problems involving equations and unit conversions. For ideal gas law problems, we suggest a slightly different procedure.

Ideal Gas Law Procedure

1. Examine the units given in the problem statement. Choose a gas constant (R) that contains as many of the units given in the problem as possible.
2. If necessary, convert all parameters into units found in the gas constant (R) that you choose.
3. Solve the ideal gas law for the variable of interest.
4. Substitute values and perform all necessary calculations.
5. If necessary, convert your final answer to the required units and apply reasonableness.

● **EXAMPLE 8-16**

A container holds 1.43 moles of nitrogen (formula: N_2) at a pressure of 3.4 atmospheres and a temperature of 500 degrees Fahrenheit. What is the volume of the container in liters?

Method	Steps
(1) Choose ideal gas constant	Given units: mol, atm, °F, L Select R: $0.08206\ \dfrac{\text{atm L}}{\text{mol K}}$
(2) Convert to units of chosen R	500°F = 533 K
(3) Solve for variable of interest	$V = \dfrac{nRT}{P}$
(4) Calculate	$V = \dfrac{1.43\ \text{mol}\ \left\vert 0.08206\ \text{atm L}\ \right\vert 533\ \text{K}\ \vert}{\vert\ \text{mol K}\ \vert\ \vert 3.4\ \text{atm}}$
(5) Convert, be reasonable	$V = 18.4\ \text{L}$

• EXAMPLE 8-17

A container holds 1.25 moles of nitrogen (formula: N_2) at a pressure of 350 kilopascals and a temperature of 160 degrees Celsius. What is the volume of the container in liters?

Method	Steps
(1) Choose ideal gas constant	Given units: mol, Pa, °C, L Select R: $8,314 \dfrac{\text{Pa L}}{\text{mol K}}$
(2) Convert to units of chosen R	$160°C = 433 \text{ K}$
(3) Solve for variable of interest	$V = \dfrac{nRT}{P}$
(4) Calculate	$V = \dfrac{1.25 \text{ mol}}{} \left| \dfrac{8,314 \text{ Pa L}}{\text{mol K}} \right| \dfrac{433 \text{ K}}{} \left| \dfrac{}{350,000 \text{ Pa}} \right.$
(5) Convert, be reasonable	$V = 13 \text{ L}$

• EXAMPLE 8-18

A gas originally at a temperature of 300 kelvin and 3 atmospheres pressure in a 3.9-liter flask is cooled until the temperature reaches 284 kelvin. What is the new pressure of gas in atmospheres?

Since the volume and the mass of the gas remain constant, we can examine the ratio between the initial condition (1) and the final condition (2) for pressure and temperature. The volume of the container (V) and the amount of gas (n) are constant, so $V_1 = V_2$ and $n_1 = n_2$.

Method	Steps
(1) Choose ideal gas constant	Given units: mol, K, L Select R: $0.08206 \dfrac{\text{atm L}}{\text{mol K}}$
(2) Convert to units of chosen R	None needed
(3) Solve for variable of interest, eliminating any variables that remain constant between the initial and final state	$\dfrac{P_1 V_1}{P_2 V_2} = \dfrac{n_1 R T_1}{n_2 R T_2}$ $\dfrac{P_1}{P_2} = \dfrac{T_1}{T_2}$
(4) Calculate	$\dfrac{3 \text{ atm}}{P_2} = \dfrac{300 \text{ K}}{284 \text{ K}}$
(5) Convert, be reasonable	$P = 2.8 \text{ atm}$

COMPREHENSION CHECK 8-10

A 5-gallon container holds 35 grams of nitrogen (formula: N_2, molecular weight = 28 grams per mole) at a temperature of 400 kelvin. What is the container pressure in units of kilopascals?

COMPREHENSION CHECK 8-11

An 8-liter container holds nitrogen (formula: N_2, molecular weight = 28 grams per mole) at a pressure of 1.5 atmospheres and a temperature of 310 kelvin. If the gas is compressed by reduction of the volume of the container until the gas pressure increases to 5 atmospheres while the temperature is held constant, what is the new volume of the container in units of liters?

8.8 ENERGY

Energy is an abstract quantity with several definitions, depending on the form of energy being discussed. You may be familiar with some of the following types of energy.

IMPORTANT CONCEPT

Work
$W = F\Delta x$

Potential Energy
$PE = mg\Delta H$

Kinetic Energy, translational
$KE_T = \frac{1}{2}m(v_f^2 - v_i^2)$

Kinetic Energy, rotational
$KE_R = \frac{1}{2}I(\omega_f^2 - \omega_i^2)$

Kinetic Energy, total
$KE = KE_T + KE_R$

Thermal Energy
$Q = mC_p\Delta T$

MOMENT OF INERTIA (*I*)

m = mass, r = radius

Object	*I* for KE_R
Cylinder: thin shell	mr^2
Cylinder: solid	$\frac{1}{2}mr^2$
Sphere: thin shell	$\frac{2}{3}mr^2$
Sphere: solid	$\frac{2}{5}mr^2$

Types of Energy

■ **Work** (*W*) is energy expended by exertion of a force (*F*) over a distance (*d*). As an example, if you exert a force on (push) a heavy desk so that it slides across the floor, which will make you more tired: pushing it 5 feet or pushing it 50 feet? The farther you push it, the more work you do.

■ **Potential energy** (PE) is a form of work done by moving a weight (*w*)—which is a force—a vertical distance (*H*). Recall that weight is mass (*m*) times gravity (*g*). Note that this is a special case of the work equation, where force is weight and distance is height.

■ **Kinetic energy** (KE) is a form of energy possessed by an object in motion. If a constant force is exerted on a body, then by $F = ma$, we see that the body experiences a constant acceleration, meaning the velocity increases linearly with time. Since the velocity increases as long as the force is maintained, work is being done on the object. Another way of saying this is that the object upon which the force is applied acquires kinetic energy, also called **energy of translational motion**. For a nonrotating body moving with some velocity (*v*) the kinetic energy can be calculated by $KE_T = (\frac{1}{2})mv^2$.

This, however, is not the entire story. A rotating object has energy whether it is translating (moving along a path) or not. If you have ever turned a bicycle upside down, spun one of the wheels fairly fast, then tried to stop it with your hand, you understand that it has energy. This is **rotational kinetic energy**, and for an object spinning in place (but not going anywhere), it is calculated by $KE_R = (\frac{1}{2})I\omega^2$. The Greek letter omega (ω) symbolizes angular velocity or the object's rotational speed, typically given in units of radians per second. The moment of inertia (*I*) depends on the mass and the geometry of the spinning object. The table shown lists the moments of inertia for a few common objects.

For an object that is rotating and translating, such as a bowling ball rolling down the lane toward the pins, the total kinetic energy is simply the sum of the two:

$$KE = KE_T + KE_R = (\frac{1}{2})mv^2 + (\frac{1}{2})I\omega^2 = (\frac{1}{2})(mv^2 + I\omega^2)$$

■ **Thermal Energy** or heat (*Q*) is energy associated with a change in temperature (ΔT). It is a function of the mass of the object (*m*) and the specific heat (C_p), which is a property of the material being heated:

$$Q = mC_p\Delta T$$

Where Does KE = ½*mv*² Come From?

If a constant force is applied to a body,

- That body will have a constant acceleration (remember $F = ma$).
- Its velocity will increase linearly with time.
- Its average velocity is the average of its initial and final values.

This is $\qquad\qquad\qquad v = (v_f + v_i)/2$

Distance traveled is $\qquad d = vt$

The work done is $\qquad\quad W = Fd$

$\qquad\qquad\qquad\qquad\quad = (ma)d$

$\qquad\qquad\qquad\qquad\quad = mavt$

Acceleration is the change in velocity over time, or $a = (v_f - v_i)/t$

Substituting for a and v in the work equation

$$W = mavt = m[(v_f - v_i)/t][(v_f + v_i)/2][t]$$

$$W = (1/2)m(v_f^2 - v_i^2)$$

This is given the name *kinetic energy*, or the energy of motion. Remember, work and energy are equivalent. This expression is for the *translation* of a body only.

Calories and BTUs and Joules—Oh My!

The SI unit of work is **joule**, defined as one newton of force acting over a distance of one meter (Table 8-10). The unit is named after James Joule (1818–1889), an English physicist responsible for several theories involving energy, including the definition of the mechanical equivalent of heat and Joule's law, which describes the amount of electrical energy converted to heat by a resistor (an electrical component) when an electric current flows through it. In some mechanical systems, work is described in units of foot pound-force [ft lb$_f$].

For energy in the form of heat, units are typically reported as British thermal units and calories instead of joules. A **British thermal unit** [BTU] is the amount of heat required to raise the temperature of one pound-mass of water by one degree Fahrenheit. A **calorie** [cal] is amount of heat required to raise the temperature of one gram of water by one degree Celsius.

Table 8-10 Dimensions of energy

Quantity	Common Units	Exponents			
		M	L	T	Θ
Work	J	1	2	−2	0
Thermal energy	BTU	1	2	−2	0
	cal	1	2	−2	0

● EXAMPLE 8-19

A 50-kilogram load is raised vertically a distance of 5 meters by an electric motor. How much work in units of joules was done on the load?

First, we must determine the type of energy. The parameters we are discussing include mass (kilograms) and height (meters). Examining the energy formulas given above, the equation for potential energy fits. Also, the words "load is raised vertically a distance" fits with our understanding of potential energy.

Step One: Convert to Base SI Units		
No conversion needed		

Step Two: Calculate		
Method		**Steps**
(1) Determine appropriate equation		$PE = mg\Delta H$
(2) Insert known quantities		$PE = \dfrac{50\ kg}{}\left|\dfrac{9.8\ m}{s^2}\right|\dfrac{5\ m}{}$
(3) Calculate, be reasonable		$PE = 2{,}450\ \dfrac{kg\ m^2}{s^2}$

This is apparently our final answer, but the units are puzzling. If the units of energy are joules and if this is a valid equation, then our final result for energy should be joules. If we consider the dimensions of energy:

Quantity	Common Units	Exponents			
		M	**L**	**T**	**Θ**
Energy	J	1	2	−2	0

A unit of energy has dimensions $E\{=\}\ M\ L^2/T^2$, which in terms of base SI units would be $E\ [=]\ kg\ m^2/s^2$. As this term occurs so frequently it is given the special name "joule." Anytime we see this term $(kg\ m^2/s^2)$, we know we are dealing with an energy, equal to a joule.

(3) Calculate, be reasonable	$PE = 2{,}450\ \dfrac{kg\ m^2}{s^2}\left|\dfrac{1\ J}{1\ \dfrac{kg\ m^2}{s^2}}\right. = 2{,}450\ J$

Step Three: Convert from Base SI Units to Desired Units
No conversion needed

● EXAMPLE 8-20

In the morning, you like to drink your coffee at a temperature of exactly 70 degrees Celsius [°C]. The mass of the coffee in your mug is 470 grams. To make your coffee, you had to raise the temperature of the water by 30 degrees Celsius. How much energy in units of British thermal units [BTU] did it take to heat your coffee? The specific heat of water is 4.18 joules per gram degree Celsius [J/(g °C)].

First, you must determine the type of energy we are using. The parameters discussed include mass, temperature, and specific heat. Examining the energy formulas given above, the equation for thermal energy fits. Also, the words "How much energy . . . did it take to heat your coffee" fits with an understanding of thermal energy.

Step One: Convert to Base SI Units	
No conversion needed	

Step Two: Calculate	
Method	**Steps**
(1) Determine appropriate equation	$Q = mC_p\Delta T$
(2) Insert known quantities	$Q = \dfrac{470\text{ g}}{} \bigg\vert \dfrac{4.18\text{ J}}{\text{g}^\circ\text{C}} \bigg\vert \dfrac{30^\circ\text{C}}{}$
(3) Calculate, be reasonable	$Q = 59{,}370\text{ J}$

Step Three: Convert from Base SI Units to Desired Units	
Method	**Steps**
(1) Term to be converted	59,370 J
(2) Conversion formula	
(3) Make a fraction	$\dfrac{59{,}370\text{ J}}{} \bigg\vert \dfrac{9.48 \times 10^{-4}\text{ BTU}}{1\text{ J}}$
(4) Multiply	
(5) Cancel, calculate, be reasonable	56 BTU

COMPREHENSION CHECK 8-12

You push an automobile with a constant force of 20 pounds-force until 1,500 joules of energy has been added to the car. How far did the car travel in units of meters during this time? You may assume that frictional losses are negligible.

COMPREHENSION CHECK 8-13

One gram of material A is heated until the temperature rises by 10 kelvin. If the same amount of heat is applied to one gram of material B, what is the temperature rise of material B in units of kelvin?

The specific heat (C_p) of material A = 4 joules per gram kelvin [J/(g K)]
The specific heat (C_p) of material B = 2 joules per gram kelvin [J/(g K)]

8.9 POWER

IMPORTANT CONCEPT

Power = energy/time

W = J/s

SI unit of power = watt

NOTE

Power is the RATE at which energy is delivered over time.

Power is defined as energy per time (Table 8-11). The SI unit of power is **watt**, named after James Watt (1736–1819), a Scottish mathematician and engineer whose improvements to the steam engine were important to the Industrial Revolution. He is responsible for the definition of **horsepower** [hp], a unit of power originally used to quantify how the steam engine could replace the work done by a horse.

Table 8-11 Dimensions of power

Quantity	Common Units	Exponents			
		M	L	T	(·)
Power	W	1	2	−3	0

To help understand the relationship between energy and power, imagine the following. Your 1,000-kilogram car has run out of gas on a level road. There is a gas station not far ahead, so you decide to push the car to the gas station. Assume that you intend to accelerate the car up to a speed of one meter per second (about 2.2 miles per hour), and then continue pushing at that speed until you reach the station. Ask yourself the following questions:

- Can I accelerate the car to one meter per second in one minute?
- On the other hand, can I accelerate it to one meter per second in one second?

Most of you would probably answer "yes" to the first and "no" to the second, but why? Well, personal experience! But that is not really an explanation. Since the change in kinetic energy is the same in each case, to accelerate the car in one second, your body would have to generate energy at a rate 60 times greater than the rate required if you accelerated it in one minute. The key word is rate, or how much energy your body can produce per second. If you do the calculations, you will find that for the one-minute scenario, your body would have to produce about $1/90$ horsepower, which seems quite reasonable. On the other hand, if you try to accomplish the same acceleration in one second, you would need to generate $2/3$ horsepower. Are you two-thirds as powerful as a horse?

As another example, assume that you attend a class on the third floor of the engineering building. When you are on time, you take 2 minutes to climb to the third floor. On the other hand, when you are late for class, you run up the three flights in 30 seconds.

- In which case do you do the most work (expend the most energy)?
- In which case do you generate the most power?

● **EXAMPLE 8-21**

A 50-kilogram load is raised vertically a distance of 5 meters by an electric motor in 60 seconds. How much power in units of watts does the motor use, assuming no energy is lost in the process?

This problem was started in Example 8-19. The energy used by the system was found to be 2,450 joules, the analysis of which is not repeated here.

Step One: Convert to Base SI Units	
No conversion needed	
Step Two: Calculate	
Method	**Steps**
(1) Determine appropriate equation	$\text{Power} = \dfrac{\text{energy}}{\text{time}}$
(2) Insert known quantities	$\text{Power} = \dfrac{2{,}450 \text{ J}}{60 \text{ s}}$
(3) Calculate, be reasonable	$\text{Power} = 41\dfrac{\text{J}}{\text{s}} \left\| \dfrac{1 \text{ W}}{1\frac{\text{J}}{\text{s}}} \right. = 41 \text{ W}$
Step Three: Convert from Base SI Units to Desired Units	
No conversion needed	

Note that since power = energy/time, energy = power * time. We pay the electric company for energy calculated this way as kilowatt-hours. If power is constant, we can obtain the total energy involved simply by multiplying the power by the length of time that power is applied. If power is *not* constant, we would usually use calculus to determine the total energy, but that solution is beyond the scope of this book.

8.10 EFFICIENCY

IMPORTANT CONCEPT

Efficiency is always less than 100%.

Efficiency (η, Greek letter eta) is a measure of how much of a quantity, typically energy or power, is lost in a process. In a perfect world, efficiency would always be 100%. All energy put into a process would be recovered and used to accomplish the desired task. We know that this can never happen, so *efficiency is always less than 100%*. If a machine operates at 75% efficiency, 25% of the energy is lost. This means you have to put in "extra" energy to complete the work.

The use of the terms "input" and "output" require some explanation. The **input** is the quantity of energy or power or whatever required by the mechanism from some source to operate and accomplish its task. The **output** is the amount of energy or power or whatever is actually applied to the task itself by the mechanism. Note that the rated power of a device, whether a light bulb, a motor, or an electric heater, refers to the input power—the power needed to operate the device—not the output power. In an ideal, 100% efficient system, the input and output would be equivalent. In an inefficient system (the real world), the input is equivalent to the sum of the output and the power or energy lost. This is perhaps best explained by way of examples.

IMPORTANT CONCEPT

Efficiency (η) = output/input
Efficiency (η) = output/(output + loss)
Input = quantity required by mechanism to operate
Output = quantity actually applied to task
Loss = quantity wasted during the application

Orders of Magnitude

The table below gives you an idea of orders of magnitude of power and energy as related to real-world objects and phenomena. All values are approximate and, in most cases, have been rounded to only one significant figure. Thus, if you actually do the calculations from power to energy, you will find discrepancies.

Unless you are already familiar with the standard metric prefixes, refer to the list on the inside back cover of the book.

A few things of possible interest:

- U.S. power consumption (all types) is one-fifth of the total world power consumption.
- The Tsar Bomba generated 1.5% of the power of the sun, but only lasted 40 nanoseconds.
- Total human power consumption on the planet is about 0.01% (1/10,000) of the total power received from the sun.

Power	"Device"	Energy per Hour	Energy per Year
10 fW	Minimum reception power for cell phone	40 pJ	300 nJ
	Single human cell	4 nJ	30 mJ
10 mW	DVD laser	40 J	300 kJ
500 mW	Cell phone microprocessor	2 kJ	15 MJ
50 W	Modern 2GHz microprocessor	200 kJ	1.5 GJ
100 W	Human at rest	400 kJ	3 GJ
500 W	Human doing strenuous work	2 MJ	15 GJ
750 W	Power per meter bright sunshine	3 MJ	25 GJ
2 kW	Maximum human power for short period	(NA)	
20 kW	Average U.S. home	80 MJ	600 GJ
100 kW	Typical automobile	400 MJ	3 TJ
150 MW	Boeing 747 jet	500 GJ	5 PJ
1 GW	Large commercial nuclear reactor	4 TJ	30 PJ
20 GW	Three Gorges Hydroelectric Dam (China)	80 TJ	600 PJ
4 TW	U.S. total power consumption	15 PJ	100 EJ
20 TW	Total human power consumption	80 PJ	600 EJ
100 TW	Average hurricane	400 PJ	(NA)
200 PW	Total power received on the Earth from the sun	1 ZJ	6 YJ
5 YW	Largest fusion bomb ever built (Russian Tsar Bomba)	(NA) Total yield 200 PJ	
400 YW	Total power of the sun	10^6 YJ	10^9 YJ

● **EXAMPLE 8-22**   A standard incandescent light bulb has an efficiency of about 5%; thus, $\eta = 0.05$. An incandescent bulb works by heating a wire (the filament) inside the bulb to such a high temperature that it glows white. About 95% of the power delivered to an incandescent bulb is discarded as heat. Only 5% results in "light" energy.

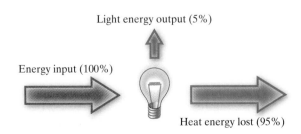

Light energy output (5%)

Energy input (100%)

Heat energy lost (95%)

If a 100-watt bulb is turned on for 15 minutes, how much energy is "lost" as heat during the 15-minute period?

Step One: Convert to Base SI Units	
No conversion needed	
Step Two: Calculate	
Method	**Steps**
(1) Determine appropriate equation	Power = energy/time
(2) Insert known quantities	Power = energy/15 min
For Unknown Quantities, Repeat the Process	
Method	**Steps**
(1) Determine appropriate equation	Input power = output power/efficiency
	"Lost" power = input power − output power
(2) Insert known quantities	100 W = output power/(0.05)
(3) Calculate, be reasonable	Output power = 5 W
	"Lost" power = 100 W − 5 W = 95 W
(2) Insert known quantities	95 W = energy/15 min
(3) Calculate, be reasonable	$E = 95\ \dfrac{J}{s}\left\|\dfrac{15\ min}{}\right\|\dfrac{60\ s}{1\ min}$
	Energy lost = 85,500 J
Step Three: Convert from Base SI Units to Desired Units	
No conversion needed	

● **EXAMPLE 8-23**

Over the past few decades, the efficiency of solar cells has risen from about 10% to the most recent technologies achieving about 40% conversion of solar energy to electricity. The losses are due to several factors, including reflectance and resistive losses, among others.

Assume you have an array of solar cells mounted on your roof with an efficiency of 28%. If the array is delivering 750 watts of electricity to your home, how much solar power is falling on the photoelectric cells?

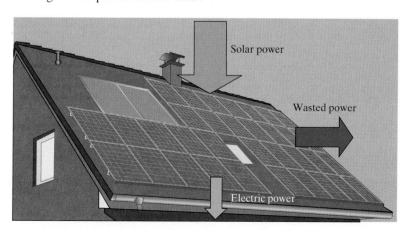

NOTE

How much power is wasted by the array of solar cells? If the array received 2,680 watts and delivered 750 watts, then the difference is 2,680 − 750 = 1,930 watts; thus, 1,930 watts are wasted.

Step One: Convert to Base SI Units	
No conversion needed	

Step Two: Calculate	
Method	**Steps**
(1) Determine appropriate equation	Input power = output power/efficiency
(2) Insert known quantities	Input power = 750 W/0.28
(3) Calculate, be reasonable	Input power = 2,680 W

Step Three: Convert from Base SI Units to Desired Units	
No conversion needed	

● **EXAMPLE 8-24**

If your microwave takes 2 minutes to heat your coffee in Example 8-20, how many watts of power does your microwave require, assuming that it is 80% efficient? Remember that our answer was 59,370 joules, before we converted the final answer to units of BTU.

Step One: Convert to Base SI Units	
No conversion needed	

Step Two: Calculate	
Method	**Steps**
(1) Determine appropriate equation	Input power = output power/efficiency
(2) Insert known quantities	Input power = output power/0.8

For Unknown Quantities, Repeat the Process	
Method	**Steps**
(1) Determine appropriate equation	Output power = energy/time
(2) Insert known quantities	Output power = $\dfrac{59{,}370\ \text{J}}{}\left\|\dfrac{}{2\ \text{min}}\right\|\dfrac{1\ \text{min}}{60\ \text{s}}\left\|\dfrac{1\ \text{W s}}{1\ \text{J}}\right.$
(3) Calculate, be reasonable	Output power = 493 W

(2) Insert known quantities	Input power = 493 W/0.8
(3) Calculate, be reasonable	Input power = 615 W

Step Three: Convert from Base SI Units to Desired Units	
No conversion needed	

COMPREHENSION CHECK 8-14

A motor with an output power of 100 watts is connected to a flywheel. How long, in units of hours, must the motor operate to transfer 300,000 joules to the flywheel, assuming the process is 100% efficient?

**COMPREHENSION
CHECK** 8-15
If a 50-kilogram load was raised 5 meters in 50 seconds, determine the minimum rated wattage of the motor needed to accomplish this, assuming the motor is 80% efficient.

● **EXAMPLE 8-25**

A simple two-stage machine is shown in the diagram below. Initially, an electric motor receives power from the power grid, accessed by being plugged into a standard electrical wall socket. The power received by the motor from the wall socket is the "input" power or the power the motor uses or requires (Point A).

The spinning drive shaft on the motor can then be used to power other devices, such as a hoist or a vacuum cleaner or a DVD drive; the power available from the spinning shaft is the "output" power of the motor (Point B). In the process of making the drive shaft spin, however, some of the input power is lost (Point C) because of both frictional and ohmic heating as well as other wasted forms such as sound.

Since some energy is being lost as heat or other unusable forms every second of operation (remember, power is energy per time!) the output power *must* be less than the input power—the efficiency of the motor is less than 100%. The power available from the spinning shaft of the motor (Point B) is then used to operate some device: the hoist or vacuum cleaner or DVD drive (Point E). In other words, the output power of the motor is the input power to the device it drives. This device will have its own efficiency, thus wasting some of the power supplied to it by the motor (Point D).

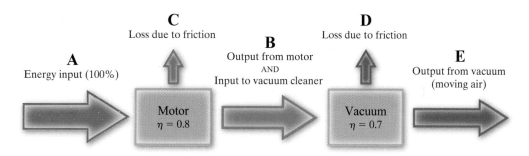

C
Loss due to friction

B
Output from motor
AND
Input to vacuum cleaner

D
Loss due to friction

A
Energy input (100%)

E
Output from vacuum
(moving air)

Motor
$\eta = 0.8$

Vacuum
$\eta = 0.7$

NOTE

The overall efficiency of two linked devices is the product of the two efficiencies.

Assume that a motor with an efficiency of 80% is used to power a vacuum cleaner that has an efficiency of 70%. If the input power to the motor (Point A) is one-half horsepower, what is the output power of the vacuum (Point E)? The output from the motor (Point B) is
$P_B = \eta P_A = 0.8 \, (0.5 \text{ horsepower}) = 0.4 \text{ horsepower}$.
This is the input to the vacuum, so the output from the vacuum (Point E) is

$$P_E = \eta P_B = 0.7 \, (0.4 \text{ hp}) = 0.28 \text{ hp}$$

What is the overall efficiency of this machine? The input power (Point A) is 0.5 horsepower, and the output power (Point E) is 0.28 horsepower; thus, the efficiency of this linked system is:

$$\eta = P_A/P_E = 0.28 \text{ hp}/0.5 \text{ hp} = 0.56, \text{ or an efficiency of } 56\%.$$

SO WHAT WAS THE ANSWER TO THE BOAT AND THE STONE QUESTION FROM THE INTRODUCTION?

This is a problem that can be analyzed without the need for any equations or formal mathematics. We will start with Archimedes' principle that states simply: "When a body is submerged in a fluid, the buoyancy force (which pushes the body upward) is equal to the weight of the fluid which the body displaces."

If an object is floating at equilibrium in a fluid, the buoyancy force is equal to the weight of the object. The upward buoyancy force and downward weight of the object (remember, weight is a force!) are the same, so the net force on the object is zero. For example, when a 500 pound-mass boat is placed in a pool, its weight acts downward (with a value of 500 pounds-force if at sea level on the Earth), and it sinks deeper and deeper, displacing more and more fluid, until it has displaced 500 pounds-mass of fluid. At that point, the buoyancy force is 500 pounds-force and this opposes the weight of the boat so that it rests at equilibrium.

If the body we place in a fluid weighs more than the fluid it displaces when completely submerged, then the body sinks, but it still experiences a buoyancy force equal to the weight of the fluid displaced. This is one reason astronauts train under water—the net downward force of themselves and of objects they are manipulating is greatly reduced although their mass is the same.

If we place the same 500 pound-mass boat in a pool, and put a 200 pound-mass person in it, the boat will sink in the fluid more deeply than before, until it displaces 700 pound-mass of fluid. At this new level of submergence, the upward and downward forces are equal and the new "system" of boat and person will rest at equilibrium. If we add a stone in the boat, the new system of the boat, the person, and the stone will sink even deeper until it displaces a mass of fluid equal to the combined masses of the boat, the person, and the stone. In each of these situations the displaced fluid has to go some-where, so the level of the liquid in the pool rises.

If we assume that the stone will sink when tossed overboard, it will displace a volume of liquid equal to the volume of the stone. Since the stone is denser than the liquid (has more mass per volume), it now displaces less liquid than it did when it was in the boat. Thus, when the stone is placed in the pool, the water level will drop.

In-Class Activities

ICA 8-1

Complete the following table, using the equation weight = (mass) (gravity) as needed:

	Pound-mass	Kilogram	Newton	Pound-force
Abbreviation		[kg]		
Example	0.33	0.15	1.47	0.33
(a)	10			
(b)			700	
(c)				25

ICA 8-2

Complete the following table, using the equation weight = (mass) (gravity) as needed:

	Slug	Grams	Newton	Pound-force
Abbreviation		[g]		
(a)		130		
(b)	50			
(c)			50	
(d)				30

ICA 8-3

Complete the following table:

	Compound	SG	Density $[lb_m/ft^3]$	Density $[g/cm^3]$	Density $[kg/m^3]$
Example	Ethyl alcohol	1.025	64.0	1.025	1,025
(a)	Acetic acid			1.049	
(b)	Octane				703
(c)	Tetrachloroethane		100		
(d)	Chloroform	1.489			

ICA 8-4

Complete the following table:

	Compound	Density			
		SG	[lb$_m$/ft^3]	[g/cm^3]	[kg/m^3]
(a)	Gallium	5.91			
(b)	Aluminum		168.5		
(c)	Zinc oxide				5,600
(d)	Sugar, powdered			0.801	
(e)	Lead	11.4			

ICA 8-5

Complete the following table. Assume you have a cube composed of each material, with "length" indicating the length of one side of the cube.

	Material	Mass [g]	Length [cm]	Volume [cm^3]	Density [g/cm^3]
Example	Tungsten	302	2.5	15.6	19.3
(a)	Zinc			25	7.14
(b)	Copper	107			8.92
(c)	Lead	1,413			11.35

ICA 8-6

Complete the following table. Assume you have a cylinder, composed of each material, with "radius" and "height" indicating the dimensions of the cylinder.

	Material	Mass [kg]	Radius [m]	Height [m]	Density [kg/m^3]
(a)	Silicon	2,750	0.50	1.5	
(b)	Aluminum		1.25	0.75	2,700
(c)	Tin	10,000	0.25		7,290
(d)	Titanium	8,000		1.0	4,540

ICA 8-7

Complete the following table:

	Compound	Boiling Temperature			
		[°F]	[°C]	[K]	[°R]
(a)	Acetic acid	180			
(b)	Ethyl alcohol				815
(c)	Octane		126		
(d)	Tetrachloroethane			419	

ICA 8-8

A **eutectic alloy** of two metals contains the specific percentage of each metal that gives the lowest possible melting temperature for any combination of those two metals. Eutectic alloys are often used for soldering electronic components to minimize the possibility of thermal damage. In the past, the most common eutectic alloy used in this application has been 63% Sn, 37% Pb, with a melting temperature of about 361 degrees Fahrenheit. To reduce lead pollution in the environment, many other alloys have been tried, including those in the table below. Complete the following table:

| | Compound | Eutectic Temperature | | | |
		[°F]	[°C]	[K]	[°R]
(a)	48% Sn, 52% In	244			
(b)	42% Sn, 58% Bi		138		
(c)	91% Sn, 9% Zn			472	
(d)	96.5% Sn, 3.5% Ag				890

ICA 8-9

In the Spring of 2004, NASA discovered a new planet beyond Pluto named Sedna. In the news report about the discovery, the temperature of Sedna was reported as "never rising above −400 degrees." What are the units of the reported temperature?

ICA 8-10

Is there a physical condition at which a Fahrenheit thermometer and Celsius thermometer will read the same numerical value? If so, what is this value?

ICA 8-11

Is there a physical condition at which a Fahrenheit thermometer and a Kelvin thermometer will read the same numerical value? If so, what is this value?

ICA 8-12

Determine the mass in units of grams of 0.025 moles of caffeine (formula: $C_8H_{10}N_4O_3$). The components are hydrogen (formula: H, amu = 1); carbon (formula: C, amu = 12); nitrogen (formula: N, amu = 14); and oxygen (formula: O, amu = 16).

ICA 8-13

Determine the mass in units of grams of 0.15 moles of the common analgesic naproxen sodium (formula: $C_{14}H_{13}NaO_3$). The components are hydrogen (formula: H, amu = 1); carbon (formula: C, amu = 12); sodium (formula: Na, amu = 23); and oxygen (formula: O, amu = 16).

ICA 8-14

Determine the amount in units of moles of 5 grams of another common analgesic acetaminophen (formula: $C_8H_9NO_2$). The components are hydrogen (formula: H, amu = 1); carbon (formula: C, amu = 12); nitrogen (formula: N, amu = 14); and oxygen (formula: O, amu = 16).

ICA 8-15

Complete the table:

	Atmosphere	Pascal	Inches of Mercury	Pound-force per Square Inch
Abbreviation	[atm]			
Example	0.030	3,000	0.886	0.435
(a)			30	
(b)				50
(c)	2.5			

ICA 8-16

Complete the table:

	Atmosphere	Pascal	Inches of Mercury	Pound-force per Square Inch
Abbreviation	[atm]			
(a)		500,000		
(b)				5
(c)	10			
(d)			25	

ICA 8-17

Complete the table, using the equation for total pressure:

	Fluid Density [kg/m³]	Height of Fluid [ft]	Surface Pressure [atm]	Total Pressure [atm]
(a)	787	10	1	
(b)		50	1	2
(c)	1,263		3	5.4
(d)	1,665	35		3.25

ICA 8-18

Complete the table, using the equation for total pressure:

	Fluid Specific Gravity	Height of Fluid [m]	Surface Pressure [atm]	Total Pressure [psi]
(a)	0.787	15	1	
(b)		75	1	30
(c)	1.263		3	80
(d)	1.665	20		48

ICA 8-19

Complete the table, using the ideal gas law:

	Compound	Mass [g]	MW [g/mol]	Amount [mol]	Pressure [atm]	Volume [L]	Temperature [K]
(a)	Oxygen (O_2)	8	32		2.0		400
(b)	Nitrogen (N_2)	5.6	28			5	373
(c)	Chlorine (Cl_2)		71	0.5	2.5	5	
(d)	Xylene	265		2.5	2.5		273

ICA 8-20

Complete the table, using the ideal gas law:

	Compound	Mass [lb_m]	MW [g/mol]	Amount [mol]	Pressure [Pa]	Volume [gal]	Temperature [°C]
(a)	Acetylene (C_2H_2)	0.1	26		303,975		−23
(b)	Benzene (C_6H_6)	0.04	78			1.32	100
(c)	Naphthalene	0.07		0.25	131,723	1.32	
(d)	Toluene	0.425		2.1	354,638		27

ICA 8-21

Complete the table for specific heat conversions:

	Compound	[cal/(g °C)]	[BTU/(lb_m °F)]	[J/(kg K)]
(a)	Neon			1,030
(b)	Benzene		0.0406	
(c)	Mercury	0.03325		
(d)	Aluminum			0.897
(e)	Ethanol			2,440
(f)	Copper		0.0912	

ICA 8-22

Complete the table for thermal conductivity conversions:

	Compound	[W/(m °C)]	[BTU/(ft h °F)]	[cal/(cm min K)]
(a)	Zinc		122	
(b)	Silver	420		
(c)	Lead	35		
(d)	Sandstone		1.387	
(e)	Copper		232	
(f)	Glass	1.1		

ICA 8-23

Complete the table. This problem involves the power required to raise a mass a given distance in a given amount of time.

	Mass [kg]	Distance [m]	Energy [J]	Time [s]	Power [W]
(a)	250	11		200	
(b)	440	52			100
(c)	60		2,940	60	
(d)		10		50	100

ICA 8-24

Complete the table. This problem involves the power required to raise a mass a given distance in a given amount of time.

	Mass [lb$_m$]	Distance [ft]	Energy [J]	Time [min]	Power [hp]
(a)	220	15		0.5	
(b)	1,875	145			0.268
(c)	200		4,000	1	
(d)		35		2	0.134

ICA 8-25

The specific heat of copper is 0.09 British thermal units per pound-mass degree Fahrenheit. Convert this value into units of joules per gram kelvin.

ICA 8-26

The specific heat of helium is 5.24 joules per gram kelvin. Convert this value into units of British thermal units per pound-mass degree Fahrenheit.

ICA 8-27

The thermal conductivity of a plastic is 0.325 British thermal units per foot hour degree Fahrenheit. Convert this value in units of watts per meter kelvin.

ICA 8-28

The heat transfer coefficient of steel is 25 watts per square meter degree Celsius. Convert this value into units of calories per square centimeter second kelvin.

ICA 8-29

One problem with solar energy is that any given point on the planet is illuminated by the sun for only half of the time at best. It would be helpful, therefore, if there were a simple, affordable, and efficient means for storing any excess energy generated on sunny days for use during the night or on cloudy days.

You are investigating the electrodes used in electrolysis cells as part of a three-stage process for solar energy collection and storage.

1. Convert sunlight to electricity with photovoltaic cells.
2. Use the electricity generated in an electrolysis cell to split water into its component elements, hydrogen and oxygen. The hydrogen can be stored indefinitely. The oxygen can simply be released into the atmosphere.
3. Use a fuel cell to recombine the stored hydrogen with oxygen from the atmosphere to generate electricity.

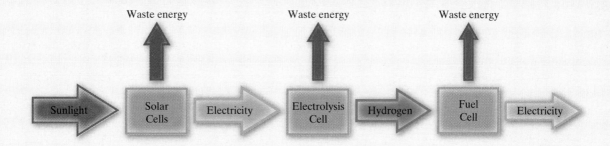

You have obtained an array of new high-efficiency, thin-film photovoltaic cells with an efficiency of 41%. The efficiency of fuel cells varies with the current demands placed on them, but the cells you have obtained yield an overall efficiency of 37% at the anticipated load.

Assume the total solar power on the solar cells is 2,000 watts. You conduct four experiments, each with a different alloy of palladium, platinum, gold, copper, and/or silver for the electrodes in the electrolysis cell. The final output power from the fuel cell is measured for each case, and the results are tabulated below. Determine the efficiency of each electrolysis cell and complete the table.

Alloy	Output Power (P_0) [W]	Electrolysis Cell Efficiency (η)
Alloy A	137	
Alloy B	201	
Alloy C	67	
Alloy D	177	

ICA 8-30

Materials

Bag of cylinders	Scale	Calipers	Ruler

Procedure

For each cylinder, record the mass, length, and diameter in the table provided.

Analysis

- Calculate the volume and density for each cylinder, recording the results in the table provided.
- Rank the rods in order of increasing density, with the least dense rod first on the list.
- Using the density, determine the material of each rod.

Data Worksheet

	Measured Values			Calculated Values	
Description	Mass	Length	Diameter	Volume	Density
Units					
Rod 1					
Rod 2					
Rod 3					
Rod 4					
Rod 5					
Rod 6					
Rod 7					
Rod 8					
Rod 9					
Rod 10					

Rank Rods Increasing Density	Density	Material

ICA 8-31

Materials

25 mL graduated cylinder   Scale   Paper towels

Unknown liquids   Basket labeled "Wash"   Water bottle

Wastewater bucket

Procedure

Record the following data in the table provided:

1. Weigh the empty graduated cylinder and record the value.
2. Pour 15 milliliters [mL] of water into the cylinder.
3. Weigh the cylinder with water and record the value.

4. Pour the water into the wastewater bucket.
5. Pour 15 milliliters [mL] of unknown liquid (UL) 1 into a cylinder.
6. Weigh the cylinder with UL 1 and record the value.
7. Pour the UL back into the original container.
8. Place the graduated cylinder in the "Wash" basket if more than one cylinder is available. If only one cylinder is being used, wash the cylinder with water so no trace of the UL is left in the cylinder.
9. Repeat Steps 5–9 with each unknown liquid provided.

Analysis

- Calculate the density and specific gravity for each fluid, recording the results in the table provided.
- Rank the fluids in order of increasing density, with the least dense fluid first on the list.
- Using the specific gravity, determine the type of liquid in each container.

Data Worksheet

	Measured			Calculated	
Description	Total Mass	Volume	Liquid Mass	Density	Specific Gravity
Units					
Empty cylinder					
Water					
UL 1					
UL 2					
UL 3					
UL 4					
UL 5					

Rank Liquids Increasing Density	Density	Liquid

Chapter 8 REVIEW QUESTIONS

1. If a person weighs 700 newtons, what is the mass of the person in units of kilograms?
2. If a person weighs 700 newtons, what is the mass of the person in units of pounds-mass?
3. If a person weighs 700 newtons, what is the mass of the person in units of slugs?
4. A football lineman weighs 300 pounds-force. What is his mass in units of kilograms?
5. A football lineman weighs 300 pounds-force. What is his mass in units of pounds-mass?
6. A football lineman weighs 300 pounds-force. What is his mass in units of slugs?
7. The weight of a can of soda on the moon (where the acceleration of gravity is 1.6 meters per second squared) is 0.6 newtons. What is the mass of the can of soda on the Earth in units of kilograms?
8. The weight of a can of soda on the moon (where the acceleration of gravity is 1.6 meters per second squared) is 0.6 newtons. What is the mass of the can of soda on the Earth in units of pounds-mass?
9. If the density of silicon is 10.5 grams per cubic centimeter, what is this in units of pounds-mass per cubic foot?
10. If the density of sulfur is 131 pounds-mass per cubic foot, what is this in units of grams per cubic centimeter?
11. If the density of sodium is 98 kilograms per cubic meter, what is this in units of slugs per gallon?
12. A basketball has a mass of approximately 624 grams and a volume of 0.25 cubic feet. Determine the density of the basketball in units of pounds-mass per cubic foot.
13. Consider the following strange, but true, units:

    1 arroba = 11.5 kilograms
    1 peck = 9 liters

    A basketball has a mass of approximately 624 grams and a volume of 0.25 cubic feet. Determine the density of the basketball in units of arroba per peck.
14. Consider the following strange, but true, units:

    1 batman = 3 kilograms
    1 hogshead = 63 gallons

    A basketball has a mass of approximately 624 grams and a volume of 0.25 cubic feet. Determine the density of the basketball in units of batman per hogshead.
15. The specific gravity of acetic acid (vinegar) is 1.049. State the density in units of pounds-mass per cubic foot.
16. The specific gravity of ethyl alcohol is 0.785. State the density in units of grams per cubic centimeter.
17. The specific gravity of chloroform is 1.489. State the density in units of kilograms per cubic meter.
18. The specific gravity of iodine is 4.927. State the density in units of slugs per liter.
19. A cube of material X, 1 inch on all sides, has a mass of 0.05 kilograms. Determine the specific gravity of material X.
20. The specific gravity of gold is 19.3. What is the length of one side of a 0.4 kilogram cube of solid gold, in units of inches?
21. The density of gasoline is 0.72 grams per cubic centimeter. What is the mass in units of kilograms of a 5-gallon container filled completely with gasoline? Ignore the mass of the container.
22. A lab reports that the density of a new element is X kilograms per cubic foot and Y grams per cubic meter. Which of the following statements is true?
    (A) $X > Y$.
    (B) $X < Y$.
    (C) $X = Y$.
    (D) Cannot be determined.

23. The Eco-Marathon is an annual competition sponsored by Shell Oil, in which participants build special vehicles to achieve the highest possible fuel efficiency. The Eco-Marathon is held around the world with events in the United Kingdom, Finland, France, Holland, Japan, and the United States.

A world record was set in Eco-marathon by a French team in 2003 called Microjoule with a performance of 10,705 miles per gallon. The Microjoule runs on ethanol. If the cars are given 100 grams of ethanol (specific gravity = 0.789) and drive until the fuel runs out, how far did the Microjoule drive in kilometers?

24. A golden bar of metal (5 centimeters by 18 centimeters by 4 centimeters) being transported by armored car is suspected of being fake, made from a less valuable metal with a thin coating of pure gold. The bar is found to have a mass of 2.7 kilograms. If the specific gravity of gold is 19.3, is the bar fake? Justify your answer.

25. A rod on the surface of Jupiter's moon Callisto has a volume of 0.3 cubic meters. Determine the weight of the rod in units of pounds-force. The density is 4,700 kilograms per cubic meter. Gravitational acceleration on Callisto is 1.25 meters per second squared.

26. A substance used to remove the few remaining molecules from a near vacuum by reacting with them or adsorbing them is called a getter. There are numerous materials used and several ways of deploying them within a system enclosing a vacuum, but here we will look at a common method used in vacuum tubes, once the workhorse of electronics but now relegated to high-end audio systems and other niche markets. In vacuum tubes, after the air is evacuated with a vacuum pump, getters are usually deposited inside the hemispherical top by flash deposition. Assume that it is desired to flash deposit 1.5×10^{-3} moles of a getter onto the hemispherical top of a vacuum tube with an inside diameter of three-quarters of an inch. For each of the following getter materials, how thick will the coating be? Report your answers using meters with an appropriately chosen prefix.

The 12AX7 is a very common dual triode vacuum tube first developed for audio applications in the mid-1940s and still in common use in guitar amplifiers.

Getter Material	Specific Gravity	Atomic Weight [g/mol]
Aluminum	2.7	26.981
Barium	3.51	137.33
Calcium	1.55	40.078
Magnesium	1.738	24.305
Sodium	0.968	22.99
Strontium	2.64	87.62

27. The largest temperature decline during a 24-hour period was 56 degrees Celsius in Browning, Montana. Express this as degrees Fahrenheit per minute.

28. The largest temperature decline during a 24-hour period was 56 degrees Celsius in Browning, Montana. Express this as degrees Rankine per second.

29. The world's lowest recorded temperature was −129 degrees Fahrenheit in Vostok, Antarctica. State this temperature in units of degrees Celsius.

30. The world's lowest recorded temperature was −129 degrees Fahrenheit in Vostok, Antarctica. State this temperature in units of kelvin.

31. If we increase the temperature in a reactor by 90 degrees Fahrenheit, how many degrees Celsius will the temperature increase?

32. We are making a cup of coffee and want the temperature to be just right, so we measure the temperature with both Fahrenheit and Celsius thermometers. The Fahrenheit meter registers 110 degrees Fahrenheit, but you prefer to it to be slightly hotter at 119 degrees Fahrenheit, so we heat it up a little. How much will the Celsius thermometer increase when we make this change?

33. Which of the following plastics has the highest melting temperature? You must prove your answer for credit!
    (A) Acrylic at 150 degrees Fahrenheit.
    (B) Polyethylene terephthalate (PET) at 423 kelvin.
    (C) High-density polyethylene (HDPE) at 710 degrees Rankine.

34. The tiles on the space shuttle are constructed to withstand a temperature of 1,950 kelvin. What is the temperature in units of degrees Rankine?

35. The tiles on the space shuttle are constructed to withstand a temperature of 1,950 kelvin. What is the temperature in units of degrees Fahrenheit?

36. The boiling point of propane is −43 degrees Celsius. What is the temperature in units of degrees Fahrenheit?

37. The boiling point of propane is −43 degrees Celsius. What is the temperature in units of kelvin?

38. We want to construct a thermometer using mercury (Hg). As the mercury in the bulb is heated, it expands and moves up the thin capillary tube connected to the bulb. The symbol used for the coefficient of volume expansion of a substance due to a temperature increase is β. It is used in the following equation:

$$\Delta V = \beta V (\Delta T)$$

Here, ΔV is the increase in volume, V is the original volume, and ΔT is the temperature increase. The value of β for mercury is 1.8×10^{-4} [1/degree Celsius]. If the bulb contains 0.2 milliliters and the tube has a diameter of 0.2 millimeters, how much will the mercury rise in the tube in units of centimeters if we increase the temperature from 30 degrees Fahrenheit to 70 degrees Fahrenheit?

39. You are designing a new thermometer using Galinstan®, an alloy of gallium, indium, and tin that is liquid at normal living temperatures. The specific alloy used has a coefficient of thermal expansion $\beta = 190 \times 10^{-6}$ [1/kelvin].

    The change in volume (ΔV) for a given change in temperature (ΔT) can be determined by

$$\Delta V = \beta V (\Delta T)$$

Here, ΔV is the increase in volume, V is the original volume, and ΔT is the temperature increase. The thermometer will contain two cubic centimeters of Galinstan®, most of which is in the "bulb" or reservoir that is connected to a capillary tube up which the liquid moves as it expands. If your design specifications are to have a 2-millimeter change in the position of the liquid in the capillary tube for each degree Fahrenheit change in temperature, what is the diameter of the capillary tube, assuming it has a circular cross section?

40. If the pressure is 250 feet of water, what is the pressure in units of inches of mercury?

41. If the pressure is 250 feet of water, what is the pressure in units of millimeters of mercury?

42. If the pressure is 250 feet of water, what is the pressure in units of pounds-force per square inch?

43. If the pressure is 100 millimeters of mercury, what is the pressure in units of feet of water?

44. If the pressure is 100 millimeters of mercury, what is the pressure in units of pascals?

45. If the pressure is 100 millimeters of mercury, what is the pressure in units of atmospheres?

46. A "normal" blood pressure is 120 millimeters of mercury (systolic reading) over 80 millimeters of mercury (diastolic reading). Convert 120 millimeters of mercury into units of pounds-force per square inch.

47. A "normal" blood pressure is 120 millimeters of mercury (systolic reading) over 80 millimeters of mercury (diastolic reading). Convert 80 millimeters of mercury into units of pascals.

48. A car tire is inflated to 30 pounds-force per square inch. If the tire has an area of 0.25 square feet in contact with the road, how much force is exerted by all four tires?

49. The force on the inside of a cork in a champagne bottle is 10 pound-force. If the cork has a diameter of 0.5 inches, what is the pressure inside the bottle in units of feet of water?

50. If a force of 15 newtons is applied to a surface and the pressure is measured as 4,000 pascals, what is the area of the surface in units of square meters?

51. A sensor is submerged in a silo to detect any bacterial growth in the stored fluid. The stored fluid has a density of 2.2 grams per cubic centimeters. What is the hydrostatic pressure felt by the sensor at a depth of 30 meters in units of atmospheres?

52. One of the National Academy of Engineering Grand Challenges for Engineering is **Develop Carbon Sequestration Methods.** The NAE defines carbon sequestration as "capturing the carbon dioxide produced by burning fossil fuels and storing it safely away from the atmosphere." The most promising storage location is underground, possibly in sedimentary brine formations. You are assigned to develop instrumentation to measure the properties of a brine formation, located 800 meters deep. Assume the instruments will feel an equivalent amount of pressure to the amount of hydrostatic pressure felt at the bottom of an 800-meter high column of brine, with a specific gravity of 1.35. To what hydrostatic pressure, in units of atmospheres, must the instrumentation be built to withstand?

53. A cylindrical tank filled to a height of 25 feet with tribromoethylene has been pressurized to 2 atmospheres ($P_{surface} = 2$ atmospheres). The total pressure at the bottom of the tank is 4 atmospheres. Determine the density of tribromoethylene in units of kilograms per cubic meter.

54. A submersible vehicle is being designed to operate in the Atlantic Ocean. Density of ocean water is 1.025 grams per cubic centimeter. For a maximum depth of 300 feet, what is the total pressure the hull of the submersible must be designed to withstand? Give your answer in units of pounds per square inch.

55. If the total pressure on the bottom of an open container is 2.5 atmospheres, what would the atmospheric pressure have to be to double the total pressure on the bottom?

56. NASA is designing a mission to explore Titan, the largest moon of Saturn. Titan has numerous hydrocarbon lakes containing a mix of methane and ethane in unknown proportions. As part of the mission, a small submersible vehicle will explore Kraken Mare, the largest of these lakes, to determine, among other things, how deep it is. Assuming that the maximum depth of Kraken Mare is less than 400 meters, how much pressure in atmospheres must the submersible be designed to withstand? Assume the surface pressure on Titan is 147 kilopascals, the surface temperature is 94 kelvin, and the gravity is 1.35 meters per second squared. The specific gravity of liquid methane is 0.415 and the specific gravity of liquid ethane is 0.546.

57. Airspeed (v), is determined from dynamic pressure using the following formula: $P_{dynamic} = \frac{1}{2}\rho v^2$. Determine the dynamic pressure, in units of pascals, for an aircraft moving at an airspeed of 600 miles per hour. Air density is 1.20 kilograms per cubic meter.

58. A 10-liter flask contains 1.3 moles of an ideal gas at a temperature of 20 degrees Celsius. What is the pressure in the flask in units of atmospheres?

59. An ideal gas, kept in a 5-liter container at 300 kelvin, exhibits a pressure of 2 atmospheres. If the volume of the container is decreased to 2.9 liters, but the temperature remains the same, what is pressure in the new container in units of atmospheres?

60. An ideal gas is kept in a 10-liter container at a pressure of 1.5 atmospheres and a temperature of 310 kelvins. If the gas is compressed until its pressure is raised to 3 atmospheres while holding the temperature constant, what is the new volume in units of liters?

61. A 5-liter container holds nitrogen (formula: N_2, molecular weight = 28 grams per mole) at a pressure of 1.1 atmospheres and a temperature of 400 kelvins. What is the amount of nitrogen in the container, in units of grams?

62. An ideal gas in a 1.25-gallon container is at a temperature of 125 degrees Celsius and pressure of 2.5 atmospheres. If the gas is oxygen (formula: O_2, molecular weight = 32 grams per mole), what is the amount of gas in the container in units of grams?

63. A 10-liter flask contains 5 moles of gas at a pressure of 15 atmospheres. What is the temperature in the flask in units of kelvin?

64. A container holding 1.5 moles of oxygen (formula: O_2, molecular weight = 32 grams per mole) at a pressure of 1.5 atmospheres and a temperature of 310 kelvins is heated to 420 kelvins, while maintaining constant volume. What is the new pressure inside the container in units of pascals?

65. The specific heat of neon is 1.03 joules per gram kelvin. Convert this value into units of calorie per kilogram degree Celsius.

66. The specific heat of copper is 0.386 joules per gram kelvin. Convert this value into units of British thermal units per pound-mass degree Fahrenheit.

67. The specific heat of octane is 0.51 calorie per gram degree Celsius. Convert this value into units of joules per gram kelvin.

68. The specific heat of tungsten is 0.134 calorie per gram degree Celsius. Convert this value into units of British thermal units per pound-mass degree Fahrenheit.

69. Which of the following requires the expenditure of more work? You must show your work to receive credit.
    (A) Lifting a 100-newton weight a height of 4 meters.
    (B) Sliding a sofa, weighing 200 pounds-force, 30 feet across a room. Assume an additional 50 pounds-force is required to overcome the resisting friction.

70. Which object—A, B, or C—has the most potential energy when held a distance H above the surface of the ground? You must show your work to receive credit.

Object A:	mass = 1 kilogram	height = 3 meters
Object B:	mass = 1 slug	height = 3 feet
Object C:	mass = 1 gram	height = 1 centimeter

71. A 10-gram rubber ball is released from a height of 6 meters above a flat surface on the moon. Gravitational acceleration on the moon is 1.62 meters per second squared. Assume that no energy is lost from frictional drag. What is the velocity, in units of meters per second, of the rubber ball the instant before it strikes the flat surface?

72. A robotic rover on Mars finds a spherical rock with a diameter of 10 centimeters. The rover picks up the rock and lifts it 20 centimeters straight up. The resulting potential energy of the rock relative to the surface is 2 joules. Gravitational acceleration on Mars is 3.7 meters per second squared. What is the specific gravity of the rock?

73. If a person weighs 200 pounds-mass, how fast must they run in units of meters per second to have a kinetic energy of 1,000 calories?

74. Measurements indicate that boat A has twice the kinetic energy of boat B of the same mass. How fast is boat A traveling if boat B is moving at 30 knots? 1 knot = 1 nautical mile per hour; 1 nautical mile = 6,076 feet.

75. If a 10-kilogram rotating solid cylinder moves at a velocity (v), it has a kinetic energy of 36 joules. Determine the velocity the object is moving in units of meters per second if the kinetic energy is given by KE = ½ mv^2 + ¼ mv^2.

76. If a ball is dropped from a height (H) its velocity will increase until it hits the ground (assuming that aerodynamic drag due to the air is negligible). During its fall, its initial potential energy is converted into kinetic energy. If the mass of the ball is doubled, how will the impact velocity change?

77. If a ball is dropped from a height (H) its velocity will increase until it hits the ground (assuming that aerodynamic drag due to the air is negligible). During its fall, its initial potential energy is converted into kinetic energy. If the ball is dropped from a height of 800 centimeters, and the impact velocity is 41 feet per second, determine the value of gravity in units of meters per second.

78. A ball is thrown vertically into the air with an initial kinetic energy of 2,500 joules. As the ball rises, it gradually loses kinetic energy as its potential energy increases. At the top of its flight, when its vertical speed goes to zero, all of the kinetic energy has been converted into potential energy. Assume that no energy is lost to frictional drag, etc. How high does the ball rise in units of meters if it has a mass of 5 kilograms?

79. The specific heats of aluminum and iron are 0.214 and 0.107 calories per gram degree Celsius, respectively. If we add the same amount of energy to a cube of each material of the same mass and find that the temperature of the aluminum increases by 30 degrees Fahrenheit, how much will the iron temperature increase in degrees Fahrenheit?

80. We go out to sunbathe on a warm summer day. If we soak up 100 British thermal units per hour of energy, how much will the temperature of 60,000-gram person increase in 2 hours in units of degrees Celsius? We assume that since our bodies are mostly water they have the same specific heat as water. The specific heat of water is 4.18 joules per gram degree Celsius.

81. A 3-kilogram projectile traveling at 100 meters per second is stopped by being shot into an insulated tank containing 100 kilograms of water. If the kinetic energy of the projectile is completely converted into thermal energy with no energy lost, how much will the water increase in temperature in units of degrees Celsius? The specific heat of water is 1 calorie per gram per degree Celsius.

82. The thermal conductivity of ice is 1.6 watts per meter kelvin. Convert this into units of calories per centimeter second degree Celsius.

83. The thermal conductivity of nickel is 91 watts per meter kelvin. Convert this into units of British thermal units per foot hour degree Fahrenheit.

84. The thermal conductivity of gold is 179 British thermal units per foot hour degree Fahrenheit. Convert this into units of watts per meter kelvin.

85. The thermal conductivity of beryllium is 126 British thermal units per foot hour degree Fahrenheit. Convert this into units of calories per centimeter second degree Celsius.

86. The heat transfer coefficient of water is 300 British thermal units per square foot hour degree Fahrenheit. Convert this into units of watts per square meter kelvin.

87. The heat transfer coefficient of stainless steel is 21 watts per square meter kelvin. Convert this into units of calories per square centimeter second degree Celsius.

88. The heat transfer coefficient of material A is 85 calories per square centimeter second degree Celsius. Convert this into units of watts per square meter kelvin.

89. The heat transfer coefficient of material X is 130 British thermal units per square foot hour degree Fahrenheit. Convert this into units of calories per square centimeter second degree Celsius.

90. When we drive our car at 100 feet per second, we measure an aerodynamic force (called drag) of 66 pounds-force that opposes the motion of the car. How much horsepower is required to overcome this drag?

91. The power required by an airplane is given by $P = Fv$, where P is the engine power, F is the thrust, and v is the plane speed. At what speed will a 500-horsepower engine with 1,000 pounds-force of thrust propel the plane?

92. The power required by an airplane is given by $P = Fv$, where P is the engine power, F is the thrust, and v is the plane speed. What horsepower is required for 1,000 pounds-force of thrust to propel a plane 400 miles per hour?

93. The power required by an airplane is given by $P = Fv$, where P is the engine power, F is the thrust, and v is the plane speed. Which of the following planes has the most power? You must show your work to receive credit.

Plane A:	Thrust = 2,000 pounds-force	Speed = 200 meters per second
Plane B:	Thrust = 13,000 newtons	Speed = 500 feet per second

94. When gasoline is burned in the cylinder of an engine, it creates a high pressure that pushes on the piston. If the pressure is 100 pound-force per square inch, and it moves the 3-inch diameter piston a distance of 5 centimeters in 0.1 seconds, how much horsepower does this action produce?

95. A 100-watt motor (60% efficient) is used to raise a 100-kilogram load 5 meters into the air. How long, in units of seconds, will it take the motor to accomplish this task?

96. A 100-watt motor (60% efficient) is available to raise a load 5 meters into the air. If the task takes 65 seconds to complete, how heavy was the load in units of kilograms?

97. You need to purchase a motor to supply 400 joules in 10 seconds. All of the motors you can choose from are 80% efficient. What is the minimum wattage on the motor you need to choose?

98. A robotic rover on Mars finds a spherical rock with a diameter of 10 centimeters. The rover picks up the rock and lifts it 20 centimeters straight up. The rock has a specific gravity of 4.75. The gravitational acceleration on Mars is 3.7 meters per second squared. If the robot's lifting arm has an efficiency of 40% and required 10 seconds to raise the rock 20 centimeters, how much power (in watts) did the arm use?

99. Consider the following strange, but true, unit:

    1 donkeypower = 0.33 horsepower

    A certain motor is rated to supply an input power of 2,500 calories per minute at an efficiency of 90%. Determine the amount of output power available in units of donkeypower.

100. When boiling water, a hot plate takes an average of 8 minutes and 55 seconds to boil 100 milliliters of water. Assume the temperature in the lab is 75 degrees Fahrenheit. The hot plate is rated to provide 283 watts. How efficient is the hot plate?

101. When boiling water, a hot plate takes an average of 8 minutes and 55 seconds to boil 100 milliliters of water. Assume the temperature in the lab is 75 degrees Fahrenheit. The hot plate is rated to provide 283 watts. If we wish to boil 100 milliliters of acetone using this same hot plate, how long do we expect the process to take? Acetone has a boiling point of 56 degrees Celsius. [*Hint:* You must determine the efficiency of the hotplate.]

102. You are part of an engineering firm on contract by the U.S. Department of Energy's Energy Efficiency and Renewable Energy task force to measure the power efficiency of home appliances. Your job is to measure the efficiency of stove-top burners. In order to report the efficiency, you will place a pan containing one gallon of room temperature water on their stove, record the initial room temperature, turn on the burner, and wait for it to boil. When the water begins to boil, you will record the time it takes the water to boil and look up the power for the burner provided by the manufacturer. After measuring the following stove-top burners, what is the efficiency of each burner?

Room Temp [°F]	Time to Boil [min]	Rated Burner Power [W]
(a) 70	20	1400
(b) 71	25	1300
(c) 72	21	1500
(d) 68	26	1400
(e) 69	18	1350

103. The power available from a wind turbine is calculated by the following equation:

$$P = \frac{1}{2}A\rho v^2$$

where P = power [watts], A = sweep area (circular) of the blades [square meters], ρ = air density [kilograms per cubic meter], and v = velocity [meters per second]. The world's largest sweep area wind turbine generator in Spain has a blade diameter of 420 feet. The specific gravity of air is 0.00123. Assuming a velocity of 30 miles per hour and an actual power of 5 megawatts, determine the efficiency of this turbine.

104. The maximum radius a falling liquid drop can have without breaking apart is given by the equation $R = \sqrt{\sigma/(g\rho)}$, where σ is the liquid surface tension, g is the acceleration due to gravity, and ρ is the density of the liquid. For bromine at 20 degrees Celsius, determine the surface tension (σ) in units of joules per meter squared if the maximum radius of a drop is 0.8 centimeter and the specific gravity of the liquid is 2.9.

105. The maximum radius a falling liquid drop can have without breaking apart is given by the equation $R = \sqrt{\sigma/g\rho}$, where σ is the liquid surface tension, g is the acceleration due to gravity, and ρ is the density of the liquid. For acetone at 20 degrees Celsius, determine the surface tension (σ) in units of joules per meter squared if the maximum radius of a drop is 1 centimeter and the specific gravity of the liquid is 0.79.

106. When a flowing fluid is stopped, its pressure increases. This is called stagnation pressure. The stagnation pressure is determined by: $P_{stagnation} = \frac{1}{2}\rho v^2 + P_{surface}$, where ρ is the fluid density, v the fluid speed, and $P_{surface}$ the atmospheric pressure. Calculate the stagnation pressure in units of atmospheres for acetone flowing at 15 feet per second. Assume the density of acetone to be 790 kilograms per cubic meter.

107. When a flowing fluid is stopped, its pressure increases. This is called stagnation pressure. The stagnation pressure is determined by $P_{stagnation} = \frac{1}{2}\rho v^2 + P_{surface}$, where ρ is the fluid density, v is the fluid speed, and $P_{surface}$ is the atmospheric pressure. If the stagnation pressure is 18 pounds-force per square inch, what is the fluid speed in units of feet per minute? Assume the fluid is methyl ethyl ketone (MEK) and the density of is 805 kilograms per cubic meter.

Chapter 9
DIMENSIONLESS NUMBERS

Recall that in Chapter 7, we discussed the concept of dimensions. A **dimension** is a measurable physical idea; it generally consists solely of a word description with no numbers. A **unit** allows us to quantify a dimension, to state a number describing how much of that dimension exists in a specific situation. Units are defined by convention and related to an accepted standard. Table 9-1 (which is the same as Table 7-1) shows the seven base units and their corresponding fundamental dimensions.

Table 9-1 Fundamental dimensions and base units

Dimension	Symbol	Unit	Symbol
Length	L	meter	m
Mass	M	kilogram	kg
Time	T	second	s
Temperature	Θ	kelvin	K
Amount of substance	N	mole	mol
Light intensity	J	candela	cd
Electric current	I	ampere	A

9.1 COMMON DIMENSIONLESS NUMBERS

Sometimes, we form the ratio of two parameters, where each parameter has the same dimensions. Sometimes, we form a ratio with two groups of parameters, where each group has the same dimensions. The final result in both cases is dimensionless.

Pi (π): One example is the parameter π, used in the calculation of a circumference or area of a circle. The reason π is dimensionless is that it is actually defined as the ratio of the circumference of a circle to its diameter:

$$\pi = \frac{C}{D} = \frac{\text{circumference}}{\text{diameter}} \; \{=\} \; \frac{\text{length}}{\text{length}} = \frac{L^1}{L^1} = L^0$$

The ratio of one length to another length yields a dimensionless ratio. We can see this in another way through reversing the process. For the circumference of a circle:

$$C = \pi D$$

and if dimensions are inserted,

$$\{L^1\} = \pi\{L^1\}$$

This equation is dimensionally correct only if π has no dimensions. The same result is obtained for the equation of the area of a circle.

$$A = \pi r^2$$

Inserting dimensions:

$$\{L^2\} = \pi\{L^1\}\ \{L^1\} = \pi\{L^2\}$$

Again, this equation is dimensionally correct only if π is dimensionless.

Specific Gravity (SG): The specific gravity is the ratio of the density of an object to the density of water.

$$\text{Specific gravity} = \frac{\text{density of the object}}{\text{density of water}} = \frac{\text{mass/volume}}{\text{mass/volume}}\ \{=\}\ \frac{\{M/L^3\}}{\{M/L^3\}} = \{M^0L^0\}$$

Mach Number (Ma): We often describe the speed at which an airplane or rocket travels in terms of the Mach number, named after Ernst Mach, an Austrian physicist. This number is the ratio of the speed of the plane compared with the speed of sound in air.

$$\text{Mach number} = \frac{\text{speed of the object}}{\text{speed of sound in air}}\ \{=\}\ \frac{\{L/T\}}{\{L/T\}} = \{L^0T^0\}$$

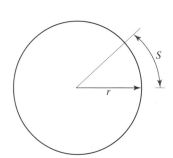

Radian [rad]: The derived unit of a radian is defined as the angle at the center of a circle formed by an arc (S) equal in length to the radius (r) of that circle. In a complete circle, there are two π radians. Since by definition a radian is a length (S) divided by a length (r), it is a dimensionless ratio. Note, however, that angles are often measured using the radian as a unit, even though it is dimensionless.

$$1 \text{ radian [rad]} = \frac{S}{r}\ \{=\}\ \frac{L}{L} = L^0$$

Table 9-2 Some common dimensionless parameters

Name	Phenomena Ratio	Symbol	Expression
Coefficient of friction	Sideways force (F)/weight of object (w) [object static or kinetic (object sliding)]	μ_{st} and μ_k	F/w
Drag coefficient	Drag force (F_d)/inertia force (ρ, density; v, speed; A body area)	C_d	$F_d/(\frac{1}{2}\rho v^2 A)$
Mach number	Object speed (v)/speed of sound (v_{sound})	Ma	v/v_{sound}
Pi	Circle circumference (C)/ circle diameter (D)	π	C/D
Poisson's ratio	Transverse contraction (ε_{trans})/ longitudinal extension (ε_{long})	v	$\varepsilon_{trans}/\varepsilon_{long}$
Radian	Arc length (S)/circle radius (r)	rad	S/r
Specific gravity	Object density/density of water	SG	ρ/ρ_{H_2O}

We must remind ourselves that it is always essential to use the appropriate dimensions and units for every parameter. Suppose that we are interested in computing the sine of an angle. This can be expressed as a dimensionless number by forming the ratio of the length of the opposite side divided by the length of the hypotenuse of a right triangle.

$$\sin(x) = \frac{\text{length opposite side}}{\text{length hypotenuse}} \{=\} \frac{L}{L} = L^0$$

In addition to the ratio of two lengths, you will know from one of your math classes that the sine can be also be expressed as an infinite series given by:

$$\sin(x) = x - \frac{x^3}{3!} + \frac{x^5}{5!} - \frac{x^7}{7!} + \cdots$$

LAW OF ARGUMENTS

Any function that can be computed using a series must employ a *dimensionless* argument.

This includes all the trigonometric functions, logarithms, and e^x, where e is the base of natural logarithms.

Let us suppose that the argument x had the units of length, say, feet. The units in this series would then read as:

$$\text{ft} - \frac{\text{ft}^3}{3!} + \frac{\text{ft}^5}{5!} - \frac{\text{ft}^7}{7!} + \cdots$$

We already know that we cannot add two terms unless they have the same units; recall the Plus law from Chapter 7. This is clearly not the case in the example above. The only way we can add these terms, all with different exponents, is if each term is dimensionless. Consequently, when we calculate $\sin(x)$, we see that the x must be dimensionless, which is why we use the unit of radians. This conclusion is true for any function that can be computed using a series form, leading to the **Law of Arguments**.

● EXAMPLE 9-1

What are the dimensions of k in the following equation, where d is distance and t is time?

$$d = Be^{kt}$$

Since exponents must be dimensionless, the product of k and t must not contain any dimensions. The dimensions of time are {T}

$$kT^1 \{=\} M^0 L^0 T^0 \Theta^0$$

Solving for k yields:

$$k\{=\}T^{-1}$$

k is expressed in dimensions of inverse time or "per time."

9.2 DIMENSIONAL ANALYSIS

Dimensionless quantities are generated as a result of a process called **dimensional analysis**. As an example, suppose we want to study rectangles, assuming that we know nothing about rectangles. We are interested in the relationship between the area of a rectangle, the width of the rectangle, and the perimeter of the rectangle. We cut out a lot of paper rectangles and ask students in the class to measure the area, the perimeter, and the width (Table 9-3).

Table 9-3 Rectangle measurements

Perimeter (P) [cm]	Area (A) [cm²]	Width (W) [cm]
4.02	1.0	1.1
8.75	4.7	1.9
6	2.3	1.55
13.1	6.0	1.1
17.75	19	5.25
10.25	1.2	0.25
12.1	3.0	5.5
6	0.3	2.9
16.25	15.4	5.1
17	7.8	1.05

If we graph the area against the perimeter, we obtain the following plot:

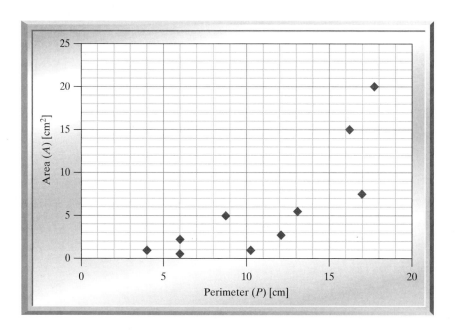

From this, we see that the data are scattered. We would not have a great deal of confidence in drawing conclusions about how the area depended on the perimeter of the rectangle (or in trying to draw a line through the data). The best we could do is to make a statement such as, "It seems that the larger the perimeter, the larger the area." However, close examination of the data table shows that as the perimeter increases from 8.75 to 10.25 centimeters and from 16.25 to 17 centimeters, the area actually decreases in each case. One reason for this problem is that our plot has omitted one important parameter: the width.

Analysis shows that one way in which to generalize plots of this type is to create dimensionless parameters from the problem variables. In this case, we have perimeter with dimension of length, width with the dimension of length, and area with the dimension of length squared. A little thought shows that we could use the ratio of P/W (or W/P) instead of just P on the abscissa. The ratio W/P has the dimensions of length/length, so it is dimensionless. It does not matter whether this is miles/miles, or centimeters/centimeters, but the ratio is dimensionless. Similarly, we could write $A/(W^2)$, and this would also be dimensionless.

These ratios are plotted and shown below. The scatter of the first plot disappears and all the data appear along a single line.

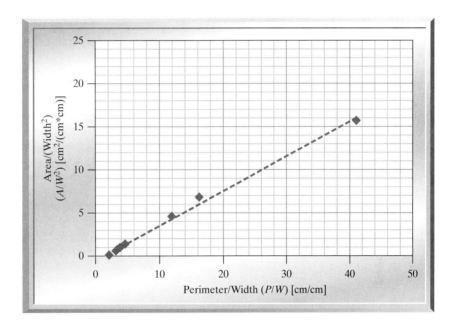

To understand how to read data from this chart, let us examine the following question. If a rectangle has a perimeter of 20 feet and a width of 2 feet, what is the area?

Step A: $P/W = (20 \text{ ft})/(2 \text{ ft}) = 10$ (with no units).

Step B: From the chart, at a P/W value of 10, we read a value from the line of $A/(W^2) = 3.5$.

Step C: Calculate A from this as $A = 3.5 * (2 \text{ ft} * 2 \text{ ft}) = 14 \text{ ft}^2$.

Some of you may be thinking that we made this problem unnecessarily difficult. After all, anyone who manages to get to college knows that the "sensible" measurements to make are length, width, and area. However, many phenomena are far more complicated than simple rectangles, and it is often not at all obvious what parameters should be measured to characterize the behavior of the system we are studying. In situations of this type, dimensionless analysis can become a powerful tool to help us understand *which* parameters affect the behavior of the system and *how* they affect it. With this in mind, let us look at a slightly more complicated example.

A not-so-famous scientist, Dr. Triticale, decided to apply his scientific skills to cooking. He had always been fascinated with the process of cooking pancakes, so it seemed reasonable that he start there. He wanted to learn how to flip the flapjacks in a graceful arc in the air and then catch them.

He spent long summer days pondering this process until he finally was able to produce a list of the parameters that he felt were important. He kept asking himself, "If I change this parameter, will the trajectory of the pancake change?" If he could answer "Yes!" or even "Probably," he then considered the parameter as important enough to include on his list. As he saw it, these parameters were:

Speed of the frying pan, U	Mass of the flapjack, m
Height of the flip, H	Gravity (it pulls the flapjack back down), g

He then wrote this dependency in equation form as $H = f:(U, m, g)$.

Dr. Triticale realized that while he felt that gravity was important, it would not be easy to change the value of gravity in his tests (he could have gone to a high mountain or the moon, but this was too hard). His plan was to do many tests (and consequently eat many pancakes). He would make many measurements for many different flipping speeds and pancake masses, and try to fit a curve to the data.

Based on his work with conversion factors and his knowledge of the Per law, he reasoned that it is acceptable to multiply parameters with different dimensions. It is also fine to raise a parameter (and its associated units) to a power. Based on his understanding of the Plus law, he knew it is not acceptable to add parameters with different dimensions.

Using this information, he decided that it would be permissible to try to "fit" the dependence of pancake flipping to the important parameters raised to different powers and multiplied together. This would create a single term like $k_1 U^{a_1} m^{b_1} g^{c_1}$.

He also knew that if this term were made to have the same dimensions as H, it just might be a legitimate expression. In fact, if this were the case, he could use many terms, each of which had the dimensions of H and add them all together. While this might not be a valid equation, at least it would satisfy the Per and Plus laws, and with many terms he would have a good chance of his equation fitting the data. So, he boldly decided to try the following series:

$$H = k_1 U^{a_1} m^{b_1} g^{c_1} + k_2 U^{a_2} m^{b_2} g^{c_2} + k_3 U^{a_3} m^{b_3} g^{c_3} + \cdots$$

He needed to determine the values of the dimensionless k constants as well as all of the exponents. He knew that all the terms on the right-hand side must have the same dimensions, or they could not be added together. He also knew that the dimensions on the left and right sides must match.

With this, he then realized that he could examine the dimensions of any term on the right-hand side since each had to be the same dimensionally. He did this by comparing a typical right-hand term with the left-hand side of the equation, or

$$H = k U^a m^b g^c$$

The next step was to select the proper values of a, b, and c, so that the dimensions of the right-hand side would match those on the left-hand side. To do this, he substituted the dimensions of each parameter:

$$L^1 M^0 T^0 = \{LT^{-1}\}^a \{M\}^b \{LT^{-2}\}^c = \{L\}^{a+c} \{M\}^b \{T\}^{-a-2c}$$

For this to be dimensionally correct, the exponents for L, M, and T on the right and left would have to match, or

$$L: 1 = a + c \quad M: 0 = b \quad T: 0 = -a - 2c$$

This yields

$$a = 2 \quad b = 0 \quad c = -1$$

From this, Dr. Triticale settled on a typical term as

$$k\, U^2 m^0 g^{-1}$$

Finally, he wrote the "curve fitting" equation (with a whole series of terms) as

$$H = \sum_{i=1}^{\infty} k_1 U^2 g^{-1} + k_2 U^2 g^{-1} + k_3 U^2 g^{-1} + \cdots = \frac{U^2}{g} \sum_{i=1}^{\infty} k_i = (K)\left(\frac{U^2}{g}\right)$$

Now, armed with this expression, he was sure that he could flip flapjacks with the best, although he knew that he would have to conduct many experiments to make sure the equation was valid (and to determine the value of K). What he did not realize was that he had just performed a procedure called "dimensional analysis."

9.3 RAYLEIGH'S METHOD

In this section we formalize the discussion presented in Example 9-2 by introducing a method of dimensional analysis devised by Lord Rayleigh, John William Strutt, the third Baron Rayleigh (1842–1919). Three detailed examples illustrate his approach to dimensionless analysis:

- Example 9-3, in which we analyze factors affecting the distance traveled by an accelerating object
- Example 9-4, in which we determine the most famous named dimensionless number, Reynolds number
- Example 9-5, in which we simplify one use of Rayleigh's method

No matter the problem, the way we solve it stays the same:

Step 1: Write each variable and raise each to an unknown exponent (use *all* the variables, even the dependent variable). Order and choice of exponent do not matter.

Step 2: Substitute dimensions of the variables into Step 1. Be sure to raise each dimension to the proper exponent groups from Step 1.

Step 3: Group by dimension.

Step 4: Exponents on each dimension must equal zero for dimensionless numbers, so form a set of equations by setting the exponent groups from Step 3 for each dimension equal to zero.

Step 5: Solve the simultaneous equations (as best as you can).

    Hint: Number of unknowns – number of equations = number of groups you will have!

Step 6: Substitute results of Step 5 back into Step 1 exponents.

Step 7: Group variables by exponent. These resulting groups are your dimensionless numbers.

Step 8: Be sure to *check* it out!! Are *all* of the ratios really dimensionless?

    Hint: If the resulting groups are *not* dimensionless, you most likely goofed in either Step 2 or Step 5!

Rayleigh's analysis is quite similar to the Buckingham Pi method, another method to determine dimensionless groups. Rayleigh's method is, however, a bit more direct and often seems less "mysterious" to those who are new to dimensional analysis. Both methods use a general form with multiplied and exponentiated variables. Any inspection of physics, engineering, and mathematical texts reveal many examples of this form of equation governing a myriad of behaviors.

● EXAMPLE 9-3

To develop an understanding of how initial velocity, acceleration, and time all affect the distance traveled by an accelerating object, we conduct some experiments and then analyze the resulting data. We asked a student to conduct a series of tests for us. She observed 25 different moving bodies with a wide range of initial speeds and different accelerations. For each, she measured the distance the bodies traveled for some prescribed time interval. Results are given in Table 9-4.

Table 9-4 Position of a body as a function of initial velocity, acceleration, and time

Test	Initial Velocity (v_0) [m/s]	Acceleration (a) [m/s^2]	Time (t) [s]	Distance (d) [m]
1	3	1	6	36
2	3	2	6	54
3	1.5	5	6	99
4	5	4	6	44
5	5	3	8	136
6	5	5	2	20
7	10	1	9	131
8	14	2	11	275
9	20	3	4	104
10	10	2	4	56
11	10	4	3	48
12	10	6	2	32
13	5	2	2	14
14	8	2	10	180
15	12	2	4	64
16	6	1	4	32
17	2	2	9	99
18	3	3	12	252
19	6	4	6	108
20	15	5	2.4	50
21	4	7	7.2	210
22	2	2	8	80
23	9	8	6.2	210
24	6.7	2	1.7	14
25	3.1	2	10	131

In addition, we would like to use this data set to help make predictions of the distance traveled by other bodies under different conditions. For example, we might want to answer the following question:

■ What is the acceleration needed to travel 4,800 meters in 200 seconds, if the initial velocity is 8 meters per second?

There are several independent variables (initial velocity, acceleration, and time), so it is not obvious what to plot. We can write the dependency as

$$d = f: (v_0, a, t)$$

We anticipate that it is difficult to draw conclusions regarding the interdependence of all of these variables. Realizing this, we plot distance against time without worrying about the initial velocity and the acceleration.

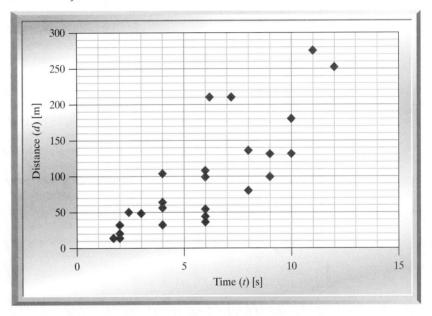

It seems that, in general, the longer one travels, the farther one goes. Upon closer inspection, however, it is obvious that this is not always the case. For example, for a travel time of 6 seconds, the distance traveled varies from about 25 to 210 meters. Since enough tests were not conducted with systematic variation of the initial velocity or acceleration, it is not possible to do much better than this. We certainly have no hope of answering the questions above with any confidence. In addition, since the values are so scattered, we realize that we have no good way to determine if any of our measurements were "bad."

With this disheartening conclusion, we perform a dimensional analysis in an attempt to place the parameters into fewer groups. The immediate problem we face is how to combine the parameters to give dimensionless groups. For rectangles, this was relatively easy to do by inspection. The parameters in this new problem are slightly more complicated, and although we might find suitable groups by inspection, as the problems become a lot harder (and they will), we need some sort of standard method, or algorithm, to define suitable dimensionless groups.

This technique is relatively simple and comprises the following eight steps:

Step 1: *Raise each variable to a different unknown power, using symbols for the power variables that do not already appear in the problem, and then multiply all of these individual* terms together. For example, since the current problem has both an a and d used as variables, we should not use a and d for the powers; hence we choose the letters p through s for the exponents. The order in which we list the variables and assign exponents is completely random.

This gives us the term: $d^p v_0^q a^r t^s$

Step 2: Substitute the correct dimensions for each variable.

$$\{L\}^p \left\{\frac{L}{T}\right\}^q \left\{\frac{L}{T^2}\right\}^r \{T\}^s$$

Step 3: Expand the expression to have each dimension as a base raised to some power.

$$\{L\}^{p+p+r} \{T\}^{-q-2r+2}$$

Step 4: For the expression in Step 3 to be dimensionless, each exponent must equal zero, or

$$p + q + r = 0 \text{ and } -q - 2r + s = 0$$

Step 5: Solve for the exponents. In this case, we have two equations and four unknowns, so it is not possible to solve for all the unknowns in terms of an actual number. We must be satisfied with finding two of the exponents in terms of the other two. This might seem problematic, but we will find that this not only is not a difficulty, but also is quite common in this type of analysis. **Note that there are many ways to do this and all will lead to two dimensionless ratios. If you do not like the plot you get from doing it one way, try solving for different exponents and see if that provides a better plot. All will be correct, but some are easier to use than others.** Although several procedures will lead to solutions, in general, we will solve for one of the variables, and then substitute into the other equations to reduce the number of variables. If a variable appears in all or most of the equations, that may be a good one to begin with.

In our example, we solve the second equation for q.

$$q = -2r + s$$

Substituting for q into the other equation gives

$$p - 2r + s + r = 0$$

thus,

$$p = r - s$$

At this point, we have defined p and q in terms of the other two variables, r and s.

Step 6: Substitute into the original expression.

$$d^{r-s} v_0^{-2r+s} a^r t^s$$

Note that all of the exponents are now expressed in terms of only two variables, r and s.

Step 7: Simplify by collecting all terms associated with the remaining exponential variables (r and s in this case).

$$\left(\frac{da}{v_0^2}\right)^r \left(\frac{tv_0}{d}\right)^s$$

Step 8: *The simplification in Step 7 gives the dimensionless ratios we are looking for. Dropping the exponents assumed in Step 1 gives the following groups:*

$$\left(\frac{da}{v_0^2}\right) \text{ and } \left(\frac{tv_0}{d}\right)$$

We need to double-check to make sure that both the groups are dimensionless. Before plotting them, we make two additional observations. (1) The variables of distance and initial velocity appear in both quantities. This may not always be desirable. (2) The initial velocity appears in the denominator of the first ratio. This may cause problems if we are examining data in which the initial velocity is very small, making the ratio very large. While dimensional analysis is much more involved than the examples given here, there are several important facts for you to remember.

First, since the results of the dimensional analysis produces dimensionless ratios, these ratios may be used as they appear above or they may be inverted. In other words, for this example, we can use $\frac{da}{v_0^2}$ or $\frac{v_0^2}{da}$ equally well.

To eliminate the problem of very small initial velocity values, the second form is preferable for our work here. As a side note, if we are interested in the behaviors at very small times, then we would prefer for time to appear in the numerator of the second ratio (and it already does).

Second, it is permissible to alter the form of one of the ratios by multiplying it by the other one or by the inverse of the other one or by the other one squared, etc. This will change the form of the first ratio and may produce results that are easier to interpret. A simple example can be used to show this. For the two ratios here, multiply the first ratio by the second ratio squared. This yields a "new" first ratio as $\left[(at^2)/d\right]$ and this could be used along with the second ratio $\left[(tv_0)/d\right]$. This result may have the advantage of initial velocity appearing in only one of the ratios. To continue this example, we create a spreadsheet with the four columns of data and then add two extra columns, one for each of the two dimensionless ratios $\left[v_0^2/(da)\right]$ and $\left[(v_0t)/d\right]$. Once this is done, it is a straightforward matter to plot one against the other. This result is shown below:

NOTE

There seems to be one "bad" data point. Would you have been able to pick out this point from the original data or from the dimensional plot?

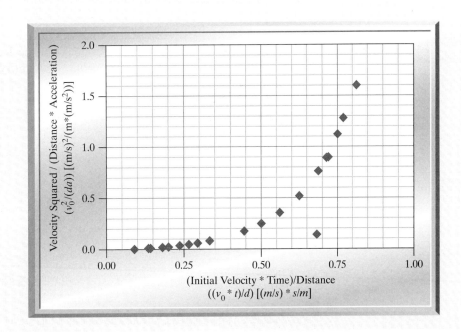

Now the scatter from the original dimensional plot is gone, and all but one data point seems to lie on a smooth curve. We can use this plot to determine the relationship between distance traveled, acceleration, time, and initial velocity. Let us see how to do this for the question we posed earlier.

What acceleration is required to go 4,800 meters in 200 seconds if the initial car velocity is 8 meters per second?

On the x-axis: [(8 m/s) (200 s)]/(4,800 m) = 0.33.
Reading from the graph: $v_0^2/(da) = 0.075$.
Solving for acceleration: a = (8 m/s)^2/(0.075) (4,800 m) = 0.18 m/s^2.

Several important conclusions can be drawn from this exercise.

- Dimensionless parameters often allow us to present data in an easily interpretable fashion when the "raw" data have no recognizable pattern.
- When we use the exponent approach to find dimensionless parameters, we need to remember that the exponents could be either positive or negative. Thus, in the case of the rectangle, P/W is just as good as W/P. You can always try both to see which gives the best-looking results. The choice is yours, and sometimes depends on whether one of the variables goes to zero; since you cannot divide by zero, that variable should not be in the denominator, if possible.
- If we "collapse" the data by using dimensionless parameters so that a single curve can fit through the resulting points, *bad data points will usually become obvious.*
- Finally, this approach can reduce literally thousands of different measurements into one simple curve. In the case of the car, the single line we obtained will work for all possible combination of times, initial velocities, accelerations, and distance. If we used dimensional plots, we would need more plots, with many lines on each plot. This would require an entire book of plots rather than the single plot with a single line that we obtained above by dimensionless ratios.

● EXAMPLE 9-4

To classify the smoothness of a flowing fluid, Osborne Reynolds (1842–1912, British, ME) developed the now famous dimensionless quantity of **Reynolds number**. His theory stated that the smoothness or roughness (a lot of eddies or swirling) of a fluid depended upon:

How fast the fluid was moving (velocity)	$v \ [=] \ m/s$
The density of the fluid	$\rho \ [=] \ kg/m^3$
The diameter of the pipe	$D \ [=] \ m$
How hard it was to move the fluid (viscosity)	$\mu \ [=] \ g/(cm\,s)$

Reynolds knew the smoothness depended upon these quantities:

$$Smoothness \ of \ the \ flow = f: (v, \rho, D, \mu)$$

But how did they depend on one another? We could write the four variables above as

$$v^a \rho^b D^c \mu^d$$

and if this was dimensionless, it would appear as $M^0 \ L^0 \ T^0 \ \Theta^0$.

To make this grouping dimensionless, first we substitute in the dimensions of the four variables to obtain:

$$\left\{\frac{L}{T}\right\}^a \left\{\frac{M}{L^3}\right\}^b \{L^c\} \left\{\frac{M}{LT}\right\}^d = M^{b+d} L^{a-3b+c-d} T^{-a-d}$$

Note we generally have dimensions of mass, length, time, and temperature in this problem; there is no temperature dimension. If this is to be dimensionless, then the exponents on all of the dimensions must equal zero, therefore:

M: $b + d = 0$
L: $a - 3b + c - d = 0$
T: $-a - d = 0$

This gives three equations in four unknowns, so we will have to solve for three of the variables in terms of the fourth. In this example, we solve for the three unknowns a, b, c in terms of d:

M: $b = -d$
T: $a = -d$
L: $c = -a + 3b + d = d - 3d + d = -d$

Substituting these back into the original parameters gives:

$$v^{-d} \rho^{-d} D^{-d} \mu^d$$

We see that there is one dimensionless group, since all the parameters have an exponent of d. We can write

$$\frac{\mu}{v \rho D} \{=\} \frac{\dfrac{M}{LT}}{\dfrac{L}{T} \dfrac{M}{L^3} L}$$

NOTE

The Reynolds number is used to describe fluid flow.

$Re < 2,000$ = laminar

$2,000 < Re < 10,000$ = transitional

$Re < 10,000$ = turbulent

Since the variables of diameter and velocity can approach zero, the Reynolds number is commonly written as follows:

$$Re = \frac{\rho D v}{\mu}$$

If the Reynolds number has a value less than 2,000, the flow is described as **laminar**, meaning it moves slowly and gently with no mixing or churning. If the Reynolds number has a value greater than 10,000, the flow is described as **turbulent**, meaning it moves quickly with much mixing and churning (lots of eddies) occurring. The region in between 2,000 and 10,000 is called the **transition region**.

● EXAMPLE 9-5

Suppose we conduct an experiment with a ball that we throw from the top of a tall tower of height H. We throw it directly downward with some initial velocity v, and then measure the elapsed time t until it hits the ground. We vary the initial height and the initial velocity. The variables of interest in this problem are H, v, and t. A little thought leads us to include g, since it is the force of gravity that causes the ball to fall in the first place.

Using Rayleigh's method, find a set of dimensionless ratios that can be used to correlate our data.

Step 1: *Write each variable and raise each to an unknown exponent (use all the variables, even the dependent variable).*

$$t^a H^b v^c g^d$$

Step 2: *Substitute dimensions of the variables into Step 1. Be sure to raise each dimension to the proper exponent from Step 1.*

$$t^a \{=\} \, \mathsf{T}^a \quad H^b \{=\} \, \mathsf{L}^b \quad v^c \{=\} \, \mathsf{L}^c \mathsf{T}^{-c} \quad g^d \{=\} \, \mathsf{L}^d \mathsf{T}^{-2d}$$

Step 3: *Group by dimension.*

$$\mathsf{L}^{b+c+d} \, \mathsf{T}^{a-c-2d}$$

Step 4: *Exponents on each dimension must equal zero for dimensionless numbers! Form a set of equations by setting the exponents for each dimension equal to zero.*

$$b + c + d = 0 \qquad a - c - 2d = 0$$

Step 5: *Solve the simultaneous equations (as best as you can).*

$$b = -c - d \qquad a = c + 2d$$

Step 6: *Substitute results of Step 5 back into Step 1 exponents.*

$$t^{c+2d} H^{-c-d} v^c g^d$$

Step 7: *Group variables by exponent. These resulting groups are your dimensionless numbers.*

$$\left[\frac{v\,t}{H} \right]^c \qquad \left[\frac{g\,t^2}{H} \right]^d$$

Step 8: *Be sure to* check *it out!! Are all of the ratios really dimensionless?*

Always remember that we initiate this procedure simply by providing a list of parameters we think are important to the situation at hand. If we omit an important parameter, our final result will not be physically correct, even if it is dimensionally correct. Consequently, if we select an improper parameter, then when tests are conducted, we will discover that it was not important to the problem and we can drop it from further consideration. *We cannot decide whether any variable is important until we conduct some experiments.*

Consequently, if we are *sure* that a parameter is important, then we know it should *not* drop from the analysis. The only way it can be retained is if at least one other parameter contains the missing dimension. In this case, we need to ask ourselves what other parameters might be important, add them to our list, and rework the analysis.

Dimensional analysis helps us organize data by allowing us to plot one-dimensionless parameter against another, resulting in one line on a single plot. This is a powerful result, and reduces a problem of multiple initial parameters to one containing only two. This discussion leads to the **Problem Simplification Rule**:

By performing dimensional analysis of the parameters, we can generally find dimensionless groupings to effectively reduce the number of parameters, facilitating the presentation of interdependencies and often simplifying the problem.

In-Class Activities

ICA 9-1

Complete the following table:

	Quantity	SI Units	Dimensions				
			M	L	T	Θ	N
Example	Acoustic impedance	(Pa s)/m	1	–2	–1	0	0
(a)	Fuel consumption	kg/(kW h)					
(b)	Heat transfer coefficient	W/(m² K)					
(c)	Ideal gas constant	(Pa L)/(mol K)					
(d)	Latent heat	BTU/lb$_m$					
(e)	Mass density	kg/m³					
(f)	Molar heat capacity	cal/(mol °C)					
(g)	Rate of drying	lb$_m$/(ft² h)					
(h)	Specific heat	J/(g °C)					
(i)	Specific weight	N/m³					
(j)	Surface tension	J/m²					
(k)	Thermal conductivity	cal/(cm s °C)					
(l)	Thermal resistance	(K m²)/W					

ICA 9-2

Calculate the numerical value of each of the dimensionless parameters listed in the table. Be sure to check that the ratio is actually dimensionless after you insert the values.

	Situation	Name	Expression	Value
(a)	Water in a pipe	Reynolds number, Re	$\dfrac{\rho v_w D_{\text{pipe}}}{\mu}$	
(b)	Hot water	Prandtl number, Pr	$\dfrac{\mu C_p}{k}$	
(c)	Air over a flat plate	Nusselt number, Nu	$\dfrac{hL}{k}$	
(d)	Sphere in air	Drag coefficient, C_d	$\dfrac{F}{\frac{1}{2}\rho v_a^2 A}$	
(e)	Water: Effect of surface tension	Weber number, We	$\dfrac{\rho v_w^2 d}{\sigma}$	
(f)	Water in a river	Froude number, Fr	$\dfrac{v_w}{\sqrt{gH}}$	
(g)	Wind making a wire "sing"	Strouhal number, St	$\dfrac{\omega D_{\text{wire}}}{v_a}$	

Properties and definitions for this problem:

Property	Symbol	Units	Air	Water
Density	ρ	slugs/ft^3	0.002378	1.94
Dynamic viscosity	μ	kg/(m s)	2×10^{-5}	4×10^{-4}
Heat transfer coefficient	h	W/(m^2 °C)	20	—
Thermal conductivity	k	W/(m K)	0.025	0.7
Surface tension	σ	dynes/cm	—	70
Specific heat	C_p	cal/(g °C)	0.24	1

Property	Symbol	Units	Value
Plate length	L	ft	2
Silhouette area of object	A	in^2	120
Water depth	H	m	3
Water speed	v_w	cm/s	210
Air speed	v_a	mph	60
Pipe diameter	D_{pipe}	in	6
Drag force on sphere	F	N	30
Depth of water film	d	cm	3
Oscillation frequency	ω	Hz (or cycles/s)	140
Wire diameter	D_{wire}	mm	20

ICA 9-3

We have encountered some equations in an old set of laboratory notes, each having two terms on the right-hand side of the equation. We realize we cannot read the final variable or variables listed in each equation. Using dimensions and the equation laws, determine the missing variable from the list below, and if that variable is multiplied or divided. In some questions, more than one variable may be required to form the necessary dimensions.

Variable choices:

Area	Acceleration	Density	Height
Mass	Speed	Time	Volume

(a) Distance = (speed) * (time) + (acceleration) ___ (___)
(b) Volume = (dimensionless constant) * (length) * (area) + (speed) ___ (___)
(c) Specific gravity = (density) * (volume)/(mass) + (height) * (length) ___ (___)
(d) Speed = (distance)/(time) + (acceleration) ___ (___)
(e) Energy = (mass) * (speed)^2 + (mass) * (height) ___ (___)
(f) Power = (mass) * (area)/(time)^3 + (mass) * (acceleration) ___ (___)
(g) Pressure = (density) * (area)/(time)^2 + (mass) * (acceleration) ___ (___)

ICA 9-4

A fluid with a specific gravity of 0.91 and a viscosity of 0.38 pascal-seconds is pumped through a 25-millimeter diameter smooth pipe at an average velocity of 2.6 meters per second. Determine the Reynolds number in the pipe for the system and indicate if the flow is laminar, transitional, or turbulent.

ICA 9-5

Brine, with a density of 1.25 grams per cubic centimeter and a viscosity of 0.015 grams per centimeter second is pumped through a 5-centimeter radius steel pipe at an average velocity of 15 centimeters per second. Determine the Reynolds number in the pipe for the system and indicate if the flow is laminar, transitional, or turbulent.

ICA 9-6

The Euler number is a function of the pressure drop, velocity, and density. Determine the form of the Euler number with Rayleigh's method.

Pressure drop	ΔP	pascal
Density	ρ	grams per cubic centimeter
Velocity	v	meters per second

ICA 9-7

When a simple turbine is used for mixing, the following variables are involved:

Power requirement	P	watt
Shaft speed	N	hertz
Blade diameter	D	meters
Blade width	W	meters
Liquid density	ρ	kilograms per meter cubed

Determine a set of dimensionless groups for the turbine, using Rayleigh's method.

ICA 9-8

We assume that the total storm water runoff (R, given in volume units) from a plot of land depends on the length of time that it rains (t), the area of the land (A), and the rainfall rate (r, given in inches per hour).

Rainfall Rate (r) [in/h]	Land Area (A) [acres]	Rainfall Duration (t) [h]	Measured Runoff (R) [ft³]
0.50	2	3.0	49
0.30	14	2.5	172
0.78	87	4.1	4,563
0.15	100	2.2	541
0.90	265	0.4	1,408
0.83	32	7.6	3,310
1.40	18	1.8	744
0.22	6	4.7	102
0.67	26	3.1	886
0.48	62	4.9	2,392

(a) Using the data, construct a plot of the runoff versus the time that it rains. You should see that this plot is of little help in understanding the relationships between the various parameters.

(b) Using this plot, estimate the total runoff from 200 acres if rain falls for 3 hours at a rate of 1.2 inches per hour. Is this even possible using this plot?

(c) Using the variables in the table, complete a dimensional analysis to help you plot the data. You should obtain two dimensionless ratios. Plot these, with the ratio containing the total runoff on the ordinate. This plot should collapse the values to a single line. Draw a smooth curve through the values.

(d) Use this line to answer question (b) again.

ICA 9-9

We are interested in analyzing the velocity of a wave in water. By drawing a sketch of the wave, and labeling it, we decide that the velocity depends on the wavelength (λ), the depth of the water (H), the density of the water (ρ), and the effect of gravity (g). We have measured wave speeds in many situations; the data are given below.

Water Depth (H) [m]	Wave Length(λ) [m]	Velocity (v) [m/s]
1.0	10.0	7.4
1.0	20.0	7.7
2.0	30.0	10.8
9.0	40.0	18.7
0.2	13.0	3.5
4.0	24.0	13.6
18.0	20.0	14.0
0.3	2.6	4.0
33.0	30.0	17.1
5.0	15.0	11.9

(a) Construct a plot of wave velocity versus either wave length or water depth. You will see substantial scatter.

(b) Using your plot, estimate the wave velocity for a wave length of 25 meters in water that is 25 meters deep. Is this even possible using this plot?

(c) Perform a dimensional analysis on the parameters (v, ρ, H, g, and λ). After calculating the new dimensionless ratio values, make a dimensionless plot.

(d) Recalculate the answer to question (b).

Chapter 9 REVIEW QUESTIONS

1. Complete the following table:

Quantity	Symbol	"Derived" from	Typical SI Units	M	L	T	Θ
(a) Dynamic viscosity	μ	Newton's law of viscosity	g/(cm s)				
(b) Mass density	ρ	$\rho = m/V$	kg/m^3				
(c) Specific heat	C_p	$Q = m\,C_p\,\Delta T$	J/(g °C)				
(d) Surface tension	σ	From theory or experiment	J/m^2				
(e) Thermal conductivity	k	Fourier's law of heat transfer	W/(m °C)				
(f) Young's modulus	E	Stress/strain	Pa				

2. The Darcy–Weisbach friction formula for pipes is used to calculate the frictional energy loss per unit mass of fluid flowing through the pipe. Examine the equations below and indicate for each if the equation is a valid or invalid equation; justify your answer for each case. In these equations,

h_L = energy loss per unit mass   [=] J/kg
f = friction factor   [=] dimensionless
L = length of pipe   [=] ft
D = diameter of pipe   [=] cm
v = average velocity   [=] cm/min
g = acceleration of gravity   [=] m/s^2

(A) $h_L = f\dfrac{L\,v}{D\,2g}$

(B) $h_L = fL\dfrac{v^2}{2g}$

(C) $h_L = f\dfrac{L\,v^2}{D\,2}$

(D) $h_L = f\dfrac{L\,2D}{v\,g}$

3. We wish to analyze the velocity (v) of a fluid exiting an orifice in the side of a pressurized tank. The tank contains a fluid to a depth (H) above the orifice. The air above the fluid in the tank is pressurized to a value of (P). We realize the greater the pressure inside, the greater the velocity. We also believe the greater the depth of fluid, the greater the velocity. Examine the equations below and indicate for each if the equation is a valid or invalid equation; justify your answer for each case. In these expressions, g is the acceleration due to gravity and ρ is the fluid density.

(A) $v = \dfrac{P}{\rho} + \sqrt{2gH}$

(B) $v = \sqrt{\dfrac{P}{\rho} + 2H}$

(C) $v = \sqrt{2gH}$

(D) $v = \sqrt{\dfrac{2P}{\rho} + 2gH}$

(E) $v = \sqrt{2P + 2gH}$

4. Wind energy uses large fans to extract energy from the wind and turn it into electric power. Examine the equations below and indicate for each if the equation is a valid or invalid equation; justify your answer for each case. In these expressions, P is the power, η is the efficiency, ρ is the density of the air, A is the area swept out by the fan blades, and v is the velocity of the wind.

   (A) $P = \eta \rho A^2 v^2$

   (B) $P = \eta \rho A v^2$

   (C) $P = \eta \rho A v^3$

   (D) $P = \eta \rho^2 A v$

   (E) $P = \eta \sqrt{\rho A v^3}$

5. One of the most famous experiments in physics is the Millikan oil-drop experiment. In it, very tiny drops are sprayed into a closed container by an atomizer. The drops tend to fall, but they are so small that the air slows them down. Sir George Stokes studied this part of the oil-drop experiment to see how fast the drops would fall in the still air in the container. He found that the fall speed (v) depended on the mass of the drop (m), the acceleration of gravity (g), the radius of the drop (r), and the viscosity of the air (μ). The equation governing the motion of the drop was determined to be

$$mg - 6\pi\mu rv = 0$$

The constant (the value 6) in the second term of the equation is dimensionless. What are the dimensions of viscosity (μ)?

6. One of the most famous experiments in physics is the Millikan oil-drop experiment. In it, very tiny drops are sprayed into a closed container by an atomizer. The drops tend to fall, but they are so small that the air slows them down. Sir George Stokes studied this part of the oil-drop experiment to see how fast the drops would fall in the still air in the container. He found that the fall speed (v) depended on the mass of the drop (m), the acceleration due to gravity (g), the radius of the drop (r), and the viscosity of the air (μ). The equation governing the motion of the drop was determined to be

$$mg - 6\pi\mu rv = 0$$

What kind of quantity is the second term ($6\pi\mu rv$)? Choose from the following and prove your answer in terms of dimensions.

   (A) Pressure
   (B) Force
   (C) Velocity
   (D) Energy

7. The friction factor (f) is a dimensionless number used in fluid flow calculations. It can be related to velocity in the pipe (v) and the fluid density (ρ) by the following equation:

$$f = \frac{2\varphi}{\rho v^2}$$

Using the laws of dimensions in equations, determine the dimensions of the variable φ.

8. The friction factor (f) is a dimensionless number used in fluid flow calculations. It can be related to velocity in the pipe (v) and the fluid density (ρ) by the following equation:

$$f = \frac{2\varphi}{\rho v^2}$$

What kind of quantity is the variable φ? Choose from the following and prove your answer in terms of dimensions:

(A) Pressure

(B) Force

(C) Velocity

(D) Energy

9. In previous examples, we discussed quantities with fundamental dimensions of mass {M}, length {L}, time {T}, and temperature {Θ}. As we study other fields, such as electricity and magnetism, we must introduce new fundamental dimensions, such as {I}, representing electrical current, which cannot be expressed in terms of mass, length, time, or temperature. Electric power can be calculated as $P = I^2R$, where R is the resistance [ohms, Ω] and I is the electrical current [amperes, A]. Assuming that electrical power must have the same dimension as mechanical power, derive the fundamental dimension of the ohm in terms of mass {M}, length {L}, time {T}, temperature {Θ}, and electrical current {I}.

10. What are the dimensions of the constant coefficient (k) in the following equations?

(a) Energy = k * (mass) * (temperature)

(b) Energy = k * (height)

(c) Energy = k * (mass) * (acceleration) * (length)

(d) Force = k * (pressure)

(e) Mass flowrate = k * (velocity) * (area)

(f) Pressure = k * (temperature)/(volume)

(g) Pressure = k * (density) * (acceleration)

(h) Power = k * (mass) * (acceleration) * (temperature)

(i) Power = k * (volume)/(time)

(j) Volume of a sphere = k * (diameter)^3

(k) Volumetric flowrate = k * (speed)

(l) Work = k * (volume)

11. A biodegradable fuel having a specific gravity of 0.95 and a viscosity of 0.04 grams per centimeter second is draining by gravity from the bottom of a tank. The drain line is a plastic 3-inch diameter pipe. The velocity is 5.02 meters per second. Determine the Reynolds number in the pipe for the system and indicate if the flow is laminar, in transition, or turbulent.

12. A sludge mixture having a specific gravity of 2.93 and a viscosity of 0.09 grams centimeter per second is pumped from a reactor to a holding tank. The pipe is a 2½-inch diameter pipe. The velocity is 1.8 meters per second. Determine the Reynolds number in the pipe for the system and indicate if the flow is laminar, in transition, or turbulent.

13. Water (specific gravity = 1.02; viscosity = 0.0102 grams per centimeter second) is pumped through a 0.5-meter diameter pipe. If the Reynolds number is 1,800 for the system, determine the velocity of the water in units of meters per second.

14. Water (specific gravity = 1.02; viscosity = 0.0102 grams per centimeter second) is pumped through 0.5 meter diameter pipe. If the Reynolds number is 5,800 for the system, determine the velocity of the water in units of meters per second.

15. The Peclet number is used in heat transfer in general and forced convection calculations in particular. It is a function of the two other dimensionless groups, the Reynolds number and the Prandlt number. Determine the functional form of these dimensionless groups, using Rayleigh's method. The problem depends on the following variables:

- Liquid density, ρ [=] kg/m^3
- Specific heat of liquid, C_p [=] J/(g °C)
- Liquid viscosity, μ [=] kg/(m s)

- Thermal diffusivity, α [=] m²/s
- Thermal conductivity of the plate, k [=] W/(m °C)
- Distance from edge of the plate, x [=] m
- Liquid velocity, v [=] m/s

16. When a fluid flows slowly across a flat plate and transfers heat to the plate, the following variables are important. Analyze this system using Rayleigh's method.
    - Liquid density, ρ [=] kg/m³
    - Specific heat of liquid, C_p [=] J/(g °C)
    - Liquid viscosity, μ [=] kg/(m s)
    - Thermal conductivity of the plate, k [=] W/(m °C)
    - Heat transfer coefficient, h [=] W/(m² °C)
    - Distance from edge of the plate, x [=] m
    - Liquid velocity, v [=] m/s

17. A projectile is fired with an initial velocity (v_0) at an angle (θ) with the horizontal plane. Find an expression for the range (R). The data are given in the table below. Use the data in the table to create one or more *dimensional* plots (e.g., launch speed on the abscissa and range on the ordinate). From these plots, answer the following questions.

Launch Angle (θ) [°]	Launch Speed (v_0) [m/s]	Measured Range (R) [m]
4	70	73
50	50	230
3	50	30
45	18	32
37	27	75
35	60	325
22	8	4.4
10	30	34
88	100	77
45	45	210

(a) If the launch speed is 83 meters per second and the launch angle is 64 degrees, what is the range? You will likely find it difficult to provide a good estimate of the range, but do the best you can.

(b) Complete a dimensional analysis of this situation. In this case, you would assume that the important parameters are θ, v_0, and R. Upon closer examination, however, it would seem that the range on Earth and on the moon would be different. This suggests that gravity is important, and that you should include g in the list of parameters. Finally, since it is not clear how to include θ, you could omit it and replace the velocity by v_x and v_z, where x is the distance downrange and z the height. You should use this information to determine dimensionless parameters. Also, you must decide how the lengths in R and g should appear. When you complete the analysis, you should find that these four parameters will be grouped into a single dimensionless ratio.

(c) Use the data from the table to calculate the numerical value of the ratio for each test. Note that $v_x = v \cos(\theta)$ and that you can find a similar expression for v_z. Insert these expressions into your dimensionless ratio.

(d) Assuming that you performed the dimensional analysis correctly, you should find that the ratio you obtained will always give the same value (at least nearly, within test-to-test error). Calculate the average value of the tests, and if it is nearly an integer, use the integer value.

(e) Finally, set this ratio equal to this integer, and then solve for the range R. Write your final equation for the range (i.e., $R = $ xxxxx). Now using this equation, answer question (a) again.

18. The drag on a body moving in a fluid depends on the properties of the fluid, the size and the shape of the body, and probably most importantly, the velocity of the body. We find that for high velocities, the fluid density is important but the "stickiness" (or viscosity) of the fluid is not. The frontal area of the object is important. You might expect that there will be more drag on a double-decker bus moving at 60 miles per hour than on a sports car moving at 60 miles per hour.

The table below gives some data for tests of several spheres placed in air and in water. The terminal velocity, the point at which the velocity becomes constant when the weight is balanced by the drag, is shown.

Object	Drag (F) [lb$_f$]	Velocity (v) [ft/s]	Diameter (D) [in]	Fluid
Table tennis ball	0.005	12	1.6	Air
Bowling ball	6	60	11	Air
Baseball	0.18	41	3	Air
Cannon ball	33	174	9	Air
Table tennis ball	0.0028	0.33	1.6	Water
Bowling ball	12.4	3.1	11	Water
Baseball	0.31	1.7	3	Water
Cannon ball	31	6.2	9	Water

(a) Plot the drag on the ordinate and the velocity of the object on the abscissa for each fluid on a separate plot. Use the graphs to answer the following question: What is the drag on a baseball in gasoline (specific gravity = 0.72) at a speed of 30 feet per second? You may struggle with this, but do the best you can.

(b) Now complete a dimensional analysis of this situation and replot the data. First, recognize that the important parameters are the ball diameter (use the silhouette area of a circle), the density of the fluid, the drag, and the velocity. You will find a single dimensionless ratio that combines these parameters.

(c) Compute the value of this ratio for the eight tests. Be sure in your analysis that you use consistent units so that the final ratio is truly unitless.

(d) Use this result to help you answer question (a) again.

Part 3

SCRUPULOUS WORKSHEETS

Scrupulous: **scroop·yə·ləs** ~ adjective;
  definition _____

LEARNING OBJECTIVES

The overall learning objectives for this part include:

- Apply basic concepts of statistics to experimental data.
- Apply basic principles from mathematical and physical sciences, such as Hooke's law, Ohm's law, and Newton's law of viscosity, to solve engineering problems.
- Communicate technical information effectively by correctly applying graphical conventions.
- Use Microsoft Excel to enhance problem solution techniques, including:
  - Entering, sorting, and formatting data;
  - Analyzing functions, including mathematical, statistical, and trigonometric;
  - Creating and formatting various types of graphs, including scatter, line, and column charts; rectilinear, semi-logarithmic, and logarithmic graphs;
  - Utilizing conditional statements, conditional formatting, data validation, built-in functions, and iteration.
- Select and interpret mathematical models to describe physical data.
- Use graphical techniques to create "proper" plots, sketch functions, and determine graphical solutions to problems.

Microsoft Excel is a worksheet computer program used internationally for an incalculable number of different applications. A **worksheet** is a document that contains data separated by rows and columns. The idea of using a worksheet to solve different types of problems originated before the advent of computers in the form of bookkeeping ledgers. The first graphical worksheet computer program for personal computers, VisiCalc, was released in 1979 for the Apple II® computer.

Figure P3-1 Comparison of VisiCalc and Excel interfaces.

Modern worksheet computer programs like Excel are significantly more powerful than earlier versions like VisiCalc; a comparison of the interface is shown in Figure P3-1. Excel contains text-formatting controls, built-in functions to perform common calculations, and a number of different plotting capabilities that make it an extremely powerful data analysis tool for engineers. Part 3 introduces the Microsoft Excel interface, the formatting controls used to create organized worksheets, and many built-in functions to assist in analyzing data or performing calculations on data contained in the worksheet.

A successful engineer must rely on knowledge of the way things work in order to develop solutions to problems, whether ameliorating climate change or trapping cockroaches. In many cases, the behavior of systems or phenomena can be described mathematically. These mathematical descriptions are often called mathematical models. The variables in the model vary with respect to one another in the same way that the corresponding parameters of the real physical system change.

As a very simple example, imagine you are driving your car on a country road at a constant speed of 30 miles per hour. You know that at this speed, you travel one-half mile every minute. If you drive at this speed for 44 minutes, you cover a distance of 22 miles.

A mathematical model for this is $d = 0.5t$, where d is distance in miles, t is time in minutes, and the value 0.5 has units of miles per minute. If you substitute *any* number of minutes in this equation for time (including 44), the distance (in miles) will be exactly half of the time numerical value. This allows you to predict what would happen in the "real world" of cars and roads without having to actually go out and drive down the road to determine what would happen if you drove 30 miles per hour for 44 minutes.

Needless to say, the mathematical descriptions for some physical systems can be extremely complicated, such as models for the weather, global economic fluctuations, or the behavior of plasma in an experimental fusion reactor.

As it turns out, a significant number of phenomena important in engineering applications can be described mathematically with only three simple types of models. Also in Part 3, we introduce these three models and their characteristics, as well as discuss the use of Excel to determine a mathematical model from a set of data determined by experimentation.

A few notes about this section of the book:

- Within the examples given in this portion of the text, note that any information you are asked to type directly into Excel will be found in quotations. Do not type the quotation marks, type only the information found within the quotation marks.
- In hardcopy, the data needed to create a chart will be shown in columns or rows, depending on the size of the data, to efficiently use space and save a few trees by using

less textbook paper. In the worksheets containing the starting data online, the data will be shown in columns.

■ Files available online are indicated by the symbol ![Excel icon].

■ ✎ This symbol indicates directions for an important process to follow. Step-by-step instructions are given once for each procedure.

■ ⌘ This symbol indicates special instructions for Mac OS users.

TIME MANAGEMENT

If you are using this text sequentially, by this point you are probably starting to feel a bit overwhelmed with all you need to do. While many introductory textbooks cover time management during the first few weeks of the semester, the authors have found it more useful to cover it a little later. In week 2 of your first semester of college, you are probably feeling like you still have things under control and do not need help. By week 10, however, you may be struggling to keep everything together and are more open to try some time management suggestions. Please note these are just suggestions and each person must develop a time management system that works best for him or her. It may take you a few attempts to find a process you can actually use, so keep making adjustments until you find your own personal solution.

There are 24 hours in each day, and 7 days in a week. Each week, you have 168 hours, or an estimated 170 hours, to spend doing something—sleeping, going to class, doing homework, or attending a football game. How do you spend all this time?

■ To get enough rest, you should sleep at least 7 hours every night, or about 50 hours every week.

■ If you spend 1 hour for each meal during the day, about 20 hours of your week will be spent eating.

■ If you allow 1 hour per day for personal hygiene and a few hours for laundry (your classmates will thank you for showering and having clean clothes), this takes about 10 hours per week.

■ Attending class is critical, and with lectures and labs you are probably in the classroom for 20 hours.

■ If you spend the maximum recommended study time on each course, this will take another 30–45 hours each week.

So what is left? Actually, quite a bit of time remains: 30 hours. While that may not seem like much, remember we assumed the maximum limits in our analysis.

■ It may only take you 30 minutes each day to get showered and dressed, saving you 3.5 hours per week.

■ Your lab may be canceled, freeing up an additional 3 hours.

■ While there are weeks when it will be necessary to study the maximum amount, this will also be balanced by weeks when you can study the minimum amount.

How, exactly, can you balance this "free" time with the "required" time? To be successful at time management, you must plan. If you carve out 1 hour each week to determine your plan for the upcoming days, you will be able to find time to work in any activities you want to do and still find time to study, eat, and sleep. Here, we present a PLAN with four steps: Prioritize, Leave time for fun, Anticipate delays, and No—learn to say it.

Prioritize

Ask:

- What must be completed this week (required assignments)?
- What can I begin to work on for next week (upcoming project, exams)?
- What would be nice to do if I have the time (recommended problems, reading)?

Rules:

- Schedule all courses in your plan. Attend every class. Be sure to include travel time, especially if you are commuting.
- Select a study time for each class and stick to it. As a general rule, plan for 2–3 hours of studying for each hour in class. For a 15-credit-hour course load, this is 30–45 hours.
- Determine when you can study best. Are you an early riser or a night owl?
- Be specific in your plan. Listing "Read Chapter 2, pages 84–97" is much better than "Read chemistry." Break down large projects into smaller tasks, each with a deadline.
- Do not study more than 2 hours at a time without a break. Pay attention to how long it takes you to become distracted easily.
- Schedule time daily to read course e-mail and check any online course management system.
- If you are working during college, do not forget to schedule in this time. As a general rule, you should not plan to work more than 10 hours per week while taking a 15-credit-hour course load. If you are working more, you may want to consult your financial aid office for advice.

Leave Time for Fun (and Chores)

Ask:

- What has to get done this week (chores)?
- What activities do I want to take part in (fun stuff you really want to do)?
- What would be nice to do if I have time (fun stuff if you have time)?

Rules:

- Schedule time for planning each week. Adopt your weekly schedule to meet the upcoming week deadlines and assignments.
- Schedule time for meals. Relax and talk with friends, read an engrossing book. Do not study during meals!
- Schedule time for sleep. Stick to this schedule — you will feel better if you go to sleep and awake each day at the same time . . . yes, even on weekends!
- Schedule time for physical activity. This can be hitting the gym, playing intramurals, or taking a walk. Staying healthy will help you stay on track.
- Schedule time for chores, such as laundry and paying bills.
- Allow time for technology on a limited basis. If you have a favorite TV show, schedule time to watch. If you want to surf on a social network, do so for a limited time each day.
- Plan outings. Colleges are wonderful resources for arts, music, theater, and athletics. Explore and find activities to enjoy, but do not compromise study time.
- Leave some open time. It is not necessary to schedule every minute of every day. Free time is a wonderful stress reliever!

Anticipate Delays

Ask:

- What can go wrong this week?
- What activities will alter my plans?

Rules:

- Plan time for "Murphy's Law": broken computers, running out of paper, getting sick, or helping a friend. If none occur, you will have extra hours in your plan.
- Leave time to proofread your work, or better yet, have someone else help you. Utilize your course teaching assistants, professor, or college academic facilities to assist you in polishing your final product.
- Plan to finish large projects 1 week before they are due to allow for any unexpected delays.

NO—Learn to Say it!

Ask:

- Will this activity help me reach my goal?
- If I do this activity, what will alter in my plan?

Rules:

- Schedule social activities around academics. Say "no" if you are not finished with your coursework.
- Remember, you are here to get an education. Employers will not care that you attended every basketball game or that you have 10,000 online friends if you have poor grades.

GOAL SETTING, BE REALISTIC

By Ms. Solange Dao, P.E., President DAO Consultants, Inc, Orlando, Florida and Vice President of Tau Beta Pi National Engineering Honor Society

Love to have fun? Many of us go out of our way to have fun or explore something intriguing. I will wait in a long line for an hour to ride a roller coaster, or detour 5 miles to drive down a new Interstate off-ramp. Fun, exploration, and intrigue naturally soak our minds with positive stimulation.

If fun could be channeled into traditionally boring things, productivity would sky-rocket. What if all Americans went to the gym regularly because it was truly fun to sweat off calories? How about smiling all day at work because the job was a pleasur-able challenge? Many New Year's resolvers give up by January 3. Why is it so hard to keep a promise to yourself just 72 hours after you have sworn to change? As an adult, you can choose to have fun in your type-A goal-oriented world. The key to mental stimulation and success is to trick yourself into having fun.

There are three main reasons why goals fail:

(1) The goal is too big to achieve quickly.

In our on-demand society, we tend to equate movement with progress. When you set a large goal (e.g., lose 50 pounds or save $1,000), it takes time to get there. The long-term goals seem more unreachable since daily movement toward them is miniscule. Since big goals are our vision of a greater life for ourselves, how can we better handle our big goals?

Breaking the big goal into small chunks is how you win. Break the "lose weight" goal down to its essentials. For example, to lose weight you have to eat less calories than you use, or burn more calories than you consume. Try to conquer each meal one at a time. This month, every breakfast will be healthy oatmeal with fruit. Then next month, add healthy choices for lunch. By focusing on the immediate task, you can win and be proud daily.

If you plan to exercise your way to a slimmer waist, think of small increments. Saying you will run 10k races every month is a big bite to chew. How about just parking in the farthest spot from the building door? Commit to taking stairs instead of the elevator, even when it is 10 flights. Stop using the TV remote and get up to change the channel. Small victories add up to a big win.

What if it is an academic goal such as getting into an honor society? Break it down to conquering one test at a time. Plan small study periods well in advance of each test instead of attempting a marathon cram session 24 hours before the tests. Micromovements are more mentally manageable.

(2) Pitfalls were not foreseen.

Minimize your disappointments in goal setting by attempting to foresee possible problems. Thinking it will be smooth sailing to your destination will only lead to bailing out. Challenges are needed to test us so we can build character. However, there is some good news: we can cheat at this obstacle course. We can cheat fate by predicting the obstacles to come. Like a football coach who studies his opponents' play tapes, he sees patterns for which he devises a counter play. You have just been assigned as Coach of your goals.

In early 2009, I decided to quit drinking sodas. I knew this was going to be a big task. I usually had a Diet Coke everyday around 3:00 p.m. to give me a little pep. I foresaw myself dragging at 3:00 p.m. without a Diet Coke, so I planned ahead: if I craved a soda, I would replace it with a noncarbonated iced tea. I still got the caffeine pep, yet the acidity was greatly reduced, not to mention the artificial additives.

What pitfall are you not foreseeing in your grand plan to achieve a goal? If it is financial, will you let a tire blow-out tumble you into despair and give up the goal of saving $1,000? As much as possible, plan your reactions to detours and missteps along the path. If the unexpected new tire costs $100, then accept it and keep going. You just extend your goal deadline by a month.

Many people look at pitfalls or setbacks as a sign from the Divine that they will never make their goal, or that they are too weak to change. The challenges are a part of the game. In the child's game Chutes and Ladders, sometimes you climb up and sometimes you slide down. Who quits Chutes and Ladders upon their first chute down? You roll the dice again and keep going. You may land on another chute, but you cannot predict that. Do not be the child who gets mad, smashes the game for being nonfun, and pouts for the rest of the day.

Challenge yourself to foresee the obstacles you will face. Brainstorm pitfalls with your friends. Make it a game by adding really goofy ones like participating in the county fair pie-eating contest. Thus, when problems arise, you can actually laugh at them by saying to yourself, "Ha! I knew this molasses and rhubarb pie was going to show up one of these days."

What if a friend invites you to a sold-out concert-of-the-decade that interferes with your planned study time for your engineering mid-term exams? If you plan some extra time into your schedule, you will be able to take advantage of unforeseen opportunities. If the study plan is too tight, any deviation will seem insurmountable.

(3) No back-up plan was devised.

It is impossible to predict everything that life will throw at you. Despite your best efforts to predict the unexpected, often a misstep occurs. You can also plan for failure. If you set several micro-goals which will lead to your bigger vision, a total failure of one small goal is much more manageable.

You ate the cake. OK, fine. Stop with the guilt and get to action. What is your back-up plan? For example, the 500-calorie cake slice means salads all day tomorrow, or it means walk a half-mile during each of your 15 minute work breaks. Make the consequences funny by putting up a sign at your desk that says "Vegan in Training."

Spent $300 on stereo equipment in a moment of insanity? No problem, that means making coffee at home. What is the back-up plan to keep you going forward?

What if your grades were not good enough to enter the honor society? There are other ways to gain recognition on campus such as joining service organizations and helping with charity events. Remember, there is always more than one path to any goal.

By preplanning micro-movement to our large goals, we can have fun predicting the inevitable, skirting around obstacles, and pulling ourselves up if we stumble.

CHAPTER 10
EXCEL WORKBOOKS

The following is an example of the level of knowledge of Excel needed to proceed with this chapter. *If you are not able to quickly recreate an Excel Worksheet similar to the one shown, including equations and formatting, please review worksheet basics in the appendix materials before proceeding.*

The wind chill temperature is calculated using the following empirical formula:

$$WC = 35.74 + 0.6215T - 35.75v^{0.16} + 0.4275Tv^{0.16}$$

where WC = wind chill temperature [degrees Fahrenheit, °F], T = actual air temperature [degrees Fahrenheit, °F], and v = wind speed [miles per hour, mph].

Create a single cell to hold the wind speed, set at 5 miles per hour, in Column A. Create a vertical series of actual air temperatures: $T = 30, 10, 0, -10$. Enter the equation for wind chill, referencing the cells and not the actual numerical values. A sample worksheet is shown below.

	A	B	C	D	E
1					
2					
3					
4					
5					
6					
7	Wind Speed	Actual Temperature (T) [deg F]			
8	(v) [mph]	30	10	0	-10
9	5	24.7	1.2	-10.5	-22.3

10.1 CELL REFERENCES

● EXAMPLE 10-1

Suppose we are given a list of *XY* coordinates in a worksheet. We want to calculate the distance between each point. We can find the distance between two *XY* coordinates by using Pythagoras' theorem:

$$d = \sqrt{(x_2 - x_1)^2 + (y_2 - y_1)^2}$$

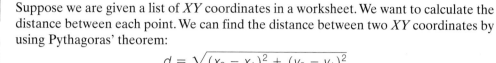

	A	B	C	D	E	F	G	H	I
2	This example demonstrates how to handle Excel's order of operations and cell references.								
3									
4									
5	Point 1			Point 2					
6	X	Y		X	Y				
7	27	20		25	10				
8	25	4		7	8				
9	4	6		24	3				
10	25	26		13	24				
11	19	24		26	1				
12	29	10		0	5				
13	7	29		13	13				
14	3	20		19	16				
15	20	7		5	17				
16	20	26		19	3				
17	13	15		13	14				
18	23	22		17	25				
19	3	27		10	22				
20	30	16		30	17				
21									

To solve this problem, we must adhere to the default behavior of Excel to properly calculate the distance between the coordinates. First, we must observe the order of operations that Excel follows to determine how we need to write our equations. Second, we must determine how to use **cell references** to translate the $x_2, x_1, y_2,$ and y_1 values in the equation shown above into locations in our worksheet.

Let us rewrite Pythagoras' theorem in the notation shown above using what we know about order of operations in Excel:

$$d = ((x_2 - x_1)\wedge2 + (y_2 - y_1)\wedge2)\wedge(1/2)$$

Let us calculate the distance between Point 1 and Point 2 in column G. In cell G7, we need to translate the equation into an equation that replaces the x_1, y_1 and x_2, y_2 variables with addresses to cells in the worksheet. Since each row represents a single calculation, we know that for the first data pair, x_1 is located in cell A7, y_1 is in B7, x_2 is in D7, and y_2 is in E7.

The equation we need to type into cell G7 becomes

$$= ((D7 - A7)\wedge2 + (E7 - B7)\wedge2)\wedge(1/2)$$

If we copy that equation down for the other pairs of XY coordinates, our sheet should now contain a column of all the distance calculations.

⊿	A	B	C	D	E	F	G	H	I
1	Distance Between XY Coordinates								
2	This example demonstrates how to handle Excel's order of operations and cell references.								
3									
4									
5		Point 1			Point 2				
6	X	Y		X	Y		Distance		
7	27	20		25	10		10.20		
8	25	4		7	8		18.44		
9	4	6		24	3		20.22		
10	25	26		13	24		12.17		
11	19	24		26	1		24.04		
12	29	10		0	5		29.43		
13	7	29		13	13		17.09		
14	3	20		19	16		16.49		
15	20	7		5	17		18.03		
16	20	26		19	3		23.02		
17	13	15		13	14		1.00		
18	23	22		17	25		6.71		
19	3	27		10	22		8.60		
20	30	16		30	17		1.00		
21									

Suppose we start off with a slightly modified worksheet that requires us to calculate the distance between all the points in the first column of XY values to a single point in the second column.

*We can calculate the distance between all the points in the first column to the single point through the use of absolute addressing. An **absolute address** allows an equation to reference a single cell that will remain constant regardless of where the equation is copied in the worksheet. An absolute reference is indicated by a dollar sign ($) in front of the row and column designators. In this example, we want to use an absolute reference on cells D7 and E7 in all distance calculations. The equation we need to type in cell G7 becomes:*

$$= ((\$D\$7 - A7)\textasciicircum 2 + (\$E\$7 - B7)\textasciicircum 2)\textasciicircum (1/2)$$

⊿	A	B	C	D	E	F	G	H	I
1	Distance Between XY Coordinates								
2	This example demonstrates how to handle Excel's order of operations and cell references.								
3									
4									
5		Point 1			Point 2				
6	X	Y		X	Y		Distance		
7	27	20		25	10		10.20		
8	25	4					6.00		
9	4	6					21.38		
10	25	26					16.00		
11	19	24					15.23		
12	29	10					4.00		
13	7	29					26.17		
14	3	20					24.17		
15	20	7					5.83		
16	20	26					16.76		
17	13	15					13.00		
18	23	22					12.17		
19	3	27					27.80		
20	30	16					7.81		
21									

Relative Addressing

- A **relative cell address** used in a formula will always *refer to the cell in the same relative position* to the cell containing the formula, no matter where the formula is copied in the worksheet. For example, if "=B2" is typed into cell C4 and then copied to cell C7, the formula in cell C7 would read "=B5". In this case, the cell reference is to call the cell two rows up and one cell to the left.
- When we insert or change cells, the formulas automatically update. This is one of a worksheet's major advantages: easily applying the same calculation to many different sets of data.

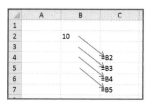
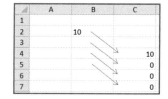

Absolute Addressing

- Absolute addressing is indicated by the presence of a dollar sign ($) immediately before both the column and row designators in the formula (e.g., $C$5; $AB$10).
- An **absolute cell address** will *always refer to the same cell* if the formula is copied to another location. For example, if "=$B$2" is typed into cell C4 and then copied to cell C7, the formula in cell C7 would read "=$B$2".

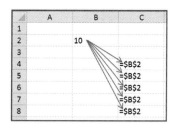
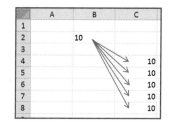

Mixed Addressing

- In **mixed addressing**, *either the row or the column designator is fixed* (by the $), but the other is relative (e.g., $C5; AB$10; $AB10).
- It may not be immediately obvious why this capability is desirable, but many problems are dramatically simplified with this approach. We will study this in more detail later.

COMPREHENSION CHECK 10-1

Type "5" in cell E22 and "9" in cell E23; type "=E22 + 4" in cell F22. Copy cell F22 to cell F23.

- Is this an example of absolute, mixed, or relative addressing?
- What is displayed in cell F23?

COMPREHENSION CHECK 10-2

Type "20" into cell G22 and "=$G$22 + 10" in cell H22. Copy cell H22 down to row 26 using the fill handle.

- Is this an example of absolute, mixed, or relative addressing?
- What is displayed in cell H26?

COMPREHENSION CHECK 10-3

Type "25" into cell A28 and "=A$28 + 5" in cell D28. Copy cell D28 down to row 30 using the fill handle. Copy cell D28 across to column F using the fill handle.

- Is this an example of absolute, mixed, or relative addressing?
- What is displayed in cell F28?

COMPREHENSION CHECK 10-4

Type "=$A28 + 5" in cell G28. Copy cell G28 down to row 30 using the fill handle. Copy cell G28 across to column J using the fill handle.

- Is this an example of absolute, mixed, or relative addressing?
- What is displayed in cell J28?

10.2 FUNCTIONS IN EXCEL

Hundreds of functions are built into Excel. Tables 10-1 through 10-4 list a few functions commonly used in engineering applications. Table 10-5 contains common error messages you may encounter. There are several things you should note when using these functions.

- You must make certain to *use the correct name of the function.* For example, the average function is written as AVERAGE and cannot be abbreviated AVE or AVG.
- *All functions must be followed by parentheses.* For example, the value of π is given as PI(), with nothing inside the parentheses.
- The *argument* of the function (the stuff in the parentheses) can include numbers, text, expressions, or cell references, as long as they are appropriate for the function.
- Many functions can *accept a list or range of cells as the argument.* These can be expressed as a list separated by commas [e.g., A6, D7, R2, F9], as a rectangular block designated by the top-left cell and bottom-right cell separated by a colon [e.g., D3:F9], or as a mixed group [e.g., A6, R2, D3:F9]. To insert cells into a formula, type the formula up to the open parenthesis and select the desired cells. You can also type in the references directly into the formula.
- Most functions will also *accept another function as the argument.* These can be fairly simple [e.g., SIN (RADIANS (90))] or more complicated [e.g., AVERAGE (SQRT(R2), COS(S4 + C4), MIN (D3:F9) + 2)].
- Some functions, such as trigonometric functions, require specific arguments. *Trigonometric functions must have an argument in units of radians, not units of degrees.* Be sure you are aware of any limitations of the functions you are using. Look up an unfamiliar function in the **HELP** menu.

■ Note that *some functions can be expressed in several different ways.* For example, raising the number 2 to the fifth power can be written as $= 2 \wedge 5$ or as POWER(2,5).

Table 10-1 Trigonometric functions in Excel

Function as Written in Excel	Definition
ACOS (cell)	Calculates the inverse cosine of a number (also ASIN)
COS (angle in radians)	Calculates the cosine of an angle (also SIN)
DEGREES (angle in radians)	Converts radians to degrees
PI()	Calculates pi (π) to about 15 significant figures
RADIANS (angle in degrees)	Converts degrees to radians

Table 10-2 Mathematical functions in Excel

Function as Written in Excel	Definition
EXP (cell)	Raises e (base of the natural log) to the power "cell"
POWER (cell, power)	Raises the cell to "power"
PRODUCT (cells)	Finds the product of a list of cells
SQRT (cell)	Finds the square root of cell
SUM (cells)	Finds the sum of a list of cells

Table 10-3 Statistical functions in Excel

Function as Written in Excel	Definition
AVERAGE (cells)	Finds the mean or average value of a list of cells
MAX (cells)	Finds the maximum value in a list of cells
MEDIAN (cells)	Finds the median value of a list of cells
MIN (cells)	Finds the minimum value in a list of cells
STDEV (cells)	Finds the standard deviation value of a list of cells
VAR (cells)	Finds the variance value of a list of cells

Table 10-4 Miscellaneous functions in Excel

Function as Written in Excel	Definition
COUNT (cells)	Counts number of cells that are not blank and that do not contain an error
COUNTIF (cells, criteria)	Counts number of cells that meet the stated criteria, such as a numerical value, text, or a cell reference
COUNTIFS (cells1, criteria1, cells2, criteria2, . . .)	Counts number of cells that meet multiple stated criteria, such as a numerical value, text, or a cell reference
INTERCEPT (y values, x values)	Calculates linear line for range of (x, y) pairs and returns the intercept value of y (where x = 0)
ROUND (cell, number of decimal places)	Rounds a number to a specific number of decimal places
SLOPE (y values, x values)	Calculates linear line for range of (x, y) pairs and returns the slope value
TRUNC (cell, number of digits)	Truncates a number to a specific number of digits

Table 10-5 Common error messages in Excel and possible solutions

Error	Explanation	Possible Fix	Example
#####	Column is not wide enough to display a number	Make column wider	−125,000,500 will not fit in a cell with a standard width
#DIV/0!	Formula has resulted in division by zero	Check values in denominator of formula contained in the cell	If cell A1 contains 12 and cell A2 is empty, the formula =A1/A2 will return #DIV/0!
#NAME?	Excel does not recognize something you have typed	Check spelling! Check operators for missing * or : Check for missing " " around text	Formula names: MXA should be MAX; PI should be PI() Range of cells: 2A:B3 should be A2:B3 Operators: (A7)(B6) should be (A7)*(B6)
#NULL!	You specify a set of cells that do not intersect	Check formulas for spaces, missing commas	= SUM(A2:A5 B4:B6) will return this error; fix as = SUM(A2:A5,B4:B6)
#VALUE!	Formula contains invalid data types	Arguments of functions must be numbers, not text Sometimes, part of a required function is missing; check for all required elements	If cell A2 contains "2 grams" and cell A3 contains 3, the formula = A2 + A3 will result in this error since A2 is text (the word grams makes the cell text, not a number) = VLOOKUP(A2:B5,2,FALSE) will result in this error since a LOOKUP function must contain four parts in the argument, not three
#N/A	Formula has called a value that is not available	Check for lookup value in data table (see Section 10.4 on LOOKUP function)	If A2 contains 11, and the data table contains values 1 to 10 in the first column, this error will appear since the value 11 is not in the first column of the data table
#REF!	Invalid cell reference	Check formula for data table size and number of column to return (see Section 10.4 on LOOKUP function)	= VLOOKUP(A2,A2:B5,3,FALSE) will return this error because there are not three columns available in the lookup table
#NUM!	Formula results invalid numeric values	Check numerical result expected is between -1×10^{307} and 1×10^{307} Check number of iterations	If the calculation results in a value outside the range given, such as 2×10^{400}, this error will appear See Chapter 11 for information on iteration

Handling Calculation Errors: IFERROR

Especially when dealing with worksheets that rely on user interaction to create meaningful information or analysis, there are often scenarios that will result in calculations that are not possible or might result in an error in a cell calculation. If you see cells in your worksheet that contain values like #DIV/0!, #N/A, or other messages that begin with the # symbol, that means that Excel was not able to calculate or look up the expression typed into the cell. The IFERROR function will allow the programmer of an Excel worksheet to specify what value should appear in a cell if there is a calculation error in the worksheet. The IFERROR function is often used when dealing with lookup statements or iterative expressions where error messages in cells might throw off the intended result of the calculation. For example, if you type = A1/A2 into cell B1 and it results in #DIV/0! you could type the following instead:

$$= \text{IFERROR(A1/A2,0)}$$

This function will check to see if A1/A2 results in an error message. If it does not generate an error, the resulting value of A1/A2 will appear in the cell, otherwise the value 0 will appear in the cell. It is worth noting that "0" in the formula above can be replaced with any valid Excel commands, including function calls, conditional statements, lookup statements, or simple hardcoding a value like 0 as shown above. For example, all of the following are valid **IFERROR** expressions:

$$= \text{IFERROR(A1/A2,MAX(A1,A2))}$$

$$= \text{IFERROR(IF(B2<3,A1/A2,B2),0)}$$

$$= \text{IFERROR(VLOOKUP(B2,A15:F20,3,FALSE),0)}$$

In the final example, if the lookup value of B2 is not found in the table located in A15:F20, the formula will return the value 0 rather than the error message #N/A.

● EXAMPLE 10-2

Assume we are studying the number of fatal accidents that occur during different times of the day. Given the data shown, we want to use Excel to analyze our data to determine the average, minimum, or maximum number of accidents, as well as a few other items that might be of significance.

	A	B
1	**Fatal Vehicular Accidents**	
2	This worksheet demonstrates the proper use of Excel Functions	
3		
4	**Week**	**Number Fatal Accidents**
5	A	190
6	B	202
7	C	179
8	D	211
9	E	160
10	F	185
11	G	172
12	H	205
13	I	177
14	J	120
15	K	235
16	L	183
17	M	177
18	N	193
19		

Total accidents: = SUM (B5:B18)
Total samples: = COUNT (B5:B18)
Mean: = AVERAGE (B5:B18)
Median: = MEDIAN (B5:B18)
Variance: = VAR (B5:B18)
Standard deviation: = STDEV (B5:B18)

*Note that decimal values appear when we calculate the mean, variance, and standard deviation of the accident data. Since it makes sense to round these values up to the nearest whole number, we need to type those functions as the argument to a rounding function. Start by modifying the equation for the mean by typing the **ROUND** function. Notice that as you start typing the ROUND function in the cell, a drop-down menu with a list of all of the functions that start with the letters ROUND appears below the cell. Note that Excel contains a function called **ROUNDUP** that will round a number up to the nearest whole value away from zero.*

NOTE
The ROUND function refers to number of decimal places, although the Excel help menu calls this "num_digits." Be sure to always read ALL the help menu file when using a new function.

2589	
14	
=round	
ⓕ ROUND　　Rounds a number to a specified number of digits	
ⓕ ROUNDDOWN	
703 ⓕ ROUNDUP	
26.51694461	

After we select the ROUNDUP function, a new box below the cell documents the arguments the function requires. Note that we need to provide the value we want to round as the first argument and the number of decimal places to which we want to round the number (in this case, 0).

Total Accidents	2589
Total Samples	14
Mean	=ROUNDUP(
Median	184
Variance	703.1483516
Standard Deviation	26.51694461

The new function we need to type ultimately becomes

= ROUNDUP (AVERAGE (B5:B18), 0)

Repeat this with the equations for calculating the median and the standard deviation.

    Suppose we want to determine how many of the samples reported accidents greater than the calculated average number of accidents. Note that the **COUNTIF** *function requires a "criteria" argument, which can take on a number of different values. For example, if we want to count the number of values greater than 200 in the range B5:B18, we need to type the criteria ">200" (in double quotes) as the 2nd argument to the COUNTIF function.*

= COUNTIF (B5:B18,">200")

In this example, we want to compare our COUNTIF result to a value calculated in a different cell. Since we cannot type cell references inside of double quotes (">E21"), we need to use the **ampersand** *operator (&) to* **concatenate** *the logical operator to the cell reference (">"&E21).*

Samples Greater than Mean: = COUNTIF (B5:B18,">"&B22)

Similarly, we could use the **COUNTIFS** *function to calculate the number of samples that have a number of accidents between (and including) 180 and 200. COUNTIFS is a special function that contains a variable number of arguments, with a minimum of two arguments required (range1, criteria1) to use the function. Since we have two criteria that must be met (>180 and <200), we must pass in four arguments to the COUNTIFS function (range1, critera1, range2, criteria2). In this example, range1 and range2 must be the same range of cells since we are enforcing the criteria on the same set of data. We will place the bounds in the worksheet as follows:*

Lower Bound in D23: 180   Upper Bound in E23: 200

Samples Between: = COUNTIFS (B5:B18,">="&D23, B5:B18, "<="&E23)

NOTE

Concatenate means to join things together. In Excel, the ampersand sign (&) will join two elements together.

=3&75 will result in 375

=3&"grams" will result in 3 grams

Your final worksheet should appear as shown.

	A	B	C	D	E
1	**Fatal Vehicular Accidents**				
2	This worksheet demonstrates the proper use of Excel Functions				
3					
4	**Week**	**Number Fatal Accidents**			
5	A	190			
6	B	202			
7	C	179			
8	D	211			
9	E	160			
10	F	185			
11	G	172			
12	H	205			
13	I	177			
14	J	120			
15	K	235			
16	L	183			
17	M	177			
18	N	193			
19					
20	Total Accidents	2589	Samples Greater than Mean		
21	Total Samples	14	6		
22	Mean	185			
23	Median	184	Samples Between	180	200
24	Variance	704	4		
25	Standard Deviation	27			

COMPREHENSION CHECK 10-5

Launch a new worksheet. Type the following Excel expressions into the specified cells. Be certain you understand *why* each of the following yields the specific result. Note that not all functions shown in this table are valid Excel functions. If the formula returns an error, how can the formula be changed to correctly display the desired result?

In Cell . . .	Enter the Formula . . .	The Cell Will Display . . .
A1	= SQRT (144)	
A2	= MAX (5, 8, 20/2, 5 + 6)	
A3	= AVERAGE (5, SQRT(100), 15)	
A4	= POWER (2, 5)	
A5	= PI()	
A6	= PI	
A7	= PRODUCT (2, 5, A2)	
A8	= SUM (2 + 7, 3 * 2, A1:A3)	
A9	= RADIANS (90)	
A10	= SIN (RADIANS (90))	
A11	= SIN (90)	
A12	= ACOS (0.7071)	
A13	= DEGREES(ACOS(0.7071))	
A14	= CUBRT(27)	

● **EXAMPLE 10-3**

The maximum height (H) an object can achieve when thrown can be determined from the velocity (v) and the launch angle with respect to the horizontal (θ):

$$H = \frac{v^2 \sin (\theta)}{2g}$$

Note the use of a cell (E7) to hold the value of the acceleration due to gravity. This cell will be referenced in the formulae instead of our inserting the actual value into the formulae. This will allow us to easily work the problem in a different gravitational environment (e.g., Mars) simply by changing the one cell containing the gravitational constant.

	A	B	C	D	E	F	G
1	Basic Examples of Trig Functions and Cell Addressing						
2	The following data is used to illustrate built-in trig functions and mixed references						
3							
4							
5							
6							
7	Planet	Earth			Gravity (g)	9.8	[m/s²]
8							
9	Velocity		Angle (θ) [degrees]				
10	(v)[m/s]	50	60	70	80		
11	10						
12	12						
13	14						
14	16						
15	18						
16	20						
17							

For the following, assume that the angle 50° is in cell B10. After setting up the column of velocities and the row of angles, we type the following into cell B11 (immediately below 50°)

= $A11^2 * SIN (RADIANS (B$10)) / (2*$E$7)

Note the use of absolute addressing (for gravity) and mixed addressing (for angle and velocity). For the angle, we allow the column to change (since the angles are in different columns) but not the row (since all angles are in row 10). For the velocity, we allow the row to change (since the velocities are in different rows) but the column is fixed (since all velocities are in column A). This allows us to write a single formula and replicate it in both directions.

The sine function requires an argument in units of radians, and the angle is given in units of degrees in the problem statement. In this example, we used the RADIANS function to convert from degrees into radians. Another method is to use the relationship 2π radians is equal to 360 degrees, or

= $A11^2 * SIN ((2 * PI() / 360) * B$10) / (2*$E$7)

We replicate the formula in cell B11 across the row to cell E11, selecting all four formulae in row 11 and replicating to row 16. If done correctly, the values should appear as shown.

Velocity	Angle (θ) [°]			
(v) [m/s]	50	60	70	80
20	3.91	4.42	4.79	5.02
12	5.63	6.36	6.90	7.24
14	7.66	8.66	9.40	9.85
16	10.01	11.31	12.27	12.86
18	12.66	14.32	15.53	16.28
20	15.63	17.67	19.18	20.10

Here, we consider the planet to be Mars with a gravity of 3.7 meters per second squared in cell E7. The worksheet should automatically update, and the values should appear as shown.

Velocity	Angle (θ) [°]			
(v) [m/s]	50	60	70	80
10	10.35	11.70	12.70	13.31
12	14.91	16.85	18.29	19.16
14	20.29	22.94	24.89	26.08
16	26.50	29.96	32.51	34.07
18	33.54	37.92	41.14	43.12
20	41.41	46.81	50.79	53.23

Now, we consider the planet to be Moon with a gravity of 1.6 meters per second squared in cell E7. The worksheet should automatically update, and the values should appear as shown.

Velocity	Angle (θ) [°]			
(v) [m/s]	50	60	70	80
10	23.94	27.06	29.37	30.78
12	34.47	38.97	42.29	44.32
14	46.92	53.04	57.56	60.32
16	61.28	69.28	75.18	78.78
18	77.56	87.69	95.14	99.71
20	95.76	108.25	117.46	123.10

10.3 LOGIC AND CONDITIONALS

Outside of the realm of computing, logic exists as a driving force for decision making. Logic transforms a list of arguments into outcomes based on a decision.

Some examples of every day decision making:

- If the traffic light is red, stop. If the traffic light is yellow, slow down. If the traffic light is green, go.

  Argument: three traffic bulbs   Decision: is bulb lit?   Outcomes: stop, slow, go

- If the milk has passed the expiration date, throw it out; otherwise, keep the milk

  Argument: expiration date   Decision: before or after?   Outcomes: garbage, keep

To bring decision making into our perspective on problem solving, we need to first understand how computers make decisions. **Boolean logic** exists to assist in the decision-making process, where each argument has a binary result and our overall outcome exhibits binary behavior. **Binary behavior**, depending on the application, is any sort of behavior that results in two possible outcomes.

In computing, we often refer to the outcome of Boolean calculations as "yes" and "no." Alternatively, we may refer to the outcomes as "true" and "false," or "1" and "0." To connect all the Boolean arguments to make a logical decision, we have a few operators that allow us to relate our arguments to determine a final outcome.

- **AND:** The AND logical operator enables us to connect two Boolean arguments and return the result as TRUE if and only if *both* Boolean arguments have the value of TRUE. In Excel, the AND function accepts more than two arguments and is TRUE if all the arguments are TRUE.
- **OR:** The OR logical operator enables us to connect two Boolean arguments and return the result as TRUE if *only one* of the Boolean arguments has the value of TRUE. In Excel, the OR function accepts two or more arguments and is TRUE if at least one of the arguments is TRUE.
- **NOT:** The NOT logical operator enables us to invert the result of a Boolean operation. In Excel, the NOT function accepts one argument. If the value of that argument is TRUE, the NOT function returns FALSE. Likewise, if the argument of the function is FALSE, the NOT function returns TRUE.

To determine the relationship between two cells (containing numbers or text), we have a few operators, listed in Table 10-6, that allow us to compare two cells to determine whether or not the comparison is true or false.

Table 10-6 Relational operators in Excel

Operator	Meaning
>	Greater than
<	Less than
> =	Greater than or equal to
< =	Less than or equal to
=	Equal to
< >	Not equal to

These relational operators are usually placed between two different cells to determine the relationship between the two values. This expression of cell–operator–cell is typically called a **relational expression**. Relational expressions can be combined by means of logical operators to create a **logical expression**. If no logical operator is required in a particular decision, then the single relational expression can be the logical expression.

Conditional statements are commands that give some decision-making authority to the computer. Specifically, the user asks the computer a question using conditional statements, and then the computer selects a path forward based on the answer to the question. Sample statements are given below:

- If the water velocity is fast enough, switch to an equation for turbulent flow!
- If the temperature is high enough, reduce the allowable stress on this steel beam!
- If the RPM level is above red line, issue a warning!
- If your grade is high enough on the test, state: You Passed!

In these examples, the comma indicates the separation of the condition and the action that is to be taken if the condition is true. The exclamation point marks the end of the statement. Just as in language, more complex conditional statements can be crafted with the use of "else" and "otherwise" and similar words. In these statements, the use of a semicolon introduces a new conditional clause, known as a nested conditional statement. For example:

- If the collected data indicate the process is in control, continue taking data; otherwise, alert the operator.
- If the water temperature is at or less than 10 degrees Celsius, turn on the heater; or else if the water temperature is at or greater than 80 degrees Celsius, turn on the chiller; otherwise, take no action.

Single Conditional Statements

In Excel, conditional statements can be used to return a value within a cell based upon specified criteria. The IF conditional statement within Excel takes the form

> = IF (logical test, value if true, value if false)

Every statement must contain three and only three parts:

1. **A logical test, or the question to be answered**
   The answer to the logical test must be TRUE or FALSE.
   Is the flow rate in Reactor #1 higher than Reactor #5?

2. **A TRUE response**, if the answer to the question is yes
   Show the number 1 to indicate Reactor #1.

3. **A FALSE response**, if the answer to the question is no
   Show the number 5 to indicate Reactor #5.

The whole statement for the above example would read:

$$= IF \ (B3 > B4, \ 1, \ 5)$$

	A	B	C
1			
2			
3	Reactor #1 Flowrate	10	[gpm]
4	Reactor #5 Flowrate	25	[gpm]
5	Maximum Flowrate in Reactor #	5	

Special Things to Note

- **To leave a cell blank, type a set of quotations with nothing in between ("").** For example, the statement = IF (C3>10, 5,"") is blank if C3 is less than 10.
- For display of a text statement, the *text must be stated within quotes* ("*text goes in here*"). For example, the statement = IF (E5 > 10, 5,"WARNING") would display the word WARNING if E5 is less than 10.

● **EXAMPLE 10-4** For the following scenarios, write a conditional statement to be placed in cell B5 to satisfy the conditions given. Below each statement are sample outcomes of the worksheet in different scenarios.

(a) Display the pressure difference between upstream station 1 (displayed in cell B3) and downstream station 2 (displayed in cell B4) if the pressure difference is positive; otherwise, display the number 1.

	A	B	C
1			
2			
3	Station #1 Pressure	2.4	[atm]
4	Station #2 Pressure	2.8	[atm]
5	Pressure Difference	1	[atm]

	A	B	C
1			
2			
3	Station #1 Pressure	3.2	[atm]
4	Station #2 Pressure	2.8	[atm]
5	Pressure Difference	0.4	[atm]

Answer: = IF ((B3 − B4) > 0, B3 − B4, 1)

(b) Display the value of the current tank pressure if the current pressure is less than the maximum tank pressure; otherwise, display the word "MAX".

	A	B	C
1			
2			
3	Maximum Tank Pressure	5	[atm]
4	Current Tank Pressure	2	[atm]
5	Pressure Status	2	[atm]

	A	B	C
1			
2			
3	Maximum Tank Pressure	5	[atm]
4	Current Tank Pressure	10	[atm]
5	Pressure Status	MAX	[atm]

Answer: = IF (B3 > B4, B4, "MAX")

(c) If the sum of the temperature values shown in cells B2, B3, and B4 is greater than or equal to 100, leave the cell blank; otherwise, display a warning to the operator that the temperature is too low.

	A	B	C
1			
2	Temperature Reading #1	25	[°C]
3	Temperature Reading #2	50	[°C]
4	Temperature Reading #3	45	[°C]
5	Cumulative Temperature		

	A	B	C
1			
2	Temperature Reading #1	25	[°C]
3	Temperature Reading #2	10	[°C]
4	Temperature Reading #3	45	[°C]
5	Cumulative Temperature	Too Low	

Answer: = IF (SUM(B2:B4) > = 100, B4, "", "Too Low")

COMPREHENSION CHECK 10-6	Are the logical expressions displayed in Example 10-4 the only logical expressions that could be used to display the required outcomes? Write out a logically equivalent expression that could be typed into cell B5 for each situation.

Nested Conditional Statements

If more than two outcomes exist, the conditional statements in Excel can be nested. The nested IF conditional statement within Excel can take the form

> = IF(logical test #1, value if #1 true, IF (logical test #2, value if #2 true, value if both false))

Note that the number of parenthesis must match (open and closed) and must be placed in the proper location. Recall that every statement must contain three and only three parts. For the first IF statement, they are:

1. **The first logical test, or the first question to be answered**
   The answer to the logical test must be TRUE or FALSE.
   Is the score for Quiz #1 less than the score for Quiz #2?

2. **A true response**, or what to do if the answer to the first question is yes
   Show the score for Quiz #1.

3. **A false response**, or what to do if the answer to the first question is no
   Proceed to the logical question for the second IF statement.

For the second IF statement, the three parts are:

1. **The second logical test, or the second question to be answered**
   The answer to the logical test must be TRUE or FALSE.
   Is the score for Quiz #2 less than the score for Quiz #1?

2. **A true response**, or what to do if the answer to the second question is yes
   Show the score for Quiz #2.

3. **A false response**, or what to do if the answer to the second question, and by default both questions, is no
   Show the text "equal".

The whole statement typed in cell B5 for the above example would read

$$= IF (B3 < B4, B3, IF (B3 > B4, B4, "equal"))$$

	A	B	C
1			
2			
3	Quiz Grade #1	70	
4	Quiz Grade #2	70	
5	Lowest Quiz Score	Equal	

	A	B	C
1			
2			
3	Quiz Grade #1	90	
4	Quiz Grade #2	70	
5	Lowest Quiz Score	70	

	A	B	C
1			
2			
3	Quiz Grade #1	50	
4	Quiz Grade #2	70	
5	Lowest Quiz Score	50	

There can be a maximum of 64 nested IF statements within a single cell. The nested IF can appear as either the true or false response to the first IF logical test. In the above example, only the false response option is shown.

● EXAMPLE 10-5

Write the conditional statement to display the state of water (ice, liquid, or steam) based upon temperature displayed in cell B4, given in degrees Celsius. Below are sample outcomes of the worksheet in different scenarios.

	A	B	C
1			
2			
3			
4	Temperature of Mixture	75	[°C]
5	State of Mixture	Liquid	

	A	B	C
1			
2			
3			
4	Temperature of Mixture	110	[°C]
5	State of Mixture	Steam	

	A	B	C
1			
2			
3			
4	Temperature of Mixture	10	[°C]
5	State of Mixture	Ice	

Here, there must be two conditional statements because there are three responses:

■ *If the temperature is less than or equal to zero, display "ice";*
■ *If the temperature is greater than or equal to 100, display "steam";*
■ *Otherwise, display "liquid".*

Answer: =IF(B4 <= 0, "ice", IF (B4 >= 100, "steam", "liquid"))

COMPREHENSION CHECK 10-7

Is the logical expression displayed in Example 10-5 the only logical expression that could be used to display the phase of the water? Write out a logically equivalent expression that could be typed into cell B5.

Compound Conditional Statements

If more than two logic tests exist for a single condition, conditional statements can be linked together by AND, OR, and NOT functions. Up to 255 logical tests can be compared in a single IF statement (only two are shown in the box below). The compound IF conditional statement takes the form

> = IF (AND (logical test #1, logical test #2), value if both tests are true, value if either test is false)
>
> = IF (OR (logical test #1, logical test #2), value if either test is true, value if both tests are false)

● EXAMPLE 10-6   Write the conditional statement that meets the following criteria:

(a) If the product has cleared all three quality checks (given in cells B2, B3, and B4) with a score of 80 or more on each check, mark the product as "OK" to ship; otherwise, mark the product as "Recycle."

	A	B	C
1			
2	Quality Check #1 Rating	90	
3	Quality Check #2 Rating	80	
4	Quality Check #3 Rating	85	
5	Mark Product	OK	

	A	B	C
1			
2	Quality Check #1 Rating	60	
3	Quality Check #2 Rating	80	
4	Quality Check #3 Rating	85	
5	Mark Product	Recycle	

Answer: = IF(AND (B2 >= 80, B3 >= 80, B4 >= 80),"OK", "Recycle")

(b) If the product has cleared all three quality checks (given in cells B2, B3, and B4) with a minimum score of 80 on each check, mark the product as "OK" to ship; otherwise, if the product scored a 50 or below on any check, mark the product as "Rejected"; otherwise, mark the product as "Rework."

	A	B	C
1			
2	Quality Check #1 Rating	90	
3	Quality Check #2 Rating	80	
4	Quality Check #3 Rating	85	
5	Mark Product	OK	

	A	B	C
1			
2	Quality Check #1 Rating	40	
3	Quality Check #2 Rating	80	
4	Quality Check #3 Rating	85	
5	Mark Product	Rejected	

	A	B	C
1			
2	Quality Check #1 Rating	60	
3	Quality Check #2 Rating	80	
4	Quality Check #3 Rating	85	
5	Mark Product	Rework	

Answer: = IF(AND (B2 >= 80, B3 >= 80, B4 >= 80), "OK", IF (OR (B2 <= 50, B3 <= 50, B4 <= 50), "Rejected", "Rework"))

COMPREHENSION CHECK 10-8   Is the logical expression displayed in Example 10-6(a) the only logical expression that could be used to display the desired outcome? Write out a logically equivalent expression that could be typed into cell B5.

10.4 LOOKUP AND DATA VALIDATION

The LOOKUP function enables Excel to locate information from a table of data in a worksheet. There are two LOOKUP functions: VLOOKUP, which searches vertically, and HLOOKUP, which searches horizontally. In the following example, we focus on VLOOKUP, but the same principles could easily be applied to HLOOKUP. To use the VLOOKUP function, we need to pass in four different arguments:

VLOOKUP (lookup_value, table_array, col_index_num, [range_lookup])

- The ***lookup_value*** argument is the value we want to look up in the table. Typically, this value is a string, but it can be a numerical value. Note that whatever we use as the *lookup_value*, Excel will perform a case-insensitive search of the data for the value, which means that any special characters used in the string, like punctuation or spaces, must appear the same in the *lookup_value* and the table, and must be a unique identifier in the table.
- The ***table_array*** is the range of cells that encapsulates the entire data table we want to search. Since we are using VLOOKUP, it is important to realize that our *table_array* must have at least two columns of data. Note that the *lookup_value* we are passing in to the VLOOKUP function will only search the first column of the *table_array*, so it might be necessary to move the data around.
- The ***col_index_num*** argument is the column number that contains the data we want as a result of our search. By default, Excel will refer to the first column where the *lookup_value* is located as the number 1, so the *col_index_num* will always be a number greater than 1.
- The last argument, [***range_lookup***], is an optional argument (so it is listed in square brackets). This argument tells the function what type of search to perform and can only take on two values: TRUE or FALSE.

  - Passing in TRUE tells Excel to conduct an approximate search of the data. That is, Excel will search the data table for the largest value that is less than the *lookup_value* and use that result as the selected value. Note that for an approximate search, the first column of the *table_array* must be sorted in ascending order.
  - Passing in FALSE tells Excel to conduct an exact search of the data. The data need not be sorted for this option. If an exact match is not found, the function returns an error.
  - If we do not specify TRUE or FALSE, Excel attempts to match the data exactly, and if a match is not found, Excel returns an approximate value. This may give undesired results. It is good practice to tell Excel which searching algorithm to use to search the *table_array*.

Assume we are given the following table of data on students. To determine what Sally's eye color is from (column C) in cell A5, we could type

$$= \text{VLOOKUP (\"Sally\", A1:D4, 3, FALSE)}$$

since the data are unsorted and we are looking for an exact match on Sally.

	A	B	C	D
1	Joe	18	Blue	EE
2	John	19	Brown	ME
3	Sally	18	Brown	IE
4	Julie	18	Blue	CE

● **EXAMPLE 10-7**

Digital audio is a relatively new medium for storing and reproducing music. Before albums were sold on CD and other digital media formats, analog recordings were commonly sold as vinyl records, 8-track tapes, and cassette tapes. We want to build a worksheet to help us compare these different media formats to observe how information storage has progressed over the past 50 years. Note the following media equivalencies:

- A 74-minute CD (44.1 kilohertz, 2 channel, 16-bit digital audio) can hold 650 MB of data.
- A single-sided, single-layer DVD can hold 4.7 GB of data (~4,813 MB, 547 minutes of 44.1 kilohertz, 2 channel, 16-bit digital audio).
- A single-sided, single-layer Blu-ray disc can hold 25 GB of data (~25,600 MB, 2,914 minutes of 44.1 kilohertz, 2 channel, 16-bit digital audio).
- A 7-inch vinyl record recorded at 45 rpm can hold 9 minutes of music.
- A 7-inch vinyl record recorded at $33\frac{1}{3}$ rpm can hold 12 minutes of music.
- A 12-inch vinyl record recorded at 45 rpm can hold 24 minutes of music.
- A 12-inch vinyl record recorded at $33\frac{1}{3}$ rpm can hold 36 minutes of music.
- An 8-track tape can hold 46 minutes of music.
- A typical cassette tape can hold 60 minutes of music.

To determine audio equivalencies between these different storage formats, we first create a worksheet. We want to allow the user to input the media type and quantity of the desired format to be converted. To complete the comparison, it would seem like each calculation requires a statement with nine questions to ask (Is it a CD? Is it a DVD? Is it a Blu-ray? . . .).

	A	B	C	D	E
1	**Digital Audio Media**				
2	This worksheet demonstrates the use of VLOOKUP and data validation				
3					
4					
5	Quantity	Format	*is equivalent to*	Quantity	Format
6	100	Blu-ray Disc		3838	CD
7				533	DVD
8				100	Blu-ray Disc
9				32378	7" @ 45 rpm
10				24284	7" @ 33 1/3 rpm
11				12142	12" @ 45 rpm
12				8095	12"@ 33 1/3 rpm
13				6335	8-track tape
14				4858	Cassette tape

*Rather than requiring the user to type the name of the media each time (CD, DVD, Blu-ray, etc), Excel can do **data validation**, so we can give the user of our worksheet a drop-down menu from which to select the media. We need to add a table that contains the name of each media type along with the length of the audio we can fit on each media. We will place this table below our initial data, in cells A17:B26.*

	Format	Length [min]
16	**Storage Information**	
17	**Format**	**Length [min]**
18	CD	74
19	DVD	547
20	Blu-ray Disc	2914
21	7" @ 45 rpm	9
22	7" @ 33 1/3 rpm	12
23	12" @ 45 rpm	24
24	12"@ 33 1/3 rpm	36
25	8-track tape	36
26	Cassette tape	60

Next, we need to calculate the quantity of each item. Since the name of the media will appear in cell B6, we use that as the lookup value in our VLOOKUP statement. To calculate the quantity for each equivalent media, we look up the length of the format specified in B6, divide that by the length of each media given in column E, and multiply that by the number of the original media provided in A6. Note that we need to round this number up since it does not make sense to have a noninteger value in our count.

For CDs, the calculation in Cell D6 should be

$$= \text{ROUNDUP (VLOOKUP (\$B\$6, \$A\$18:\$B\$26, 2, FALSE)/}$$
$$\text{VLOOKUP (E6, \$A\$18:\$B\$26, 2, FALSE) *\$A\$6, 0)}$$

A sample calculation using 100 Blu-ray discs is shown.

The next step to finish our worksheet is to include a drop-down menu of the different media formats. To insert data validation on the media format, we click Cell B6 and go to **Data > Data Tools > Data Validation.**

The **Data Validation** *window is displayed. Under the* **Settings** *tab, the* **Allow:** *menu lets us specify the type of data that can be provided in the cell we selected. Since we want to restrict the data to a list of values, we select* **List***.*

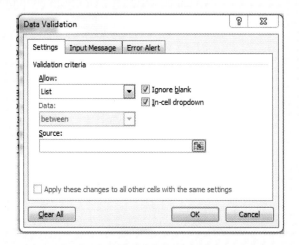

Under the **Source:** *option, we select the range of all of the media types, $A$18:$A$26, and click* **OK** *to close the Data Validation window.*

Notice the drop-down handle next to cell B6. When the user of the worksheet clicks B6, a drop-down menu appears that lists all of the possible media types so that the user can quickly select an item from the list. Furthermore, this feature prevents the user from typing items that are not on the list, making a typo, or entering any other information that will cause an error in calculations that rely on the value in B6.

6	100	CD
7		CD
8		DVD
		Blu-ray Disc
9		7" @ 45 rpm
		7" @ 33 1/3 rpm
10		12" @ 45 rpm
11		12"@ 33 1/3 rpm
		8-track tape

In addition to controlling the input type to a cell, it is also possible to give feedback to the person using the worksheet using pop up messages. In this example, the quantity cannot be a negative number, so we need to bring up the Data Validation window again and restrict the input to only allow whole numbers that are greater than or equal to zero.

Next, we need to click the input Message tab to type in a message that will appear below the cell when the person using our worksheet clicks on the cell to type in a quantity.

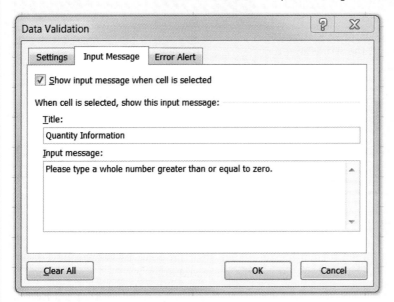

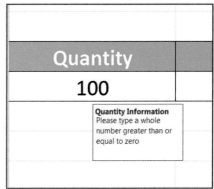

Finally, we need to click the Error Alert tab to provide the message that should pop up when an invalid number is typed into the cell.

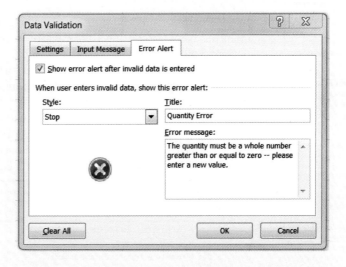

COMPREHENSION CHECK 10-9

Open the worksheet you created in Example 10-3 and modify it to use VLOOKUP and data validation to allow the user of the worksheet to select the planet and automatically fill in the gravity for each planet.

Planet	Gravity (g) [m/s^2]
Earth	9.8
Jupiter	24.8
Mars	3.7
Mercury	3.7
Moon	1.6
Neptune	11.2
Pluto	0.7
Saturn	10.4
Uranus	8.9
Venus	8.9

10.5 CONDITIONAL FORMATTING

You can use conditional formatting to change the font color or background of a cell based upon the values found in that cell. As an example:

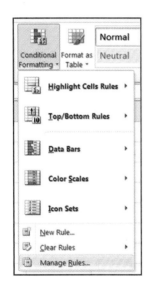

- On a blank worksheet, type the value of 20 in cell A4, a value of 30 in cell B4, and a value of 50 in cell C4.
- Select cells A4 to C4.
- Select the **Conditional Formatting** drop-down menu in **Home > Styles** and click the **Manage Rules** item.
- In the Conditional Formatting Rules Manager window, click the New Rule button to open the New Formatting Rule wizard. In this window, you can now select several options. Under Select a Rule Type, click the option **Format only cells that contain** to bring up the interface for creating custom formats based on certain conditions. In the first drop-down menu, you can choose among **Cell Value** and a few other options. For this activity, use only Cell Value.
- In the second drop-down menu, you can choose various conditional statements such as "between" or "less than." To begin this problem, choose "less than" from this list.
- The choice of "less than" will combine the next two boxes into a single box. You can enter a number or formula, or reference a cell within the worksheet.

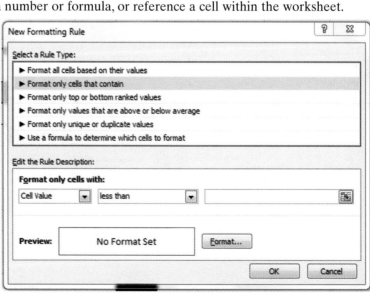

For this example, enter the value "25." Note: If you enter a formula, the same rules apply for absolute and relative referencing. In addition, if you select a cell within the worksheet, the program automatically defaults to an absolute reference.

- Click the **Format** button. Select the formatting you want to apply when the cell value meets the condition or the formula returns the value TRUE. You can change the font, border, or background of the cell. For this example, choose a green background on the Fill tab. When you are finished, click **OK**.

- To add another condition, click **New Rule**. Additionally, in this window, you can edit or delete existing conditional formatting options. For this condition, make it greater than 40, with a font of white, bolded on a red background. Click **OK**. To apply the conditions in the Conditional Formatting Rules Manager window, click the **Apply** button, then click **OK**.

Your worksheet should now look like the one shown. If none of the specified conditions are TRUE, the cells keep their existing formats.

	A	B	C
1			
2			
3			
4	20	30	50
5			
6			

● EXAMPLE 10-8

Let us assume we want to build an interactive worksheet that changes the format of a cell to model the behavior of a traffic light. We want the user to input the number of seconds it takes for a light (which is initially green) to turn red. In addition, the user must also be able to provide the "warning" so that the light can switch from green to yellow and then to red.

- The green light (bottom) will only be lit if the time remaining is greater than the warning time.
- The yellow light (middle) will only be lit if the time remaining is greater than 0 seconds, but less than the warning time.
- The red light (top) will only be lit if the time remaining is 0 seconds.

Before we set up the conditional formatting for each cell, we need to write IF statements in the light cells that will be used as a trigger for conditional formatting.

For Cell E5 (the red light): $= IF(A5 = 0, "R", "")$
For Cell E9 (the yellow light): $= IF(AND (A5 <= B5, A5 > 0), "Y", "")$
For Cell E13 (the green light): $= IF(A5 > B5, "G", "")$

Next, we add a set of conditional formatting rules for each cell.

For the red light, we click E5 and select the Manage Rules item from the Conditional Formatting menu to create a rule that turns the fill color red when the cell has the letter "R" in the text.

For the yellow light, we click E9 and repeat this process to turn the fill color yellow when cell has the letter "Y" in the text. For the green light, we click E13 and repeat this process to turn the fill color green when cell has the letter "G" in the text.

The final worksheet should appear as shown. Note that cell formats should change when the time remaining changes.

	A	B	C	D	E	F	G	H
1	Modeling a Traffic Light							
2	This worksheet demonstrates the use of conditional formatting and interation in Excel							
3								
4	Time Remaining [s]	Warning Times [s]						
5	2	5						
6								
7								
8								
9								
10								
11								
12								
13								
14								
15								

COMPREHENSION CHECK 10-10

Open the worksheet you created in Example 10-3 and modify it to highlight all heights greater than 100 meters with a light blue background and all heights less than 25 meters with a dark blue background with a white font.

10.6 SORTING AND FILTERS

Excel provides a number of built-in tools for sorting and filtering data in a worksheet. This section describes how to use these tools effectively without causing unintended side effects.

Each year, the federal government publishes a list of fuel economy values. The complete lists for recent years can be found at www.fueleconomy.gov/feg. A partial list of 2009 vehicles is shown below. In the table, MPG = miles per gallon.

Make	Model	MPG City	MPG Highway	Annual Fuel Cost
Jeep	Liberty 4WD	15	21	$1,835
BMW	X5 xDrive 30i	15	21	$1,927
Honda	Civic Hybrid	40	45	$743
Volkswagen	Jetta 2.5L	20	29	$1,301
Ford	Mustang 5.4L	14	20	$2,166
Bentley	Continental GTC	10	17	$2,886
Honda	Fit	27	33	$1,039

Given this information, assume you are to present it with some sort of order. What if you want to sort the data on text values (Make or Model) or numerical values (MPG City, MPG Highway, Annual Fuel Cost), or what if you want to view only certain vehicles that meet a certain condition?

Sorting Data in a Worksheet

- Select the cells to be sorted. You can select cells in a single column or row, or in a rectangular group of cells.
- Select **Home > Editing > Sort & Filter**. By default, two commonly used sorting tools (Sort A to Z and Sort Z to A) appear, in addition to a button for Custom Sort. With a group of cells selected, the common sorting tools will sort according to the values in the leftmost column. If the leftmost column contained numerical values, the options would have read Sort Smallest to Largest/Largest to Smallest. Since it is often desired to involve multiple sorting conditions, click **Custom Sort**.
- The sorting wizard is displayed as shown. If your selected group of cells had a header row (a row that displays the names of the columns and not actual data) the "My data has headers" checkbox should be selected.

By default, Excel automatically detects whether the top row of your selected data is a header or a data row. Since you selected the data including the header rows, the "Sort by" drop-down menu will contain the header names. If you had not included the header row, the "Sort by" drop-down menu would show the column identifiers as options. It is good practice to select the headers in addition to the data to make sorting easier to understand.

- Assume you want to sort the list alphabetically (A to Z) by the make, then by smallest-to-largest annual fuel cost. Click the **Add Level** button to add two levels of sorting since there are two conditions. In the sorting wizard, the topmost sorting level will be the sort applied first, and then the next level will sort each data group that forms from the first sort. In the example, there is more than one Honda vehicle, so the second level will place the Civic Hybrid above the Fit, since the Civic Hybrid has a smaller annual fuel cost.

The resulting sorted data appear as shown.

	A	B	C	D	E
1	Fuel Economy of Vehicles				
2	This worksheet demonstrates the use of sorting and filtering in excel				
3					
4	**Make**	**Model**	**MPG City**	**MPG Highway**	**Annual Fuel Cost**
5	Bentley	Continental GTC	10	17	$2,885
6	BMW	X5 xDrive 30i	15	21	$1,927
7	Ford	Mustang 5.4L	14	20	$2,165
8	Honda	Civic Hybrid	40	45	$743
9	Honda	Fit	27	33	$1,039
10	Jeep	Liberty 4WD	15	21	$1,835
11	Volkswagon	Jetta 2.5L	20	29	$1,301
12					

NOTE

To "undo" a sort, either choose the "Undo" arrow button on the top menu or use CTRL-Z.

It is important to be sure to select all of the data when using the sort functions because it is possible to corrupt your data set. To demonstrate, select only the first three columns (Make, Model, MPG City) and sort the data smallest to largest on the MPG City column.

Notice after sorting that the last two columns (MPG Highway, Annual Fuel Cost) are not the correct values for the vehicle. There is no way to recover the original association if you were to save the file and open it at a later time, so it is critical that when using the built-in sorting functions, you verify the correctness of your data before saving your workbook. In this case, you can click Excel's Undo button or CTRL-Z to unapply the last sort.

	A	B	C	D	E
1	**Fuel Economy of Vehicles**				
2	This worksheet demonstrates the use of sorting and filtering in excel				
3					
4	**Make**	**Model**	**MPG City**	**MPG Highway**	**Annual Fuel Cost**
5	Bentley	Continental GTC	10	17	$2,885
6	Ford	Mustang 5.4L	14	21	$1,927
7	BMW	X5 xDrive 30i	15	20	$2,165
8	Jeep	Liberty 4WD	15	45	$743
9	Volkswagon	Jetta 2.5L	20	33	$1,039
10	Honda	Fit	27	21	$1,835
11	Honda	Civic Hybrid	40	29	$1,301
12					

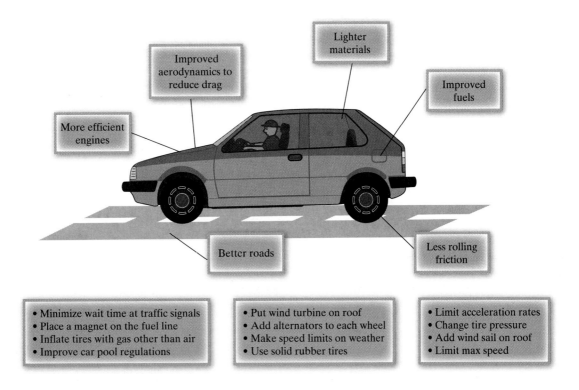

Lighter materials

Improved aerodynamics to reduce drag

Improved fuels

More efficient engines

Better roads

Less rolling friction

- Minimize wait time at traffic signals
- Place a magnet on the fuel line
- Inflate tires with gas other than air
- Improve car pool regulations

- Put wind turbine on roof
- Add alternators to each wheel
- Make speed limits on weather
- Use solid rubber tires

- Limit acceleration rates
- Change tire pressure
- Add wind sail on roof
- Limit max speed

Improving automotive gas mileage, while keeping costs under control, is a complex puzzle, involving many different types of engineers. Above are some ways to possibly improve fuel efficiency. Some really work, some are false claims, and some are fictitious. Can you tell the difference? What other ways can you think of to improve today's automobiles?

COMPREHENSION
CHECK 10-11

In 1980, the Environmental Protection Agency (EPA) began the Superfund Program to help cleanup highly polluted areas of the environment. There are over 1,300 Superfund sites across the country. Not all Superfund sites are from deliberate pollution. Some sites are old factories, where chemicals were dumped on the ground; landfills where garbage was dumped along with other poisonous waste; remote places where people secretly dumped hazardous waste because they did not know what to do with it; or old coal, iron ore, or silver mines.

According to the EPA (http://www.epa.gov/superfund/index.htm), the following groundwater contaminants were found in South Carolina Superfund sites in Greenville, Pickens, Oconee, and Anderson counties.

- Sort by city in ascending order. Examine the result: Which city appears first?
- Sort again: first by city in descending order, then by site name in descending order. Examine the results: Which site name now appears first?
- Sort again by contaminant in ascending order, then by site name in ascending order. Examine the results: Which site name appears last?

Contaminants	Site Name	City
Polycyclic aromatic hydrocarbons	Sangamo Weston	Pickens
Volatile organic compounds	Beaunit Corporation	Fountain Inn
Polycyclic aromatic hydrocarbons	Beaunit Corporation	Fountain Inn
Polycyclic aromatic hydrocarbons	Para-Chem Southern, Inc.	Simpsonville
Volatile organic compounds	Golden Strip Septic Tank Service	Simpsonville
Volatile organic compounds	Para-Chem Southern, Inc.	Simpsonville
Metals	Para-Chem Southern, Inc.	Simpsonville
Polycyclic aromatic hydrocarbons	Rochester Property	Travelers Rest
Volatile organic compounds	Sangamo Weston	Pickens
Polychlorinated biphenyl	Sangamo Weston	Pickens
Metals	Rochester Property	Travelers Rest
Metals	Golden Strip Septic Tank Service	Simpsonville
Metals	Beaunit Corporation	Fountain Inn
Volatile organic compounds	Rochester Property	Travelers Rest

Filtering Data in a Worksheet

Assume you want to look only at a specific portion of the data set and hide all the other rows of data. For example, you might want to look only at Honda vehicles or all vehicles that have an MPG City rating between 10 and 15 MPG. Excel has a built-in filtering capability by which you can conditionally display rows in a data set.

- Select the header row for a data set and click the **Sort & Filter** button in the **Home > Editing** ribbon. Click the **Filter** option to enable filtering for each column of data. Each column label contains a drop-down menu with various sorting options, as well as a number of different approaches for filtering.

  - For data sets that contain a small number of options, use the checkboxes in the drop-down filter to manually check certain options to display.
  - For numerical values, use the Number Filters submenu to filter on certain conditional expressions. The Custom Filter option in the Number Filters submenu lets you combine up to two logical expressions to filter a single column of data.

Assume you want to revisit your fuel economy data set and add in a number of statistical functions to assist in analysis.

	A	B	C	D	E
1	Fuel Economy of Vehicles				
2	This worksheet demonstrates the use of sorting and filtering in excel				
3					
4	**Make**	**Model**	**MPG City**	**MPG Highway**	**Annual Fuel Cost**
5	Bentley	Continental GTC	10	17	$2,886
6	BMW	X5 xDrive 30i	15	21	$1,927
7	Ford	Mustang 5.4L	14	20	$2,165
8	Honda	Civic Hybrid	40	45	$743
9	Honda	Fit	27	33	$1,039
10	Jeep	Liberty 4WD	15	21	$1,835
11	Volkswagon	Jetta 2.5L	20	29	$1,301
12					
13		Average:	20	27	$1,700
14		Min:	10	17	$743
15		Max:	40	45	$2,886
16					

Suppose you filter the data set to look only at the Honda vehicles.

	A	B	C	D	E
1	Fuel Economy of Vehicles				
2	This worksheet demonstrates the use of sorting and filtering in excel				
3					
4	**Make**	**Model**	**MPG City**	**MPG Highway**	**Annual Fuel Cost**
8	Honda	Civic Hybrid	40	45	$743
9	Honda	Fit	27	33	$1,039
12					
13		Average:	20	27	$1,700
14		Min:	10	17	$743
15		Max:	40	45	$2,886
16					

Notice that the statistical calculations at the bottom are still referencing the entire data set, even though, because of the filter, only a subset of the data is displayed. For data comparisons, this will be a valuable side effect; however, if you want the calculations to apply only to the visible data, you will need to use built-in functions other than the traditional functions (AVERAGE, MIN, MAX).

Using the SUBTOTAL Function

The **SUBTOTAL** function allows the worksheet to dynamically recalculate expressions generated with a filtered list. In the example where only Honda vehicles are selected, only the two visible vehicles will be used in the calculations, if you modify your worksheet to use the SUBTOTAL function instead of the traditional statistical functions. To use the SUBTOTAL function, pass in two different arguments:

> = SUBTOTAL (function_num, range)

- The *function_num* argument is a number associated to various built-in Excel functions. Table 10-7 lists the available functions for use with the SUBTOTAL function.
- The *range* argument is the range of cells to which the function should be applied.

Table 10-7 Available functions in SUBTOTAL

function_num	Function	Definition
1	AVERAGE	Computes the average value of the range
2	COUNT	Counts the number of cells in the range that contain numbers
3	COUNTA	Counts the number of nonempty cells in the range
4	MAX	Calculates the maximum value of the range
5	MIN	Calculates the minimum value of the range
6	PRODUCT	Calculates the product of each number in the range
7	STDEV	Calculates the standard deviation of the numbers in the range ("$n - 1$" method)
8	STDEVP	Calculates the standard deviation of the numbers in the range ("n" method)
9	SUM	Calculates the sum of all of the numbers in the range
10	VAR	Calculates the variance of the numbers in the range ("n" method)
11	VARP	Calculates the variance of the numbers in the range ("$n - 1$" method)

In the example, use the following calculation in cell C13 to calculate the average of MPG City:

> = AVERAGE(C5:C11)

The AVERAGE function corresponds to function_num 1, so the resulting calculation in cell C13 using the SUBTOTAL function would appear as follows:

> = SUBTOTAL (1, C5:C11)

After you modified all of the statistical calculations in the worksheet to use the SUBTOTAL function, the sheet should appear as shown in the examples below. Note that the values recalculate automatically according to the filtered data.

Filter on Make: Honda Only

	A	B	C	D	E
1	**Fuel Economy of Vehicles**				
2	This worksheet demonstrates the use of sorting and filtering in excel				
3					
4	**Make**	**Model**	**MPG City**	**MPG Highway**	**Annual Fuel Cost**
8	Honda	Civic Hybrid	40	45	$743
9	Honda	Fit	27	33	$1,039
12					
13		Average:	34	39	$891
14		Min:	27	33	$743
15		Max:	40	45	$1,039
16					

Filter on Annual Fuel Cost: Less than $1,500

	A	B	C	D	E
1	**Fuel Economy of Vehicles**				
2	This worksheet demonstrates the use of sorting and filtering in excel				
3					
4	**Make**	**Model**	**MPG City**	**MPG Highway**	**Annual Fuel Cost**
8	Honda	Civic Hybrid	40	45	$743
9	Honda	Fit	27	33	$1,039
11	Volkswagon	Jetta 2.5L	20	29	$1,301
12					
13		Average:	29	36	$1,028
14		Min:	20	29	$743
15		Max:	40	45	$1,301
16					

In-Class Activities

ICA 10-1

You want to set up a worksheet to investigate the oscillatory response of an electrical circuit. Create a worksheet similar to the one shown, including the proper header information.

4			
5			
6			
7	Neper Frequency (α_0)	25	[rad/s]
8	Resonant Frequency (ω_0)	400	[rad/s]
9	Initial Voltage (V_0)	15	[V]
10			
11	Damped Frequency (ω_d)		[rad/s]
12			
13			
14	Time (t) [s]	Voltage (V) [V]	
15			

First, calculate another constant, the damped frequency ω_d, which is a function of the neper frequency (α_0) and the resonant frequency (ω_0). This can be calculated with the formula

$$\omega_d = \sqrt{\omega_0^2 - \alpha_0^2}$$

Next, create a column of times (beginning in A15) used to calculate the voltage response, ranging from 0 to 0.002 seconds at an increment of 0.0002 seconds.

In column B, calculate the voltage response with the following equation, formatted to one decimal place:

$$V = V_0 e^{-\alpha_0 t} \cos(\omega_d t)$$

(a) Change neper frequency to 200 radians per second, resonant frequency to 800 radians per second, and initial voltage to 100 volts. At a time of 0.0008 seconds, what is the voltage?

(b) Change neper frequency to 100 radians per second, resonant frequency to 600 radians per second, and initial voltage to 100 volts. At a time of 0.0008 seconds, what is the voltage?

(c) Change neper frequency to 200 radians per second, resonant frequency to 400 radians per second, and initial voltage to 75 volts. At a time value of 0.0008 seconds, what is the voltage?

ICA 10-2

Some alternate energy technologies, such as wind and solar, produce more energy than needed during peak production times (windy and sunny days), but produce insufficient energy at other times (calm days and nighttime). Many schemes have been concocted to store the surplus energy generated during peak times for later use when generation decreases. One scheme is to use the energy to spin a massive flywheel at very high speeds, then use the rotational kinetic energy stored to power an electric generator later.

The worksheet shown below was designed to calculate how much energy is stored in flywheels of various sizes. The speed of the flywheel (revolutions per minute) is to be entered in cell B2, and the density of the flywheel in cell B4. A formula in cell B3 converts the speed into units of radians per second. There are 2π radians per revolution of the wheel.

To simplify the computations, the stored energy was calculated in three steps. The first table calculates the volumes of the flywheels, the second table uses these volumes to calculate the masses of the flywheels, and the third table uses these masses to determine the stored rotational kinetic energy.

Note that in all cases, changing the values in cells B2 and/or B4 should cause all appropriate values to be automatically recalculated.

	A	B	C	D	E	F	G	H	I
1									
2	**Speed (v) [rpm]**	15000		**Volume (V) [m³]**			**Height (H) [m]**		
3	**Speed (ω) [rad / s]**	1571		**Diameter (D) [m]**	**0.3**	**0.6**	**0.9**	**1.2**	**1.5**
4	**Density (ρ) [kg / m³]**	8000		0.2	0.009	0.019	0.028	0.038	0.047
5				0.4	0.038	0.075	0.113	0.151	0.188
6				0.6	0.085	0.170	0.254	0.339	0.424
7				0.8	0.151	0.302	0.452	0.603	0.754
8				1.0	0.236	0.471	0.707	0.942	1.178
9									
10				**Mass (m) [kg]**			**Height (H) [m]**		
11				**Diameter (D) [m]**	**0.3**	**0.6**	**0.9**	**1.2**	**1.5**
12				0.2	75	151	226	302	377
13				0.4	302	603	905	1206	1508
14				0.6	679	1357	2036	2714	3393
15				0.8	1206	2413	3619	4825	6032
16				1.0	1885	3770	5655	7540	9425
17									
18				**Kinetic Energy (KE) [J]**			**Height (H) [m]**		
19				**Diameter (D) [m]**	**0.3**	**0.6**	**0.9**	**1.2**	**1.5**
20				0.2	1.9E+06	3.7E+06	5.6E+06	7.4E+06	9.3E+06
21				0.4	3.0E+07	6.0E+07	8.9E+07	1.2E+08	1.5E+08
22				0.6	1.5E+08	3.0E+08	4.5E+08	6.0E+08	7.5E+08
23				0.8	4.8E+08	9.5E+08	1.4E+09	1.9E+09	2.4E+09
24				1.0	1.2E+09	2.3E+09	3.5E+09	4.7E+09	5.8E+09
25				**Average KE [J]**	3.6E+08	7.3E+08	1.1E+09	1.5E+09	1.8E+09
26				**Max KE - Min KE [J]**	4.3E+09				

(a) What should be typed in cell B3 to convert revolutions per minute in cell B2 into radians per second?

(b) What should be typed into cell E4 that can then be copied through the rest of the first table to calculate the flywheel volumes? Assume the shape of the flywheel to be a cylinder.

(c) What should be typed into cell E12 that can then be copied through the rest of the second table to calculate the flywheel masses?

(d) What should be typed into cell E20 that can then be copied through the rest of the third table to calculate the kinetic energies stored in the flywheels? The rotational kinetic energy is given by the formula: $KE_{rot} = I\omega^2 = (mr^2)\omega^2$

(e) What should be typed into cell E25 that can then be copied through Row 25 to determine the average kinetic energy at each height (in each column)?

(f) What should be typed into cell E26 to determine the difference between the maximum kinetic energy and 800 times the minimum kinetic energy given in the table?

ICA 10-3

Refer to the following worksheet. The following expressions are typed into the Excel cells indicated. Write the answer that appears in the cell listed. If the cell will be blank, write "BLANK" in the answer space. If the cell will return an error message, write "ERROR" in the answer space.

	A	B	C	D	E	F	G	H
1								
2								
3	**Fluid Type**	Benzene			**Fluid Type**	Olive Oil		
4	**Density (ρ)**	0.879	[g / cm^3]		**Density (ρ)**	0.703	[g / cm^3]	
5	**Viscosity (μ)**	6.47E-03	[g / (cm s)]		**Viscosity (μ)**	1.01	[g / (cm s)]	
6								
7	**Velocity (v)**	15	[cm / s]		**Velocity (v)**	50	[cm / s]	
8								
9	**Pipe Diameter**	**Reynolds Number**			**Pipe Diameter**	**Reynolds Number**		
10	**(D) [cm]**	**(Re) [--]**			**(D) [cm]**	**(Re) [--]**		
11	1.27	2,588			1.27	44		
12	2.54	5,176			2.54	88		
13	3.81	7,764			3.81	133		
14	5.08	10,352			5.08	177		
15	6.35	12,940			6.35	221		
16	7.62	15,529			7.62	265		

	Expression	Typed into Cell
(a)	= IF (B4 > F4, B3, "F3")	D4
(b)	= IF (B7/2 > F7/10, " ", B7*2)	H7
(c)	= IF (B11 < F11, "B11", IF (B11 > F11, SUM(B11, F11), F11))	D11
(d)	= IF (AND(B4 < F4,B5 < F5), B3, MAX(F11:F16))	D9
(e)	= IF(OR(E16/2^2 > E15*2,E11+E12 < E14),F4*62.4,F4*1000)	H16

ICA 10-4

Use the following worksheet to answer the questions.

	A	B	C
1	Temperature Limit of Phase B	250	
2	Temperature Limit of Phase D	350	
3			
4			
5			
6	Amount of Type S Polymer (S) [wt %]	Temperature (T) [K]	Phase
7	92	387	Phase D
8	86	125	Phase E
9	17	407	Phase D
10	51	323	Phase C
11	12	73	Phase A
12	53	174	Phase B
13	73	79	Phase B
14	75	275	Phase C
15	19	51	Phase B
16	47	300	Phase C
17	45	79	Phase B

(a) If the following is typed into cell C1, what will appear when the statement is executed?

$$= IF (B1 > 200,\ IF (B1 < 300, 200 + 50, \text{"High"}), MAX(B1, B2))$$

(b) If the following is typed into cell C2, what will appear when the statement is executed?

$$= IF (OR(B1 > 200, B2 < 500), \text{""}, 300)$$

(c) The axes for a phase diagram of a polymer blend are shown on the answer sheet. Use the following worksheet to sketch the phase diagram, containing 5 phases A–E, that was used to create the conditional statement that appears in cell C7. The following appears in cell C7, typed on a single line:

$$= IF (A7 < 15, \text{"Phase A"}, IF (AND (A7 < 80, B7 < \$B\$1), \text{"Phase B"}, IF (B7 > \$B\$2,$$
$$\text{"Phase D"} IF (A7 > 80, \text{"Phase E"}, \text{"Phase C"})))$$

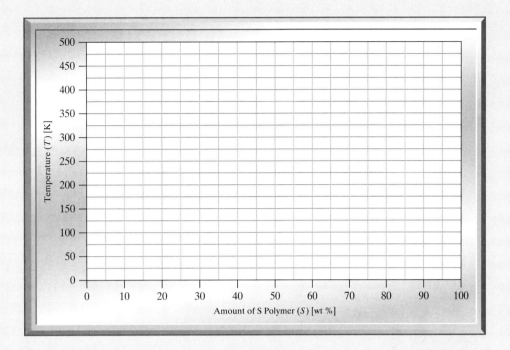

ICA 10-5

A bioengineer conducts clinical trials on stressed-out college students to see if a sleep aid will help them fall asleep faster. She begins the study by having 20 students take a sleep aid for seven days and records through biofeedback the time when they fall asleep. To analyze the data, she sets up the following worksheet. Evaluate the expressions below; state what will appear in the cell when the command is executed. Column I contains the average time each student took to fall asleep during the seven-day trial. Column J contains any adverse reactions the students experienced (H = headache; N = nausea).

(a) Column K will contain the rating of the time it took the student to fall asleep compared with the control group, who did not take the medication. The statement as it appears in cell K14 is given below. What will appear in cell K14 when this statement is executed?

> = IF > (I14 > $I$2 + $I$3, "MORE", IF (I14 < $I$2 − $I$3, "LESS", ""))

(b) Column L groups the participants into three groups according to their reaction to the drug and the time it took them to fall asleep. Assume the statement for part (a) is executed in Column K. The statement as it appears in cell L7 is given below. What will appear in cell L7 when this statement is executed?

> = IF (AND (K7 = "MORE", J7 = "H"), "MH", IF (AND (K7 = "MORE", J7 = "N"), "MN", ""))

(c) Suppose the formula in Column L was changed to regroup the participants. The statement as it appears in cell L9 is given below. In Excel, this statement would appear as a continuous line, but here it is shown on two lines for space. What will appear in cell L9 when this statement is executed?

> = IF (AND (K9 = "MORE", OR (J9 = "H", J9 = "N")), "SEVERE",
>
> IF (OR (J9 = "H", J9 = "N"), "MILD", IF (K9 = "LESS", "HELPFUL", "")))

(d) Suppose the formula in part (c) was copied into cell L16. What would appear in cell L16 when this statement is executed?

(e) Suppose the formula in part (c) was copied into cell L18. What would appear in cell L18 when this statement is executed?

(f) Does the expression given in part (c) accurately identify all severe reaction conditions? If not, how would you change the expression to include all severe reactions?

	A	B	C	D	E	F	G	H	I	J	K	L
1						Control Group Data						
2						Overall Average			35	[min]		
3						Standard Deviation			4	[min]		
4												
5					Number of Minutes to Fall Asleep							
6	Patient	Day 1	Day 2	Day 3	Day 4	Day 5	Day 6	Day 7	Average	Reaction	Time	Group
7	A	45	39	83	47	39	25	42	46	H		
8	B	35	75	15	36	42	12	29	35			
9	C	42	32	63	45	37	34	31	41	N		
10	D	14	25	65	38	53	33	32	37	H		
11	E	14	71	48	18	29	14	24	31			
12	F	14	25	29	24	18	24	15	21	H N		
13	G	31	14	42	19	28	17	21	25			
14	H	12	24	32	42	51	12	16	27	H N		
15	I	28	29	44	15	43	15	22	28	N		
16	J	21	19	35	41	34	25	18	28	H		
17	K	44	36	51	39	30	26	25	36			
18	L	38	43	36	59	14	34	18	35	N		
19	M	19	15	63	50	55	27	31	37	H		

ICA 10-6

Refer to the worksheet shown, set up to calculate the displacement of a spring. Hooke's law states the force (F, in newtons) applied to a spring is equal to the stiffness of the spring (k, in newtons per meter) times the displacement (x, in meters): $F = kx$.

	A	B	C	D	E	F	G	H
1								
2	Spring Code	Stiffness [N/m]	Maximum Displacement [mm]			Spring Code	Stiffness [N / m]	Maximum Displacement [mm]
3	3-Blue	50	20			1-Blue	10	40
4						1-Black	25	60
5	Mass [g]	Displacement [cm]	Warning			2-Blue	30	25
6	25	0.49				2-Black	40	60
7	50	0.98				2-Red	20	30
8	75	1.47				3-Blue	50	20
9	100	1.96				3-Red	40	30
10	125	2.45	Too Much Mass			3-Green	60	10
11	150	2.94	Too Much Mass					
12	175	3.43	Too Much Mass					
13	200	3.92	Too Much Mass					
14	225	4.41	Too Much Mass					
15	250	4.90	Too Much Mass					
16	275	5.39	Too Much Mass					
17	300	5.88	Too Much Mass					
18								

Cell A3 contains a data validation list of springs. The stiffness (cell B3) and maximum displacement (cell C3) values are found using a VLOOKUP function linked to the table shown at the right side of the worksheet. These data are then used to determine the displacement of the spring at various mass values. A warning is issued if the displacement determined is greater than the maximum displacement for the spring. Use this information to determine the answers to the following questions.

(a) Write the expression, in Excel notation, that you would type into cell B6 to determine the displacement of the spring. Assume you will copy this expression to cells B7 to B17, so be sure and watch your absolute and relative references—and your units!

(b) A LOOKUP function in Excel contains four parts. Fill in the following information in the VLOOKUP function used to determine the maximum displacement in cell C3 based on the choice of spring in cell A3.

= VLOOKUP(___(a)___ , ____(b)____ , ____(c)____ , ____(d)____)

(c) An IF statement in Excel contains three parts. Fill in the following information in the IF function used to determine the warning given in cell C6, using the maximum displacement in cell C3. Assume you will copy this expression to cells C7 to C17, so be sure and watch your absolute and relative references—and your units!

= IF(___(a)___ , ____(b)____ , ____(c)____)

ICA 10-7

You are interested in analyzing different implant parts being made in a bioengineering production facility. The company has the ability to make 13 different parts from 17 different metal alloys.

	A	B	C
1			
2	MATERIAL	WC-2	
3	Cost / lbm	17	
4	Specific Gravity	11.88	
5			
6	Part Number	JB3	
7	Material Weight [lbm]	0.3	
8	Part Volume [cin]	3.5	
9			
10	Number of Parts	6000	Too Many
11			
12	Amount of Mat'l to Order [lbm]	1800	Check $
13			
14	Size of Shipping Package		OK
15	Length [in]	24	
16	Width [in]	5	
17	Height [in]	2	
18			
19	Amount of Boxes Needed	44	
20			

	D	E	F	G
1		Material	Specific Gravity	Material Cost / lbm
2		AuZn-4	19.8	$ 15.00
3		WC-2	11.88	$ 17.00
4		CuAg-5	6.6	$ 24.00
5		PdSi-3	15.84	$ 18.00
6		PdSi-5	15.84	$ 15.00
7		ZnCd-2	6.6	$ 16.00
8		CoNi-7	7.92	$ 16.00
9		CoAg-12	13.2	$ 24.00
10		CdAl-2	7.92	$ 14.00
11		PtZn-4	7.92	$ 13.00
12		PtC-9	19.8	$ 22.00
13		MnPd-8	14.52	$ 5.00
14		WTi-3	11.88	$ 6.00
15		ScCo-4	6.6	$ 18.00
16		ZrW-8	5.28	$ 18.00
17		MnRh-5	6.6	$ 7.00
18		PdCd-7	7.92	$ 9.00

	H	I	J	K	L	M
1		Part Number	Material Weight [lbm]	Number / Box	Part Volume [cin]	Energy Cost / Part
2		JB2	0.1	12	7.5	0.1
3		JB3	0.3	76	3.5	0.03
4		JB5	0.1	48	7.5	0.08
5		JB6	0.4	66	1	0.1
6		JB8	0.1	96	6.5	0.07
7		KA9	0.1	4	1.5	0.05
8		KA11	0.45	85	4	0.04
9		KA2	0.05	96	7.5	0.01
10		KA5	0.45	5	3.5	0.04
11		DS3	0.3	86	3	0.03
12		DS7	0.5	92	3	0.08
13		DS8	0.05	51	1	0.05
14		DS12	0.1	47	5	0.08
15						

On the worksheet shown, you have created a place for the user to choose the material, linked to Column E, and the part number, linked to Column I, using data validation lists. Once these two values are chosen, the cost per pound, specific gravity, material weight of the part, and part volume all automatically adjust using VLOOKUP functions. LOOKUP functions in Excel contain four parts.

= VLOOKUP(__(a)__ , __(b)__ , __(c)__ , __(d)__)

(a) Fill in the following information in the VLOOKUP function used to determine the cost per pound of material in cell B3 based on the choice of material in cell B2.

(b) Fill in the following information in the VLOOKUP function used to determine the part volume in cell B8 based on the choice of part number in cell B6.

An IF statement in Excel contains three parts. Fill in the following information in the IF function used to determine the following conditions.

$$= IF(\_\_(a)\_\_,\_\_(b)\_\_,\_\_(c)\_\_)$$

(c) In cell B10, the user can enter the number of parts needed in production. If this value is more than 2,500 parts, a warning will appear in cell C10 telling the user the quantity is too high; otherwise, the cell remains blank.

(d) In cell B12, the worksheet determines the amount of material to order. If the cost of the materials (determined by the amount in cell B12 times the cost per pound in cell B3) is more than $1,000, a warning will appear in cell C12 telling the user to check with supervision first; otherwise, the cell displays the cost of the material to order.

(e) In cells B15–B17, the user enters the size of the shipping box being used once the pieces have been made and are ready to ship to the hospitals. The worksheet checks to ensure the box dimensions are not larger than 24 inches long by 12 inches wide by 12 inches high. If the user enters values outside this range, the program will warn the user in cell C14 to resize the box; otherwise, the cell will indicate the dimensions are ok.

ICA 10-8

You have a large stock of several values of inductors and capacitors, and are investigating how many possible combinations of a single capacitor and a single inductor chosen from the ones you have in stock will give a resonant frequency between specified limits.

Create two cells to hold a minimum and maximum frequency the user can enter. If the value entered for the maximum frequency (f_{MAX}) is less than the minimum frequency (f_{MIN}), the maximum frequency cell should be shaded dark grey with white, bold text.

Incorrect Data:

Allowable Range	
f_{min} [Hz]	f_{max} [Hz]
2,500	1,000

Correct Data:

Allowable Range	
f_{min} [Hz]	f_{max} [Hz]
2,500	7,777

Calculate the resonant frequency (f_R) for all possible combinations of one inductor and one capacitor, rounded to the nearest integer. For a resonant inductor/capacitor circuit, the resonant frequency in hertz [Hz] is calculated by

$$f_R = \frac{1}{2\pi\sqrt{LC}}$$

Here, L is the inductance in units of henry [H] and C is the capacitance in units of microfarads [μF]. Automatically format each result to indicate its relation to the minimum and maximum frequency values as listed below.

- $f_R > f_{MAX}$: The cell should be shaded white with light grey text and no border.
- $f_R < f_{MIN}$: The cell should be shaded light grey with dark grey text and no border.
- $f_{MIN} < f_R < f_{MAX}$: The cell should be shaded white with bold black text and a black border.

If done properly, the table should appear similar to the table below for $f_{MIN} = 2,500$ and $f_{MAX} = 7,777$.

After you have this working properly, modify the frequency input cells to use data validation instead of conditional formatting to warn the user of an invalid value entry.

Resonant Frequency (f_R) [Hz]	Capacitance (C) [µF]							
Inductance (L) [H]	0.0022	0.0082	0.05	0.47	0.82	1.5	3.3	10
0.0005	151748	78601	31831	10382	7860	5812	3918	2251
0.002	75874	39301	15915	5191	3930	2906	1959	1125
0.01	33932	17576	7118	2322	1758	1299	876	503
0.05	15175	7860	3183	1038	786	581	392	225
0.068	13012	6740	2729	890	674	498	336	193
0.22	7234	3747	1517	495	375	277	187	107
0.75	3918	2029	822	268	203	150	101	58

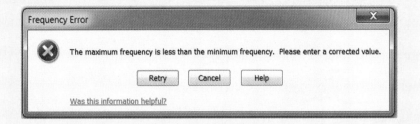

Frequency Error ☒

The maximum frequency is less than the minimum frequency. Please enter a corrected value.

Retry   Cancel   Help

Was this information helpful?

ICA 10-9

We accidentally drop a tomato from the balcony of a high-rise apartment building. As it falls, the tomato has time to ponder some physics and says, "You know, the distance I have fallen equals $\frac{1}{2}$ gravity times the time I have fallen squared." Create a worksheet to solve the question of when the tomato goes splat.

- The user will input the initial balcony height in units of feet. Use data validation to set a limit for the height of 200 feet.
- Place the acceleration due to gravity in a cell under the balcony height and not within the formulas themselves. *Be sure to watch the units for this problem!*
- Column A will be the distance the tomato falls, starting at a distance of zero up to a distance of 200 feet, in 5-foot increments.
- Column B will show the calculated time elapsed at each distance fallen.
- Column C will display the status of the tomato as it falls.
  - If the tomato is still falling, the cell should display the distance the tomato still has to fall.
  - If the tomato hits the ground, the cell should display "SPLAT" on a red background.
  - SPLAT should appear once; the cells below are blank.

Use the worksheet just created to test the following parameters.

	Balcony Height [ft]
(a)	200
(b)	120
(c)	50
(d)	25

ICA 10-10

You are interested in calculating the best place to stand to look at a statue. Where should you stand so that the angle subtended by the statue is the largest?

At the top of the worksheet, input the pedestal height (P) and the statue height (S).

In Column A, create a series of distances (d) from the foot of the statue, from 2 feet to 40 feet by 2-foot increments.

In Column B, calculate the subtended angle in radians using the following equation:

$$\theta = \tan^{-1}\left(\frac{P + S}{d}\right) - \tan^{-1}\left(\frac{P}{d}\right)$$

In Column C, write a function to change the angles in Column B from radians to degrees. At the bottom of Column C, insert a function to display the maximum value of all the angles.

Photo courtesy of E. Stephan

In Column D, use a conditional statement whose output is blank except at the single distance where the angle is a maximum; at the maximum, print "Stand Here."

Use the worksheet just created to test the following parameters

	Pedestal Height (P) [ft]	Statue Height (S) [ft]
(a)	20	10
(b)	10	10
(c)	15	20
(d)	30	20

ICA 10-11

Many college students have compact refrigerator-freezers in their dorm room. The data set provided is a partial list of energy efficient models less than 3.6 cubic feet [cft], according to the American Council for an Energy Efficient Economy (www.aceee.org). Complete the analysis below.

(a) We would like to compute the cost to run each model for a year. Assume that it costs $0.086 per kilowatt-hour [kWh]. Create a new column, "Annual Energy Cost [$/year]," that calculates the annual energy cost for each refrigerator.

(b) First, we will sort the first table by energy usage, with the model with the highest kilowatt-hour rating listed first. Which model appears first?

(c) Next, we will sort by the volume in ascending order and the annual energy cost in ascending order. Which model appears first?

(d) Assume we want to restrict our selection to refrigerators that can contain more than 2.5 cubic feet. Which models appear in the list?

(e) Assume we want to restrict our selection to refrigerators that can contain more than 2.5 cubic feet and only require between 0 and 300 kilowatt-hours per year. Which models appear in the list?

ICA 10-12

The complexity of video gaming consoles has evolved over the years. The data set provided is a list of energy usage data on recent video gaming consoles, according to the Sust-It consumer energy report data (www.sust-it.net). Complete the analysis below.

(a) Compute the cost to run each gaming console for a year, including the purchase price. Assume that it costs $0.086 per kilowatt-hour [kWh]. Create a new column, "Cost + Energy [$/yr]," that calculates the total (base + energy) cost for each gaming console.

(b) On average, a consumer will own and operate a video gaming console for four years. Calculate the total carbon emission [kilograms of carbon dioxide, or kg CO_2] for each gaming console over the average lifespan; put the result in a column labeled "Average Life Carbon Emission [kg CO_2]."

(c) First, sort the table by total cost, with the console with the highest total cost listed first. Which console appears first?

(d) Next, sort by the original cost in ascending order and the average life carbon emission in ascending order. Which console appears last?

(e) Restrict your selection to video game consoles that originally cost $300. Which models appear in the list?

(f) Restrict your selection to video game consoles that originally cost less than or equal to $300 and have an average life carbon emission less than or equal to 25 kg CO_2. Which models appear in the list?

Chapter 10 REVIEW QUESTIONS

1. With current rocket technology, the cost to lift one kilogram of mass to geosynchronous orbit (GSO) is about $20,000. Several other methods of lifting mass into space for considerably less cost have been envisioned, including the Lofstrom loop, the orbital airship, and the space elevator.

   In space elevators, a cargo compartment (climber) rides up a slender tether attached to the Earth's surface and extending tens of thousands of miles into space. Many designs provide power to the climber by beaming it to a collector on the climber using a laser of maser.

   The leftmost column of the table should contain efficiencies from 0.5% to 2% in 0.25% increments. The top row of the table should list electricity prices from 4 cents to 14 cents per kilowatt-hour with 2 cent increments. Each row of the table thus represents a specific efficiency and each column represents a specific electricity cost. The intersection of each row and column should contain the corresponding total cost of the electricity used to lift one kilogram to GSO.

   Assume that the total change in the potential energy of an object lifted from sea level to GSO is 50 megajoules per kilogram.

   Any constants and conversion factors used should appear as properly labeled constants in individual cells, and your formulae should reference these. Conversions and constants should NOT be directly coded into the formulae. You are expected to use absolute, relative, and mixed cell addressing as appropriate. As an example (not within the table) so you can check your formulae, if electricity costs 18 cents per kilowatt-hour and the conversion efficiency is 3%, the electricity to lift one kilogram to GSO would cost $83.33.

2. A history major of your acquaintance is studying agricultural commerce in nineteenth century Wales. He has encountered many references to "hobbits" of grain, and thinking that this must be some type of unit similar to a bushel (rather than a diminutive inhabitant of Middle Earth), he has sought your advice because he knows you are studying unit conversions in your engineering class.

   He provides a worksheet containing yearly records for the total number of hobbits of three commodities sold by a Mr. Thomas between 1817 and 1824, and has asked you to convert these to not only cubic meters, but also both U.S. and imperial bushels.

	A	B	C	D
1	Grain sold by Mr. Thomas			
2				
3			Hobbits Sold	
4	Year	Barley	Wheat	Oats
5	1817	106	154	203
6	1818	118	145	187
7	1819	98	167	167
8	1820	137	124	199
9	1821	102	105	210
10	1822	142	168	147
11	1823	93	132	186
12	1824	117	136	193

   After a little research, you find that the hobbit was equal to two and a half imperial bushels, the imperial bushel equals 2,219 cubic inches, and the U.S. bushel equals 2,150 cubic inches.

   First, you create a table showing the conversion factors from hobbits to the other units, including comments documenting the conversion. You then use these calculated conversion factors to create the rest of the table. (See example, including embedded comment for conversion to cubic meters.)

	A	B	C	D	E	F	G	H	I	J	K	L	M
1	Grain sold by Mr. Thomas												
2													
3			Barley				Wheat				Oats		
4	Year	Hobbits	Imp. Bushels	US Bushels	Cubic Meters	Hobbits	Imp. Bushels	US Bushels	Cubic Meters	Hobbits	Imp. Bushels	US Bushels	Cubic Meters
5	1817	106	265	273.5	9.6	154	385	397.4	14.0	203	507.5	523.8	18.5
6	1818	118	295	304.5	10.7	145	362.5	374.1	13.2	187	467.5	482.5	17.0
7	1819	98	245	252.9	8.9	167	417.5	430.9	15.2	167	417.5	430.9	15.2
8	1820	137	342.5	353.5	12.5	124	310	319.9	11.3	199	497.5	513.5	18.1
9	1821	102	255	263.2	9.3	105	262.5	270.9	9.5	210	525	541.8	19.1
10	1822	142	355	366.4	12.9	168	420	433.5	15.3	147	367.5	379.3	13.4
11	1823	93	232.5	240.0	8.5	132	330	340.6	12.0	186	465	479.9	16.9
12	1824	117	292.5	301.9	10.6	136	340	350.9	12.4	193	482.5	498.0	17.5
13													
14		Hobbit	Imp. Bushels	US Bushels	Cubic Meters								
15		1	2.5	2.58023	0.09091	[Imp. Bushel] [2219 cu. in./Imp. Bushel] [2.54 cm/in]^3 [meter/100 cm]^3							
16													

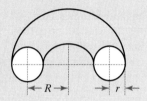

3. In the 1950s, a team at Los Alamos National Laboratories built several devices they called "Perhapsatrons," thinking that PERHAPS they might be able to create controllable nuclear fusion. After several years of experiments, they were never able to maintain a stable plasma and abandoned the project.

   The perhapsatron used a toroidal (doughnut-shaped) plasma confinement chamber, similar to those used in more modern Tokamak fusion devices. You have taken a job at a fusion research lab, and your supervisor asks you to develop a simple spreadsheet to calculate the volume of a torus within which the plasma will be contained in a new experimental reactor.

   (a) Create a simple calculator to allow the user to type in the radius of the tube (r) in meters and the radius of the torus (R) in meters and display the volume in cubic meters.

   (b) Data validation should be used to assure that $R > r$.

   (c) Create a table that calculates the volumes of various toruses with specific values for r and R. The tube radii (r) should range from 5 centimeters to 100 centimeters in increments of 5 centimeters. The torus radii (R) should range from 1.5 meters to 3 meters in increments of 0.1 meters.

   The volume of a torus can be determined using $V = 2\pi^2 R r^2$.

4. A phase diagram for carbon and platinum is shown. Assuming the lines shown are linear, we can say the mixture has the following characteristics:

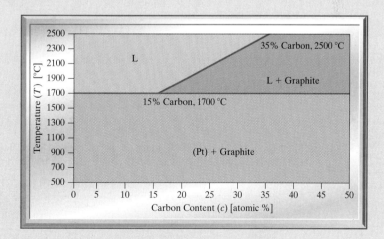

   - Below 1,700°C, it is a mixture of solid platinum and graphite.
   - Above 1,700°C, there are two possible phases: a liquid (L) phase and a liquid (L) + graphite phase. The endpoints of the division line between these two phases are labeled on the diagram. Develop a worksheet to determine the phase of a mixture, given the temperature and carbon content. A sample worksheet is shown below.

	A	B	C	D	E
1					
2					
3	Maximum Temperature for Pt + G			1700	[°C]
4					
5					
6					
7	Temperature (T) [°C]	Carbon Content (c) [%]	Temp between L & L+G	Phase	
8					
9	1220	30			
10	1300	34			
11	1150	45			
12	1310	56			

(a) Write the equation to describe the temperature of the dividing line between the liquid (L) region and the liquid (L) + graphite region in Column C. Reference the carbon content found in Column B as needed. Add any absolute reference cells you feel are needed to complete this calculation.

(b) Write the conditional statement to determine the phase in Column D. For simplicity, call the phases Pt + G, L, and L + G. For points on the line, YOU can decide which phase they are included in.

(c) Use conditional formatting to indicate each phase. Provide a color key.

5. A simplified phase diagram for cobalt and nickel is shown. Assuming the lines shown are linear, we can say the mixture has the following characteristics:

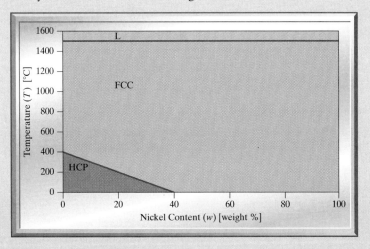

- Above 1,500°C, it is a liquid.
- Below 1,500°C, there are two possible phases: face-centered cubic (FCC) phase and hexagonal close-packed (HCP) phase. Develop a worksheet to determine the phase of a mixture, given the temperature and nickel content. A sample worksheet is shown below.

	A	B	C	D	E	F
1						
2						
3		Minimum Temperature for Liquid			1500	[°C]
4						
5	Temperature (T) [°C]	Nickel Content (w) [%]	Temp Between HCP & FCC	Phase		
6						
7	250	12				
8	1015	43				
9	125	49				
10	360	27				
11	108	68				
12	200	40				

(a) Write the mathematical equation to describe the dividing line between the HCP region and the FCC region in Column C. Reference the nickel content found in Column B as needed. Add any absolute reference cells you feel are needed to complete this calculation.

(b) Write the conditional statement to determine the phase in Column D. For simplicity, call the phases HCP, FCC, and L. For points on the line, YOU can decide which phase they are included in.

(c) Use conditional formatting to indicate each phase. Provide a color key.

6. You enjoy drinking coffee but are particular about the temperature (T) of your coffee. If the temperature is greater than or equal to 70 degrees Celsius [°C], the coffee is too hot to drink; less than or equal to 45°C is too cold by your standards. Your coffee pot produces coffee at the initial temperature (T_0). The cooling of your coffee can be modeled by the equation below, where time (t) and the cooling factor (k) are in units per second:

$$T = T_0 e^{-kt}$$

(a) At the top of the worksheet, create an area where the user can modify four properties of the coffee:

- Initial temperature (T_0); for the initial problem, set to 80°C.
- Cooling factor (k); set to 0.001 per second [s^{-1}].
- Temperature above which coffee is "Too Hot" to drink (T_{hot}); set to 70°C.
- Temperature below which coffee is "Too Cold" to drink (T_{cold}); set to 45°C.

(b) Create a temperature profile for the coffee:

- In column A, generate a time range of 0–300 seconds, in 15-second intervals.
- In column B, generate the temperature of the coffee, using the equation given and the input parameters set by the user (T_0 and k).

(c) In column B, the temperature values should appear on a red background if the coffee is too hot to drink, and a blue background if it is too cold using conditional formatting.

(d) In column C, create a warning next to each temperature that says "Do not Drink" if the calculated temperature in column B is too hot or too cold in comparison with the temperature values the user enters.

7. The phase diagram below for the processing of a polymer relates the applied pressure to the raw material porosity.

- Region A or B = porosity is too high or too low for the material to be usable.
- Region C = combinations in this region yield material with defects, such as cracking or flaking.
- Region D = below a pressure of 15 pound-force per square inch [psi] the polymer cannot be processed.
- Region E = optimum region to operate.

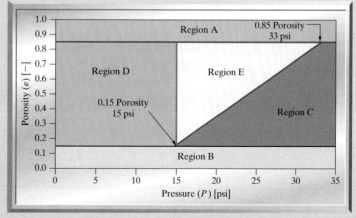

There are often multiple ways to solve the same problem; here we look a few alternative ways to determine the phase of the material and the processability of the material.

	A	B	C	D	E
1					
2					
3	Porosity Upper Limit [%]			85	
4	Porosity Lower Limit [%]			15	
5	Pressure Limit [psi]			15	
6					
7	Pressure	Porosity	Porosity between	Phase	Is Material Able
8	(P) [psi]	(ε) [%]	C and E		to be Processed?
9	13	68			
10	26	39			
11	10	43			
12	15	82			
13	22	47			
14	16	84			
15	7	73			
16	25	45			

Method One: Conditional Statements — (a)–(d)

(a) Begin on the worksheet titled Conditionals. In Column C, develop the equation for the line dividing the phases of Region E and Region C. Assume it was written in cell C9 and copied to Column C.

(b) In Column D, write an expression to determine the phase of the material (Phase A–Phase E).

(c) In Column E, write an expression to determine if the material is processible.

(d) When the conditions of Phase E are met, the cell should be highlighted by conditional formatting. Provide a color key.

Method Two: Data Validation — (e)–(i)

(e) Begin on the worksheet titled Validation. In Column A and Column B, use data validation to restrict the user from entering values outside the valid parameter ranges—pressure: 0–35 psi and porosity: 0–100%.

(f) In Column C, develop the equation for the line dividing the phases of Region E and Region C.

(g) In Column D, write an expression to determine the phase of the material (Phase A–Phase E).

(h) In Column E, write an expression to determine if the material is processible.

(i) When the conditions of Phase E are met, the cell should be highlighted by conditional formatting.

(j) Can you write an expression in Column F to tell the user why the material was rejected? For example, under the conditions of pressure = 25 psi and porosity = 40%, the statement might say "Porosity too low."

(k) Can you write an expression in Column F to tell the user why the material was rejected and how to fix the problem to make the product useful? For example, under the conditions of pressure = 25 psi and porosity = 40%, the statement might say "Porosity too low, increase to at least a value of 55%."

	A	B	C	D	E	F
1						
2						
3	Porosity Upper Limit [%]			85		
4	Porosity Lower Limit [%]			15		
5	Pressure Limit [psi]			15		
6						
7	Pressure	Porosity	Porosity between			
8	(P) [psi]	(ε) [%]	C and E			
9	50	68				
10						
11						
12						

Microsoft Excel

The value you entered is not valid.

A user has restricted values that can be entered into this cell.

Retry   Cancel   Help

Was this information helpful?

8. The following phase diagram is for salt water. There are four possible phases, which depend on the temperature and the sodium chloride content (NaCl).

- Ice and SC = Mixed ice and salt crystals.
- Ice and SW = Ice and saltwater.
- SW = Saltwater.
- SW and SC = Saltwater and salt crystals.

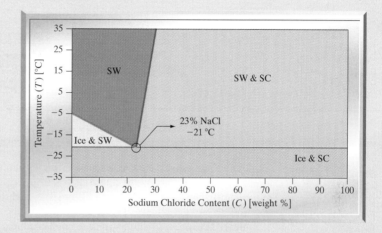

There are often multiple ways to solve the same problem; here we look a few alternative ways to determine the phase of the mixture.

Method One: Conditional Statements—(a)–(d)

(a) Begin on the worksheet titled Conditionals. In Column C, develop the equation for the line dividing the phases of the ice–saltwater mix and the saltwater. Assume it was written in cell C11 and copied down.

(b) In Column D, develop the equation for the line dividing the phases of the saltwater and the saltwater–salt crystals mix. Assume it was written in cell D11 and copied down.

(c) In Column E, write an expression to determine the phase of the mixture.

(d) Use conditional formatting to highlight the various phases. Provide a color key.

	A	B	C	D	E
5					
6					
7	Upper Limit of Mixed Ice and Salt Crystals			-21	[°C]
8					
9			Dividing Temp [°C]		
10	NaCl [%]	Temp [°C]	Ice and SW to SW	SW to SW and SC	Phase
11	83	-25			
12	73	2			
13	30	12			
14	56	-12			
15	90	31			
16	39	23			
17	33	-16			
18	35	5			

Method Two: Data Validation—(e)–(i)

(e) Begin on the worksheet titled Validation. In Column A and Column B, use data validation to restrict the user from entering values outside the valid parameter ranges: NaCl (%): 0–100%; Temp [°C]: −35°C to 35°C.

(f) In Column C, develop the equation for the line dividing the phases of the ice–saltwater mix and the saltwater.

(g) In Column D, develop the equation for the line dividing the phases of the saltwater and the saltwater–salt crystals mix.

(h) In Column E, write an expression to determine the phase of the mixture.

(i) Use conditional formatting to highlight the various phases. Provide a color key.

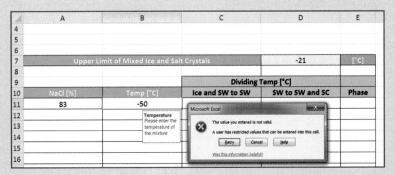

9. Guitars are constructed in a manner to allow musicians to place their fingers on different arrangements of strings to create chords. Chords are arrangements of three or more notes (or pitches) played at the same time. (There is some disagreement among musicologists concerning whether two notes played simultaneously should be classified as a chord.) On a standard electric guitar, there are typically 6 strings strung from the nut down the neck of the guitar with 21 frets set in the neck terminating at the bridge on the guitar. On a 6-string, 21-fret guitar, there are 126 different locations that can generate a pitch plus the 6 pitches generated by playing a string without pressing a fret. However, each fretted string on a guitar can take on 1 of 12 different semitones. As a guitarist moves their finger from lower to higher frets on a string on a guitar, each note changes by increasing 1 semitone.

The 12-tone semitone transitions are shown in the chart below:

From	To	From	To
A	A#/Bb	D#/Eb	E
A#/Bb	B	E	F
B	C	F	F#/Gb
C	C#/Db	F#/Gb	G
C#/Db	D	G	G#/Ab
D	D#/Eb	G#/Ab	A

In other words, if we are on the 5th string playing a B on the 10th fret, moving to the 11th fret would play a C.

Fill in the provided Excel worksheet using LOOKUP statements to the semitone transition table for a standard tuning guitar (E-A-D-G-B-E) to show where each note will be located on the guitar fretboard. The worksheet should be set up so that it can be easily modified for different guitar tunings:

(a) What will the resulting notes on the 5th fret be if an "open tuning," typical of blues and slide guitar music, if the open G tuning were used (G-B-D-G-B-D)?

(b) What will the resulting notes on the 8th fret be if the D-A-D-G-A-D tuning were used (e.g., Led Zeppelin's "Kashmir")?

(c) Use conditional formatting to highlight the sharps and flats (notes with # or b) in light yellow.

(d) Use conditional formatting to highlight all entries with the same note name as one of the base tuning notes within the column by making the text white on a black background. For standard tuning, this would highlight all occurrences of E in the first column, A in the second column, etc.

(e) Use data validation to allow the user to choose the notes from a drop-down list.

10. When liquid and vapor coexist in a container at equilibrium, the pressure is called vapor pressure. Several models predict vapor pressure. One, called the **Antoine equation**, first introduced by Ch. Antoine in 1888, yields vapor pressure in units of millimeters of mercury [mm Hg].

$$P = 10^{(A - \frac{B}{T+C})}$$

The constants A, B, and C are called the *Antoine constants*; they depend on both fluid type and temperature. Note that "B" and "C" must be in the same units as temperature and "A" is a dimensionless number, all determined by experiment.

	A	B	C	D	E	F	G
6							
7			Constants			Validity Range	
8	Compound	A	B [°C]	C [°C]	T_{min}[°C]	T_{max}[°C]	
9	Ethanol	8.204	1642.89	230.3	-57	80	
10							
11							
12							
14							
15	Temperature (T) [°C]	Pressure (P) [mm Hg]					
16	-100						
17	-95						
18	-90						
19	-85						
20	-80						
21	-75						
22	-70						
23	-65						

Create a worksheet using the provided template. The Antoine constants should automatically fill in after the user selects one from a drop-down menu in Cell A9 of the compounds shown below. (*Hint:* Use data validation and LOOKUP expressions.) The data in the table below has been provided in the online workbook.

Compound	Constants			Validity Range	
	A	B [°C]	C [°C]	T_{min} [°C]	T_{max} [°C]
Acetic acid	7.5596	1,644.05	233.524	17	118
Acetone	7.1327	1,219.97	230.653	−64	70
Cyclohexane	6.85146	1,206.47	223.136	7	81
Ethanol	8.20417	1,642.89	230.3	−57	80
Hexadecane	7.0287	1,830.51	154.45	150	321
Methanol	8.08097	1,582.27	239.7	15	100
m-Xylene	7.22319	1,585.83	226.5	−45	140
Tetrahydrofuran	6.99515	1,202.29	226.254	23	100

Next, create a column of temperature (T) beginning at -100 degrees Celsius and increasing in increments of 5 degrees Celsius until a temperature of 400 degrees Celsius.

In column **B**, calculate the vapor pressure (P, in millimeters of mercury, [mm Hg]) using the Antoine equation, formatted to four decimal places. If the equation is outside the valid temperature range for the compound, the pressure column should be blank.

11. The ideal gas law assumes that molecules bounce around and have negligible volume themselves. This is not always true. To compensate for the simplifying assumptions of the ideal gas law, the Dutch scientist Johannes van der Waals developed a "real" gas law that uses several factors to account for molecular volume and intermolecular attraction. He was awarded the Nobel Prize in 1910 for his work. The **van der Waals equation** is as follows:

$$\left(P + \frac{an^2}{V^2}\right)(V - bn) = nRT$$

P, V, n, R, and T are the same quantities as found in the ideal gas law. The constant "a" is a correction for intermolecular forces [atm L^2/mol^2], and the constant "b" accounts for molecular volume [L/mol]. Each of these factors must be determined by experiment.

⬛	A	B	C
1			
2			
3			
4			
5			
6			
7	Type of Gas	Argon	🔽
8	Quantity [g]	4	
9	Temperature (T) [°C]	15	
10			
11			
12	Molecular Weight (MW) [g/mol]	40	
13	vdW Constant "a" [atm L²/mol²]	1.345	
14	vdW Constant "b" [L/mol]	0.0322	
15			
16	Ideal Gas Constant (g) [(atm l)/mol K)]	0.08206	
17			
18			
19		Ideal Gas	van der Waals
20	Volume (V) [L]	Pressure (P) [atm]	
21	0.5	4.73	4.70
22	0.6	3.94	3.92
23	0.7	3.38	3.36
24	0.8	2.95	2.95
25	0.9	2.63	2.62
26	1.0	2.36	2.36

Create a worksheet using the provided template. The molecular weight, "a," and "b" should automatically fill in after the user selects the type of gas in cell B7. (*Hint:* Use data validation and LOOKUP expressions.) The user will also set the quantity of gas and the temperature of the system. The data in the table below has been provided online.

Next, create a column of volume beginning in A21 at 0.5 liters and increasing in increments of 0.1 liters to a volume of 5 liters.

In column B, calculate the pressure (P, in atmospheres [atm]) using the ideal gas law.

In column C, calculate the pressure (P, in atmospheres [atm]) using the van der Waals equation.

Compound	a [atm L^2/mol^2]	b [L/mol]	MW [g/mol]
Acetic acid	17.587	0.1068	60
Acetone	13.906	0.0994	58
Ammonia	4.170	0.0371	17
Argon	1.345	0.0322	40
Benzene	18.001	0.1154	78
Chlorobenzene	25.433	0.1453	113
Diethyl ether	17.380	0.1344	74
Ethane	5.489	0.0638	30
Ethanol	12.021	0.0841	46
Hexane	24.387	0.1735	86
Methanol	9.523	0.0670	32
Neon	0.211	0.0171	20
Oxygen	1.360	0.0318	32
Pentane	19.008	0.1460	72
Propane	8.664	0.0845	44
Toluene	24.061	0.1463	92
Water	5.464	0.0305	18
Xenon	4.194	0.0511	131

NOTE

The astronomical unit (AU) is the average distance from the Earth to the Sun.

12. One of the NAE Grand Challenges for Engineering is **Engineering the Tools of Scientific Discovery.** According to the NAE website: "Grand experiments and missions of exploration always need engineering expertise to design the tools, instruments, and systems that make it possible to acquire new knowledge about the physical and biological worlds."

Solar sails are a means of interplanetary propulsion using the radiation pressure of the sun to accelerate a spacecraft. The table below shows the radiation pressure at the orbits of the eight planets.

Planet	Distance from Sun (d) [AU]	Radiation Pressure (P) [mPa]
Mercury	0.46	43.3
Venus	0.72	17.7
Earth	1	9.15
Mars	1.5	3.96
Jupiter	5.2	0.34
Saturn	9.6	0.099
Uranus	19.2	0.025
Neptune	30.1	0.01

Create a table showing the area in units of square meters of a solar sail needed to achieve various accelerations for various spacecraft masses at the distances from the sun of the various planets. Your solution should use data validation and VLOOKUP to select a planet and the corresponding radiation pressure. The columns of your table should list masses of the spacecraft (including the mass of the sail) ranging from 100 to 1,000 kilograms in increments of 100 kilograms. The rows should list accelerations from 0.0001 to 0.001 g in increments of 0.001 g, where "g" is the acceleration of Earth's gravity, 9.81 meters per second squared. All constants and conversion factors should be placed in individual cells using appropriate labels, and all formulae should reference these cells and NOT be directly coded into the formulae. You should use absolute, relative, and mixed addressing as appropriate.

13. A hands-on technology museum has hired you to do background research on the feasibility of a new activity to allow visitors to assemble their own ferrite core memory device—a technology in common use until the 1970s, and in specialized applications after that. The computers onboard the early space shuttle flights used core memory due to their durability, non-volatility, and resistance to radiation—core memory recovered from the wreck of the Challenger still functioned.

Ferrite core memory comprises numerous tiny ferrite rings ("cores") in a grid, each of which has either two or three wires threaded through it in a repeating pattern and can store a single bit, or binary digit—a 0 or a 1. Since the cores were typically on the order of one millimeter in diameter, workers had to assemble these under microscopes.

After investigating ferrite materials, you find several that would be suitable for fabrication of the cores. The museum staff has decided to have the visitors assemble a 4 × 4 array (16 cores—actual devices were MUCH larger) and anticipate that 2,500 people will assemble one of these over the course of the project. Assuming that the cores are each cylindrical rings with a hole diameter half that of the outside diameter of the ring and a thickness one-fourth the outside diameter, you need to know how many grams of ferrite beads you need to purchase with 10% extra beyond the specified amount for various core diameters and ferrite materials. You also wish to know the total cost for the beads.

Using the provided online worksheet that includes a table of different ferrite material densities and costs, use data validation to select one of the materials from the list, then create a table showing the number of pounds of cores for core diameters of 1.2 to 0.7 millimeter in 0.1 millimeter increments as well as the total cost. For cores with a diameter less than 1 millimeter, there is a 50% manufacturing surcharge, thus the smallest cores cost more per gram. Include table entries for individual core volume and total volume of all cores. Your worksheet should resemble the example below.

Cost of Ferrite Cores for Hands-on Museum

Ferrite Compound	Specific Gravity	Cost per Gram [$/g]			Ferrite Compound	Specific Gravity	Cost per gram [$/g]
CMP C	3.73	$ 32.50			CMP A	8.39	$ 39.41
					CMP B	7.47	$ 27.32
Sets needed	2500				CMP C	3.73	$ 32.50
Cores per set	16				CMP D	4.01	$ 27.80
Total Cores	40000				CMP E	7.27	$ 35.56
Total + 10%	44000				CMP F	3.92	$ 22.51

	Core Outside Diameter (D) [mm]					
	0.7	0.8	0.9	1	1.1	1.2
Volume of one core [mm³]	0.0505	0.0754	0.1074	0.1473	0.1960	0.2545
Volume of all Cores [mm³]	2222	3318	4724	6480	8624	11197
Mass of Cores [g]	8.29	12.37	17.62	24.17	32.17	41.76
Cost of Cores [$]	$ 404.11	$ 603.21	$ 858.87	$ 785.43	$ 1,045.41	$ 1,357.23

Note: Volume of ring: $V = \pi (D/2)^2 H - \pi (0.5 (D/2))^2 H = \pi (3/64) D^3$ assuming that the center hole is half the outside diameter (D) and the thickness (H) is one fourth the outside diameter.

14. A substance used to remove the few remaining molecules from a near vacuum by reacting with them or adsorbing them is called a getter. There are numerous materials used and several ways of deploying them within a system enclosing a vacuum, but here we will look at a common method used in vacuum tubes, once the workhorse of electronics but now relegated to high-end audio systems and other niche markets. In vacuum tubes, after the air is evacuated with a vacuum pump, getters are usually deposited on the inside of the tube, often at the top, by flash deposition.

Assume we are investigating getter materials for use in vacuum tubes with various inside diameters and hemispherical tops. The getter will be flash deposited on this hemispherical area.

Photo courtesy of W. Park

We wish to set up a worksheet that will allow the user to select a getter material from a menu using data validation, and produce a table showing the number of moles of that material and the thickness of the deposited film for various masses of material from 20 to 300 milligram with 20 milligram increments and various tube inside diameters from 0.6 to 1.2 inches by 0.1 inch. Your final worksheet should appear similar to the example shown below. A starting worksheet including the table of possible materials and their specific gravities and atomic weights is available online.

Getter Material	Specific Gravity	Atomic Weight
Sodium	0.968	22.99

Getter Material	Specific Gravity	Atomic Weight
Barium	3.51	137.33
Aluminum	2.7	26.981
Sodium	0.968	22.99
Strontium	2.64	87.62
Calcium	1.55	40.078
Magnesium	1.738	24.305

Getter Thickness [μm]		Vacuum Tube Inside Diameter (D) [in]						
Mass [mg]	Moles	0.6	0.7	0.8	0.9	1	1.1	1.2
20	8.7E-04	56.6	41.6	31.9	25.2	20.4	16.8	14.2
40	1.7E-03	113.3	83.2	63.7	50.3	40.8	33.7	28.3
60	2.6E-03	169.9	124.8	95.6	75.5	61.2	50.5	42.5
80	3.5E-03	226.5	166.4	127.4	100.7	81.6	67.4	56.6
100	4.3E-03	283.2	208.0	159.3	125.8	101.9	84.2	70.8
120	5.2E-03	339.8	249.6	191.1	151.0	122.3	101.1	84.9
140	6.1E-03	396.4	291.3	223.0	176.2	142.7	117.9	99.1
160	7.0E-03	453.1	332.9	254.8	201.4	163.1	134.8	113.3
180	7.8E-03	509.7	374.5	286.7	226.5	183.5	151.6	127.4
200	8.7E-03	566.3	416.1	318.6	251.7	203.9	168.5	141.6
220	9.6E-03	623.0	457.7	350.4	276.9	224.3	185.3	155.7
240	1.0E-02	679.6	499.3	382.3	302.0	244.7	202.2	169.9
260	1.1E-02	736.2	540.9	414.1	327.2	265.0	219.0	184.1
280	1.2E-02	792.9	582.5	446.0	352.4	285.4	235.9	198.2
300	1.3E-02	849.5	624.1	477.8	377.5	305.8	252.7	212.4

15. Create an Excel worksheet that will allow the user to type in the radius of a sphere and the standard abbreviation for the units used.

Standard Unit Abbreviations						
Unit	meter	centimeter	millimeter	yard	foot	inch
Abbreviation	m	cm	mm	yd	ft	in

The volume of the sphere should then be calculated and expressed by the following units: cubic meters, cubic centimeters, cubic millimeters, liters, gallons, cubic yards, cubic feet, and cubic inches. Your worksheet should appear similar to the sample shown below, although you will probably need additional information in the worksheet not shown here.

Enter Radius Value Here	Enter Radius Units Here	Volume							
		m^3	cm^3	mm^3	liters	gallons	yd^3	ft^3	in^3
1.7	in	3.372E-04	3.372E+02	3.372E+05	3.372E-01	8.910E-02	4.416E-04	1.191E-02	2.058E+01

16. Most resistors are so small that the actual value would be difficult to read if printed on the resistor. Instead, colored bands denote the value of resistance in ohms. Anyone involved in constructing electronic circuits must become familiar with the color code, and with practice, one can tell at a glance what value a specific set of colors means. For the novice, however, trying to read color codes can be a bit challenging.

You are to design a worksheet similar to the one shown, allowing the user to enter a resistance value and automatically show the color code for that resistance. Note the cells below "First Digit," "Second Digit," and "Number of Zeros" should actually take on the appropriate colors in addition to showing the numerical value. The cells below should use the LOOKUP function to show the color names. The colors assigned to each digit are on the right of the worksheet for convenience.

NOTE

Without giving too much away, try typing **"Round"** into Excel help, click on the link to the ROUND function, then scroll down to the list of related functions. One or more of those might be useful.

To enter the resistance, use data validation along with the table on the left. The resistance will be entered in two parts, the first two digits, and a power of 10 by which those digits will be multiplied. Include an appropriate input message when the cell is selected and an appropriate warning if the user enters an invalid number. The user should be able select from a drop-down menu. Note that you must use the two input values to calculate the total resistance, the first digit, the second digit, and the number of zeros.

For example, a resistance of 4,700 ohms [Ω] has first digit 4 (yellow), second digit 7 (violet), and 2 zeros following (red). A resistance of 56 Ω would be 5 (green), 6 (blue), and 0 zeros (black); 1,000,000 Ω is 1 (brown), 0 (black), and 5 zeros (green). Particularly note that if the second digit is zero, it does not count in the multiplier value. There are numerous explanations of the color code on the web if you need further information or examples.

The worksheet shown is available online except for the nine cells in the center containing the desired resistance values and the color code for that resistance.

Resistor Color Code Calculator

Resistor Values			Enter Values Below							Color Code	
Standard Values	Multipliers		First Two Digits	Multiplier	Resistance	First Digit	Second Digit	Number of zeros		Value	Color
10	1		68	1,000	68000	6	8	3		0	Black
12	10						Blue	Grey	Orange	1	Brown
15	100		Please enter first							2	Red
18	1,000		two digits of							3	Orange
22	10,000		resistance value or							4	Yellow
27	100,000									5	Green
33	1,000,000									6	Blue
39										7	Violet
47										8	Grey
56										9	White
68											
82											

17. Download the starting file, and complete the following commands using the data provided.

(a) Indicate the following using conditional formatting commands of your choice. Each condition below should appear in a unique format.
- Length shown in Column B is greater than 6 inches or less than 4 inches.
- Width shown in Column C is less than 2.5 inches.
- Inner radius shown in Column D is above average for the inner radius values.
- Outer radius shown in Column E is below average for the outer radius values.
- Volume shown in Column F is less than 10 cubic inches or greater than 20 cubic inches.

(b) For the following conditions, in Column H use an IF statement to indicate the Status:
- If length is less than 4 inches or width is less than 2.5 inches, list the status as "Too Small."
- Otherwise, if twice the inner radius is greater than the outer radius, list the status as "Off Center."
- Otherwise, if the volume is greater than 20 cubic inches or the mass is greater than 3,000 grams, list the status as "Too Large."
- Otherwise, if none of these conditions are true, leave the cell blank.

(c) For the following conditions, in Column J use an IF statement to indicate the Action Code:
- If the status is "Too Small" or "Too Large," list as action code as a numerical value of one.
- If the status is "Off Center," list as action code as a numerical value of two.
- If none of these conditions are met, list as action code as a numerical value of three.

(d) Use a conditional formatting icon set in Column J to indicate the following:
- Status as green for action code 3.
- Status as yellow for action code 2.
- Status as red for action code 1.

(e) Count the following items, showing the results somewhere above the data table. Be sure to indicate each counted item with an appropriate label.

- Indicate the number of items classified as each action code, such as how many items are listed as 1.
- Indicate number of parts that meet the following conditions:
  - Length is greater than 6 inches.
  - Volume is less than 10 cubic inches or greater than 20 cubic inches. As a hint, use two "COUNT" functions and add them together.

(f) Sort the worksheet in the following order: Length, increasing and simultaneously then Outer Radius, decreasing. Be careful to select only the data and not the entire worksheet.

(g) Set the worksheet controls to be filtered in the header row. Filter the worksheet so only parts of length 2.80, 5.20, and 7.15 inches are shown.

CHAPTER 11
GRAPHICAL SOLUTIONS

Often, the best way to present technical data is through a "picture." But if not done properly, it is often the worst way to display information. As an engineer, you will have many opportunities to construct such pictures. *The Visual Display of Quantitative Information* by Edward Tufte contains one of the best discussions on this topic. If technical data are presented properly in a graph, it is often possible to explain a point in a concise and clear manner that is impossible any other way.

11.1 GRAPHING TERMINOLOGY

Abscissa is the horizontal axis; **ordinate** is the vertical axis. Until now, you have probably referred to these as "x" and "y." This text uses the terms *abscissa* and *ordinate*, or *horizontal* and *vertical*, since x and y are only occasionally used as variables in engineering problems.

The **independent** variable is the parameter that is controlled or whose value is selected in the experiment; the **dependent** variable is the parameter that is measured corresponding to each set of selected values of the independent variable. Convention usually shows the independent variable on the abscissa and the dependent variable on the ordinate.

Data sets given in tabular form are commonly interpreted and graphed with the leftmost column or topmost row as the independent variable and the other columns as the dependent variable(s). For the remainder of this text, if not specifically stated, assume that the abscissa variable is listed in the leftmost column or topmost row in a table of data values.

Time	Distance (*d*)[m]	
(*t*) [s]	Car 1	Car 2
Abscissa	Ordinate	Ordinate

Time (*t*) [s]		Abscissa
Distance	Car 1	Ordinate
(*d*)[m]	Car 2	Ordinate

11.2 PROPER PLOTS

We call graphs constructed according to the following rules **proper plots**:

- Caption with a **brief description**. The restating of "*d* versus *t*" or "distance versus time" or even "the relationship between distance and time" does not constitute a proper caption. The caption should give information about the graph to allow the graph to stand alone, without further explanation. It should include information about the problem that does not appear elsewhere on the graph. For example, instead of stating "distance versus time," better choices would be "Lindbergh's Flight across the Atlantic," "The Flight of Voyager I," or "Walking between Classes across Campus, Fall 2008." When including a graph as part of a written report, place the caption below the graph.
- **Label both axes clearly.** Three things are required unless the information is unavailable: category (e.g., Time), symbol used (*t*), and units [s]. Units should accompany all quantities when appropriate, enclosed in square brackets [ ].
- **Select scale increments (both axes) that are easy to read and interpolate between.** With a few exceptions, base your scale on increments of 1, 2, 2.5, and 5. You can scale each value by any power of 10 as necessary to fit the data. Avoid unusual increments (such as 3, 7, 15, or 6.5).

Increment	Sequence				
1	0	10	20	30	40
5	0.05	0.10	0.15	0.20	0.25
2.5	−2,500	0	2,500	5,000	7,500
2	6×1^{-5}	8×10^{-5}	1×10^{-4}	1.2×10^{-4}	1.4×10^{-4}

In this final case, reading is easier if the axis is labeled something like Time (*t*) [s] $\times 10^{-4}$ so that only the numbers 0.6, 0.8, 1.0, 1.2, and 1.4 show on the axis.

- **Provide horizontal and vertical gridlines** to make interpolation easier to aid the reader in determining actual numerical values from the graph.

  When minor gridlines are present, the reader should be able to easily determine the value of each minor increment. For example, examine the graphs shown in Figure 11-1. In

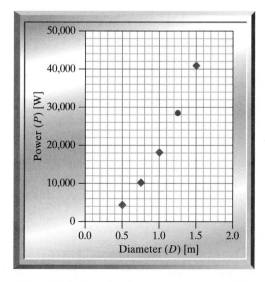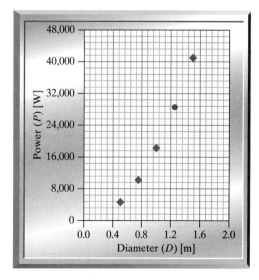

Figure 11-1 Example of importance of minor gridline spacing.

which graph is it easier to determine the abscissa value for the blue point? In the graph on the left, the increment can easily be determined as 0.1 meters. In the graph on the right, it is more difficult to determine the increment as 0.08 meters.

- **Provide a clear legend** describing each data set of multiple data sets shown. Do not use a legend for a single data set. Legends may be shown in a stand-alone box or captioned next to the data set. Both methods are shown in Figure 11-2.

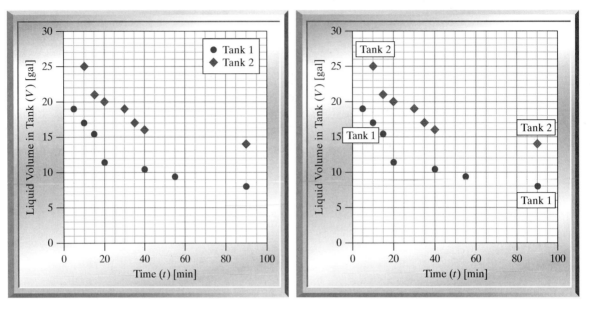

Figure 11-2 Options for displaying legends.

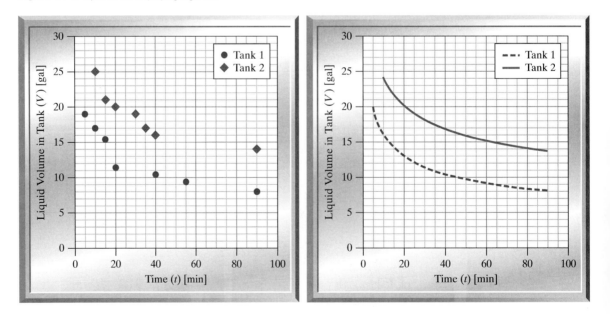

Figure 11-3 Illustration of experimental data (shown as points) versus theoretical data (shown as lines).

- **Show measurements as symbols. Show calculated or theoretical values as lines.** Do not display symbols for calculated or theoretical values. A symbol shown on a graph indicates that a measurement has been made (see Figure 11-3).

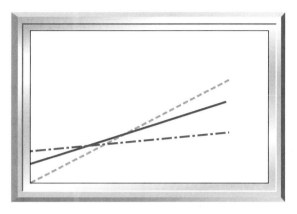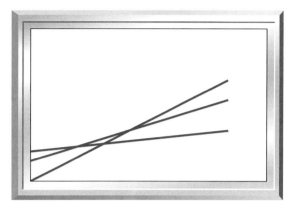

Figure 11-4 Example of importance of different line types.

- **Use a different symbol shape or color** for each experimental data set and a different line style and color for each theoretical data set. **Never use yellow and other light pastel colors** for either symbols or lines. Remember that when graphs are photocopied, all colored lines become black lines. Some colors disappear when copied and are hard to see in a projected image. For example, in Figure 11-4, left, it is much easier to distinguish between the different lines than in the figure on the right.

- **Produce graphs in portrait orientation** whenever possible within a document. Portrait orientation does not necessarily mean that the graph is distorted to be taller than it is wide; it means that readers can study the graph without turning the page sideways.

- **Be sure the graph is large enough to be easily read.** The larger the graph, the more accurate the extracted information.

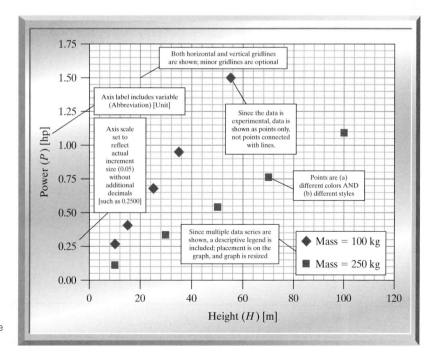

Figure 11-5 Example of a proper plot, showing multiple experimental data sets.

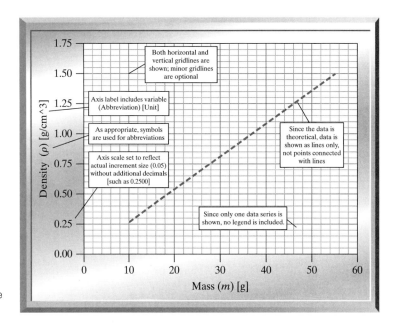

Figure 11-6 Example of a proper plot, showing a single theoretical data set.

Below is an example of a poorly constructed plot. Some problems with this plot are listed below:

- It is a plot of distance versus time, but is it the distance of a car, a snail, or a rocket? What are the units of distance—inches, meters, or miles? What are the units of time—seconds, days, or years? Is time on the horizontal or vertical axis?
- Two data sets are shown, or are there three? Why is the one data set connected with a line? Is it a trendline? Is the same data set shown in the triangles? What do the shaded and open triangles represent—different objects, different trials of the same object, or modifications to the same object?
- Lack of gridlines and strange axis increments makes it difficult to interpolate between values.

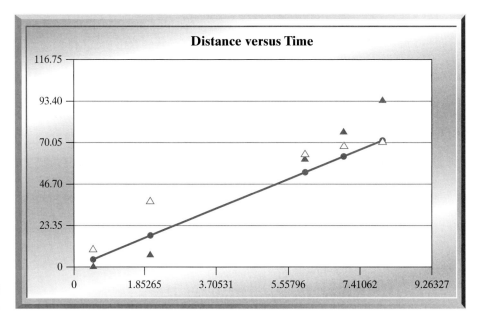

Figure 11-7 Example of poorly constructed graph.

● **EXAMPLE 11-1**   When attempting to stop a car, a driver must consider both the reaction time and the braking time. The data are taken from www.highwaycode.gov.uk. Create a proper plot of these data.

Vehicle Speed (v) [mph]	Distance	
	Reaction (d_r) [m]	Braking (d_b) [m]
20	6	6
30	9	14
40	12	24
50	15	38
60	18	55
70	21	75

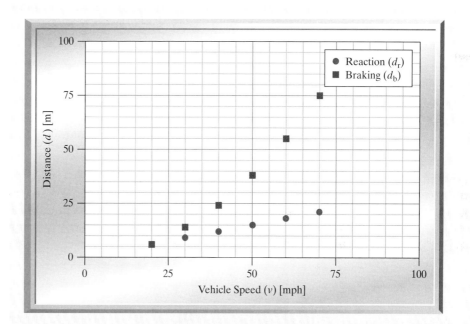

Figure 11-8 At various speeds, the necessary reaction time and braking time needed to stop a car.

● **EXAMPLE 11-2**   Ohm's law describes the relationship between voltage, current, and resistance within an electrical circuit, given by the equation $V = IR$, where V is the voltage [V], I is the current [A], and R is the resistance [Ω]. Construct a proper plot of the voltage versus current, determined from the equation, for the following resistors: 3,000, 2,000, and 1,000 Ω. Allow the current to vary from 0 to 0.05 amperes.

Note that while the lines were probably generated from several actual points along each line for each resistor, the points are not shown; only the resulting line is shown since the values were developed from theory and not from experiment. If you create a plot like this by hand, you would first put in a few points per data set, then draw the lines and erase the points so that they are not shown on the final graph.

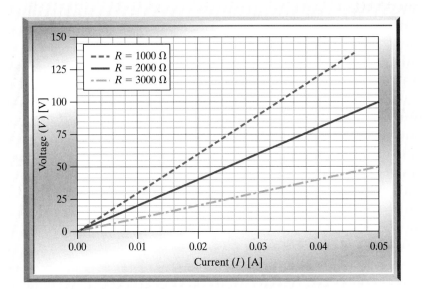

Figure 11-9 Ohm's law determined for a simple circuit to compare three resistor values within a current range of 0–0.05 amperes.

COMPREHENSION CHECK 11-1

In the following plot, identify violations of the proper plot rules.

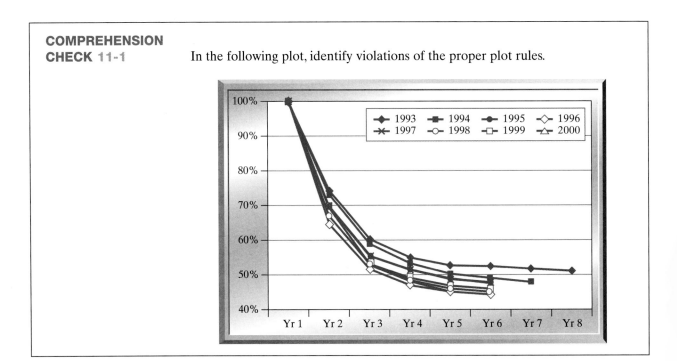

COMPREHENSION CHECK 11-2

In the following plot, identify violations of the proper plot rules.

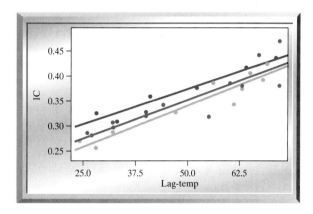

COMPREHENSION CHECK 11-3

In the following plot, identify violations of the proper plot rules.

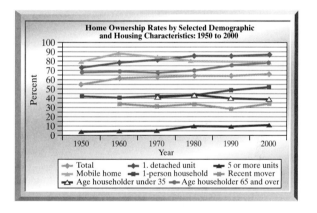

COMPREHENSION CHECK 11-4

In the following plot, identify violations of the proper plot rules.

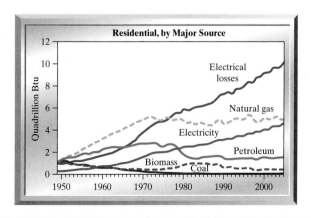

11.3 AVAILABLE GRAPH TYPES IN EXCEL

The following is an example of the level of knowledge of Excel needed to proceed. *If you are not able to quickly recreate the following exercise in Excel, please review graphing basics in the appendix materials before proceeding.*

Two graphs are given here; they describe the draining of tanks through an orifice in the bottom. When the tank contains a lot of liquid, the pressure on the bottom is large and the tank empties at a higher rate than when there is less liquid. The first graph shows actual data obtained from two different tanks. These data are given in the table below. The second plot shows curves (developed from theoretical equations) for two tanks. The equations for these curves are also given.

Experimental data for first graph:

Time (t) [min]	5	10	15	20	40	55	90
Volume Tank #1 (V1) [gal]	19.0	17.0	15.5	11.5	10.5	9.5	8.0

Time (t) [min]	10	15	20	30	35	40	90
Volume Tank #2 (V2) [gal]	25	21	20	19	17	16	14

Theoretical equations for second graph (with t in minutes):

- Tank 1: Volume remaining in tank 1 [gal] $V = 33\, t^{-0.31}$
- Tank 2: Volume remaining in tank 2 [gal] $V = 44\, t^{-0.26}$

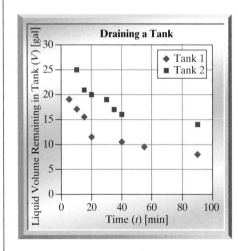

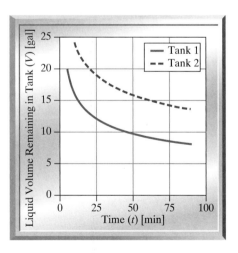

You can create many different types of charts in Excel. Usually, you will only be concerned with a few main types, shown in Table 11-1.

● EXAMPLE 11-5

What if the parameter plotted on the vertical axis was not a simple straight line or straight-line segments? For example, the flow rate of liquid out of a pipe at the bottom of a cylindrical barrel follows an exponential relationship. Assume the flow rate out of a tank is given by $F = 4\,e^{-t/8}$ gallons per minute. A graph of this is shown in Figure 11-12.

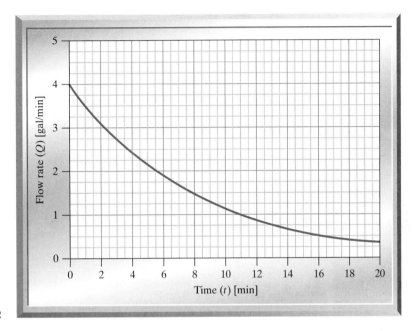

Figure 11-12

Although we might be able to make a reasonable estimate of the area under the curve (the total volume of water that has flowed out of the tank) simple algebra is insufficient to arrive at an accurate value. Those of you who have already studied integral calculus should know how to solve this problem. However, some students using this text may not have progressed this far in math, so we will have to leave it at that. It is enough to point out that there are innumerable problems in many engineering contexts that require calculus to solve. To succeed in engineering, you must have a basic understanding of calculus.

● EXAMPLE 11-6

From the past experience of driving an automobile down a highway, you should understand the concepts relating acceleration, velocity, and distance. As you slowly press the gas pedal toward the floor, the car accelerates, causing both the speed and the distance to increase. Once you reach a cruising speed, you turn on the cruise control. Now, the car is no longer accelerating and travels at a constant velocity while increasing in distance. These quantities are related through the following equations:

$$\text{velocity} = (\text{acceleration})(\text{time}) \quad v = (a)(t)$$
$$\text{distance} = (\text{velocity})(\text{time}) \quad\;\; d = (v)(t)$$

If we create a graph of velocity versus time, the form of the equation tells us that acceleration is the slope of the line. Likewise, a graph of distance versus time has velocity as the slope of the line.

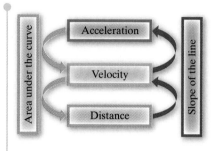

However, if we had a graph of velocity versus time and we wanted to determine distance, how can we do this? The distance is determined by how fast we are traveling times how long we are traveling at that velocity; we can find this by determining the area under the curve of velocity versus time. Likewise, if we had a graph of acceleration versus time, we could determine the velocity from the area under the curve. In technical terms, the quantity determined by the slope is referred to as the **derivative**; the quantity determined by the area under the curve is referred to as the **integral**.

In the graph shown in Figure 11-13, we drive our car along the road at a constant velocity of 60 miles per hour [mph]. After 1.5 hours, how far have we traveled?

IMPORTANT CONCEPT

Be certain that you understand the difference between a variable increasing or decreasing and a variable changing at an increasing or decreasing rate.

This is directly related to one of the fundamental concepts of calculus, and will arise many times during your engineering studies.

The area under the curve, shown by the rectangular box, is:

Area of the rectangle = (height of rectangle) (width of rectangle)

$$= (60 - 0 \text{ mph}) (1.5 - 0 \text{ h})$$

$$= 90 \text{ miles}$$

Can you explain, in your own words and without resorting to calculus terminology, why the area under the graph of acceleration and time gives velocity?

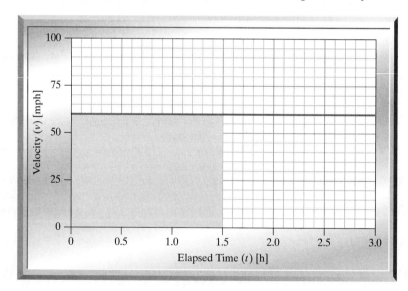

Figure 11-13 Example of distance calculation from area under velocity versus time graph.

11.5 MEANING OF LINE SHAPES

In addition to the value of the slope of the line, the shape of the line contains useful information. In Figure 11-13, the speed is shown as a horizontal line. This implies that it has a constant value; it is not changing over time. The slope of this line is zero, indicating that the acceleration is zero. Table 11-2 contains the various types of curve shapes and their physical meanings.

Table 11-2 What do the lines on a graph mean?

If the graph shows a . . .	. . . it means that the dependent variable . . .	Sketch
horizontal line	is not changing. Here, the derivative of the variable is zero. The integral of the variable is changing at a constant rate (a straight line).	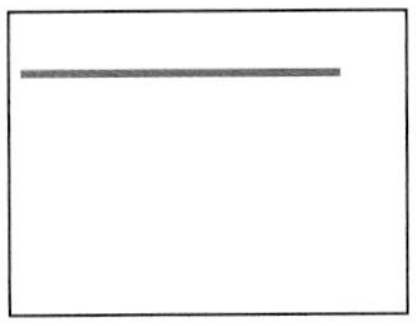
vertical line	has changed "instantaneously."	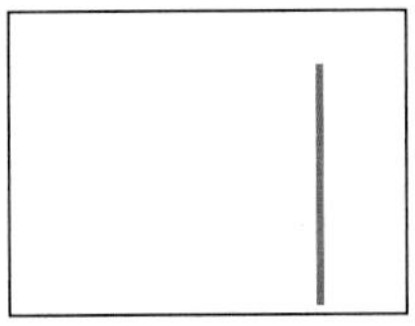
straight line positive or negative slope neither horizontal nor vertical	is changing at a constant rate. Here, the derivative of the variable is not changing (equal to a constant value). The integral of the variable is changing at a varying rate.	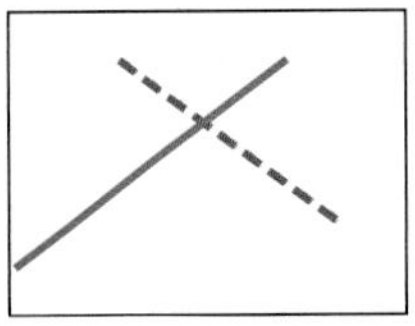
curved line concave up, increasing trend	is changing at an increasing rate. Here, the derivative of the variable is positive and increasing.	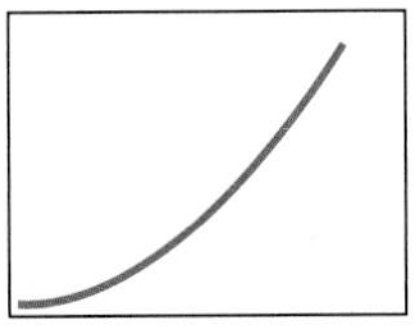
curved line concave down, increasing trend	is changing at a decreasing rate. Here, the derivative of the variable is positive and decreasing.	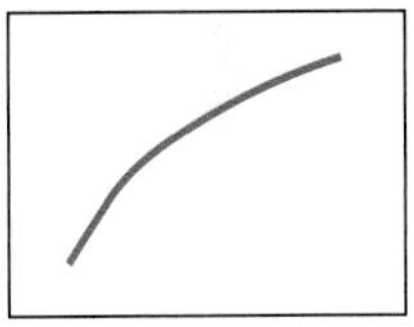
curved line concave up, decreasing trend	is changing at a decreasing rate. Here, the derivative of the variable is negative and its magnitude is decreasing.	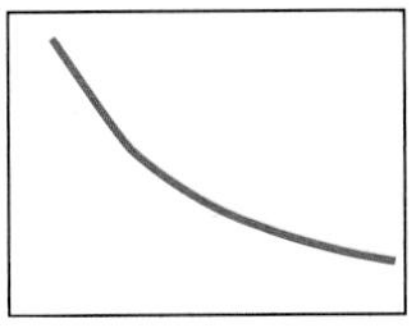
curved line concave down, decreasing trend	is changing at an increasing rate. Here, the derivative of the variable is negative and its magnitude is increasing.	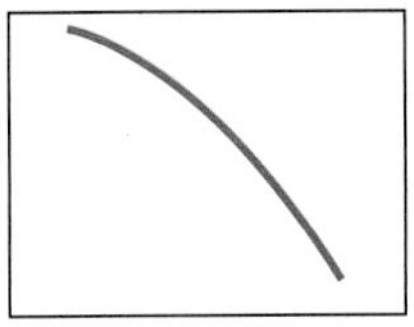

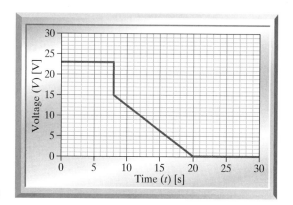

Figure 11-14

In Figure 11-14, the voltage is constant, as indicated by the horizontal line at 23 volts, from time = 0 to 8 seconds. At time = 8 seconds, the voltage changes instantly to 15 volts, as indicated by the vertical line. Between time = 8 seconds and 20 seconds, the voltage decreases at a constant rate, as indicated by the straight line, and reaches 0 volts at time = 20 seconds, where it remains constant.

In Figure 11-15, the force on the spring increases at an increasing rate from time = 0 until time = 2 minutes, then remains constant for 1 minute, after which it increases at a decreasing rate until time = 5 minutes, after which it remains constant at about 6.8 newtons.

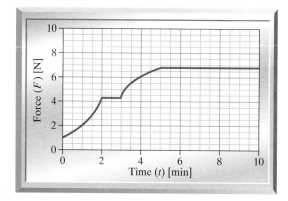

Figure 11-15

The height of a blimp is shown in Figure 11-16. The height decreases at an increasing rate for 5 minutes, then remains constant for 2 minutes, after which its height decreases at a decreasing rate until $t = 10$ minutes, after which its height remains constant at 10 meters.

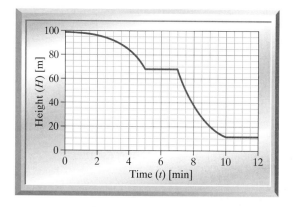

Figure 11-16

COMPREHENSION CHECK 11-5

Use the graph to answer the following questions. Choose from the following answers:

NOTE

The rate of change (derivative) of acceleration is called JERK.

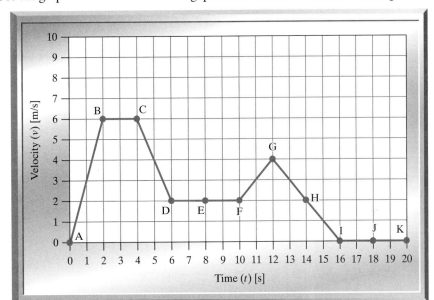

- Equal to a constant, positive value
- Equal to zero
- Equal to a constant, negative value
- Decreasing at a constant rate
- Increasing at a constant rate
- Increasing at a decreasing rate
- Increasing at an increasing rate
- Decreasing at a decreasing rate
- Decreasing at an increasing rate

(a) Between points (A) and (B), the acceleration is ___
(b) Between points (B) and (C), the acceleration is ___
(c) Between points (C) and (D), the acceleration is ___
(d) Between points (D) and (E), the distance is ___
(e) Between points (F) and (G), the distance is ___
(f) Between points (G) and (H), the distance is ___

● EXAMPLE 11-7

As a club effort to raise money for charity, we are going to push a university fire truck along a flat section of highway. The fire truck weighs 29,400 newtons. Using teams, we will exert a force of 10 newtons for the first 2 minutes, 20 newtons for the next 3 minutes, and 5 newtons for the next 4 minutes. We notice a billboard nearby that reads, "Newton says that 'FORCE = MASS × ACCELERATION'!"

Graph the acceleration and speed versus time on separate graphs, using the same timescale for each. In your analysis, ignore the effect of friction and assume the acceleration changes instantaneously.

Step 1: *Calculate acceleration based on Newton's law. Plot acceleration versus time; the results are shown in Figure 11-17a. The details of this calculation are left for the reader, but as a hint, for the first 2 minutes, a force of 10 newtons is applied to the 3,000 kilogram mass truck, giving an acceleration of 0.0033 meters per second squared. Note that in the graph, the horizontal lines indicate that acceleration is constant.*

The vertical lines imply that the acceleration instantaneously increased in value. While this is not exactly accurate, no information was given on how the acceleration changes between values. In such cases, an instantaneous change is often assumed.

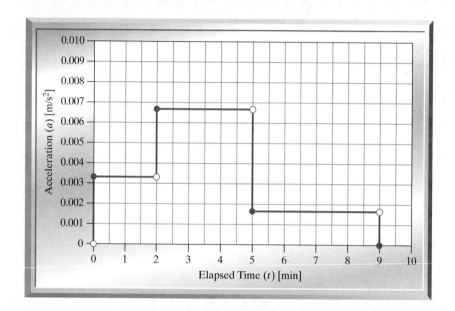

Figure 11-17a
Acceleration of a fire truck.

Step 2: *Calculate speed based on the area under the curve generated in Step 1. Plot speed versus time. The details are left to the reader. Results are shown in Figure 11-17b.*

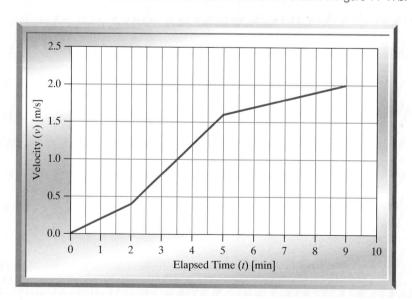

Figure 11-17b
Velocity determined from area under the acceleration curve.

COMPREHENSION CHECK 11-6

The graph gives an acceleration profile for a car that begins to move from an initial speed of zero. Draw the corresponding velocity and the corresponding distance profile.

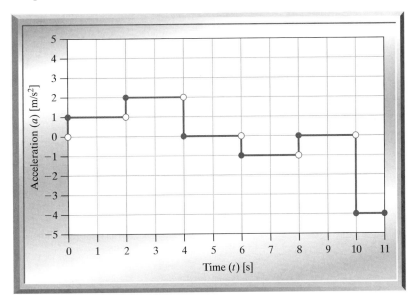

11.6 GRAPHICAL SOLUTIONS

When you have two equations containing the same two variables, it is sometimes desirable to find values of the variables that satisfy both equations. Most of you have studied methods for solving simultaneous linear equations (there are many techniques); however, most of these methods apply only to linear equations and do not work if one or both of the equations is nonlinear. It also becomes problematic if you are working with experimental data.

For systems of two equations (or data sets in two variables), you can use a graphical method to determine the value or values that satisfy both. Essentially, graph the two equations and visually determine where the curves intersect. This may be nowhere, at one point, or at several points.

● **EXAMPLE 11-8**

We assume that the current through two electromagnets is given by the following equations

$$\text{Electromagnet A: } I = 5t + 6$$
$$\text{Electromagnet B: } I = -3t + 12$$

We want to determine when the value of the current through the electromagnets is equal.

Graphing both equations gives Figure 11-18. We know not to show the points when plotting data derived from equations.

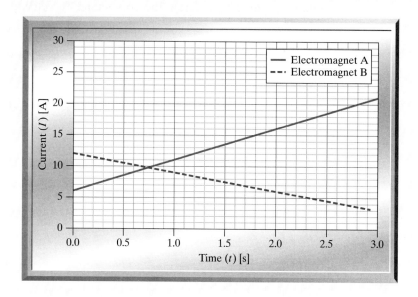

Figure 11-18

The two lines cross at time 0.75 seconds (approximately), and the current at this time is approximately 9.7 amperes. The larger we make this graph and the more gridlines we include, the more accurately we can determine the solution.

Solution: t = 0.75 seconds, I = 9.7 amperes.

Economic Analysis

Breakeven analysis determines the quantity of product a company must make before they begin to earn a profit. Two types of costs are associated with manufacturing: fixed and variable. **Fixed costs** include equipment purchases, nonhourly employee salaries, insurance, mortgage or rent on the building, etc., or "money we must spend just to open the doors." **Variable costs** depend on the production volume, such as material costs, hourly employee salaries, and utility costs. The more product produced, the higher the variable costs become.

Total cost = Fixed cost + Variable cost * Amount produced

The product is sold at a **selling price**, creating **revenue**.

Revenue = Selling price * Amount sold

Any excess revenue remaining after all production costs have been paid is **profit**. Until the company reaches the breakeven point, they are operating at a **loss** (negative profit), where the money they are bringing in from sales does not cover their expenses.

Profit = Revenue − Total Cost

The **breakeven point** occurs when the revenue and total cost lines cross, or the point where profit is zero (not negative or positive). These concepts are perhaps best illustrated through an example.

● **EXAMPLE 11-9**

Let the amount of product we produce be G [gallons per year]. Consider the following costs:

- Fixed cost: $1 million
- Variable cost: 10 cents/gallon of G
- Selling price: 25 cents/gallon of G

Plot the total cost and the revenue versus the quantity produced. Determine the amount of G that must be produced to breakeven. Assume we sell everything we make.

The plot of these two functions is shown in Figure 11-19. The breakeven point occurs when the two graphs cross, at a production capacity of 6.7 million gallons of G.

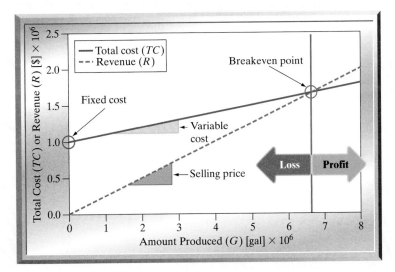

Figure 11-19 Breakeven analysis definitions.

COMPREHENSION CHECK 11-7

You are working for a tire manufacturer, producing wire to be used in the tire as a strengthening agent. You are considering implementing a new machining system, and you must present a breakeven analysis to your boss. You develop the graph, showing two possible machines that you can buy.

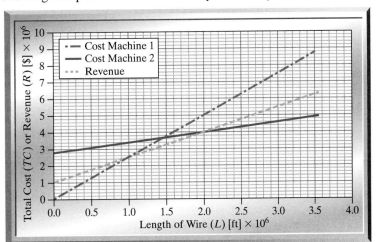

(a) Which machine has a higher fixed cost?
(b) Which machine has a lower variable cost?
(c) How much wire must be produced on Machine 1 to breakeven?
(d) If you make 3 million feet of wire, which machine will yield the highest profit?
(e) Which machine has the lower breakeven point?

COMPREHENSION CHECK 11-8

You want to install a solar panel system on your home. According to one source, if you install a 40-square foot system, the cost curve is shown in the graph.

(a) List the fixed cost and the variable cost for this system.
(b) If the source claims that you can breakeven in 3.5 years, how much savings are you generating per year (or, what is the slope of the savings curve or the "revenue" that you generate by installing the system)? Draw the "revenue" curve on the graph and use it to answer this question.
(c) If you receive a Federal Tax Credit for "going green," you can save 30% on the initial fixed cost. With this savings, how long does it take to breakeven? Draw this operating cost curve, labeled "Credit Cost" on the graph and use it to answer this question.

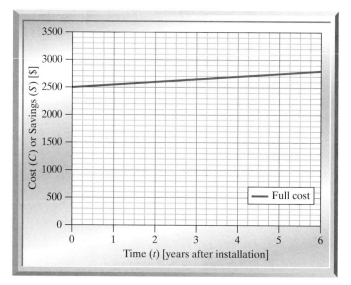

● EXAMPLE 11-10

NOTE

This example demonstrates the graphical solution of simultaneous equations when one of the equations is nonlinear. We do not expect you to know how to perform the involved circuit analyses. Those of you who eventually study electronics will learn these techniques in considerable detail.

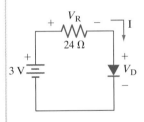

The semiconductor diode is sort of like a one-way valve for electric current: it allows current to flow in one direction, but not the other. In reality, the behavior of a diode is considerably more complicated. In general, the current through a diode can be found with the **Shockley equation**,

$$I = I_0 \left(e^{\frac{V_D}{nV_T}} - 1 \right)$$

where I is the current through the diode in amperes; I_0 is the saturation current in amperes, constant for any specific diode; V_D is the voltage across the diode in volts; and V_T is the thermal voltage in volts, approximately 0.026 volts at room temperature. The emission coefficient, n, is dimensionless and constant for any specific diode; it usually has a value between 1 and 2.

The simple circuit shown has a diode and resistor connected to a battery. For this circuit, the current through the resistor can be given by:

$$I = \frac{V - V_D}{R}$$

where I is the current through the resistor in milliamperes [mA], V is the battery voltage in volts [V], V_D is the voltage across the diode in volts, and R is the resistance in ohms [Ω].

In this circuit, the diode and resistor are in series, which implies that the current through them is the same. We have two equations for the same parameter (current), both of which are a function of the same parameter (diode voltage). We can find a solution to these two equations, and thus the current in the circuit, by graphing both equations and finding the point of intersection. For convenience of scale, the current is expressed in milliamperes rather than amperes.

Plot these two equations for the following values and determine the current.

$$I_0 = 0.01 \text{ mA}$$
$$V = 3 \text{ V}$$
$$R = 24 \ \Omega$$
$$nV_T = 0.04 \text{ V}$$

The point of intersection shown in Figure 11-20 is at $V_D = 0.64$ V and $I = 100$ mA; thus, the current in the circuit is 100 mA or 0.1 A.

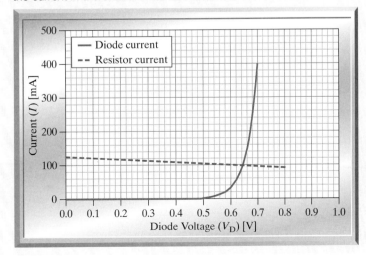

Figure 11-20

11.7   AUTOMATED CALCULATIONS: ITERATION

Traditional representations of experimental and theoretical data involve the creation of static graphs that visualize some phenomena. This section presents an alternative method that allows you to create rich, interactive graphs in Excel through the use of **iteration**. Iteration is the idea of repeatedly increasing the value of some number until some condition is true. This chapter addresses two methods for creating iterative graphs in Excel: automatic cell iteration and slider bars. Each method has its own advantages. In this section, we use some of the capabilities of Excel to show in a continuous manner how things change as "time passes" or as some other variable varies. This is best explained through examples.

Iteration — Incrementing a Variable

✍ *To enable iteration in Excel*:

- Click the **Office** button in the upper left corner of Excel.
- Click **Excel Options**.
- Choose the **Formulas** tab on the left menu of the Excel Options window.
- Click the **Enable iterative calculations** checkbox under the "Calculation Options" section.
- Under the checkbox, set the **Maximum Iterations** to **1**.
- When you are finished, click **OK** to return to the worksheet.

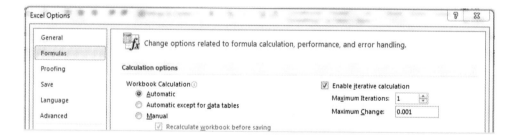

In addition to enabling iterative calculations, make sure that your workbook is set to automatic calculations.

- Go to **Formulas > Calculation** and click the **Calculation Options** drop-down menu. Be sure that your workbook is set to manually recalculate by checking the **Manual** option. By default, Excel is set to calculate automatically, but for iteration, it is preferable to control when the calculations are conducted.

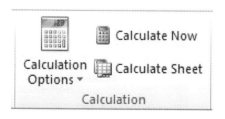

This means that the worksheet will now recalculate one time for every time you press the **F9 key**. If you hold down the F9 key, the worksheet will continue to recalculate (over and over very rapidly) until you release the key. By default, iteration is not enabled in a new workbook, so you will need to run through these instructions for each new workbook that requires iteration.

● **EXAMPLE 11-11**

Suppose you want to create an integrative graph in order to visualize the motion of a projectile. Assume that you know the launch angle and launch speed of some projectile. From physics, you know how to calculate the horizontal and vertical positions of a projectile in motion, assuming no horizontal forces and gravity as the only vertical force, using the following equations:

$$x = v \cos (\theta) t$$
$$y = -\tfrac{1}{2} g t^2 + v \sin (\theta) t$$

where

t	time	[s]
x	horizontal position at time t	[m]
y	vertical position at time t	[m]
v	launch speed of the projectile	[m/s]
θ	launch angle of the projectile	[°]
g	acceleration due to gravity	[m/s^2]

To visualize the motion of the projectile, you iterate on time (since it is the value changing in the calculation of horizontal and vertical positions) and graph the horizontal position on the x-axis and the vertical position on the y-axis.

To make this worksheet "iteration ready," you must incorporate at least one cell that will recalculate every time you press the F9 key. Since you are iterating over the time variable, you need to type a special equation into B6, which represents the time in seconds. To take the existing value of time and add the time increment stored in Cell B5, type the following equation into B6: = B5 + B6.

In a traditional worksheet, this equation would cause a circular reference error, but since iteration is enabled, this equation is acceptable.

Whenever a horizontal and vertical speed and time are inserted, the horizontal and vertical positions of the projectile will appear in cells B10 and B11. This corresponds to the (x, y) coordinates of the projectile.

*Create a plot of this single point. For x range, select only the single cell B10. For the y range, select only the single cell B11. Label the ordinate as Elevation (y) [m] and the abscissa as Horizontal Position (x) [m]. For the graph location, select the default choice of "**As object in**:" to insert the graph on the current worksheet. Drag the graph and place it below the cells containing the values. Now as you change values of the angle, velocity, or time, the projectile will move to a different location and you can watch it change location as you select different input values.*

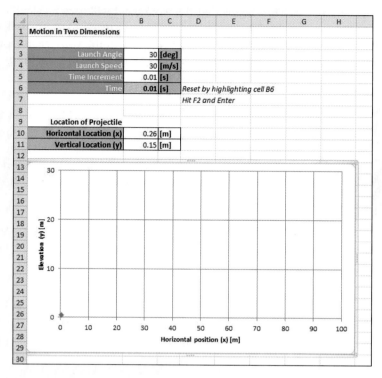

Notice, however, that Excel automatically selects the scale on both axes, so that as you change values, the scales will change. This makes it difficult to track the projectile; it would be best if the scales remain the same.

✎ To "freeze" the axis scales

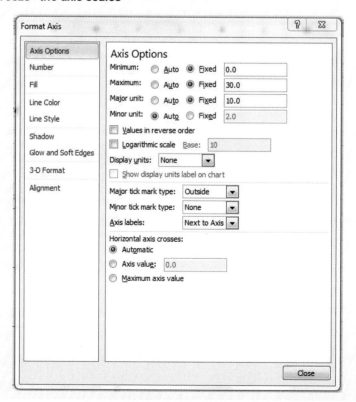

- *Right-click one of the axes and select the **Format Axis** option. The **Format Axis** window will appear.*
- *On the **Axis Options** tab, select the **Fixed** buttons for the **Minimum**, **Maximum**, and **Major unit** selections. Specify the minimum as zero, the maximum as 30, and the major unit as 10.*
- *Repeat for the other axis.*

Now when you change values, the scale will remain the same.

✎ To change the size of the symbol

*Since you are only plotting a single symbol, using a larger symbol would allow the visualization to be a little easier to follow. Right-click the symbol and select **Format Data Series**. On the **Marker Options** tab, in the **Size** box increase the value.*

✎ To test your completed worksheet

You should now be able to watch the visualization run by holding down the F9 key. To reset the time for a new run, select the cell that calculates the time (Cell B6) and press F2 to edit the equation. Do not change the equation and press the Enter key to reset the calculation.

> **NOTE**
>
> To "run" the iteration: press F9.
>
> To reset the iteration: choose the iteration cell and press F2.

11.8 AUTOMATED CALCULATIONS: SLIDER BARS

The second method for varying a parameter continuously in an Excel worksheet is through the use of **slider bars**. With slider bars, the user of a worksheet can click and drag a bar in a slider and move it to alter some parameters in the worksheet. Unlike iterative calculations, multiple slider bars can be used in a single worksheet, so more than one "variable" can be included. Example 11-12 revisits Example 11-11 and shows how a slider bar can be used in place of variable iteration in an Excel worksheet.

✎ **To enable the creation of slider bars** in a worksheet, enable the developer tools in Excel:

- Under the **File** tab, click **Options.**
- Choose the **Customize Ribbon** tab on the right side.
- On the left side, under **Customize the Ribbon**, check the box for **Developer**. Then click **OK**.

The slider bar as well as all the other special controls will now be available under the **Developer** ribbon in the **Controls** box.

● **EXAMPLE 11-12**

Example 11-11 demonstrated that when a projectile is launched from ground level at a set speed and angle with the horizontal, it rises for some distance as it moves down-range, and then gravity gradually pulls it back down to ground level. Rework this example by inserting a slider bar.

✎ *To add a slider bar*

■ *Click the **Developer** tab.*

■ *If the **Design Mode** icon is active, click it to activate the Design Mode. If it is not active, click "Properties" to engage the Design Mode icon and then click the icon to enter Design Mode.*

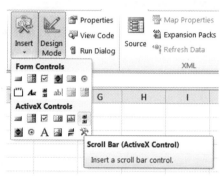

■ *Click **Insert** and under ActiveX Controls, click the scroll bar icon. By default, the scroll bar icon is the top-right ActiveX control.*

■ *In the worksheet, click and drag in an empty space on the worksheet to draw the slider bar. After you have drawn the slider bar, you will be able to resize it and to move it around the worksheet while Design Mode is enabled.*

■ *While still in Design Mode, right-click the scroll bar and click **Format Control**. Go to the **Cell Link** box and type in the cell address of the value that is going to be changed. In this example, you will change the time, so type B5. Set the maximum value to 100 (this sets the maximum value of time) and the minimum value to 0. Close the box.*

■ *Click the Design Mode icon to exit this mode.*

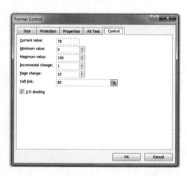

You should now be able to drag the slider bar and watch the value of cell B5 increases from 0 to 100. Next, modify B7 (time) to calculate according to the current value of the slider bar and the time increment in B6. By scaling the current value of the slider bar by the size of each time increment B7, you can calculate time by typing

$$= B5 * B6$$

As you move the slider bar, notice that the distances between the calculated data points are pretty far apart. You can increase the resolution of the calculations in your interactive graph by using a smaller time increment. Change the value of the time increment in B6 to 0.01.

Notice that the slider bar does not calculate enough values to display each horizontal and vertical position of the flight of the projectile. You can fix this by modifying the properties of the slider bar and changing the Max value to something much larger than 100 (for this example, 350) to allow the projectile to complete its return to the ground. The maximum value and axis properties require adjustment for different angles and launch speeds to display properly.

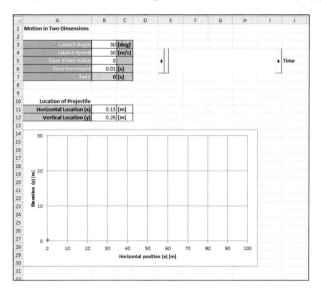

Modify the work completed in Example 11-12 to include an additional slider bar to control the launch angle.

Copy the worksheet containing Example 11-12 and insert a new slider bar. Since you now have two slider bars in your current worksheet, label and color-code the controls so that it is apparent to the user of the worksheet which controls will change each variable.

The angle slider bar should be linked to Cell B3 and should vary from 0 to 90 degrees. Using an angle increment of 1 degree is acceptable, so you need not calculate a scaled angle as you did with the time. The resulting worksheet should appear as shown below.

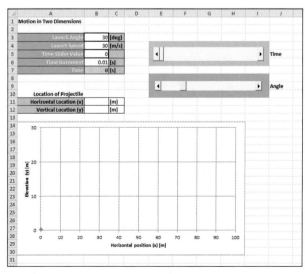

In-Class Activities

ICA 11-1

Using the list provided, you may be assigned a topic for which to create a graph. You must determine the parameters to graph and imagine a set of data to show on the chart or on a copy.

1. Air temperature
2. Airplane from airport to airport
3. Baking bread
4. Baseball
5. Bird migration
6. Birthday party
7. Boiling water in a whistling teapot
8. Bouncing a basketball
9. Brushing your teeth
10. Burning a pile of leaves
11. Burning candle
12. Climbing a mountain
13. A coastal river
14. Cooking a Thanksgiving turkey
15. Cow being picked up by a tornado
16. D.O.R. possum ("dead on the road")
17. Daily electric power consumption
18. A day in the life of a chicken
19. Detecting a submarine by using sonar
20. Diet of a parakeet
21. Diving into a swimming pool
22. Drag racing
23. Driving home from work
24. Dropping ice in a tub of warm water
25. Eating a stack of pancakes
26. Eating at a fast-food restaurant
27. Email messages
28. Engineer's salary

29. Exercising
30. Feedback from an audio system
31. Feeding birds at a bird feeder
32. Firing a bullet from a rifle
33. Fishing
34. Flight of a hot air balloon
35. Football game crowd
36. Forest fire
37. Formation of an icicle
38. A glass of water in a moving vehicle
39. Hair
40. Hammering nails
41. Hiking the Appalachian Trail
42. Homework
43. Kangaroo hopping along
44. Leaves on a tree
45. Letting go of a helium balloon
46. Marching band
47. Moving a desk down a staircase
48. Oak tree over the years
49. Oil supply
50. People in Florida
51. People in Michigan
52. Person growing up
53. Playing with a yo-yo
54. Plume from a smokestack
55. Political affiliations
56. Pony Express
57. Popping corn
58. Pouring water out of a bottle
59. Power consumption of your laptop
60. Power usage on campus
61. Pumping air into a bicycle tire
62. Rabbit family
63. Rain filling a pond
64. Recycling
65. River in a rainstorm
66. Running of the bulls
67. Skipping a stone on water
68. Sleeping
69. Slipping on a banana peel
70. Snoring
71. Snow blowing over a roof
72. Solar eclipse
73. Sound echoing in a canyon
74. Space station
75. Spinning a hula-hoop
76. Strength of concrete
77. Student attention span during class
78. Studying for an exam
79. Taking a bath in a tub
80. Talking on a cell phone
81. The moon
82. Throwing a ball
83. Thunderstorm
84. Tiger hunting
85. Traffic at intersections
86. Train passing through town
87. Typical day
88. Using a toaster
89. Using Instant Messenger
90. Washing clothes

ICA 11-2

Joule's first law relates the heat generated to current flowing in a conductor. It is named after James Prescott Joule, the same person for whom the unit of Joule is named. The Joule effect states that electric power can be calculated as $P = I^2R$, where R is the resistance [in ohms, Ω] and I is the electrical current [in amperes, A]. Use the following experimental data to create a scatter graph of the power (P, on the ordinate) and current (I, on the abscissa).

Current (I) [A]	0.50	1.25	1.50	2.25	3.00	3.20	3.50
Power (P) [W]	1.20	7.50	11.25	25.00	45.00	50.00	65.00

ICA 11-3

There is a large push in the United States currently to convert from incandescent light bulbs to compact fluorescent bulbs (CFLs). The lumen [lm] is the SI unit of luminous flux (LF), a measure of the perceived power of light. LF is adjusted to reflect the varying sensitivity of the human eye to different wavelengths of light. To test the power usage, you run an experiment and measure the following data. Create a proper plot of these experimental data, with electrical consumption (EC) on the ordinate and LF on the abscissa.

Luminous Flux [lm]	Electrical Consumption [W]	
	Incandescent 120 V	Compact Fluorescent
80	16	
200		4
400	38	8
600	55	
750	68	13
1,250		18
1,400	105	19

ICA 11-4

Your team has designed three tennis ball launchers, and you have run tests to determine which launcher best meets the project criteria. Each launcher is set to three different launch angles, and the total distance the ball flies through the air is recorded. These experimental data are summarized in the table. Plot all three sets of data on a scatter plot, showing one data set for each of the three launchers on a single graph. Launch angle should be plotted on the horizontal axis.

Launcher 1		Launcher 2		Launcher 3	
Launch Angle (θ) [°]	Distance (d) [ft]	Launch Angle (θ) [°]	Distance (d) [ft]	Launch Angle (θ) [°]	Distance (d) [ft]
20	5	10	10	20	10
35	10	45	25	40	20
55	12	55	18	50	15

ICA 11-5

You may be assigned several of the following functions. Plot them, all on the same graph. The independent variable (angle) should vary from 0 to 360 degrees.

(a) $\sin \theta$
(b) $2 \sin \theta$
(c) $-2 \sin \theta$
(d) $\sin 2\theta$

(e) $\sin 3\theta$
(f) $\sin (\theta) + 2$
(g) $\sin (\theta) - 3$
(h) $\sin (\theta + 90)$

(i) $\sin (\theta - 45)$
(j) $2 \sin (\theta) + 2$
(k) $\sin (2\theta) + 1$
(l) $3 \sin (2\theta) - 2$

ICA 11-6

You may be assigned several of the following functions. Plot them, all on the same graph. The independent variable (angle) should vary from 0 to 360 degrees.

(a) $\cos \theta$	**(e)** $\cos 3\theta$	**(i)** $\cos(\theta - 45)$
(b) $2\cos \theta$	**(f)** $\cos(\theta) + 2$	**(j)** $2\cos(\theta) + 2$
(c) $-2\cos \theta$	**(g)** $\cos(\theta) - 3$	**(k)** $\cos(2\theta) + 1$
(d) $\cos 2\theta$	**(h)** $\cos(\theta + 90)$	**(l)** $3\cos(2\theta) - 2$

ICA 11-7

You need to create a graph showing the relationship of an ideal gas between pressure (P) and temperature (T). The ideal gas law relates the pressure, volume, amount, and temperature of an ideal gas using the relationship: $PV = nRT$. The ideal gas constant (R) is 0.08207 atmosphere liter per mole kelvin. Assume the tank has a volume of 12 liters and is filled with nitrogen (formula: N_2, molecular weight: 28 grams per mole). Allow the initial temperature to be 270 kelvin at a pressure of 2.5 atmospheres. Create the graph, showing the temperature on the abscissa from 270 to 350 kelvin.

ICA 11-8

The decay of a radioactive isotope can be modeled using the following equation, where C_0 is the initial amount of the element at time zero and k is the half-life of the isotope. Create a graph of the decay of Isotope A [$k = 1.48$ hours]. Allow time to vary on the abscissa from 0 to 5 hours with an initial concentration of 10 grams of Isotope A.

$$C = C_0 e^{-t/k}$$

ICA 11-9

In researching alternate energies, you find that wind power is calculated by the following equation:

$$P = \frac{1}{2} A \rho v^3$$

where

- P = power [watts]
- A = sweep area (circular) of the blades [square meters]
- ρ = air density [kilograms per cubic meter]
- v = velocity [meters per second]

The specific gravity of air is 0.00123 and the velocity is typically 35 meters per second. Create a graph of the theoretical power (P, in units of watts) as a function of the blade diameter (D, in units of meters). Allow the diameter to be graphed on the abscissa and vary from 0.5 to 1.5 meters.

ICA 11-10

Using the following graph and answer choices, complete the following sentences:

- Equal to a constant, positive value
- Equal to zero
- Equal to a constant, negative value
- Increasing at a constant rate
- Decreasing at a constant rate
- Increasing at a decreasing rate
- Increasing at an increasing rate
- Decreasing at a decreasing rate
- Decreasing at an increasing rate

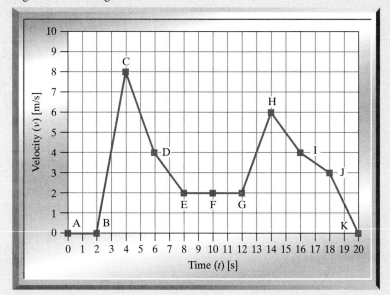

(a) Between points (A) and (B), the acceleration is ___

(b) Between points (B) and (C), the acceleration is ___

(c) Between points (C) and (D), the acceleration is ___

(d) Between points (F) and (G), the acceleration is ____

(e) Between points (G) and (H), the acceleration is ____

(f) Between points (I) and (J), the acceleration is ____

(g) Between points (A) and (B), the distance is ___

(h) Between points (B) and (C), the distance is ___

(i) Between points (C) and (D), the distance is ___

(j) Between points (F) and (G), the distance is ____

(k) Between points (G) and (H), the distance is ____

(l) Between points (I) and (J), the distance is ____

Use the graph to determine the following numerical values:

(m) Between points (A) and (B), the acceleration is ___ meters per second squared.

(n) Between points (B) and (C), the acceleration is ___ meters per second squared.

(o) Between points (C) and (D), the acceleration is ___ meters per second squared.

(p) Between points (F) and (G), the acceleration is ___ meters per second squared.

(q) Between points (I) and (J), the acceleration is ___ meters per second squared.

(r) At point (B), the distance is ___ meters.

(s) At point (C), the distance is ___ meters.

(t) At point (E), the distance is ___ meters.

(u) At point (G), the distance is ___ meters.

(v) At point (I), the distance is ___ meters.

(w) At point (K), the distance is ___ meters.

ICA 11-11

The music industry in the United States has had a great deal of fluctuation in profit over the past 20 years due to the advent of new technologies such as peer-to-peer file sharing and mobile devices such as the iPod and iPhone. The following graph displays data from a report published by eMarketer in 2009 about the amount U.S. consumers spend on digital music files and physical music formats (CDs, records, cassette tapes, etc.), where the values for 2009–2013 are reported as projections and for 2008 is reported using actual U.S. spending measurements.

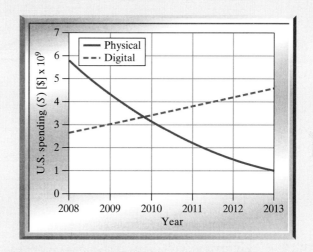

(a) According to the study, when will the sale of physical media be equivalent to the sale of digital audio files?

(b) When will the sales of digital audio files exceed that of physical media by $1 billion?

(c) If the physical media sales were $2 billion higher than the trend displayed on the graph, when would the sale of digital audio files exceed physical media?

(d) If the digital audio file sales were $1 billion lower than the trend displayed on the graph, when would the sale of digital audio files exceed physical media?

(e) According to this graph, how much do U.S. consumers spend on digital audio files per year?

(f) If you were to create a third data set on this graph that displayed the total amount U.S. consumers spent on digital and physical music sales, would the trend be increasing, decreasing, or stay the same?

ICA 11-12

You are working for a chemical manufacturer, producing solvents used to clean lenses for micro-scopes. You are working on determining the properties of three different solvent blends. You develop the following chart, showing the evaporation of the three blends.

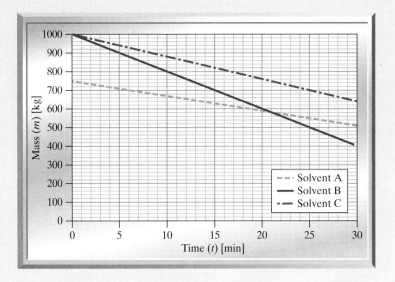

(a) Which solvent evaporates at the slowest rate?
(b) Which solvent evaporates at the fastest rate?
(c) What is the initial mass of Solvent A? Be sure to include units.
(d) What is the rate of solvent evaporation of Solvent A? Be sure to include units.
(e) What is the rate of solvent evaporation of Solvent B? Be sure to include units.
(f) What is the rate of solvent evaporation of Solvent C? Be sure to include units.

ICA 11-13

In this exercise and the one that follows, you will develop two worksheets, one for a standing wave and the other for a progressive (stationary) wave. In the real world, both of these types of waves occur in lakes, rivers, and oceans. Before starting each of these exercises, first place your worksheet into manual iteration mode.

When you throw a rock in a lake, a series of waves emanates from the impact point. Depending on conditions, these waves may move across the water surface as "progressive waves." The equation of such waves is:

$$y = A \sin(cx - kt)$$

In other words, if you "rode the crest of a wave," you would find yourself moving horizontally, much as a surfer. Alternatively, if you positioned yourself at a fixed location (you anchored the boat), as time passed, the water would move up and down at that location.

(a) Create a worksheet similar to the one shown. Create a graph with water height (y) on the ordinate and position (x) on the abscissa embedded in the worksheet. Set the horizontal scale at a minimum of 0 to a maximum of 25 with a major unit of 5. Set the vertical scale at a minimum of -10 to a maximum of 10 with a major unit of 5 and change the value x-axis crosses at to -10. Enable data validation to prevent an amplitude -10 from being selected. Hold the F9 key down to make the waves move.

(b) As an alternative to part (a), create a slider bar to control the value of A and k.

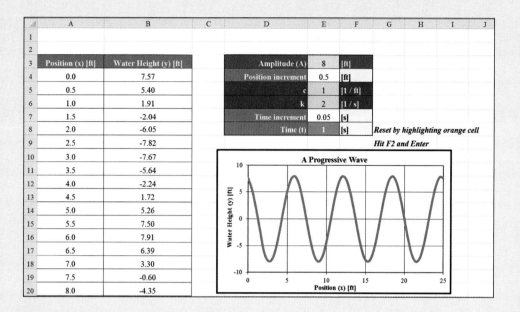

	A	B	C	D	E	F	G	H	I	J
1										
2										
3	Position (x) [ft]	Water Height (y) [ft]		Amplitude (A)	8	[ft]				
4	0.0	7.57		Position increment	0.5	[ft]				
5	0.5	5.40		c	1	[1 / ft]				
6	1.0	1.91		k	2	[1 / s]				
7	1.5	-2.04		Time increment	0.05	[s]				
8	2.0	-6.05		Time (t)	1	[s]	*Reset by highlighting orange cell*			
9	2.5	-7.82					*Hit F2 and Enter*			
10	3.0	-7.67								
11	3.5	-5.64								
12	4.0	-2.24								
13	4.5	1.72								
14	5.0	5.26								
15	5.5	7.50								
16	6.0	7.91								
17	6.5	6.39								
18	7.0	3.30								
19	7.5	-0.60								
20	8.0	-4.35								

ICA 11-14

In this exercise and the one preceding, you will develop two worksheets, one for a standing wave and the other for a progressive (stationary) wave. In the real world, both of these types of waves occur in lakes, rivers, and the ocean. Before starting each of these, first place your worksheet into manual iteration mode.

When a wave is created in a water body with solid boundaries, the waves can reflect and move back in the opposite direction. In such a situation, standing waves may be created where the crest and troughs do not move along the surface but simply move up and down in one location. The equation describing standing waves is:

$$y = A \sin(c\,x) \sin(k\,t)$$

In this equation, A is the amplitude, x is the position, and t is the time. The variable c is the frequency of the standing wave, and controls how far apart the peaks are. The variable k determines how rapidly the standing wave rises and falls.

We can position ourselves at a fixed location (we anchor the boat), and then as time passes, the water moves up and down at that location. If, however, we position ourselves at a different place, it may be that the water surface remains constant.

(a) Create a worksheet similar to the one shown. Create a graph with water height (y) on the ordinate and position (x) on the abscissa embedded in the worksheet. Set the horizontal scale at a minimum of 0 to a maximum of 25 with a major unit of 5. Set the vertical scale at a minimum of −10 to a maximum of 10 with a major unit of 5 and change the value x-axis crosses at to −10. Enable data validation to prevent selecting amplitude −10. Hold the F9 key down to make the waves move.

(b) As an alternative to part **(a)**, create a slider bar to control the values of A, c, and k.

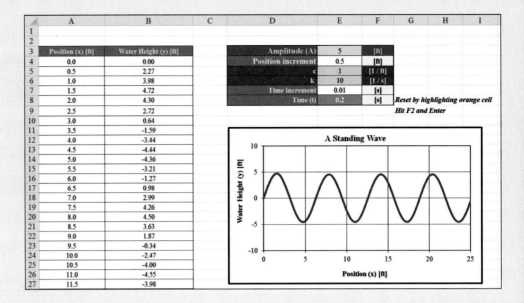

ICA 11-15

Materials

Balloons (2)   Stopwatch (2)   String (40 inches)   Tape Measure

Part I: Blowing Up a Balloon

One team member is to inflate one balloon, a second team member is to time the inhalation stage (how long it takes to inhale a single breath), and a third team member is to time the exhalation stage (how long it takes to exhale a single breath into the balloon). A fourth team member is to measure the balloon size at the end of each inhale/exhale cycle, using the string to measure the balloon circumference.

Record the observations on a worksheet similar to the following one for three complete inhale/exhale cycles or until the balloon appears to be close to maximum volume, whichever occurs first. Repeat the entire balloon inflation process for a second balloon; average the times from the balloons to obtain the time spent at each stage and the average circumference at each stage. Calculate the balloon volume at each stage, assuming the balloon is a perfect sphere.

Balloon	Stage	Inhale Time	Exhale Time	Circumference
1	1			
	2			
	3			
2	1			
	2			
	3			

	Stage	Inhale Time	Exhale Time	Circumference	Volume
Average Balloon	1				
	2				
	3				

Part II: Analysis

Graph the balloon volume (V, ordinate) versus time (t, time) on the following graph or on a copy. Allow the process to be continuous, although in reality it was stopped at various intervals for measurements. The resulting graph should contain only the time elapsed in the process of inhaling and exhaling, not the time required for recording the balloon size. For this procedure, assume that the air enters the balloon at a constant rate and the balloon is a perfect sphere.

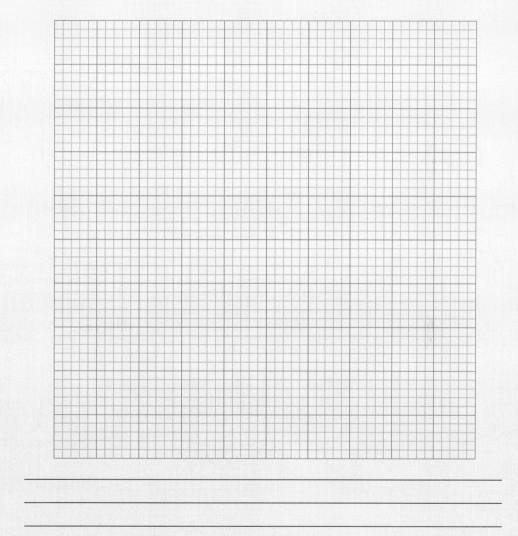

_____

_____

_____

_____

(a) What does the assumption of the air entering the balloon at a constant rate indicate about the slope?

(b) Calculate the following graphically.

■ The rate at which the air enters the balloon in the first stage.
■ The rate at which the air enters the balloon in the third stage.

(c) On the same graph, sketch the balloon volume (V, ordinate) versus time (t, time) if you were inflating a balloon that contained a pinhole leak.

(d) On the same graph, sketch the balloon volume (V, ordinate) versus time (t, time) if you were inflating a balloon from a helium tank.

1. A computer engineer has measured the power dissipated as heat generated by a prototype microprocessor running at different clock speeds. The data are shown in the following table. Create a proper plot of the following experimental data set.

Speed (S) [GHz]	0.8	1.3	1.8	2.5	3.1
Power dissipated as heat (P) [W]	135	217	295	405	589

2. Due to increased demand, an industrial engineer is experimenting with increasing the speed (S) of a machine used in the production of widgets. The machine is normally rated to produce five widgets per second, and the engineer wants to know how many defective parts (D) are made at higher speeds, measured in defective parts per thousand. The following table indicates the data collected. Create a proper plot of the following experimental data set.

Speed (S) [parts/min]	5.5	5.9	6.5	7.2	8.0
Defects in parts per thousand (D)	1	3	7	13	21

3. An engineer is conducting tests of two prototype toothbrush sanitizers that use ultraviolet radiation to kill pathogenic organisms while the toothbrush is stored. The engineer is trying to determine the minimum power needed to reliably kill pathogens on toothbrushes. Several toothbrushes are treated with a mix of bacteria, fungi, and viruses typically found in the human mouth, and then each is placed in one of the sanitizers for 6 hours at a specific power level (P). After 6 hours in the sanitizers, the viable pathogens remaining (R) on each toothbrush is assayed. The data are presented in the following table. Create a proper plot of the following experimental data set.

Power (P) [W]		10	18	25	40
Pathogens remaining (R) [%]	Sanitizer A	46	35	14	2
	Sanitizer B	58	41	21	7

4. Several reactions are carried out in a closed vessel. The following data are taken for the concentration (C) in units of grams per liter of solvent processed for compounds A and B as a function of time (t). Create a proper plot of the following experimental data set.

	Concentration [g/L]	
Time (t) [min]	A(C_A)	B(C_B)
36	0.145	0.160
65	0.120	0.155
100	0.100	0.150
160	0.080	0.140

5. The following experimental data are collected on the current (symbolized by I, in units of milliamperes) in the positive direction and voltage (symbolized by V, in units of volts) across the terminals of two different thermionic rectifiers. Create a proper plot of the following experimental data set.

	Current (I) [mA]	
Voltage (V) [V]	Rectifier A	Rectifier B
18	5	15
30	18	26
40	24	34
45	30	50

6. Plastic bottles and containers are often "blow molded" to form the shapes needed in a particular application. This requires energy. Additional energy is also required to operate the robots and other machine movements that complete the process. Consequently, for any particular blow-molding machine, we can measure the amount of energy or power needed for processing (melting) the plastic as well as the total energy or power required for the overall process.

The following data were obtained for 10 different machines. Using this information, create an appropriate graph. Experiment with several different types of graphs, and justify your final choice.

Machine Model	Energy for Extrusion [W/kg]	Total Energy [W/kg]
KLS6	269	494
KLS6D	219	381
KLS8	270	487
KLS8D	221	409
KLS10	236	433
KLS10D	223	395
KLS12	237	444
KLS12D	226	405
KLS14	218	420
KLS14D	227	392

Data provided by Kautex Machines

7. You are asked to relate the power consumption to the mass flow rate of two plastic molding processing machines. The plastic can be melted by the extruder either with an AC or a DC motor (E60 AC water cooled or E60 DC air cooled). The following data were acquired.

You need to produce a graphical representation that will allow a user to quickly judge the efficiency between the AC and DC motors. Experiment with several different types of graphs, and justify your final choice.

E60 AC Water-Cooled Drive		E60 DC Air-Cooled Drive	
Power Consumption [kW]	Mass Flow Rate [kg/h]	Power Consumption [kW]	Mass Flow Rate [kg/h]
3.2	8	12	31
4.3	11	14	42
14	54	17	54
17	65	23	63
22	76	26	74
24	88	31	85
28	99	35	95
33	108	39	104

Data provided by Kautex Machines

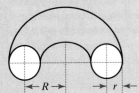

8. In the 1950s, a team at Los Alamos National Laboratories built several devices called "Perhapsatrons," thinking that PERHAPS they might be able to create controllable nuclear fusion. After several years of experiments, they were never able to maintain stable plasma and abandoned the project.

   The perhapsatron used a toroidal (doughnut-shaped) plasma confinement chamber, similar to those used in more modern Tokamak fusion devices. You have taken a job at a fusion research lab, and your supervisor asks you to develop a simple spreadsheet to calculate the volume of a torus within which the plasma will be contained in a new experimental reactor.

   (a) Create a simple calculator to allow the user to type in the radius of the tube (r) in meters and the radius of the torus (R) in meters and display the volume in cubic meters.

   (b) Create a table that calculates the volumes of various toruses with specific values for r and R. The tube radii (r) should range from 5 to 100 centimeters in increments of 5 centimeters. The torus radii (R) should range from 1.5 to 3 meters in increments of 0.1 meters.
   The volume of a torus can be determined using $V = 2\pi^2 R r^2$.

   (c) Using the table of volumes, create a graph showing the relationship between volume (independent variable) and tube radius (r) (dependent variable) for torus radii (R) of 1.5, 2, 2.5, and 3 meters.

   (d) Using the table of volumes, create a graph showing the relationship between volume (independent variable) and torus radius (R) (dependent variable) for tube radii (r) of 10, 40, 70, and 100 centimeters.

9. Below is a graph of the vertical position of a person bungee jumping, in meters.

   (a) What is the closest this person gets to the ground?

   (b) When this person stops bouncing, how high off the ground will the person be?

   (c) If the person has a mass of 70 kilograms, how would the graph change for a jumper of 50 kilograms? Approximately sketch the results on the graph provided or on a copy.

   (d) If the person has a mass of 70 kilograms, how would the graph change for a jumper of 80 kilograms? Approximately sketch the results on the graph provided or on a copy.

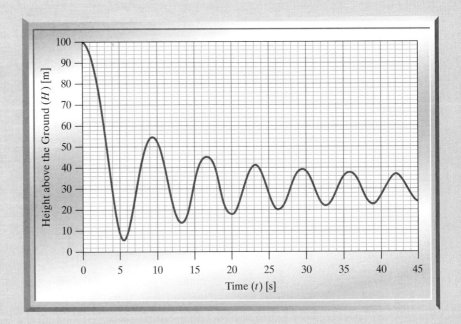

10. Shown are graphs of the altitude in meters, and velocity in meters per second, of a person skydiving.

(a) When does the skydiver reach the ground?

(b) How fast is he moving when he reaches the ground?

(c) What is the fastest he ever falls, and when does this occur?

(d) At what altitude does he open the parachute?

(e) Terminal velocity is the velocity at which the acceleration of gravity is exactly balanced by the drag force of air. How long does it take him to reach his terminal velocity without the parachute open?

(f) How long after his parachute opens does it take him to reach open parachute terminal velocity?

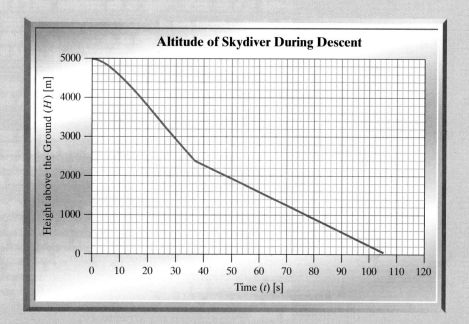

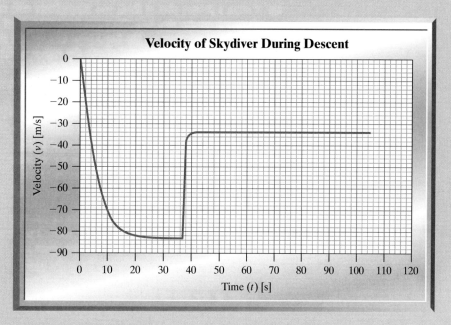

11. Generally, when a car door is opened, the interior lights come on and turn off again when the door is closed. Some cars turn the interior lights on and off gradually. Suppose that you have a car with 25 watts of interior lights. When a door is opened, the power to the lights increases linearly from 0 to 25 watts over 2 seconds. When the door is closed, the power is reduced to zero in a linear fashion over 5 seconds.

   (a) Create a proper plot of power (P, on the ordinate) and time (t, on the abscissa).
   (b) Using the graph, determine the total energy delivered to the interior lights if the door to the car is opened and then closed again 10 seconds later.

12. One of the 22 named, derived units in the metric system is the volt, which can be expressed as 1 joule per coulomb ($V = J/C$). A coulomb is the total electric charge on approximately 6.24×10^{18} electrons. The voltage on a capacitor, an electrical device that stores electric charge, is given by $V = \Delta Q/C + V_0$ volts, where ΔQ is the change in charge [coulombs] stored, V_0 is the initial voltage on the capacitor, and C is the capacitance [farads].

   (a) Create a proper plot of voltage (V, on the ordinate) and total charge (ΔQ, on the abscissa) for a 5 farad capacitor with an initial voltage of 5 volts for $0 < \Delta Q < 20$.
   (b) Using the graph, determine the total energy stored in the capacitor for an addition of 15 coulombs.

13. In a simple electric circuit, the current (I) must remain below 40 milliamps ($I < 40$ mA), and must also satisfy the function $I > 10^{-6} e^{25V}$, where V is the voltage across a device called a diode.

   (a) Create a proper plot of these two inequalities with current on the ordinate and voltage on the abscissa. The values on the vertical axis should range from 0 to 50 milliamperes, and the values on the horizontal axis should range from 0 to 1 volt.
   (b) If graphing part (a) by hand, shade the region of the graph where *both* inequalities are satisfied.
   (c) Graphically determine the maximum allowable voltage across the diode.

14. In a hard drive design, the faster the disk spins, the faster the information can be read from and written to the disk. In general, the more information to be stored on the disk, the larger the diameter of the disk must be. Unfortunately, the larger the disk, the lower the maximum rotational speed must be to avoid stress-related failures. Assume the minimum allowable rotational speed (S) of the hard drive is 6,000 revolutions per minute [rpm], and the rotational speed must meet the criterion $S < 12{,}000 - 150\,D^2$, where D is the diameter of the disk in inches.

   (a) Create a proper plot of these two inequalities with rotational speed on the ordinate and diameter on the abscissa. The values on the vertical axis should range from 0 to 12,000 rpm, and the values on the horizontal axis should range from 0 to 7 inches.
   (b) If graphing part (a) by hand, shade the region of the graph where both inequalities are satisfied.
   (c) Graphically determine the range of allowable rotational speeds for a 4-inch diameter disk.
   (d) Graphically determine the largest diameter disk that meets the design criteria.

15. We have obtained a contract to construct metal boxes (square bottom, rectangular sides, no top) for storing sand. Each box is to contain a specified volume and all edges are to be welded. Each box will require the following information: a volume (V, in units of cubic inches), the length of one side of the bottom (L, in units of inches), the box height (H, in units of inches), and the material cost (M, in units of dollars per square inch). To determine the total cost to manufacture a box, we must include not only the cost of the material, but also the cost of welding all the edges. Welding costs depend on the number of linear inches that are welded (W, in units of dollars per inch). The client does not care what the box looks like, but it should be constructed at the minimum cost possible.

   (a) Construct a worksheet that will depict the cost of the material for one box, the welding cost for one box, and the total cost for the box as a function of the box side length. First, create at the top of your worksheet a section to allow the user to specify as absolute references the variables $V, H, M,$ and W. Next, create a column for length ranging from 0 to 30 inches in increments of 2 inches. Finally, determine the material cost per box, welding cost per box, and total cost.

(b) Create a proper plot of the material cost, welding cost, and total cost (all shown as oridiante values) versus the box length.

(c) For the following values, determine the box shape for minimum cost: $V = 500$ cubic inches, $M = \$1.00$ per square inch, and $W = \$1.50$ per inch.

(d) For the following values, determine the box shape for minimum cost: $V = 300$ cubic inches, $M = \$1.00$ per square inch, and $W = \$1.00$ per inch.

(e) For the following values, determine the box shape for minimum cost: $V = 900$ cubic inches, $M = \$2.00$ per square inch, and $W = \$5.00$ per inch.

16. Your company has developed a new high-mileage automobile. There are two options for manufacturing this new vehicle.

- Process A: The factory can be completely retooled and workers trained to use the new equipment.
- Process B: The old equipment can be modified.

A graph of the costs of each process and the revenues from sales of the vehicles is shown.

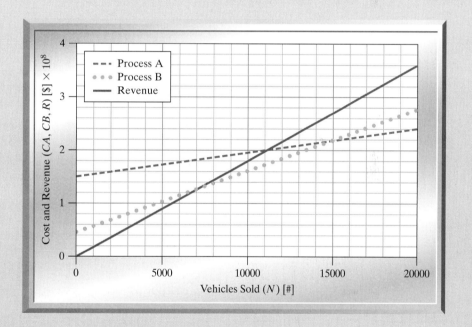

Use the chart to answer the following questions.

(a) What is the sales price per vehicle?

(b) What is the breakeven point (number of vehicles) for each of the two processes?

(c) Which process yields the most profit if 18,000 vehicles are sold? How much profit is made in this case?

(d) If the sales price per vehicle is reduced by $2,000 with a rebate offer, what is the new breakeven point (number of vehicles) for each of the two processes?

17. One of the fourteen Grand Challenges for Engineering as determined by the National Academy of Engineering committee is **Make Solar Energy Economical.** According to the NAE website: The solar "share of the total energy market remains rather small, well below 1 percent of total energy consumption, compared with roughly 85 percent from oil, natural gas, and coal." "... today's commercial solar cells ... typically convert sunlight into electricity with an efficiency of only 10 percent to 20 percent." "Given their manufacturing costs, modules of today's cells . . . would produce electricity at a cost roughly 3 to 6 times higher than current prices." "To make solar economically competitive, engineers must find ways to improve the efficiency of the cells and to lower their manufacturing costs."

The following graph shows a breakeven analysis for a company planning to manufacture modular photoelectric panels.

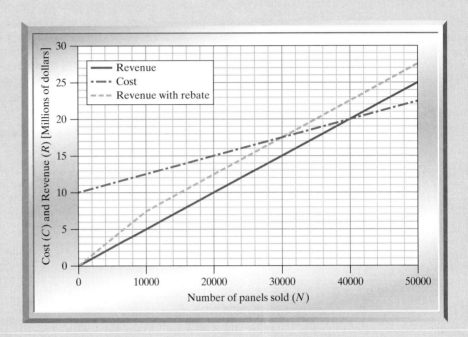

(a) What is the fixed cost incurred in manufacturing the photoelectric panels?
(b) How much does it cost to manufacture each photoelectric panel?
(c) What is the sales price of one photoelectric panel?
(d) If the company makes and sells 30,000 panels, is there a net loss or profit, and how much?
(e) While the company is still in the planning stages, the government starts a program to stimulate the economy and encourage green technologies. In this case, the government agrees to reimburse the company $250 for each of the first 10,000 units sold. Sketch a modified revenue curve for this situation.
(f) Using this new revenue curve, how many units must the company make to break even? Be sure to clearly indicate this point on the graph.
(g) Also using the new revenue curve, how many units must the company make and sell to make a profit of $1,500,000? Be sure to clearly indicate this point on the graph.

18. You are an engineer for a plastics manufacturing company. In examining cost-saving measures, your team has brainstormed the following ideas (labeled Idea A and Idea B). It is your responsibility to evaluate these ideas and recommend which one to pursue. You have been given a graph of the current process.

(a) What is the selling price of the product?

Current Cost: The current process has been running for a number of years, so there are no initial fixed costs to consider.
In the operating costs, the process requires the following:

- Material cost: $2.00/pound-mass of resin
- Energy cost: $0.15/pound-mass of resin
- Labor cost: $0.10/pound-mass of resin

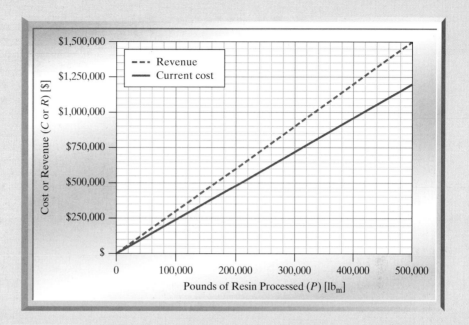

(b) There is also a cost associated with taking the scrap material to the landfill. Using the total cost determined from the graph, find the cost of landfill, in dollars per pounds-mass of resin.

Idea (A): Your customer will allow you to use regrind (reprocessed plastic) in the parts instead of 100% virgin plastic. Your process generates 10% scrap. Evaluate using all your scrap materials as regrind, with the regrind processed at your plant.

(c) You will need to purchase a regrind machine to process the plastic, estimated at a cost of $100,000. Using the regrind will alter the following costs, which account for using 10% scrap material:

- Material cost: $1.80/pound-mass of resin
- Energy cost: $0.16/pound-mass of resin
- Labor cost: $0.11/pound-mass of resin

This idea will eliminate the landfill charge required in the current process (see part **(b)**). Draw the total cost curve for Idea (A) on the graph or on a copy.

(d) How long (in pounds of resin processed) before the company reaches breakeven on Idea A?

(e) At what minimum level of production (in pound-mass of resin processed) will Idea (A) begin to generate more profit than the current process?

Idea (B): Your customer will allow you to use regrind (reprocessed plastic) in the parts instead of 100% virgin plastic. Evaluate using 25% regrind purchased from an outside vendor.

(f) Using the regrind from the other company will alter the following costs, which account for using 25% scrap material purchased from the outside vendor:

- Material cost: $1.85/pound-mass of resin
- Energy cost: $0.15/pound-mass of resin
- Labor cost: $0.11/pound-mass of resin

This idea will eliminate the landfill charge required in the current process (see part **(b)**) and will not require the purchase of a regrind machine as discussed in Idea (A). Draw the total cost curve for Idea (B) on the graph or on a copy.

(g) At what minimum level of production (in pound-mass of resin processed) will Idea (B) begin to generate more profit than the current process?

(h) At a production level of 500,000 pound-mass of resin, which Idea (A, B, or neither) gives the most profit over the current process?

(i) If the answer to part **(h)** is neither machine, list the amount of profit generated by the current process at 500,000 pound-mass of resin. If the answer to part **(h)** is Idea A or Idea B, list the amount of profit generated by that idea at 500,000 pound-mass of resin.

19. We have decided to become entrepreneurs by raising turkeys for the Thanksgiving holiday. We already have purchased some land in the country with buildings on it, so that expense need not be a part of our analysis. A study of the way turkeys grow indicates that the mass of a turkey (m) from the time it hatches (at time zero) until it reaches maturity is:

$$m = K(1 - e^{-bt})$$

Here, we select values of K and b depending on the breed of turkey we decide to raise. The value (V) of our turkey is simply the mass of the turkey times the value per pound-mass (S) when we sell it, or:

$$V = Sm$$

Here, S is the value per pound-mass (in dollars). Finally, since we feed the turkey the same amount of food each day, the cumulative cost (C) to feed the bird is:

$$C = Nt$$

Here, N is the cost of one day's supply of food [$/day].

Create a graph of this situation, showing three lines: cumulative food cost, bird value, and profit on a particular day. For the graph, show the point after which you begin to lose money, and show the time when it is most profitable to sell the bird, indicating the day on which that occurs. Use values of $K = 21$ pound-mass, $b = 0.03$ per day, $S = \$1$ per pound-mass, and $N = \$0.12$ per day.

20. As an engineer, suppose you are directed to design a pumping system to safely discharge a toxic industrial waste into a municipal reservoir. The concentrated wastewater from the plant will be mixed with freshwater from the lake, and this mixture is to be pumped into the center of the lake. You realize that the more water you mix with the waste, the more dilute it will be and thus will have a smaller impact on the fish in the lake. On the other hand, the more water you use, the more it costs in electricity for pumping. Your objective is to determine the optimum amount of water to pump so the overall cost is a minimum.

■ Assume that the cost of pumping is given by the expression $C_{pump} = 10\ Q^2$. The cost C_{pump} [$/day] depends on the pumping rate Q [gallons per minute, or gpm] of the water used to dilute the industrial waste.

■ Now, suppose that some biologists have found that as more and more water dilutes the waste, the fish loss C_{fish} [$/day] can be expressed as $C_{fish} = 2,250 - 150\ Q$.

(a) With this information, construct a graph, with pumping rate on the abscissa showing the pumping cost, the fish-loss cost, and total cost on the ordinate. For the scale, plot 0 to 15 gallons per minute for flow rate.

(b) Determine both the minimum cost and the corresponding flow rate.

21. In this problem, we model the trajectory of a skydiver's jump from an airplane. The skydiver will free fall until reaching a specified height above the ground. At that point the skydiver will pull the rip cord to release the parachute slowing the skydiver's descent. Create a worksheet as follows.

■ Cells B5–B10 and B16–B18 contain the numbers shown in the figure below.

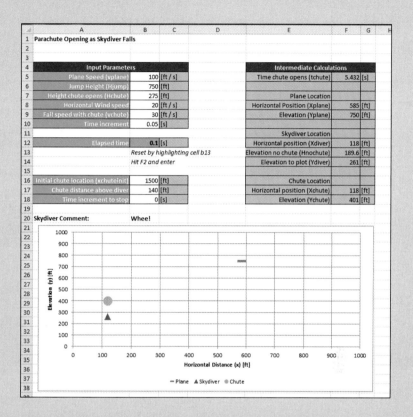

■ Cell B12 contains the time elapsed since the skydiver jumped from the airplane. To increment the value in Cell B12, we hold down the F9 button. That repeatedly adds the time increment value in B10 to the previous value in B12. We can reset the value in B12 by selecting B12, pressing the F2 key, and then Enter. This resets the "clock."

■ Cell B20 contains the skydiver's comments. Upon reaching the ground, the skydiver's comment should change from "Whee!" to "Down Safely."

■ Cell F5 contains an equation to calculate the time for the parachute to open (t_{chute}), which is a function of the difference between the jump height (H_{jump}) and the height at which the parachute opens (H_{chute}), as well as of the acceleration due to gravity.

$$t_{chute} = \sqrt{\frac{2(H_{jump} - H_{chute})}{g}}$$

■ Cell F8 contains an equation to calculate the airplane's horizontal position. This can be calculated very simply from the airplane's speed and the elapsed time.

■ Cell F9 contains the airplane's elevation, which remains the same as the skydiver's jump height, that is, the airplane continues in level flight after the skydiver jumps.

■ Cell F12 contains an equation to calculate the skydiver's horizontal position. Since we are neglecting the effects of wind drag on the skydiver's horizontal travel, this can be calculated very simply from the skydiver's horizontal speed, which is the same as the airplane's speed, and the time elapsed since leaving the airplane.

- Cell F13 contains an equation to calculate the skydiver's elevation above the ground assuming that the parachute has not opened. This is calculated with the following free-fall trajectory formula.

$$H_{nochute} = H_{jump} - \frac{g\,t^2}{2}$$

- Cell F14 contains an equation to calculate the skydiver's elevation above the ground. This will equal the height calculated in F13 whenever that height is greater than or equal to the height to open the parachute, specified in B7. After the height in F13 falls below the height specified in B7, the height to plot is calculated with the following formula.

$$Y_{diver} = H_{chute} - v_{chute}(t - t_{chute})$$

- Cell F17 contains an equation to calculate the horizontal position of the parachute. This will equal the horizontal position of the skydiver if the diver's height is less than or equal to the height to open the parachute. Before the diver's height reaches the height to open the chute, the horizontal position of the parachute is set to its initial value typed into B16 (which is off the chart, set at 1,500). This requires that the axes scales be set to a maximum of 1,000 feet.
- Cell F18 contains an equation to calculate the elevation of the parachute. This will equal the elevation of the skydiver plus the height of the parachute above the skydiver after the parachute has been opened. Until the elevation to plot falls below the height at which the chute opens, the elevation of the parachute is set to its initial value typed into B16 (which is off the chart).
- The proper plot chart contains a legend and three data series. The legend is used as a text box. The three series are:
  - The skydiver position with x = F12 and y = F14. The marker option is set as a solid, gray triangle (size 12). The series name is set to cell B20. This will display in the Legend whatever text is in cell B20, which will be "Whee!" or "Down Safely" depending on whether the skydiver has reached the ground.
  - The parachute position with x = F17 and y = F18. The marker option is set to a hollow, black circle (size 20). The default series name, Series 2, should be deleted from the legend.
  - The airplane position with x = F8 and y = F9. The marker option is set to a large (size 40) gray horizontal dash. The default series name, Series 3, should be deleted from the legend.
  - Set your worksheet to calculate iteratively so that holding down the F9 button will cause the elapsed time to advance, causing the skydiver and airplane (and later the parachute) to move across the graph.

22. Rework Question 21, using slider bars instead of iteration.
23. We throw an egg horizontally from the top of a cliff. We want to watch as the egg drops and finally goes splat. The equations of motion for the egg (neglecting aerodynamic drag) are:

$$x = vt \quad \text{and} \quad y = y_0 - \frac{1}{2}g\,t^2$$

Here, y_0 is the initial height (height of the cliff), g is the acceleration due to gravity, t is the elapsed time, and v is the initial velocity (horizontal).

If we iterate on time, we can calculate the position of the egg at any time (x and y positions or coordinates). Create a worksheet like the one shown below. Fill in all cells except B10 and B16, which require conditional (IF) statements. Use data validation to limit the initial height to 100 feet.

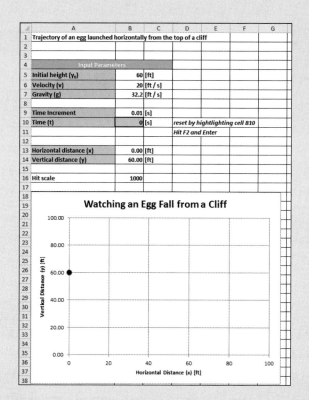

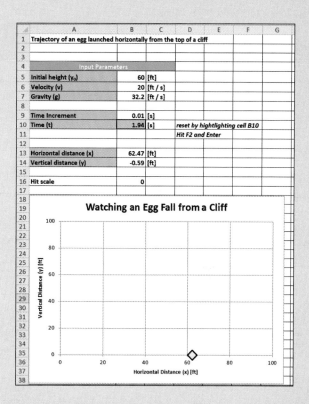

The program will start with just a single data point on the plot (maybe a circle point to represent the egg). Once the egg hits the ground, we want a large diamond for the yolk to show up. This will be a second point.

In cell B10, write an IF statement to show the sum of B9 + B10 until the egg hits the ground (i.e., the y value goes negative), and then it will use B10. This means that after the egg has reached the ground, the actual value of time will cease to change.

Construct a graph to plot the single x and y values for the egg. Fix both axis scales as 0–100 feet with a major unit of 20. This will keep the scale the same as the egg falls.

Now add one more data series to the graph where the egg hits the ground. It will have a y coordinate of zero and an x coordinate of the horizontal position when the egg hits (and time ceases to change). Write an IF statement in cell B16 to check whether the egg has hit. If it has not hit the ground yet, specify an x coordinate of 1,000, so that the point does not appear on the graph. If the egg has hit, specify the value in cell B13. Select a large diamond marker shape.

Now you should be able to run the program and watch the egg fall and go splat. Try different speeds and different cliff heights (initial height).

24. Rework Question 23, using slider bars instead of iteration.

CHAPTER 12
MODELS AND SYSTEMS

A **model** is an abstract description of the relationship between variables in a system. A model allows the categorization of different types of mathematical phenomena so that general observations about the variables can be made for use in any number of applications.

For example, if we know that $t = v + 5$ and $M = z + 5$, any observations we make about v with respect to t also apply to z with respect to M. A specific model describes a *system* or *function* that has the same *trend* or *behavior* as a generalized model. In engineering, many specific models within different subdisciplines behave according to the same generalized model.

This section covers three general models of importance to engineers: **linear, power,** and **exponential**. It is worth noting that many applications of models within these three categories contain identical math but apply to significantly different disciplines.

Linear models occur when the dependent variable changes in direct relationship to changes in the independent variable. We discuss such systems, including springs, resistive circuits, fluid flow, and elastic materials, in this chapter by relating each model to Newton's generalized law of motion.

Power law systems occur when the independent variable has an exponent not equal to 1 or 0. We discuss these models by addressing integer and rational real exponents.

Exponential models are used in all engineering disciplines in a variety of applications. We discuss these models by examining the similarities between growth and decay models.

The following is an example of the level of knowledge of Excel needed to proceed. *If you are not able to quickly recreate the following exercise in Excel, including trend-lines and formatting, please review trendline basics in appendix materials before proceeding.*

Energy (E) stored in an **inductor** is related to its inductance (L) and the current (I) passing through it by the following equation.

$$E = \frac{1}{2}LI^2$$

The SI unit of inductance, **henry** [H], is named for Joseph Henry (1797–1878), credited with the discovery of self-inductance of electromagnets.

Three inductors were tested and the results are given here. Create a proper plot of the data and add a properly formatted power law trendline to each data set.

Current (*I*) [A]	2	6	10	14	16
Energy of Inductor 1 (*E1*) [J]	0.002	0.016	0.050	0.095	0.125
Energy of Inductor 2 (*E2*) [J]	0.010	0.085	0.250	0.510	0.675
Energy of Inductor 3 (*E3*) [J]	0.005	0.045	0.125	0.250	0.310

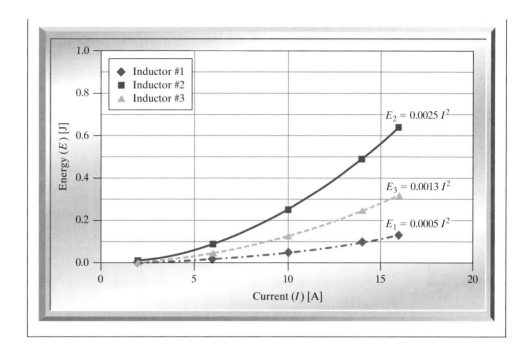

Figure 12-1 is an example of a properly formatted graph, showing an experimental data series with linear trendlines.

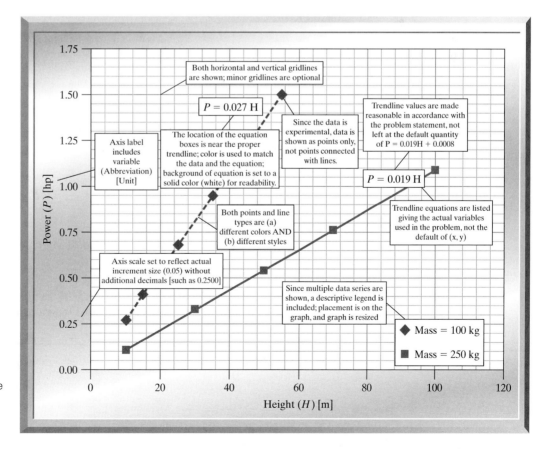

Figure 12-1

Example of a proper plot, showing multiple experimental data sets with linear trendlines.

12.1 LINEAR FUNCTIONS

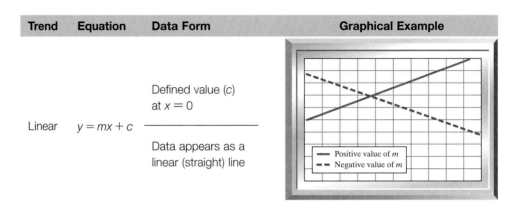

Trend	Equation	Data Form	Graphical Example
Linear	$y = mx + c$	Defined value (c) at $x = 0$ ——— Data appears as a linear (straight) line	

One of the most common models is **linear**, taking the form $y = mx + c$, where the ordinate value (y) is a function of the abscissa value (x) and a constant factor called the **slope** (m). At an initial value of the abscissa ($x = 0$), the ordinate value is equal to the **intercept** (c). Examples include

- Distance (d) traveled at constant velocity (v) over time (t) from initial position (d_0):

$$d = vt + d_0$$

- Rate of rotation (ω) as a function of time (t) and angular acceleration (α) from initial rotational rate (ω_i):

$$\omega = \alpha t + \omega_i$$

- Total pressure (P_{total}), relating density (ρ), gravity (g), liquid height (H), and the pressure above the surface ($P_{surface}$):

$$P_{total} = \rho g H + P_{surface}$$

- Newton's second law, relating force (F), mass (m), and acceleration (a):

$$F = ma$$

Note that the intercept value (c) is zero in the last example.

General Model Rules

Given a linear system of the form $y = mx + c$ and assuming $x \geq 0$:

- When $m = 1$, the function is equal to $x + c$.
- When $m = 0$, $y = c$, regardless of the value of x (y never changes).
- When $m > 0$, as x increases, y increases, regardless of the value of c.
- When $m < 0$, as x increases, y decreases, regardless of the value of c.

● **EXAMPLE 12-1**

We want to determine the effect of depth of a fluid on the total pressure felt by a submerged object. Recall that the total pressure is

$$P_{total} = P_{surface} + P_{hydro} = P_{surface} + \rho g H$$

where P_{total} = total pressure [atm]; $P_{surface}$ = pressure at the surface [atm]; ρ = density [kg/m^3]; g = gravity [m/s^2]; H = depth [m]. We enter the lab, take data, and create the following chart.

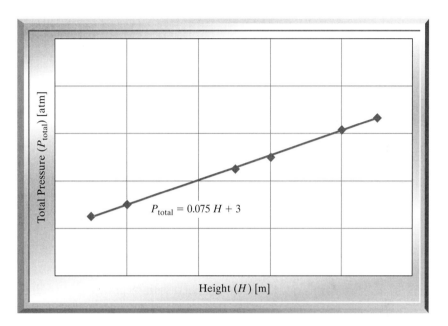

Determine the density of the fluid, in units of kilograms per cubic meter.

*We can determine the parameters by matching the trendline generated in Excel with the theoretical expression. In theory: total pressure = density * gravity * height of fluid + pressure on top of the fluid*
*From graph: total pressure = 0.075 * height + 3*
*By comparison: density * gravity = 0.075 [atm/m]*

$$\frac{0.075 \text{ atm}}{\text{m}} \left| \frac{101{,}325 \text{ Pa}}{1 \text{ atm}} \right| \frac{1 \text{ kg/(ms}^2)}{1 \text{ Pa}} = \rho(9.8 \text{ m/s}^2)$$

$$7{,}600 \text{ kg/(m}^2\text{s}^2) = \rho(9.8 \text{ m/s}^2)$$

$$\rho = 7{,}600 \text{ kg/(m}^2\text{s}^2) \left| \frac{\text{s}^2}{9.8 \text{ m}} \right. = 775 \text{ kg/m}^3$$

Determine if the tank is open to the atmosphere or pressurized, and determine the pressure on the top of the fluid in units of atmospheres.

*Once again, we can compare the Excel trendline to the theoretical expression. In theory: total pressure = density * gravity * height of fluid + pressure on top of the fluid*
*From graph: total pressure = 0.075 * height + 3*
By comparison, the top of the tank is pressurized at 3 atm.

Increasingly, engineers are working at smaller and smaller scales. Tiny beads made of glass are on the order of 50 micrometers in diameter. They are manufactured so that they become hollow, allowing the wall thickness to be a few nanometers. The compositions of the glass were engineered, so when processed correctly, they would sustain a hollow structure and the glass walls would be infiltrated with hundreds of thousands of nanometer-sized pores. These beads can possibly revolutionize the way fluids and gases are stored for use. The pores are small enough that fluids and even gases could be contained under normal conditions. However, if activated properly, the pores would allow a path for a gas to exit the "container" when it is ready to be used.

S4800 5.0kV 10.2mm x2.20k SE(M)   20.0um

Photo courtesy of K. Richardson

COMPREHENSION CHECK 12-1

The graph shows the ideal gas law relationship ($PV = nRT$) between pressure (P) and temperature (T).

(a) What are the units of the slope (0.0087)?

(b) If the tank has a volume of 12 liters and is filled with nitrogen (formula, N_2; molecular weight, 28 grams per mole), what is the amount of gas in the tank (n) in units of grams?

(c) If the tank is filled with 48 grams of oxygen (formula, O_2; molecular weight, 32 grams per mole), what is the volume of the tank (V) in units of liters?

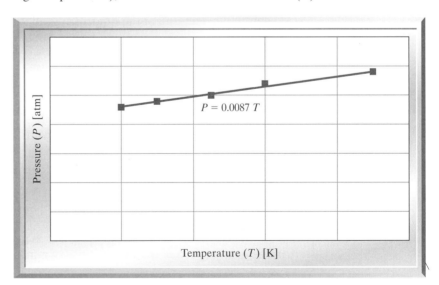

$P = 0.0087\ T$

Pressure (P) [atm]

Temperature (T) [K]

12.2 LINEAR RELATIONSHIPS

Most physics textbooks begin the study of motion ignoring how that object came to be moving in the first place, since that information is not needed to figure out what will happen to the object in the future. This is appropriate to the way physicists study the world, by observing the world as it is. Engineering is about changing the way things are, such as making things move that are stationary and stopping things that are moving. As a result, engineers are concerned with forces and the changes those forces cause. While physicists study how far a car travels through the air when hit by a truck, engineers focus on stopping the truck before it hits the car or on designing the car so it does not fly as far or on designing an air-bag system or crush-proof doors. Engineering has many diverse branches because of the many different kinds of forces and ways to apply them.

Another Way of Looking at Newton's First Law

NOTE

A system keeps doing what it is doing unless the forces acting on the system change.

Newton's first law is given as "An object at rest remains at rest and an object in motion will continue in motion with a constant velocity unless it experiences a net external force." As we consider variables other than motion, we want to expand this definition: A system keeps doing what it is doing unless the forces acting on the system change.

Another Way of Looking at Newton's Second Law

NOTE

When a force influences a change to a system parameter, the system opposes the change according to its internal resistance.

Newton's second law is given as "The acceleration of an object is directly proportional to the net force acting on it and inversely proportional to its mass." This is summarized by the familiar expression $F = ma$, which can be interpreted as follows: When an external force acts on a system to cause acceleration, the system resists that acceleration according to its mass. Mass, therefore, becomes a resistance to acceleration. Expanding Newton's second law to generalize it for use with variables other than motion: When a force influences a change to a system parameter, the system opposes the change according to its internal resistance. In generalizing these relationships, we can start to establish a pattern observed in a wide variety of phenomena, summarized in Table 12-1.

Table 12-1 Generalized Newton's second law

When a "system"...	... is acted upon by a "force"...	... to change a "parameter" ...	... the "system" opposes the change by a "resistance"	Equation
Physical object	External push or pull (F)	Acceleration (a)	Object mass (m)	$F = ma$

Springs

When an external force (F), such as a weight, is applied to a spring, it will cause the spring to stretch a distance (x), according to the following expression:

$$F = kx$$

This equation is called **Hooke's law**, named for Robert Hooke (1635–1703), an English scientist. Among other things, he is credited with creating the biological term "cell." The comparison of Hooke's Law and Newton's Second Law is shown in Table 12-2.

Table 12-2 Generalized second law . . . applied to springs

When a "system" . . .	. . . is acted upon by a "force" . . .	. . . to change a "parameter". . .	. . . the "system" opposes the change by a "resistance"	Equation
Physical object	External push or pull (F)	Acceleration (a)	Object mass (m)	$F = ma$
Spring	External push or pull (F)	Elongation (x)	Spring stiffness (k)	$F = kx$

Property	Symbol	Units
Spring Constant	k	N/m

The variable k is the **spring constant**, a measure of the stiffness of the spring. Stiff springs are hard to stretch and have high k values; soft springs are easy to stretch and have low k values. The constant k is a material property of the spring, determined by how it is made and what material it is made from.

● **EXAMPLE 12-2**   Two springs were tested; a weight was hung on one end and the resulting displacement measured. The results were graphed. Using the graph shown on the following page, give the spring constant of each spring and determine which spring is stiffer.

Spring 1 has a linear trendline of F = 66x. The slope of the line is the spring constant:

$$k_1 = 66 \, N/m$$

Spring 2 has a linear trendline of F = 8x, which corresponds to:

$$k_2 = 8 \, N/m$$

Spring 1 is stiffer since it has a higher spring constant.

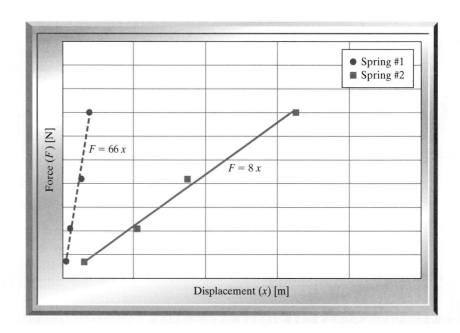

Basics of Electric Circuits

Electric charge (Q) is a property of some subatomic particles, notably the electron and the proton. On a small scale, charge can be measured in terms of the **elementary charge** (e). The magnitude of the elementary charge on the electron or proton equals 1, and by convention, the electron has a negative charge, $e = -1$; the proton has a positive charge $e = +1$.

On larger scales, the elementary charge is far too small to be useful for most practical problems, so a new unit was defined in terms of two base SI units, the ampere and the second. The new unit of charge was called the **coulomb** [C], named for the French physicist Charles-Augustin de Coulomb (1736–1806), who first described and quantified the attractive and repulsive electrostatic force, today known as Coulomb's law. In terms of elementary charges, one coulomb is approximately $6.24 \times 10^{18}e$. Thus, 6.24 quintillion electrons is one coulomb of charge.

Electric current (I) is a measure of how many charges (normally electrons) flow through a wire or component in a given amount of time. This is analogous to measuring water flowing through a pipe as amount per time, whether the units are tons per hour, gallons per minute, or molecules per second.

The unit of electric current is the **ampere** [A], one of the seven base SI units. It is named for Andre-Marie Ampere (1775–1836), a French physicist who is credited with discovering electromagnetism. How the ampere is defined as one of the base units is beyond the scope of this book, but can be found in basic physics books. Of more practical use to the engineer is a definition in terms of charges moving through a conductor per unit time. From this perspective, the ampere is one coulomb per second, or approximately 6.24×10^{18} elementary charges per second. Thus, if 6.24 quintillion electrons move through a wire per second, the wire is carrying one ampere of current.

Voltage (V) is the "force" that pushes the electrons around. Although its effects on charged particles are similar to those of a true force, voltage is quite different dimensionally. The unit of voltage is the **volt** [V], described as the potential difference (voltage) across a conductor when a current of one ampere dissipates one watt of power. (Yes, that probably sounds like gibberish to most readers. We make this a bit more intuitive below.) The volt is named for Italian physicist Alessandro Volta (1745–1827), who perhaps invented the first chemical battery, called the voltaic pile.

Resistance (R) is a measure of how difficult it is to push electrons through a substance or device. When a voltage is applied to a circuit, a current is generated. This current depends on the equivalent resistance of the circuit. Resistance has units of volts per ampere, which is given the special name **ohm** [Ω]. It is named for Georg Ohm (1789–1854), the German physicist who developed the theory, called **Ohm's Law**, to explain the relationship between voltage, current, and resistance. The similarities between Ohm's Law and Newton's Second Law are given in Table 12-3.

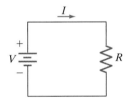

$$V = IR$$

Table 12-3 Generalized second law . . . applied to circuits

When a "system" . . .	. . . is acted upon by a "force" . . .	. . . to change a "parameter" . . .	. . . the "system" opposes the change by a "resistance"	Equation
Physical object	External push or pull (F)	Acceleration (a)	Object mass (m)	$F = ma$
Electrical circuit	Circuit voltage (V) (electromotive force)	Circuit current (I)	Circuit resistance (R)	$V = IR$

NOTE

A volt times an ampere
is a watt.

$V = J/C$

$A = C/s$

$V A = J/s = W$

Power in an electric circuit is simply the product of voltage and current:

$$P = VI$$

If one ampere is moving through a potential difference of one volt, the power is one watt. Some electrical devices can use power (converting the energy to heat, light, sound, motion, etc.) and some can provide power (e.g., a battery). Some devices, such as a rechargeable battery can do both; when recharging, it is using or absorbing energy (converting it into chemical energy); when powering a device, it is providing or delivering energy.

We all learned in grade school that "opposite charges attract and like charges repel." This well-known phenomenon underlies the idea of voltage. Imagine an accumulation of electrons (negative charges), all jammed together in a small space and somehow made immovable. Then imagine a similar group of protons (positive charges) a short distance away. Now imagine that a single electron is placed between these two opposite charge accumulations. What happens to the single electron when it is released? It is repelled by the negative charges and attracted by the positive charges; thus, it accelerates toward the positive charges. Recall Newton's second law: a force is required to accelerate a mass ($F = ma$), even one as tiny as a single electron. Somehow, the accumulated charge is causing a force to be applied to this single electron.

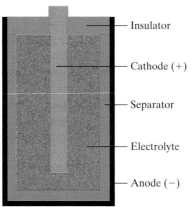

Now imagine a device that could somehow pull electrons out of one piece of material and deposit them on a different piece of material. The material losing the electrons is left with a positive charge since it has more protons than electrons. This is called the cathode. Similarly, the piece of material receiving electrons has an overall negative charge and is called the anode. Now, a charge separation exists, just as in the hypothetical case with the accumulations of charges. This, however, describes a real device, known as a battery, or, more properly, a single Galvanic cell. One can buy these almost anywhere, the supermarket, the drug store, etc., and they power all sorts of devices from singing greeting cards and flashlights to laptop computers and electric automobiles.

In the Galvanic cell, a chemical reaction is responsible for relocating the electrons. When the cell is first constructed, the chemical reaction starts moving the electrons from cathode to anode, but as this charge separation increases, the accumulated charges start trying to make the electrons being moved go the other way (they "push back" against the chemical reaction). Very quickly, a stalemate is achieved, with the accumulated charge trying to move the electrons backward just as hard as the chemical reaction is trying to move them forward. In this condition, the battery can sit in its package on the shelf for months or years without going bad. The charge separation and the relative "push" being exerted by that accumulation against the chemical reaction can be quantified in terms of voltage.

But we still have not really quantified the volt. The question to ask is "how much energy was needed to separate the charges?" If one joule of energy was necessary to separate one coulomb of charge, then that charge separation is quantified as a potential difference of one volt; thus, 1 volt = 1 joule/coulomb: [$V = J/C$]. Thus, in a standard alkaline D cell, with a voltage of 1.5 volts, for each coulomb of charge that is moved from cathode to anode, the chemical reaction must provide 1.5 joules of energy. Similarly, if such systems were 100% efficient, one joule of energy could be extracted from one coulomb of charge separated by a potential difference of one volt.

Labels for figure: Insulator; Cathode (+); Separator; Electrolyte; Anode (−)

Simplified Galvanic Cell

NOTE

Technically, a battery is a group of single cells connected in series to yield a higher voltage. The common AA battery is actually a single cell, whereas the 9-volt rectangular battery is really six single cells in series.

Materials through which electrons can move easily (e.g., most metals) have a low resistance and are called conductors. Materials through which electrons move only with great difficulty have a high resistance and are called insulators (e.g., glass and many other ceramics). Commercially available resistors are available in standard values from less than 1 Ω to over 10 MΩ.

Conductance is the reciprocal of resistance ($G = 1/R$), and in certain types of problems is much easier to use than resistance. The old unit for conductance was the mho (ohm spelled backward), but it has been supplanted in more recent times by the **siemens** [S], named after Ernst Werner von Siemens, the German inventor who, among other things, built the first electric elevator and founded the company known today as Siemens AG.

Capacitance (C) is a measure of the ability of a device called a capacitor to store an electric charge. Capacitance is measured in **farads** [F], named for the English scientist Michael Faraday (1791–1867). Among Faraday's numerous contributions to both electromagnetism and chemistry was the development of the principles for the first practical electric motor. We will define the farad after we explain how a capacitor functions.

In its simplest form, a capacitor is made from two flat conducting plates separated by an insulator. When a current enters one plate of the capacitor, the electrons making up the current begin to accumulate on that plate. Since the two plates are separated by an insulator, the electrons cannot cross over to the other plate. With accumulating negative charge on one plate, the electrons in the other plate are repelled, leaving behind an overall positive charge. This superficially gives the appearance that the current is going through the capacitor. However, the electrons entering one plate and leaving the other are different electrons, and a charge separation is accumulated on the plates of the capacitor.

Note that the larger the accumulated charge, the greater the voltage. The voltage developed across the plates of a capacitor for a given amount of charge depends primarily on the total area of the plates and the distance between them. If one coulomb of charge is stored on the plates of a capacitor, and the resulting voltage is one volt, then by definition, the capacitance is one farad: a one-farad capacitor requires one coulomb of stored charge to have a potential difference across the plates of one volt [F = C/V].

Capacitors are used in literally hundreds of diverse applications, a few being audio filters, timing circuits, power supplies, and AC motors (efficiency improvement). Commercially available capacitors range from about one picofarad to over one farad. Since the farad is far too large to conveniently describe the majority of practical devices, **microfarads** [μF] are much more commonly used, although picofards or nanofarads, and, rarely, millifarads are also used.

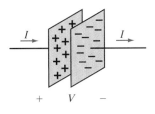

NOTE

In the figure, the current (*I*) is *conventional current*, which assumes the charges moving are positive. In reality, electrons are moving in the opposite direction, but the result is identical. Essentially all circuit analysis is done with conventional current, which is why we use it here instead of electron current. A more complete explanation is beyond the scope of this text.

Table 12-4 Summary of electrical properties

Property	Symbol	Typical Units	Equivalent Units
Charge	Q	coulomb [C]	C = A s
Electric current	I	ampere [A]	A = C/s
Voltage	V	volt [V]	V = J/C
Electric power	P	watt [W]	W = J/s
Resistance	R	ohm [Ω]	Ω = V/A
Conductance	G	siemens [S]	S = A/V
Capacitance	C	microfarad [μF]	F = C/V

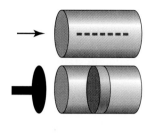

Fluid Flow

For motion in a fluid to be created, a force must be applied over an area of the fluid. While both liquids and gases can be defined as fluids, we focus on liquids in this section. Imagine a section of fluid-filled pipe placed on the desk in front of you. While in reality the liquid would flow freely from both sides of the pipe, for this example, imagine that it stays in place. If we apply a force at a single point in the fluid, only the particles at that point will move. To move the entire fluid uniformly, we must apply the force at all points at the pipe entrance simultaneously. Applying a force over the cross-sectional area of the pipe results in the application of a pressure to the fluid. The pressure that results in fluid flow has a special name: **shear stress** (τ, Greek letter tau).

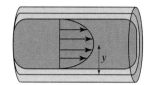

As the fluid moves, we find that the fluid molecules in contact with the wall adhere to the wall and do not move. The motion of the fluid can be visualized as occurring in layers; as the distance from the wall increases, the fluid moves faster. The fluid moves fastest at the farthest point from the wall, which is the center of the pipe. Since the velocity changes depend on the location in the pipe from the wall, the parameter we are changing cannot be expressed as a simple velocity, but rather as a **velocity gradient**, given as ($\Delta v/\Delta y$ or $\dot\gamma$). This is sometimes called the *shear rate* or *strain rate*.

Not all fluids respond equally to an applied pressure. The fluid property that represents the resistance of a fluid against flow is called the **dynamic viscosity** (μ, Greek letter mu). The relationship between shear stress and the velocity profile of a fluid is called **Newton's law of viscosity**, named after Isaac Newton. Fluids that behave in this way are called **Newtonian fluids** (e.g., water and oil). The comparison between Newton's Law of Viscosity and Newton's Second Law is given in Table 12-5.

$$\tau = \mu \frac{\Delta v}{\Delta y}$$

Table 12-5 Generalized second law . . . applied to fluid flow

When a "system" . . .	. . . is acted upon by a "force" . . .	. . . to change a "parameter" . . .	. . . the "system" opposes the change by a "resistance"	Equation
Physical object	External push or pull (F)	Acceleration (a)	Object mass (m)	$F = ma$
Fluid	Shear stress (τ)	Shear rate ($\Delta v/\Delta y$)	Dynamic viscosity (μ)	$\tau = \mu \dfrac{\Delta v}{\Delta y}$

Sometimes, a fluid must have a certain amount of stress (called the **yield stress**, τ_0) applied before it will begin to move like a Newtonian fluid. These fluids are called **Bingham plastics**, named after Eugene Bingham, a chemist who made many contributions to the field of **rheology** (the science of deformation and flow of matter, a term he, along with Markus Reiner, is credited in creating). Examples of Bingham plastics include toothpaste and slurries.

$$\tau = \mu \frac{\Delta v}{\Delta y} + \tau_0$$

Common units of dynamic viscosity are **centipoise** [cP], named after the French physician Jean Louis Poiseuille (1799–1869) who studied the flow of blood in tubes. Dynamic viscosity is a function of temperature. In most instances, viscosity decreases with increasing temperature; as the fluid heats up, it becomes easier to move.

Property	Symbol	Typical Units	Equivalent Units
Dynamic viscosity	μ	cP	P = g/(cm s)
Kinematic viscosity	ν	St	St = cm^2/s

Another useful term in describing a fluid is **kinematic viscosity** (ν, Greek letter nu). The kinematic viscosity is the ratio of dynamic viscosity to density and is given the unit of **stokes** [St], named after George Stokes (1819–1903), the Irish mathematician and physicist who made important contributions to science, including Stokes' law, optics, and physics. Several values of dynamic and kinematic viscosity are given in Table 12-6.

$$\nu = \frac{\mu}{\rho}$$

Table 12-6 Summary of material properaties for several liquids

Liquid	Specific Gravity	Dynamic Viscosity (μ) [cP]	Kinematic Viscosity (ν) [cSt]
Acetone	0.791	0.331	0.419
Benzene	0.879	0.647	0.736
Corn syrup	1.36	1,380	1,015
Ethanol	0.789	1.194	1.513
Glycerin	1.260	1,490	1,183
Honey	1.36	5,000	3,676
Mercury	13.600	1.547	0.114
Molasses	1.400	8,000	5,714
Olive oil	0.703	101	143
SAE 30W oil	0.891	290	325
Toluene	0.865	0.590	0.682
Water	1.000	1.000	1.000

COMPREHENSION CHECK 12-2

Fluid A has a dynamic viscosity of 0.5 centipoise and a specific gravity of 1.1. What is the density of Fluid A in units of pound-mass per cubic foot?

COMPREHENSION CHECK 12-3

Fluid A has a dynamic viscosity of 0.5 centipoise and a specific gravity of 1.1. What is the dynamic viscosity of Fluid A in units of pound-mass per foot/second?

COMPREHENSION CHECK 12-4

Fluid A has a dynamic viscosity of 0.5 centipoise and a specific gravity of 1.1. What is the kinematic viscosity of Fluid A in units of stokes?

Elastic Materials

Elasticity is the property of an object or material that causes it to be restored to its original shape after distortion. A rubber band is easy to stretch and snaps back to near its original length when released, but it is not as elastic as a piece of piano wire. The piano wire is harder to stretch, but would be said to be more elastic than the rubber band because of the precision of its return to its original length. The term elasticity (**Young's modulus** or **modulus of elasticity**, E) is described as the amount of deformation resulting from an applied force. Young's modulus is named for Thomas Young (1773–1829), a British scientist, who contributed to several fields: material elongation theory; optics, with his "double slit" optical experiment that led to the deduction that light travels in waves; and fluids, with the theory of surface tension and capillary action.

Property	Symbol	Units
Young's modulus	E	Pa

Like fluids, elastic materials accept a force applied over a unit area rather than a point force. **Stress** (σ, Greek letter sigma) is the amount of force applied over a unit area of the material, which has units of pressure [Pa]. The **strain** (ε, Greek letter epsilon) is the ratio of the elongation to the original length, yielding a dimensionless number. Since the modulus values tend to be large, Young's modulus is usually expressed in units of Gigapascals [GPa]. A comparison between Young's Modulus and Newton's Second Law is shown in Table 12-7.

$$\sigma = E\varepsilon$$

Table 12-7 Generalized second law . . . applied to elastic materials

When a "system" . . .	. . . is acted upon by a "force" . . .	. . . to change a "parameter" . . .	. . . the "system" opposes the change by a "resistance"	Equation
Physical object	External push or pull (F)	Acceleration (a)	Object mass (m)	$F = ma$
Elastic object	Stress (σ)	Strain (ε)	Young's modulus (E)	$\sigma = E\varepsilon$

From this discussion, you can see examples from many areas of engineering that are similar to Newton's second law. We often want to change something and find that it resists this change; this relationship is often linear. In all of these situations, we discover a coefficient that depends on the material encountered in the particular situation (mass, spring stiffness, circuit resistance, fluid viscosity, or Young's modulus).

Many other examples are not discussed here, such as Fourier's law of heat transfer, Fick's law of diffusion, and Darcy's law of permeability. You can enhance your understanding of your coursework by attempting to generalize the knowledge presented in a single theory to other theories that may be presented in other courses. Many different disciplines of engineering are linked by common themes, and the more you can connect these theories across disciplines, the more meaningful your classes will become.

Combinations of Springs and Circuits

When connected, both springs and circuits form a resulting system that behaves like a single spring or single resistor. In a combination of springs, the system stiffness depends on the stiffness of each individual spring and on the configuration, referred to as the effective spring constant (k_{eff}). In a network of circuits, the system resistance depends on the value of the individual resistors and on the configuration, referred to as the effective resistance (R_{eff}).

Springs in Parallel

When springs are attached in *parallel*, they must *displace the same distance* even though they may have different spring constants. The derivation below shows how this leads to an effective spring constant, that is, the sum of the individual spring constants in the system. Each spring is responsible for supporting a proportional amount of the force.

Writing Hooke's law for two springs each displacing the same distance (x):

$$F_1 = k_1 x \tag{a}$$
$$F_2 = k_2 x \tag{b}$$

NOTE

Springs in parallel both displace the same distance.

Solve for F_1 in terms of F_2 since the displacement is the same:

$$F_1 = k_1 \frac{F_2}{k_2} = F_2 \frac{k_1}{k_2} \tag{c}$$

Writing Hooke's law as applied to the overall system:

$$F = k_{eff} x \tag{d}$$

The total force applied to the configuration (F) is the sum of the force supported by each spring:

$$F = F_1 + F_2 \tag{e}$$

Eliminating force (F) from Equation (e) with Equation (d):

$$k_{eff} x = F_1 + F_2 \tag{f}$$

Eliminating displacement (x) with Equation (b):

$$k_{eff} \frac{F_2}{k_2} = F_1 + F_2 \tag{g}$$

Substituting for F_1 with Equation (c):

$$k_{eff} \frac{F_2}{k_2} = F_2 \frac{k_1}{k_2} + F_2 \tag{h}$$

Dividing Equation (h) by F_2:

NOTE

A system of two springs in parallel will always be stiffer than either spring individually.

$$\frac{k_{eff}}{k_2} = \frac{k_1}{k_2} + 1 \tag{i}$$

Multiplying Equation (i) by k_2 gives:

$$k_{eff} = k_1 + k_2 \tag{j}$$

Springs in Series

When two springs are attached in *series*, the *force is the same for both springs*. The effective spring constant is derived below. The applied force affects each spring as though the other spring did not exist, and each spring can stretch a different amount.

Writing Hooke's law for two springs each under the same applied force (F):

$$F = k_1 x_1 \tag{k}$$

$$F = k_2 x_2 \tag{l}$$

Solve for x_1 in terms of x_2 since the force is the same:

$$x_1 = \frac{k_2}{k_1} x_2 \tag{m}$$

Writing Hooke's law as applied to the overall system:

$$F = k_{\text{eff}} x \tag{n}$$

NOTE

Springs in series are acted on by the same force for both springs.

The total distance stretched by the configuration (x) is the sum of the distance stretched by each spring:

$$x = x_1 + x_2 \tag{o}$$

Eliminating force (F) from Equation (n) with Equation (l):

$$k_2 x_2 = k_{\text{eff}} x \tag{p}$$

Eliminating displacement (x) with Equation (o):

$$k_2 x_2 = k_{\text{eff}} (x_1 + x_2) \tag{q}$$

Substituting for x_1 with Equation (m):

$$k_2 x_2 = k_{\text{eff}} \left(\frac{k_2}{k_1} x_2 + x_2 \right) \tag{r}$$

Dividing Equation (r) by x_2

$$k_2 = k_{\text{eff}} \left(\frac{k_2}{k_1} + 1 \right) \tag{s}$$

Dividing Equation (s) by k_2 gives

$$1 = k_{\text{eff}} \left(\frac{1}{k_1} + \frac{1}{k_2} \right) \tag{t}$$

NOTE

A system of two springs in series will always be less stiff than either spring individually.

Thus,

$$k_{\text{eff}} = \frac{1}{\left(\frac{1}{k_1} + \frac{1}{k_2} \right)} = \left(\frac{1}{k_1} + \frac{1}{k_2} \right)^{-1} \tag{u}$$

These forms for springs in parallel and series generalize to any number of springs. For N springs in parallel, the effective spring constant is

$$k_{\text{eff}} = k_1 + k_2 + \cdots + k_{N-1} + k_N \tag{v}$$

For N springs in series, the effective spring constant is

$$k_{\text{eff}} = \left(\frac{1}{k_1} + \frac{1}{k_2} + \cdots + \frac{1}{k_{N-1}} + \frac{1}{k_N} \right)^{-1} \tag{w}$$

When Are Components Connected in Series, Parallel, or Neither?

Note that in each diagram, the lines with one end loose indicate where the circuit or spring configuration is connected to other things.

Series

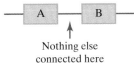

Nothing else connected here

When one end (but not both) of each of two components is connected together with NOTHING ELSE CONNECTED AT THAT POINT, they are in series.

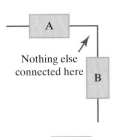

Nothing else connected here

Note that they do not necessarily have to be in a straight line as shown. Electrical components can be physically mounted in any position relative to one another, and as long as a wire connects one end of each together (with nothing else connected there), they would be in series. Two springs can be connected by a string, so that the string makes a right angle direction change over a pulley, and the two springs would be in series.

Parallel

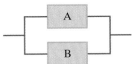

When each end of one component is connected to each of the two ends of another component, they are in parallel.

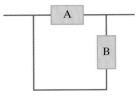

Similar to the series connection, the components do not have to be physically mounted parallel to each other or side by side, as long both ends are connected directly together with no intervening components. This is simple to do with electrical components since wire can be easily connected between any two points. Can you determine a method to physically connect two springs in parallel so that one is vertical and the other horizontal?

Sample Combinations

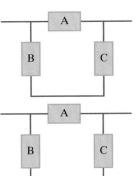

In the figure at left, B and C are in series. A is neither in series nor parallel with B or C since the lines extending to the left and right indicate connection to other stuff. A is, however, in parallel with the series COMBINATION of B and C.

In the figure at left, no components are in series or parallel with anything since the lines extending to the left and right indicate connection to other stuff. Note the extra line at lower right.

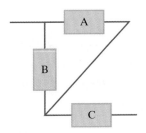

In the figure at left, A and B are in parallel. C is neither in series nor parallel with A or B. C is, however, in series with the parallel combination of A and B.

12.3 POWER FUNCTIONS

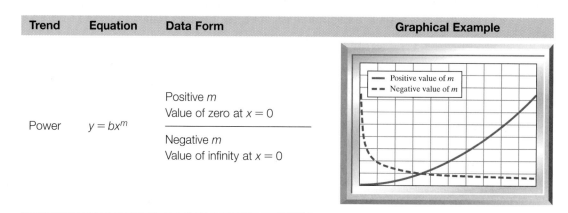

Trend	Equation	Data Form	Graphical Example
Power	$y = bx^m$	Positive m Value of zero at $x = 0$ ――― Negative m Value of infinity at $x = 0$	

Power models take the form $y = bx^m$. Examples include

■ Many geometric formulae involving areas, volumes, etc., such as the volume of a sphere (V) as a function of radius (r):

$$V = 4/3\pi r^3$$

■ Distance (d) traveled by a body undergoing constant acceleration (a) over time (t), starting from rest:

$$d = at^2$$

■ Energy calculations in a variety of contexts, both mechanical and electrical, such as the kinetic energy (KE) of an object as a function of the object's velocity (v), where the constant (k) depends upon the object shape and type of motion:

$$KE = kmv^2$$

■ Ideal gas law relationships, such as Boyle's law, relating volume (V) and pressure (P) of an ideal gas, holding temperature (T) and quantity of gas (n) constant:

$$V = (nRT)P^{-1}$$

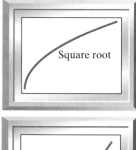

General Model Rules

Given a power system of the form $y = bx^m + c$, assuming $x \geq 0$:

■ When $m = 1$, the model is a linear function.
■ When $m = 0$, $y = b + c$, regardless of the value of x.
■ When m is rational, the function will contain a rational exponent or may be described with a radical symbol ($\sqrt{\ \ }$). Certain rational exponents have special names ($1/2$ is "square root," $1/3$ is "cube root").
■ When m is an integer, the function will contain an integer exponent on the independent variable. Certain exponents have special names (2 is "squared," 3 is "cubed").
■ When $0 < |m| < 1$ and $x < 0$, the function may contain complex values. In this chapter, we will only consider power law models, where c is zero. In the next chapter we will discuss ways of dealing with data when the value of c is non-zero.

● **EXAMPLE 12-5**

The volume (V) of a cone is calculated in terms of the radius (r) and height (H) of the cone. The relationship is described by the following equation:

$$V = \frac{\pi r^2 H}{3}$$

Given a height of 10 centimeters, calculate the volume of the cone when the radius is 3 centimeters.

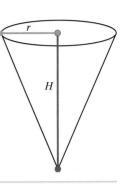

$$V = \frac{\pi (3\ cm)^2 (10\ cm)}{3} \approx 94.2\ cm^3$$

What is the volume of the cone when the radius is 8 centimeters?

$$V = \frac{\pi (8\ cm)^2 (10\ cm)}{3} \approx 670\ cm^3$$

NOTE

With a positive integer exponent, the volume of the cone increases as the independent variable (radius) increases. This observation is true with any power model with a positive integer exponent.

● **EXAMPLE 12-6**

The resistance (R [g/(cm^{4}s)]) of blood flow in an artery or vein depends upon the radius (r [cm]), as described by **Poiseuille's equation**:

$$R = \frac{8\mu L}{\pi} r^{-4}$$

The viscosity of blood (μ [g/(cm s)]) and length of the artery or vein (L [cm]) are constants in the system. In studying the effects of a cholesterol-lowering drug, you mimic the constricting of an artery being clogged with cholesterol. You use the data you collect to create the following graph.

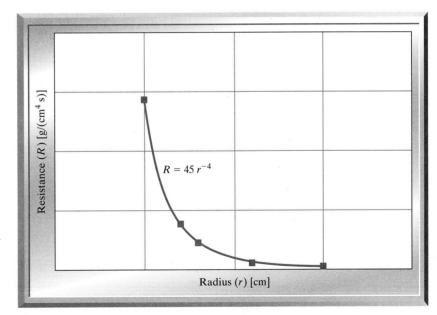

NOTE

With a negative integer exponent, the dependent variable (resistance) decreases as the independent variable (radius) increases. This trend is true for any power model with a negative integer exponent.

If the length of the artificial artery tested was 505 centimeters, what is the viscosity of the sample used to mimic blood, in units of grams per centimeter second [g/(cm s)]?

The constant "45" has physical meaning, found by comparison to the theoretical expression. In theory: $R = \dfrac{8\mu L}{\pi} r^{-4}$ and from graph: $R = 45r^{-4}$

By comparison:

$$45 \frac{g}{s} = \frac{8\mu L}{\pi} = \frac{8\mu (505 \text{ cm})}{\pi}$$

$$\mu = 0.035 \text{ g/(cm s)}$$

COMPREHENSION CHECK 12-7

The graph shows the ideal gas law relationship ($PV = nRT$) between pressure (P) and volume (V). If the tank is at a temperature of 300 kelvin and is filled with nitrogen (formula, N_2; molecular weight, 28 grams per mole), what is the amount of gas in the tank (n) in units of grams?

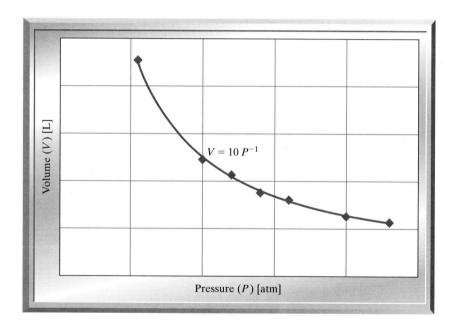

$V = 10\,P^{-1}$

Volume (V) [L]

Pressure (P) [atm]

COMPREHENSION CHECK 12-8

The graph above shows the ideal gas law relationship ($PV = nRT$) between pressure (P) and volume (V). If the tank is filled with 10 grams of oxygen (formula, O_2; molecular weight, 32 grams per mole), what is the temperature of the tank in units of degrees Celsius?

12.4 EXPONENTIAL FUNCTIONS

Trend	Equation	Data Form	Graphical Example
Exponential	$y = be^{mx}$	Defined value b ($b \neq 0$) at $x = 0$ Positive m: asymptotic to 0 for large negative values of x Negative m: asymptotic to 0 at large positive values of x	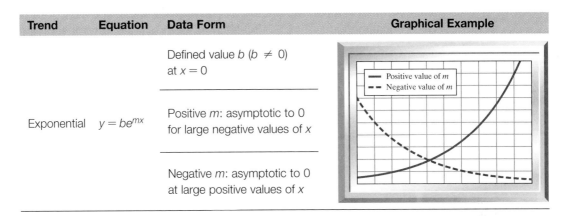

Exponential models take the form $y = be^{mx} + c$. Examples include

- The voltage (V) across a capacitor (C) as a function of time (t), with initial voltage (V_0) discharging its stored charge through resistance (R):

$$V = V_0 e^{-t/(RC)}$$

- The number (N) of people infected with a virus such as smallpox or H1N1 flu as a function of time (t), given the following: an initial number of infected individuals (N_0), no artificial immunization available and dependence on contact conditions between species (C):

$$N = N_0 e^{Ct}$$

- The transmissivity (T) of light through a gas as a function of path length (L), given an absorption cross-section (s) and density of absorbers (N):

$$T = e^{-sNL}$$

- The growth of bacteria (C) as a function of time (t), given an initial concentration of bacteria (C_0) and depending on growth conditions (g):

$$C = C_0 e^{gt}$$

Note that all exponents must be dimensionless, and thus unitless. For example, in the first equation, the quantity RC must have units of time.
Note that the intercept value (c) is zero in all of the above examples.

General Model Rules

Given an exponential system of the form $y = be^{mx} + c$:

- When $m = 0$, $y = b + c$ regardless of the value of x.
- When $m > 0$, the model is a **growth function**. The minimum value of the growth model for $x \geq 0$ is $b + c$. As x approaches infinity, y approaches infinity.
- When $m < 0$, the model is a **decay function**. The value of the decay model approaches c as x approaches infinity. When $x = 0$, $y = b + c$.

What Is "e"?

The **exponential constant** "e" is a transcendental number, thus also an irrational number, that can be rounded to 2.71828. It is defined as the base of the natural logarithm function. Sometimes, e is referred to as **Euler's number** or the **Napier constant**. The reference to Euler comes from the Swiss mathematician Leonhard Euler (pronounced "oiler," 1707–1783), who made vast contributions to calculus, including the notation and terminology used today. John Napier (1550–1617) was a Scottish mathematician credited with inventing logarithms and popularizing the use of the decimal point.

Growth Functions

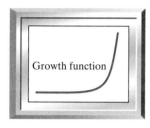

Growth function

An exponential *growth function* is a type of function that increases without bound with respect to an independent variable. For a system to be considered an exponential growth function, the exponential growth model ($y = be^{mx} + c$) requires that m be greater than zero.

A more general exponential growth function can be formed by replacing the Napier constant with an arbitrary constant, or $y = ba^{mx} + c$. In the general growth function, a must be greater than 1 for the system to be a growth function. The value of a is referred to as the *base*, m is the *growth rate*, b is the *initial value*, and c is a *vertical shift*. Note that when $a = 1$ or $m = 0$, the system is reduced to $y = b + c$, which is a constant.

● EXAMPLE 12-7

In 1965, Gordon E. Moore, co-founder of Intel Corporation, claimed in a paper that the number of transistors on an integrated circuit will double every 2 years. This idea by Moore was later referred to as **Moore's law**. The Intel 4004 CPU was released in 1971 as the first commercially available microprocessor. The Intel 4004 CPU contained 2,300 transistors. This system can be modeled with the following growth function.

$$T = T_0 2^{t/2}$$

In the equation, T_0 represents the initial number of transistors, and t is the number of years since T_0 transistors were observed on an integrated circuit. Predict the number of transistors on an integrated circuit in 1974 using the Intel 4004 CPU as the initial condition.

$$t = 1974 - 1971 = 3 \text{ years}$$
$$T = T_0 2^{1/2} = 2,300\left(2^{3/2}\right) = 2,300\left(2^{1.5}\right) \approx 6,505 \text{ transistors}$$

In 1974, the Intel 8080 processor came out with 4,500 transistors on the circuit.

Predict the number of transistors on integrated circuits in 1982 using the Intel 4004 CPU as the initial condition.

$$t = 1982 - 1971 = 11 \text{ years}$$
$$T = T_0 2^{t/2} = 2,300\left(2^{11/2}\right) = 2,300\left(2^{5.5}\right) \approx 104,087 \text{ transistors}$$

In 1982, the Intel 286 microprocessor came out with 134,000 transistors in the CPU.

Predict the number of transistors on integrated circuits in 2007 using the Intel 4004 CPU as the initial condition.

$$t = 2007 - 1971 = 36 \text{ years}$$
$$T = T_0 2^{t/2} = 2,300\left(2^{36/2}\right) = 2,300\left(2^{18}\right) \approx 603,000,000 \text{ transistors}$$

In 2007, the NVIDIA G80 came out with 681,000,000 transistors in the CPU.

No one really knows how long Moore's law will hold up. It is perhaps interesting to note that claims have consistently been made for the past 30 years that Moore's law will only hold up for another 10 years. Although many prognosticators are still saying this, some are not. There is, however, a limit to how small a transistor can be made. Any structure has to be at least one atom wide, for example, and as they become ever smaller, quantum effects will probably wreak havoc. Of course, chips can be made larger, multilayer structures can be built, new technologies may be developed (the first functional memristor was demonstrated in 2008), and so forth.

● **EXAMPLE 12-8**   An environmental engineer has obtained a bacteria culture from a municipal water sample and allowed the bacteria to grow. After several hours of data collection, the following graph is created.

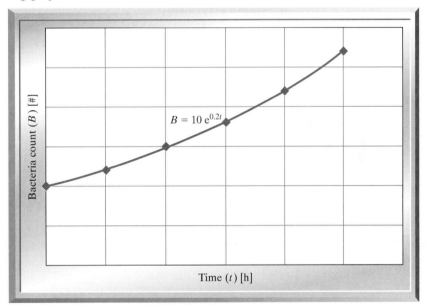

What was the initial concentration of bacteria?

In theory: $B = B_0 e^{gt}$ and from graph: $B = 10 e^{0.2t}$

By comparison: $B_0 = 10$ bacteria

What was the growth constant (g) of this bacteria strain?

In theory: $B = B_0 e^{gt}$ and from graph: $B = 10 e^{0.2t}$

By comparison: $g = 0.2$ per hour. Recall that exponents must be unitless, so the quantity of ($g\,t$) must be a unitless group. To be unitless, g must have units of inverse time.

The engineer wants to know how long it will take for the bacteria culture population to grow to 30,000.

To calculate the amount of time, plug in 30,000 for B and solve for t:

$$30{,}000 = 10e^{0.2t}$$

$$3{,}000 = e^{0.2t}$$

$$\ln(3{,}000) = \ln(e^{0.2t}) = 0.2t$$

$$t = \frac{\ln(3{,}000)}{0.2\left[\frac{1}{h}\right]} = 40 \text{ h}$$

Decay Functions

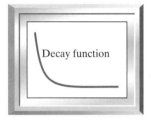
Decay function

A *decay function* is a type of function that decreases and asymptotically approaches a value. In the exponential decay model $(y = be^{-mx} + c)$, m is a positive value that represents the **decay rate**.

● **EXAMPLE 12-9**

An electrical engineer wants to determine how long it will take for a particular capacitor in a circuit to discharge. The engineer wired a voltage source across a *capacitor* (C, farads) and a resistor (R, ohms) connected in series. After the capacitor is fully charged, the circuit is completed between the capacitor and resistor, and the voltage source is removed from the circuit. The product of R and C in a circuit like this is called the "time constant" and is usually denoted by the Greek letter tau ($\tau = RC$).

$$\tau = \frac{1}{RC}$$

The following equation can be used to calculate the voltage across a discharging capacitor at a particular time.

$$V = V_0 e^{-\tau} = V_0 e^{-\frac{t}{RC}}$$

Assuming a resistance of 100 kiloohms [kΩ], a capacitance of 100 microfarads [μF], and an initial voltage (V_0) of 20 volts [V], determine the voltage across the capacitor after 10 seconds.

$$V = 20 \text{ [V] } e^{-\frac{10 \text{ s}}{(100 \text{ k}\Omega)(100 \text{ }\mu\text{F})}}$$

$$= 20 \text{ [V] } e^{-\frac{10 \text{ s}}{(100 \times 10^3 \Omega)(100 \times 10^{-6}\text{F})}} \approx 7.36 \text{ V}$$

Assuming a resistance of 200 kiloohms [kΩ], a capacitance of 100 microfarads [μF], and an initial voltage (V_0) of 20 volts [V], determine the voltage across the capacitor after 20 seconds.w

$$V = 20e^{-\frac{20 \text{ s}}{(200 \text{ k}\Omega)(100 \text{ }\mu\text{F})}} \approx 7.36 \text{ V}$$

Note that doubling the resistance in the circuit doubles the amount of time required to discharge the capacitor. In RC circuits, it is easy to increase the discharge time of a capacitor by increasing the resistance in the circuit.

> **NOTE**
>
> Exponential models are often given in the form $Y = be^{-t/\tau} + C$, where t is time; thus τ also has units of time. In this case, the constant τ is often called the **time constant**.
>
> Basically, the time constant is a measure of the time required for the response of the system to go approximately two-thirds of the way from its initial value to its final value, as t approaches infinity. The exact value is not $2/3$, but $1 - e^{-1} \approx 0.632$ or 63.2%.

COMPREHENSION CHECK 12-9

The decay of a radioactive isotope was tracked over a number of hours, resulting in the following data. The decay of a radioactive element is modeled by the following equation, where C_0 is the initial amount of the element at time zero, and k is the decay constant of the isotope.

$$C = C_0 e^{-kt}$$

Determine the initial concentration and decay constant of the isotope, including value and units.

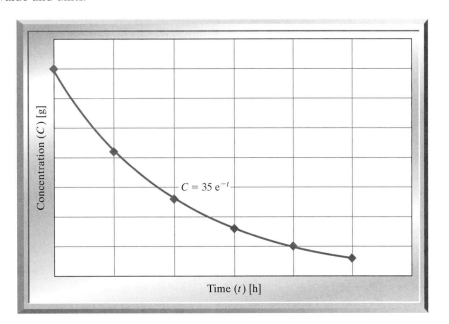

$$C = 35\,e^{-t}$$

Picture of a single mortar shot. The creation of fireworks involves knowledge of chemistry (what materials to include to get the desired colors), physics and dynamics (what amounts of combustible charge should be included to launch the object properly), and artistry (what colors, shapes, patterns, and sounds the firework should emit such that it is enjoyable to watch). This picture is a close-up of the instant when a firework is detonating.

Photo courtesy of E. Fenimore

In-Class Activities

ICA 12-1

The graph shows the ideal gas law relationship ($PV = nRT$) between volume (V) and temperature (T).

(a) What are the units of the slope (0.0175)?
(b) If the tank has a pressure of 1.2 atmospheres and is filled with nitrogen (formula, N_2; molecular weight, 28 grams per mole), what is the amount of gas in the tank (n) in units of grams?
(c) If the tank is filled with 10 grams of oxygen (formula, O_2; molecular weight, 32 grams per mole), what is the pressure of the tank (P) in units of atmospheres?

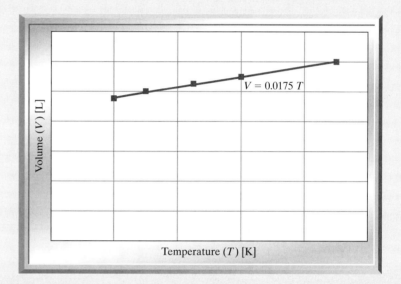

ICA 12-2

An inductor is an electrical device that can store energy in the form of a magnetic field. In the simplest form, an inductor is a cylindrical coil of wire, and its inductance (L), measured in henrys [H], can be calculated by

$$L = \frac{\mu_0 n^2 A}{\ell}$$

where

$\mu_0 =$ permeability of free space $= 4\pi \times 10^{-7}$ [newtons per ampere squared, N/A²]
$n \ =$ number of turns of wire [dimensionless]
$A =$ cross-sectional area of coil [square meters, m²]
$\ell \ =$ length of coil [meters, m]
$L =$ inductance [henrys, H] = [J/A²] (One henry is one joule per ampere squared.)

Several inductors were fabricated with the same number of turns of wire (n) and the same length (ℓ), but with different diameters, thus different cross-sectional areas (A). The inductances were measured and plotted as a function of cross-sectional area, and a mathematical model was developed to describe the relationship, as shown on the graph below.

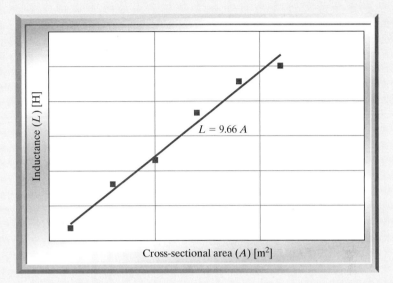

(a) What are the units of the slope (9.66)?

(b) For an inductor fabricated as described above, what is its diameter if its inductance is 0.2 henrys? Give your answer in centimeters.

(c) If the length of the coil (ℓ) equals 0.1 meter, how many turns of wire (n) are in the inductor?

ICA 12-3

Mercury has a dynamic viscosity of 1.55 centipoises and a specific gravity of 13.6.

(a) What is the density of mercury in units of kilograms per cubic meter?

(b) What is the dynamic viscosity of mercury in units of pound-mass per foot second?

(c) What is the dynamic viscosity of mercury in units of pascal seconds?

(d) What is the kinematic viscosity of mercury in units of stokes?

ICA 12-4

SAE 10W30 motor oil has a dynamic viscosity of 0.17 kilograms per meter second and a specific gravity of 0.876.

(a) What is the density of the motor oil in units of kilograms per cubic meter?

(b) What is the dynamic viscosity of the motor oil in units of pound-mass per foot second?

(c) What is the dynamic viscosity of the motor oil in units of centipoise?

(d) What is the kinematic viscosity of the motor oil in units of stokes?

ICA 12-5

You have two springs each of stiffness 1 newton per meter [N/m] and one spring of stiffness 2 newtons per meter [N/m].

(a) List all possible stiffness combinations that can be formed with these springs. Your solution should include drawings and calculations of each unique configuration.

(b) How many unique combinations can be formed?

(c) What is the stiffest configuration?

(d) What is the least stiff configuration?

ICA 12-6

You have three resistors of resistance 30 ohm [Ω].

(a) List all possible resistance combinations that can be formed with these resistors. Your solution should include drawings and calculations of each unique configuration.
(b) How many unique combinations can be formed?
(c) What is the greatest resistance configuration?
(d) What is the least resistance configuration?

ICA 12-7

Four springs were tested, with the results shown graphically below. Use the graph to answer the following questions. Be sure to justify your answers as to *why* you are making the choice.

(a) Which spring is the stiffest?
(b) Which spring, if placed in parallel with Spring C, would yield the stiffest combination?
(c) Which spring, if placed in series with Spring C, would yield the stiffest combination?
(d) Rank the following combinations in order of stiffness:

Spring A and Spring D are hooked in parallel
Spring B and Spring C are hooked in series, then connected with Spring D in parallel
Spring A
Spring D

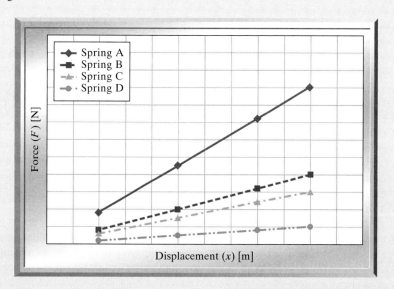

ICA 12-8

Four circuits were tested, with the results shown graphically below. Use the graph to answer the following questions. Be sure to justify your answers as to *why* you are making the choice.

(a) Which resistor gives the most resistance?
(b) What is the resistance of Resistor A?
(c) Which resistor, if placed in parallel with Resistor C, would yield the highest resistance?
(d) Which resistor, if placed in series with Resistor C, would yield the highest resistance?

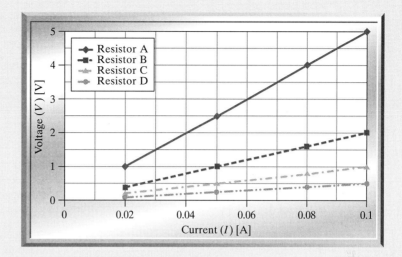

ICA 12-9

The resistance of a wire (R [ohm]) is a function of the wire dimensions (A = cross-sectional area, L = length) and material (ρ = resistivity) according to the relationship

$$R = \frac{\rho L}{A}$$

The resistance of three wires was tested. All wires had the same cross-sectional area.

Length (L) [m]	0.01	0.1	0.25	0.4	0.5	0.6
Wire 1	8.00E-05	8.00E-04	2.00E-03	3.50E-03	4.00E-03	4.75E-03
Wire 2	4.75E-05	4.80E-04	1.00E-03	2.00E-03	2.50E-03	3.00E-03
Wire 3	1.50E-04	1.70E-03	4.25E-03	7.00E-03	8.50E-03	1.00E-02

(a) Plot the data and fit a linear trendline model to each wire.
(b) From the following chart, match each wire (1, 2, and 3) with the correct material according to the results of the resistivity determined from the trendlines, assuming a 0.2-centimeter diameter wire was used.

Material	Resistivity (ρ) [Ωm] $\times 10^{-8}$
Aluminum	2.65
Copper	1.68
Iron	9.71
Silver	1.59
Tungsten	5.60

ICA 12-10

A piano, much like a guitar or a harp, is a stringed instrument. The primary difference is that a piano has a keyboard containing 88 keys that, when pressed, cause a small felt hammer to strike the strings (mostly 3) associated with that key. Each hammer can hit the strings harder or softer depending on the force applied to the key on the keyboard. In order to create the different pitches of sound, the strings for each key are of different lengths as well as having different mass per unit length.

The equation for the fundamental frequency of a vibrating string is given by

$$f = \frac{\sqrt{T/\mu}}{2L}$$

where

f = frequency [Hz]
T = string tension [N]
μ = mass per unit length [kg/m]
L = string length [m]

In avant-garde music, the piano is sometimes played much like a harp, where the strings are directly manipulated (e.g., plucked and strummed) without the use of the keys on the keyboard. Assume we construct a special piano for use by an avant-garde musician in which each string has the same mass per unit length and the same length on the soundboard. The only difference is the string tension on each string on the soundboard. To test our design, we collect data using a single string on the sound board and create a graph of the observed frequency at different string tensions:

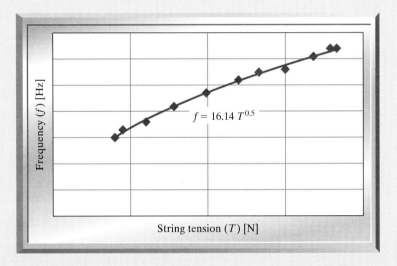

(a) What are the units of the coefficient (16.14)?
(b) If the observed frequency is 450 hertz, what is the string tension in newtons?
(c) If mass per unit length is 4.5 grams per meter, what is the length of the string in meters?
(d) If the length of the string is 0.7 meters, what is the mass per unit length in kilograms per meter?

ICA 12-11

To assist airplane manufacturers, NASA and other similar agencies around the world regularly release atmosphere models of Earth that are updated based on climate change around the world. Consider the graphs (Figures 1–2) of an atmosphere model of two different atmospheric layers (troposphere and lower stratosphere) and answer the following questions. These models are based on http://www.grc.nasa.gov/WWW/K-12/airplane/atmos.html.

(a) In Figure 1, what are the units of the value "2180" given in the trendline?
(b) In Figure 2, what are the units of "–5E-05" given in the trendline?
(c) What is the pressure at an altitude of 0 feet? How does this compare to the definition of standard atmospheric pressure of 1 atmosphere?
(d) What is the altitude at an air pressure of 150 pounds-force per square foot?

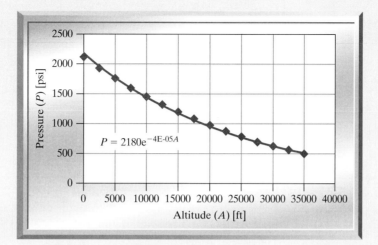

Figure 1 Earth Atmosphere Model: Troposphere

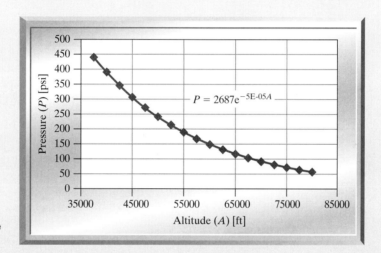

Figure 2 Earth Atmosphere Model: Lower Stratosphere

ICA 12-12

Eutrophication is a process whereby lakes, estuaries, or slow-moving streams receive excess nutrients that stimulate excessive plant growth. This enhanced plant growth, often called an algal bloom, reduces dissolved oxygen in the water when dead plant material decomposes and can cause other organisms to die. Nutrients can come from many sources, such as fertilizers; deposition of nitrogen from the atmosphere; erosion of soil containing nutrients; and sewage treatment plant discharges. Water with a low concentration of dissolved oxygen is called hypoxic. A biosystems engineering models the algae growth in a lake. The concentration of algae (C), measured in grams per milliliter [g/mL], can be calculated by

$$C = C_0 e^{\left(\frac{kt}{r}\right)}$$

where

C_0 = initial concentration of algae [?]

k = multiplication rate of the algae [?]

r = estimated nutrient supply amount [mg of nutrient per mL of sample water]

t = time [days]

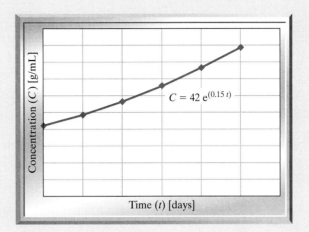

(a) For the exponential model shown, list the value and units of the parameters m and b. You do not need to simplify any units. Recall that an exponential model has the form:

    Exponential $y = be^{mx}$ m = exponent b = constant

(b) What is the initial concentration of the algae (C_0)?

(c) What are the units on the multiplication rate of the algae (k)?

(d) If the algae are allowed to grow for 10 days and an estimated nutrient supply of 3 milligrams of nutrient per milliliter of water sample, what is the multiplication rate of the algae (k)?

ICA 12-13

Select the data series from the options shown on the graph below that represent each of the following model types. You may assume that power and exponential models do not have a constant offset. You may also assume that only positive values are shown on the two axes (only first quadrant shown). For each match, write "Series X," where X is the appropriate letter, A through F. If no curve matches the specified criterion, write "No Match." If more than one curve matches a given specification, list both series.

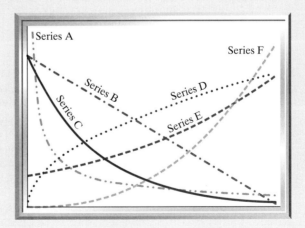

(a) Exponential, negative numeric value in exponent.

(b) Power, negative numeric value in exponent.

(c) Linear, negative slope.

(d) Exponential, positive numeric value in exponent.

(e) Power, positive numeric value in exponent.

Chapter 12 REVIEW QUESTIONS

1. A capacitor is an electrical device that can store energy in the form of an electric field. A capacitor is simply two conducting plates (usually metal) separated by an insulator, with a wire connected to each of the two plates. For a simple capacitor with two flat plates, the capacitance (C) [F] can be calculated by

$$C = \frac{\varepsilon_r \varepsilon_0 A}{d}$$

where

$\varepsilon_0 = 8.854 \times 10^{-12}$ [F/m] (the permittivity of free space in farads per meter)
ε_r = relative static permittivity, a property of the insulator [dimensionless]
A = area of overlap of the plates [m^2]
d = distance between the plates [m]

Several experimental capacitors were fabricated with different plate areas (A), but with the same inter-plate distance ($d = 1.2$ mm) and the same insulating material, and thus the same relative static permittivity (ε_r). The capacitance of each device was measured and plotted versus the plate area. A mathematical model to describe the data was then obtained. The graph and equation are shown below. The numeric scales were deliberately omitted.

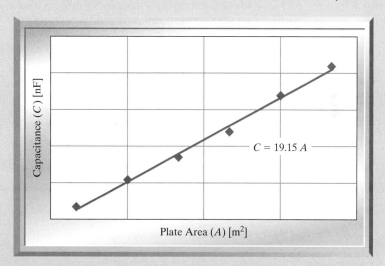

(a) What are the units of the slope (19.15)?
(b) If the capacitance is 2 nanofarads [nF], what is the area (A) of the plates?
(c) What is the relative static permittivity of the insulating layer?
(d) If the distance between the plates were doubled, how would the capacitance be affected?

2. When rain falls over an area for a sufficiently long time, it will run off and collect at the bottom of hills and eventually find its way into creeks and rivers. A simple way to estimate the maximum discharge flow rate (Q, in units of cubic feet per second [cfs]) from a watershed of area (A, in units of acres) with a rainfall intensity (i, in units of inches per hour) is given by an expression commonly called the Rational Method, as

$$Q = CiA$$

Values of C vary between about 0 (for flat rural areas) to almost 1 (in urban areas with a large amount of paved area).

A survey of a number of rainfall events was made over a 10-year period for three different watersheds. The data that resulted is given in the table below. Watershed A is 120 acres, B is 316 acres, and C is 574 acres.

Storm event	Watershed	Rainfall Intensity (i) [in/h]	Maximum Runoff (Q) [cfs]
1	A	0.5	30
2	A	1.1	66
3	A	1.6	96
4	A	2.1	126
5	B	0.3	47
6	B	0.7	110
7	B	1.2	188
8	B	1.8	283
9	C	0.4	115
10	C	1	287
11	C	1.5	430
12	C	2.4	690

(a) Create a graph containing all three watersheds, with flowrate on the ordinate and fit trendlines to obtain a simple model for each watershed.

From the information given and the trendline model obtained, answer the following:

(b) What is the value and units of the coefficient C?

(c) What would the maximum flow rate be from a watershed of 400 acres if the rainfall intensity was 0.6 inches per hour?

(d) How long would it take at this flowrate to fill an Olympic sized swimming pool that is 50 meters long, 20 meters wide, and 2 meters deep?

3. When we wish to generate hydroelectric power, we build a dam to back up the water in a river. If the water has a height (H, in units of feet) above the downstream discharge, and we can discharge water through the turbines at a rate (Q, in units of cubic feet per second [cfs]), the maximum power (P, in units of kilowatts) we can expect to generate is:

$$P = CHQ$$

For a small "run of the river" hydroelectric facility, we have obtained the following data.

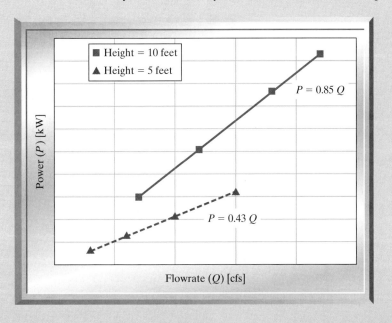

(a) Using the trendline results, and examining the general equation above, determine the value and units of the coefficient C for a height of 10 feet and a height of 5 feet.

(b) If the flowrate was 20 cubic feet per second and the height is 10 feet, what would the power output be in units of kilowatts?

(c) If the flowrate was 4,000 gallons per minute and the height 5 feet, what would the power output be in units of kilowatts?

(d) If the power output was 10 horsepower and the height was 10 feet, what would the flowrate be in cubic feet per second?

(e) If the power output was 8 horsepower and the height was 5 feet, what would the flowrate be in cubic feet per second?

(f) If the flow rate was 15 cubic feet per second and the height is 3 meters, what would the power output be in units of horsepower?

(g) If the flow rate was 10 cubic feet per second and the height is 8 meters, what would the power output be in units of horsepower?

4. You are experimenting with several liquid metal alloys to find a suitable replacement for the mercury used in thermometers. You have attached capillary tubes with a circular cross-section and an inside diameter of 0.3 millimeters to reservoirs containing 5 cubic centimeters of each alloy. You mark the position of the liquid in each capillary tube when the temperature is 20 degrees Celsius, systematically change the temperature, and measure the distance the liquid moves in the tube as it expands or contracts with changes in temperature. Note that negative values correspond to contraction of the material due to lower temperatures. The data you collected for four different alloys is shown in the table below.

Alloy G1		Alloy G2		Alloy G3		Alloy G4	
Temperature (T) [°C]	Distance (d) [cm]	Temperature (T) [°C]	Distance (d) [cm]	Temperature (T) [°C]	Distance (d) [cm]	Temperature (T) [°C]	Distance (d) [cm]
22	1.05	21	0.95	24	2.9	25	5.1
27	3.05	29	7.65	30	7.2	33	13.8
34	6.95	33	10.6	34	9.8	16	−4.3
14	−3.5	17	−2.6	19	−0.6	13	−7.05
9	−5.1	3	−14.8	12	−6.15	6	−14.65
2	−8.7	−2	−19.8	4	−11.5	−2	−22.15
−5	−11.7	−8	−25.4	−5	−18.55	−6	−26.3
−11	−15.5					−12	−32.4

(a) In Excel, create two new columns for each compound to calculate the change in temperature (ΔT) relative to 20°C (for example, 25°C gives $\Delta T = 5$°C) and the corresponding change in volume (ΔV).

(b) Plot the change in volume versus the change in temperature; fit a linear trendline to each data set.

(c) From the trendline equations, determine the coefficient of thermal expansion, β. Note that $\Delta V = \beta V \Delta T$, where V is the initial volume.

(d) There is a small constant offset (C) in each trendline equation ($\Delta V = \beta V \Delta T + C$). What is the physical origin of this constant term? Can it be safely ignored? In other words, is its effect on the determination of β negligible?

5. In order to determine the value of an unknown capacitor, it is charged with a constant current of 50 microamperes. The voltage across the capacitor was measured at several times and the results are shown below. It is known that the voltage across an initially discharged capacitor being charged by a constant current is given by $V = It/C$, where V is voltage in volts. I is the current in amperes, t is time in seconds, and C is capacitance in farads.

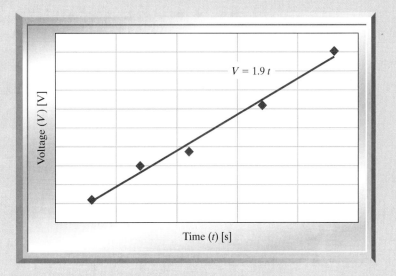

(a) What are the units on the constant (1.9)?

(b) What is the value of the capacitor?

6. Solid objects, such as your desk or a rod of aluminum, can conduct heat. The magnitude of the thermal diffusivity of the material determines how quickly the heat moves through a given amount of material. The equation for thermal diffusivity (α) is given by:

$$\alpha = \frac{k}{\rho\, C_\mathrm{p}}$$

Experiments are conducted to change the thermal conductivity (k) of the material while holding the specific heat (C_p) and the density (ρ) constant. The results are shown graphically.

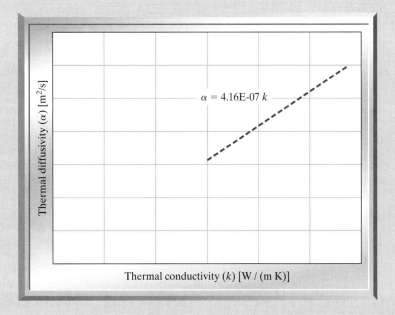

(a) What are the units of the constant 4.16×10^{-7}? Simplify your answer.

(b) If the specific heat of the material is 890 joules per kilogram kelvin, what is the density of the material?

(c) If the material has a density of 4,500 kilograms per cubic meter, what is the specific heat of the material in units of joules per kilogram kelvin?

7. Use the figure shown to answer the following questions.

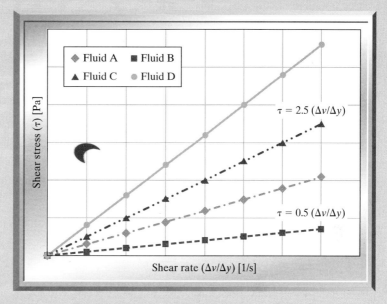

(a) Which fluid has the lowest dynamic viscosity?
(b) What is the dynamic viscosity of Fluid B?
(c) What is the dynamic viscosity of Fluid B in units of centipoise?
(d) If the specific gravity of Fluid B is 1.3, what is the kinematic viscosity of Fluid B in units of stokes?
(e) What is the dynamic viscosity of Fluid C?
(f) What is the dynamic viscosity of Fluid C in units of centipoise?
(g) If the specific gravity of Fluid C is 0.8, what is the kinematic viscosity of Fluid C in units of stokes?

8. You are given four springs, one each of 2.5, 5, 7.5, and 10 newtons per meter [N/m].

(a) What is the largest equivalent stiffness that can be made using these four springs? Draw a diagram indicating how the four springs are connected and what the value of each is.
(b) What is the smallest equivalent stiffness that can be made using these four springs? Draw a diagram indicating how the four springs are connected and what the value of each is.
(c) What is the largest equivalent stiffness that can be made using only three of these springs? Draw a diagram indicating how the three springs are connected and what the value of each is.
(d) What is the smallest equivalent stiffness that can be made using only three of these springs? Draw a diagram indicating how the three springs are connected and what the value of each is.
(e) How close an equivalent stiffness to the average of the four springs (6.25 newtons per meter) can you make using only these springs? You may use all four springs to do this, but you may use less if that will yield an equivalent stiffness closer to the average.

9. You are given four resistors, each of 7.5, 10, 15, and 20 kiloohms [kΩ].

(a) What is the largest equivalent resistance that can be made using these four resistors? Draw a diagram indicating how the four resistors are connected and what the value of each is.
(b) What is the smallest equivalent resistance that can be made using these four resistors? Draw a diagram indicating how the four resistors are connected and what the value of each is.

(c) What is the largest equivalent resistance that can be made using only three of these resistors? Draw a diagram indicating how the three resistors are connected and what the value of each is.

(d) What is the smallest equivalent resistance that can be made using only three of these resistors? Draw a diagram indicating how the three resistors are connected and what the value of each is.

(e) How close an equivalent resistance to the average of the four resistors (13.125 kΩ) can you make using only these resistors? You may use all four resistors to do this, but you may use less if that will yield an equivalent resistance closer to the average.

10. Use the diagrams shown to answer the following questions.

(a) Determine the equivalent stiffness of four springs connected as shown.

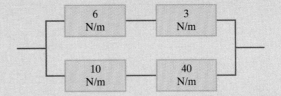

(b) Determine the equivalent stiffness of four springs connected as shown.

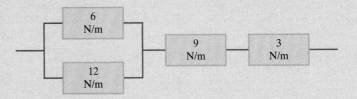

(c) Determine the equivalent resistance of four resistors connected as shown.

(d) Determine the equivalent resistance of four resistors connected as shown.

(e) Determine a SIMPLE function for the equivalent stiffness of N springs, each with a spring constant k, all connected in series. You MUST show how you derived this for credit. You may NOT use Σ or Π notation in your final answer.

(f) Determine a SIMPLE function for the equivalent resistance of N resistors, each with resistance R, all connected in series. You MUST show how you derived this for credit. You may NOT use Σ or Π notation in your final answer.

11. You have three springs. You conduct several tests and determine the following data.

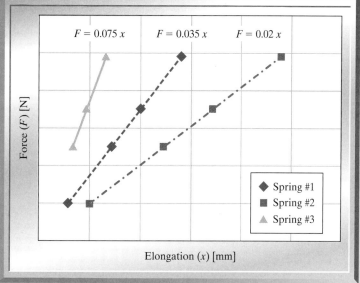

(a) Determine the stiffness of Spring 1.
(b) Determine the stiffness of Spring 2.
(c) Determine the stiffness of Spring 3.

Choose one correct spring or spring combination that will meet the following criteria as closely as possible. Assume you have one of each spring available for use. List the spring or spring combination and the resulting spring constant.

(d) You want the spring or spring system to hold 95 grams and displace approximately 1 centimeter.
(e) You want the spring or spring system to displace approximately 4 centimeter when holding 50 grams.
(f) You want the spring or spring system to displace approximately 5 millimeter when holding 75 grams.
(g) You want the spring or spring system to hold 20 grams and displace approximately 1 centimeter.

12. You have three resistors. You conduct several tests and determine the following data.

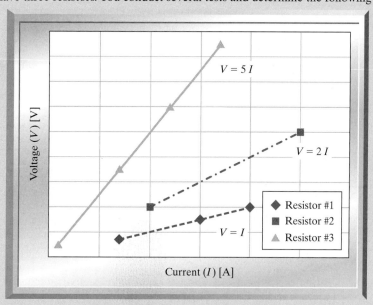

(a) Determine the resistance of Resistor 1.
(b) Determine the resistance of Resistor 2.
(c) Determine the resistance of Resistor 3.

Choose one correct resistor or resistor combination that will meet the following criteria as closely as possible. Assume you have one of each resistor available for use. List the resistor or resistor combination and the resulting resistor constant.

(d) You want the resistor or resistor system to provide approximately 20 amperes when met with 120 volts.
(e) You want the resistor or resistor system to provide approximately 46 amperes when met with 30 volts.
(f) You want the resistor or resistor system to provide approximately 15 amperes when met with 120 volts.
(g) You want the resistor or resistor system to provide approximately 33 amperes when met with 45 volts.

13. A piano, much like a guitar or a harp, is a stringed instrument. The primary difference is that a piano has a keyboard containing 88 keys that, when pressed, cause a small felt hammer to strike the strings (mostly 3) associated with that key. Each hammer can hit the strings harder or softer depending on the force applied to the key on the keyboard. In order to create the different pitches of sound, the strings for each key are of different lengths as well as having different mass per unit length.

In avant-garde music, the piano is sometimes played much like a harp, where the strings are directly manipulated (plucked, strummed, etc.) without the use of the keys on the keyboard. Assume we construct a special piano for use by an avant-garde musician in which each string has the same mass per unit length and the same tension on the soundboard. The only difference is the length of each string on the soundboard. To test our design, we collect the following data on the observed frequency at different string lengths:

Length (L) [M]	0.41	0.48	0.61	0.69	0.86	0.97	1.1	1.3	1.5
Frequency (f) [Hz]	1,500	1,300	1,000	880	740	650	600	490	400

The equation for the fundamental frequency of a vibrating string is given by

$$f = \frac{\sqrt{T/\mu}}{2L}$$

where

f = frequency [Hz]
T = string tension [N]
μ = mass per unit length [kg/m]
L = string length [m]

(a) Using the equation above, express the unit hertz [Hz] in terms of base SI units. You must prove this relationship using the equation given.
(b) Is the relationship between frequency and length linear, power, or exponential?
(c) Create a graph of the observed frequency data, including the trendline and equation generated by Excel.
(d) If the tension was reduced to half of its original value, would the frequency increase or decrease and by how much?
(e) If the tension on the strings is 200 newtons, what is the mass per unit length in grams per meter?
(f) If the mass per length of the string is 0.5 grams per meter, what is the tension in newtons?

14. Solid objects, such as your desk or a rod of aluminum, can conduct heat. The magnitude of the thermal diffusivity of the material determines how quickly the heat moves through a given amount of material. The equation for thermal diffusivity (α) is given by:

$$\alpha = \frac{k}{\rho \, C_p}$$

Experiments are conducted to change the specific heat (C_p) of the material while holding the thermal conductivity (k) and the density (ρ) constant. The results are shown graphically.

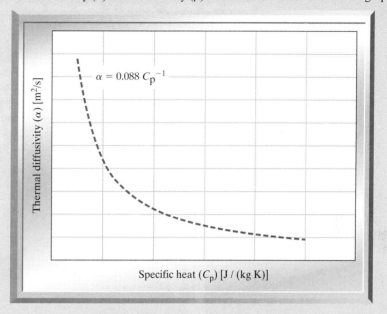

$$\alpha = 0.088\, C_p^{-1}$$

Thermal diffusivity (α) [m²/s]

Specific heat (C_p) [J / (kg K)]

(a) What are the units of the constant 0.088? Simplify your answer.

(b) If the thermal conductivity of the material is 237 watts per meter kelvin, what is the density of the material?

(c) If the material has a density of 4,500 kilograms per cubic meter, what is the thermal conductivity of the material in units of watts per meter kelvin?

15. Your supervisor has assigned you the task of designing a set of measuring spoons with a "futuristic" shape. After considerable effort, you have come up with two geometric shapes that you believe are really interesting.

You make prototypes of five spoons for each shape with different depths and measure the volume each will hold. The table below shows the data you collected.

Depth (d) [cm]	Volume (V_A) [mL] Shape A	Volume (V_B) [mL] Shape B
0.5	1	1.2
0.9	2.5	3.3
1.3	4	6.4
1.4	5	7.7
1.7	7	11

Use Excel to plot and determine appropriate power models for this data. Use the resulting models to determine the depths of a set of measuring spoons comprising the following volumes for each of the two designs:

Volume Needed (V) [tsp or tbsp]	Depth of Design A (d_A) [cm]	Depth of Design B (d_B) [cm]
¹/₄ tsp		
¹/₂ tsp		
³/₄ tsp		
1 tsp		
1 tbsp		

16. A piano, much like a guitar or a harp, is a stringed instrument. The primary difference is that a piano has a keyboard containing 88 keys that, when pressed, cause a small felt hammer to strike the strings (mostly 3) associated with that key. Each hammer can hit the strings harder or softer depending on the force applied to the key on the keyboard. In order to create the different pitches of sound, the strings for each key are of different lengths as well as having different mass per unit length.

In avant-garde music, the piano is sometimes played much like a harp, where the strings are directly manipulated (e.g. plucked and strummed) without the use of the keys on the keyboard. Assume we construct a special piano for use by an avant-garde musician in which each string has the same mass per unit length and the same tension on the soundboard. The only difference is the length of each string on the soundboard. To test our design, we collect the following data on the observed frequency at different string lengths.

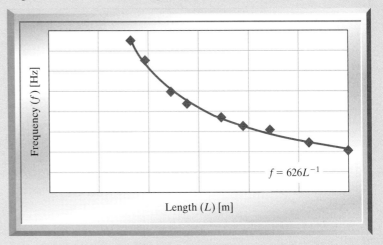

The equation for the fundamental frequency of a vibrating string is given by

$$f = \frac{\sqrt{T/\mu}}{2L}$$

where

f = frequency [Hz]
T = string tension [N]
μ = mass per unit length [kg/m]
L = string length [m]

(a) Is the relationship between frequency and length linear, power, or exponential?
(b) What are the units of the coefficient (626)?
(c) If the tension on the strings is 200 newtons, what is the mass per unit length in grams per meter?
(d) If the mass per length of the string is 0.5 grams per meter, what is the tension in newtons?

17. It is extremely difficult to bring the internet to some remote parts of the world. This can be inexpensively facilitated by installing antennas tethered to large helium balloons. To help analyze the situation, assume we have inflated a large spherical balloon. The pressure on the inside of the balloon is balanced by the elastic force exerted by the rubberized material. Since we are dealing with a gas in an enclosed space, the Ideal Gas Law will be applicable.

$$PV = nRT$$

where

P = pressure
V = volume
n = quantity of gas [moles]

R = ideal gas constant [0.08206 (atm L)/(mol K)]
T = temperature (absolute units)

If the temperature increases, the balloon will expand and/or the pressure will increase to maintain the equality. As it turns out, the increase in volume is the dominate effect, so we will treat the change in pressure as negligible.

The circumference of an inflated spherical balloon is measured at various temperatures; the resulting data are shown in the graph below.

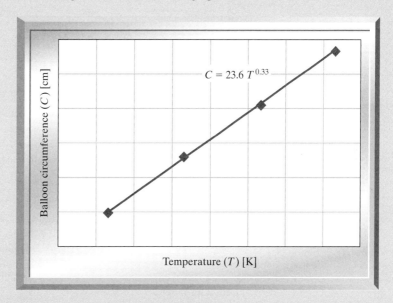

(a) What are the units of the constant 0.33?
(b) What are the units of the constant 23.6?
(c) What would the temperature of the balloon be if the circumference was 162 centimeters?
(d) If a circle with an area of 100 square centimeters is drawn on the balloon at 20 degrees Celsius, what would the area be at a temperature of 100 degrees Celsius?
(e) If the pressure inside the balloon is 1.2 atmospheres, how many moles of gas does it contain?
(f) If the balloon holds 50 grams of nitrogen, with a molecular weigth of 28 grams per mole, what is the pressure inside the ballon in units of atmospheres?

18. When a buoyant cylinder of height H, such as a fishing cork, is placed in a liquid and the top is depressed and released, it will bob up and down with a period T. We can conduct a series of tests and see that as the height of the cylinder increases, the period of oscillation also increases. A less dense cylinder will have a shorter period than a denser cylinder, assuming of course all the cylinders will float. A simple expression for the period is:

$$T = 2\pi\sqrt{\frac{\rho_{cylinder}}{\rho_{liquid}}\frac{H}{g}}$$

where g is the acceleration due to gravity [9.8 meters per second squared], $\rho_{cylinder}$ is the density of the material, and ρ_{liquid} is the density of the fluid. By testing cylinders of differing heights, we wish to develop a model for the oscillation period, shown in the graph below.

(a) What are the units of the coefficient (0.104) shown in the model?
(b) What is the oscillation period in units of seconds of a cylinder that is 4-inches tall?
(c) If the oscillation period is 45 seconds, what is the height of the cylinder in units of inches?
(d) We will conduct a series of tests with a new plastic (polystretchypropylene) that has a specific gravity of 0.6. What is the specific gravity of the fluid?

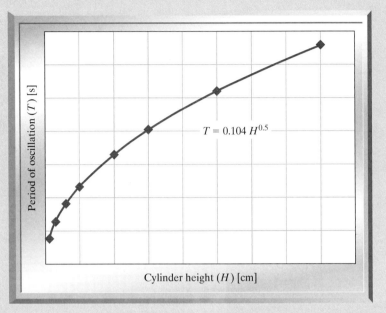

(e) We will conduct a series of tests with cylinders that have a specific gravity of 0.9. What is the density of the fluid in units of kilograms per cubic meter?

(f) Assume the tests were conducted using citric acid as the fluid, which has a specific gravity of 1.67. What is the specific gravity of the material of the cylinder?

(g) Assume the tests were conducted using acetone as the fluid, which has a specific gravity of 0.785. What is the density of the material of the cylinder in units of grams per cubic centimeter?

19. Experiments have shown that the visible flame from the back of a rocket can be estimated from the simple equation:

$$L = \sqrt{\frac{F}{f}}$$

where L is the flame length in feet, F is the thrust of the rocket in pounds-force, and f is an empirical factor found through experimentation.

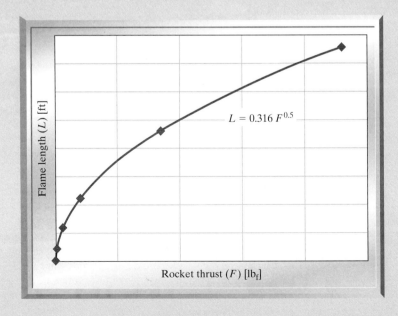

(a) What are the value and units of the factor f?

(b) If a rocket has visible flame of 30 meters, what is an estimate of the thrust in units of pounds-force?

(c) The empirical factor (f) shown is specific for the units system used in the figure. What should the value of the empirical factor (f) be if we wish to use the units of meters for length and newtons for the thrust?

20. One of the NAE Grand Challenges for Engineering is **Engineering the Tools of Scientific Discovery**. According to the NAE website: "Grand experiments and missions of exploration always need engineering expertise to design the tools, instruments, and systems that make it possible to acquire new knowledge about the physical and biological worlds."

Solar sails are a means of interplanetary propulsion using the radiation pressure of the sun to accelerate a spacecraft. The table below shows the radiation pressure at the orbits of several planets.

Planet	Distance from Sun (d) [AU]	Radiation Pressure (P) [mPa]
Mercury	0.46	43.3
Venus	0.72	17.7
Earth	1	9.15
Mars	1.5	3.96
Jupiter	5.2	0.34

NOTE

The astronomical unit (AU) is the average distance from the Earth to the Sun.

(a) Plot this data and determine the power law model for radiation pressure as a function of distance from the sun.

(b) What are the units of the exponent in the trendline?

(c) What are the units of the other constant in the trendline?

(d) What is the radiation pressure at Uranus (19.2 AU from sun)?

(e) At what distance from the sun is the radiation pressure 5 μPa?

21. The data shown graphically below was collected during testing of an electromagnetic mass driver. The energy to energize the electromagnets was obtained from a bank of capacitors. The capacitor bank was charged to various voltages, and for each voltage, the exit velocity of the projectile was measured when the mass driver was activated.

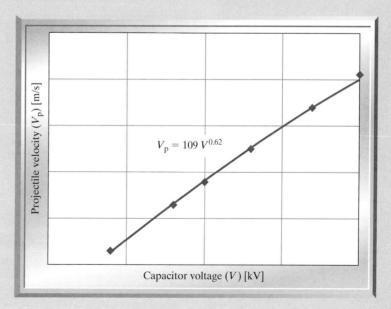

$V_p = 109\ V^{0.62}$

Projectile velocity (V_p) [m/s]

Capacitor voltage (V) [kV]

(a) What would the velocity be if the capacitors were charged to 1,000 volts?

(b) What would the velocity be if the capacitors were charged to 100,000 volts?

(c) What voltage would be necessary to accelerate the projectile to 1,000 meters per second?

(d) Assume that the total capacitance is 5 farads. If the capacitors are initially charged to 10,000 volts and are discharged to 2,000 volts during the launch of a projectile, what is the mass of the projectile if the overall conversion of energy stored in the capacitors to kinetic energy in the projectile has an efficiency of 0.2? Recall that the energy stored in a capacitor is given by $E = 0.5\,CV^2$, where C is capacitance in farads and V is voltage in volts.

(e) Assuming that the capacitors are initially charged to 10,000 volts and are discharged to 2,000 volts during the launch of a projectile, what is the total capacitance in farads if the projectile has a mass of 500 grams and the overall conversion of energy stored in the capacitors to kinetic energy in the projectile has an efficiency of 0.25?

(f) Assume the capacitors are discharged to 20% of their initial voltage when a projectile is launched. Is the energy conversion efficiency a constant, or does it depend on the initial capacitor voltage? If efficiency depends on initial capacitor voltage, determine a mathematical model for efficiency (η) as a function of voltage.

22. The Ramberg–Osgood Relationship can be used to describe the relationship between stress and strain in a material near its yield strength—that is when the stress on the material is great enough to cause it to permanently deform.

This relationship is given by

$$\varepsilon = \frac{\sigma}{E} + \alpha \frac{\sigma_0}{E}\left(\frac{\sigma}{\sigma_0}\right)^n$$

where

σ is the strain [dimensionless]
ε is the stress [pascals]
σ_0 is the yield strength [pascals]
E is Young's modulus
α and n are parameters that depend on the material.

A graph of the stress/strain relationship for an aluminum bar is shown below.

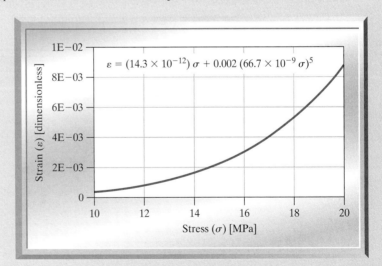

(a) What is Young's Modulus (E) for this aluminum bar? Be sure to include units with an appropriate metric prefix.

(b) What is the yield strength (σ_0) for this aluminum bar? Be sure to include units with an appropriate metric prefix.

(c) What is the value of the parameter n and its dimensions?

(d) What is the value of the parameter α and its dimensions?

If we can measure a length reliably to several millimeters, it seems reasonable that the longer a beam, the smaller would be the percentage error in length measurement. To test this hypothesis, we measure beams of differing lengths. The results are given in the figure.

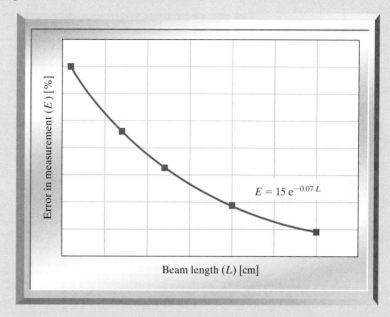

(e) What are the value and units of the coefficients in your model?

(f) How much error would you expect to have if you measured a beam that was 2 meters long?

(g) How long would a beam have to be to have a measurement error of less than 1%?

(h) Rewrite the model assuming that the measuring apparatus gives lengths in inches.

23. The total quantity (mass) of a radioactive substance decreases (decays) with time as

$$m = m_0 e^{-\frac{t}{\tau}}$$

where

$\quad t =$ time
$\quad \tau =$ time constant
$m_0 =$ initial mass (at $t = 0$)
$\quad m =$ mass at time t

A few milligrams each of three different isotopes of uranium were assayed for isotopic composition over a period of several days to determine the decay rate of each. The data was graphed and a mathematical model derived to describe the decay of each isotope.

(a) What are the units of τ if time is measured in days?

(b) What is the initial amount of each isotope at $t = 0$?

(c) How many days are required for only 1 milligram of the original isotope to remain in each sample?

(d) How many days are required for 99.99% of the original isotope in each sample to decay?

(e) What is the half-life, $T_{1/2}$, of each isotope? Half-life is the time required for half of the mass to decay into a different isotope.

(f) Four isotopes of uranium are shown in the table with their half-lives. Which isotope most likely matches each of the three samples?

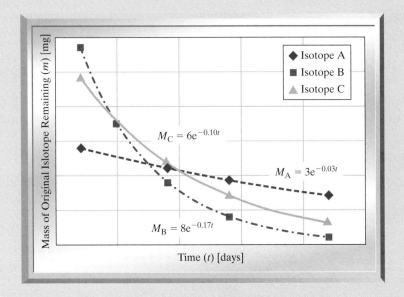

Isotope	Half-life [days]
230U	20.8
231U	4.2
237U	6.75
240U	0.59

24. When volunteers build a Habitat for Humanity house, it is found that the more houses that are completed, the faster each one can be finished since the volunteers become better trained and more efficient. A model that relates the building time and the number of homes completed can generally be given by

$$t = t_0 \, e^{-N/\nu} + t_M$$

where

t = time required to construct one house [days]
t_0 = a constant related to (but not equal) the time required to build the first house
N = the number of houses built previously [dimensionless]
ν = a constant related to the decrease in construction time as N increases
t_M = another constant related to construction time

A team of volunteers has built several houses, and their construction time was recorded for four of those houses. The construction time was then plotted as a function of number of previously built houses and a mathematical model derived as shown below. Using this information, answer the following questions:

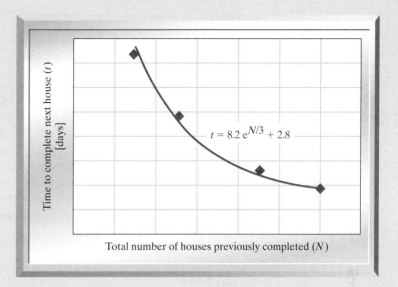

(a) What are the units of the constants 8.2, 3, and 2.8?
(b) If the same group continues building houses, what is the minimum time to construct one house that they can expect to achieve?
(c) How long did it take for them to construct the first house?
(d) How much quicker [days] was the third house built compared to the second?
(e) How many days (total) were required to build the first five houses?

25. As part of an electronic music synthesizer, you need to build a gizmo to convert a linear voltage to an exponentially related current. You build three prototype circuits, make several measurements of voltage and current in each, and graph the results as shown below.

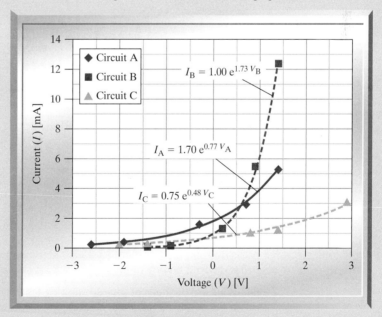

Assume that each circuit is modeled by the equation

$$I_X = A_X e^{(R_M/(R_X V_T))^{V_X}}$$

where

I_X is the current in circuit X [milliamperers, mA]
A_X is a scaling factor associated with circuit X
R_M is a master resistor, and has the same value in all circuits [ohms, Ω]
R_X is a resistor in circuit X whose value is different in each circuit [ohms, Ω]
V_T is the thermal voltage, and has a value of 25.7 volts
V_X is the voltage in circuit X [volts, V]

(a) What are the units of A_X?
(b) If you wish $I_X = 1$ mA when $V_X = 0$, what should the value of A_X be?
(c) Using the trendline models, if $R_M = 10$ kΩ, what is the value of R_A?
(d) Using the trendline models, if $R_M = 10$ kΩ, what is the value of R_B?
(e) Using the trendline models, if $R_M = 10$ kΩ, what is the value of R_C?

26. The graph below shows the relationship between current and voltage in a 1N4148 small signal diode (a semiconductor device that allows current to flow in one direction but not the other).

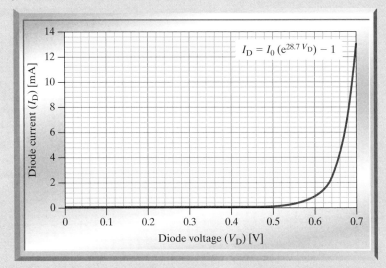

$$I_D = I_0 \left(e^{28.7\,V_D}\right) - 1$$

Semiconductor diodes can be characterized by the Shockley Equation:

$$I_D = I_0 \left(e^{\frac{qV_D}{nkT}} - 1\right)$$

where

I_D is the diode current [amperes]
I_0 is the reverse saturation current, constant for any specific diode
q is the charge on a single electron, 1.602×10^{-19} coulombs
V_D is the voltage across the diode [volts]
n is the emission coefficient, having a numerical value typically between 1 and 2, and constant for any specific device.
k is Boltzmann's Constant, 1.381×10^{-23} joules per kelvin
T is the temperature of the device [kelvin]

(a) What are the units of the -1 following the exponential term? Justify your answer.
(b) If the device temperature is 100 degrees Fahrenheit, what are the units of the emission coefficient, n, and what is its numerical value? (*Hint:* Electrical power [W] equals a volt times an ampere: $P = VI$. One ampere equals one coulomb per second.)
(c) What is the numerical value and units of the reverse saturation current, I_0? Use an appropriate metric prefix in your final answer.

27. Essentially all manufactured items are made to some "tolerance," or how close the actual product is to the nominal specifications. For example, if a company manufactures hammers, one customer might specify that the hammers should weigh 16 ounces. With rounding, this means that the actual weight of each hammer meets the specification if it weighs between 15.5 and 16.5 ounces. Such a hammer might cost 10 dollars. However, if the U.S. military, in its quest for perfection, specifies that an essentially identical hammer should have a weight of 16.000 ounces, then in order to meet specifications, the hammer must weigh between 15.9995 and 16.0005 ounces. In other words, the weight must fall within a range of one-thousandth of an ounce. Such a hammer might cost $1,000.

You have purchased a "grab bag" of 100 supposedly identical capacitors. You got a really good price, but there are no markings on the capacitors. All you know is that they are all the same nominal value. You wish to discover not only the nominal value, but the tolerance: are they within 5% of the nominal value, or within 20%? You set up a simple circuit with a known resistor and each of the unknown capacitors. You charge each capacitor to 10 volts, and then use an oscilloscope to time how long it takes for each capacitor to discharge to 2 volts. In a simple RC (resistor–capacitor) circuit, the voltage (V_C) across a capacitor (C) discharging through a resistor (R) is given by:

$$V_C = V_0 e^{-t/RC}$$

where t is time in seconds and V_0 is the initial voltage across the capacitor.

After measuring the time for each capacitor to discharge to from 10 to 2 V, you scan the list of times, and find the fastest and slowest. Since the resistor is the same in all cases, the fastest time corresponds to the smallest capacitor in the lot, and the slowest time to the largest. The fastest time was 3.3 microseconds and the slowest was 3.7 microseconds. For the two capacitors, you have the two pairs of data points.

(a) Plot these points in Excel, the pair for C_1 and the pair for C_2, on the same graph, using time as the independent variable. Fit exponential trendines to the data.

Time for C_1 (s)	Voltage of C_1	Time for C_2 (s)	Voltage of C_2
0	10	0	10
3.3×10^{-6}	2	3.7×10^{-6}	2

(b) Referring to the known form of the response, explain why the faster time corresponded to the smaller capacitor.
(c) The trendline for each capacitor had an R^2 value of 1, a perfect fit of trendline to data. Explain why this happened.
(d) Assuming you chose a precision resistor for these measurements that had a value of $R = 1,000.0$ ohms, determine the capacitance of the largest and smallest capacitors.
(e) You selected the fastest and slowest discharge times from a set of 100 samples. Since you had a fairly large sample set, it is not a bad assumption, according to the Laws of Large Numbers, that these two selected data sets represent capacitors near the lower and higher end of the range of values within the tolerance of the devices. Assuming the nominal value is the average of the minimum and maximum allowable values, what is the nominal value of the set of capacitors?
(f) What is the tolerance, in percent, of these devices? As an example, if a nominal 1 μF (microfarad) capacitor had an allowable range of 0.95 μF $< C <$ 1.05 μF, the tolerance would be 5%.

If standard tolerances of capacitors are 5%, 10%, and 20%, to which of the standard tolerances do you think these capacitors were manufactured? If you pick a smaller tolerance than you calculated, justify your selection. If you picked a higher tolerance, explain why the tolerance is so much larger than the measured value.

CHAPTER 13
MATHEMATICAL MODELS

As we have already seen, a large number of phenomena in the physical world obey one of the three basic mathematical models.

- Linear: $y = mx + b$
- Power: $y = bx^m + c$
- Exponential: $y = be^{mx} + c$

As we have mentioned previously, Excel can determine a mathematical model (trendline equation) for data conforming to all three of these model types, with the restriction that the constant c in the power and exponential forms must be 0.

Here, we consider how to determine the best model type for a specific data set, as well as learning methods of dealing with data that fit a power or exponential model best but have a nonzero value of c.

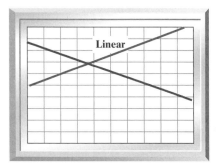
Linear

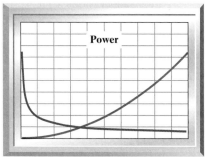
Power

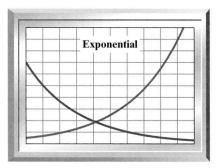
Exponential

Except as otherwise noted, the entire discussion in this chapter assumes that the data fits one of the three trendlines models: linear, power, or exponential. You should always keep this in mind when using the techniques discussed here.

13.1 SELECTING A TRENDLINE TYPE

When you determine a trendline to fit a set of data, in general you want the line, which may be straight or curved, to be as close as is reasonable to most of the data points. ***The objective is* not *to ensure that the curve passes through every point.***

To determine an appropriate model for a given situation, we use five guidelines, presented in general order of importance:

1. Do we already know the model type that the data will fit?
2. What do we know about the behavior of the process under consideration, including initial and final conditions?
3. What do the data look like when plotted on graphs with logarithmic scales?
4. How well does the model fit the data?
5. Can we consider other model types?

Guideline 1: Determine if the Model Type Is Known

If you are investigating a phenomenon that has already been studied by others, you may already know which model is correct or perhaps you can learn how the system behaves by looking in appropriate technical literature. In this case, all you need are the specific values for the model parameters since you already know the form of the equation. As we have seen, Excel is quite adept at churning out the numerical values for trendline equations.

If you are certain you know the proper model type, you can probably skip guidelines 2 and 3, although it might be a good idea to quantify how well the model fits the data as discussed in guideline 4. For example, at this point you should know that the extension of simple springs has a linear relationship to the force applied. As another example, from your study of the ideal gas law, you should know that pressure is related to volume by a power law model (exponent $= -1$).

At other times, you may be investigating situations for which the correct model type is unknown. If you cannot determine the model type from experience or references, continue to Guideline 2.

Guideline 2: Evaluate What Is Known About the System Behavior

The most important thing to consider when selecting a model type is whether the model makes sense in light of your understanding of the physical system being investigated. Since there may still be innumerable things with which you are unfamiliar, this may seem like an unreasonable expectation. However, by applying what you *do* know to the problem at hand, you can often make an appropriate choice without difficulty.

When investigating an unknown phenomenon, we typically know the answer to at least one of three questions:

1. How does the process behave in the initial state?
2. How does the process behave in the final state?
3. What happens to the process between the initial and the final states—if we sketch the process, what does it look like? Does the parameter of interest increase or decrease? Is the parameter asymptotic to some value horizontally or vertically?

● EXAMPLE 13-1

Suppose we do not know Hooke's law and would like to study the behavior of a spring. We hang the spring from a hook, pull downward on the bottom of the spring with varying forces, and observe its behavior. We know initially the spring will stretch a little under its own weight even before we start pulling on it, although in most cases this is small or negligible. As an extreme case, however, consider what would happen if you hang one end of a Slinky® from the ceiling, letting the other end fall as it will.

As we pull on the spring, we realize the harder we pull, the more the spring stretches. In fact, we might assume that in a simple world, if we pull twice as hard, the spring will stretch twice as far, although that might not be as obvious. In words we might say,

The distance the spring stretches (x) is directly proportional to the pulling force (F), or we might express the behavior as an equation:

$$x = kF + b$$

where b is the amount of stretch when the spring is hanging under its own weight. This is what we mean by using an "expected" form. Always remember, however, that what you "expect" to happen may be in error.

In addition, suppose we had tested this spring by hanging five different weights on it and measuring the stretch each time. After plotting the data, we realize there is a general trend that as the weight (force) increases, the stretch increases, but the data points do not lie exactly on a straight line. We have two options:

- *If we think our assumption of linear behavior may be in error, we can try nonlinear models.*
- *Or we can use a linear model, although the fit may not be as good as one or more of the nonlinear models.*

To bring order to these questions, we should ask the following sequence of questions:

Is the system linear?

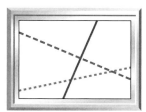

Linear systems have the following characteristics. If any of these is not true, then the system is not linear.

1. As the independent variable gets larger, the dependent variable continues to increase (positive slope) or decrease (negative slope) without limit. (See item 4 below.)
2. If the independent variable becomes negative, as it continues negative, the dependent variable continues to decrease (positive slope) or increase (negative slope) without limit unless one of the variables is constant. (See item 4 below.)
3. The rate of increase or decrease is constant; in other words, it will not curve upward or downward, but is a straight line.
4. There are no horizontal or vertical asymptotes unless the dependent variable is defined for only *one* value of the independent variable *or* if the dependent variable is the same value for *all* values of the independent variable.

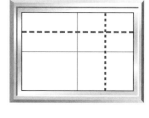

Examples illustrating if a system is linear:

- You are driving your car at a constant speed of 45 miles per hour [mph]. The longer you drive, the farther you go, without limit. In addition, your distance increases by the same amount each hour, regardless of total time elapsed. This is a linear system.
- You observe the temperature of the brake disks on your car to be slowly decreasing. If it continued to decrease without limit, the temperature would eventually be less than absolute zero (an impossibility); thus, it is not linear. Also, it seems reasonable that the temperature will eventually approach the surrounding air temperature; thus, there is a horizontal asymptote.

If the system is not linear, is there a vertical asymptote?

If there *is* a vertical asymptote, it will also have a horizontal asymptote. This is a power law model with a negative exponent. REMEMBER: We are assuming that our data fit one of the three models being considered here, and the previous statement is certainly not true for all other models. For example, $y = \tan x$ has multiple vertical asymptotes, but no horizontal asymptote.

If there is not *a vertical asymptote, is there a horizontal asymptote?*

If there is a horizontal asymptote (but not a vertical one), then the model is exponential. If the horizontal asymptote occurs for positive values of the independent variable, then the exponent is negative. If the horizontal asymptote occurs for negative values of the independent variable, then the exponent is positive.

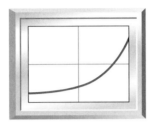 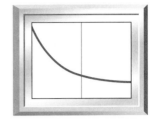

What if there is not *a horizontal asymptote or a vertical asymptote?*

It is a power law model with a positive exponent. Such models can have a variety of shapes.

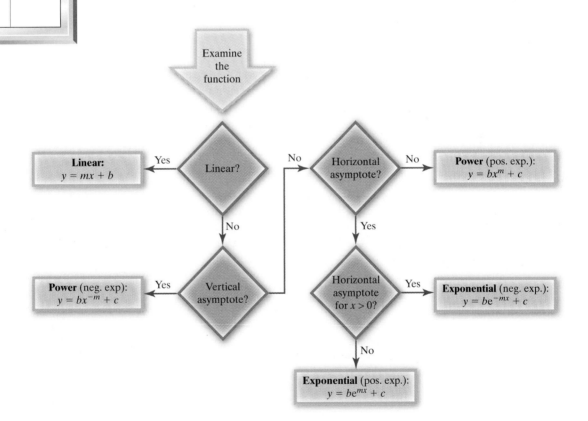

This sequence of questions can be represented pictorially as shown above. Remember, this is only valid if we assume the data fits one of the three models being discussed.

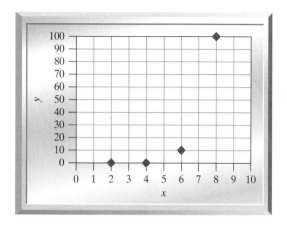

Guideline 3: Convert Axes to a Logarithmic Scale

If the logarithm of the dependent or independent variable is plotted instead of the variable itself, do the modified data points appear to lie on a straight line?

To see how logarithmic axes are constructed, let us consider a simple case. Plotting the data points below gives the graph shown to the left.

x	2	4	6	8
y	0.1	1	10	100

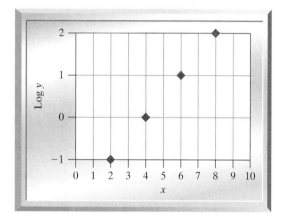

Now let us take the logarithm (base 10) of the independent (vertical) variable and plot.

x	2	4	6	8
y	0.1	1	10	100
log y	−1	0	1	2

Finally, we simply change the labels on the vertical axis of the second chart back to the original values of y from which each logarithmic value was obtained.

Note that the vertical positions of the data points are determined by the logarithms of the values, but the numeric scale uses the actual data values.

A note about the use of logarithmic scales:

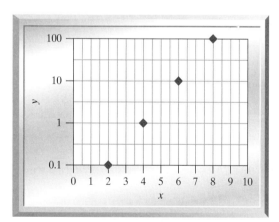

- The original data would fit an exponential model ($y = 0.01e^{1.15x}$), and when plotted on a logarithmic vertical axis, the data points appear in a straight line.
- The logarithmic axis allows us to more easily distinguish between the values of the two lowest data points, even though the data range over three orders of magnitude. On the original graph, 0.1 and 1 were almost in the same vertical position.
- Note that you *do not* have to calculate the logarithms of the data points. You simply plot the actual values on a logarithmic scale.

Logarithm graphs are discussed in more detail in Section 13.2. How does this help us determine an appropriate model type?

- Plot the data using normal (linear) scales for both axes. If the data appear to lie more or less in a straight line, a linear model is likely to be a good choice.
- Plot the data on a logarithmic vertical scale and a normal (linear) horizontal scale. If the data then appear to lie more or less in a straight line, an exponential model is likely to be a good choice.

- Plot the data with logarithmic scales for *both* axes. If the data then appear to lie more or less in a straight line, a power law model is likely to be a good choice.
- Although not covered in this course, you could plot the data on a logarithmic horizontal scale and a normal (linear) vertical scale. If the data then appear to lie more or less in a straight line, a logarithmic model is likely to be a good choice.

REMEMBER, this is only valid if we assume the data fits one of the three models being discussed.

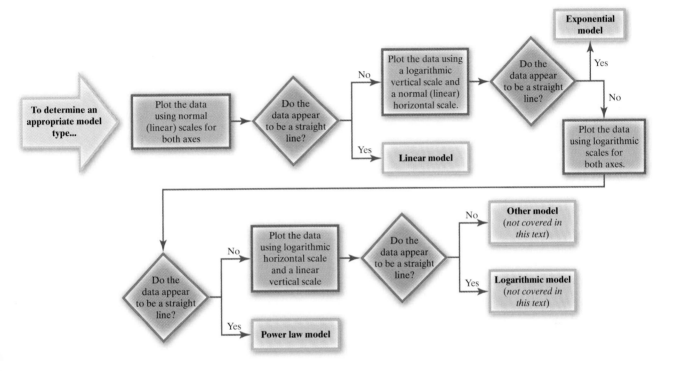

Guideline 4: Consider the R^2 Value

When a trendline is generated in Excel, the program can automatically calculate an **R^2 value**, sometimes called the **coefficient of determination**. The R^2 value is an indication of the variation of the actual data from the equation generated—in other words, it is a measure of how well the trendline fits the data. The value of R^2 varies between 0 and 1. If the value of R^2 is exactly equal to 1, a perfect correlation exists between the data and the trendline, meaning that the curve passes exactly through all data points. The farther R^2 is from 1, the less confidence we have in the accuracy of the model generated. When fitting a trendline to a data set, we always report the R^2 value to indicate how well the fit correlates with the data.

In reality, a fit of $R^2 = 1$ is rare, since experimental data are imprecise in nature. Human error, imprecision in instrumentation, fluctuations in testing conditions, and natural specimen variation are among the factors that contribute to a less-than-perfect fit. **The best R^2 value is not necessarily associated with the best model and should be used as a guide only**. Once again, making such decisions becomes easier with experience.

When displaying the equation corresponding to a trendline, you may have already noticed how to display the R^2 value.

↳ *To display an R^2 value:*

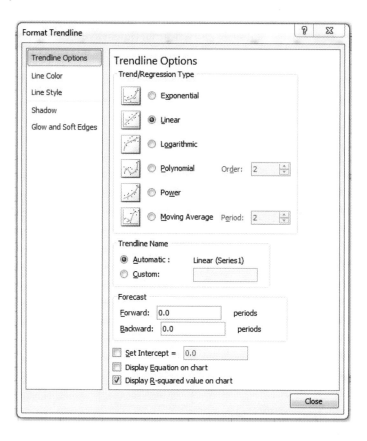

- Right-click the trendline or choose the trendline, then choose **Format > Format Selection**.
- In the **Format Trendline** window that opens, from the **Trendline** Options tab, check the box for **Display *R*-squared value on chart**. Click **Close**.

⌘ **Mac OS:** To show the R^2 value on a Mac, double-click the trendline. In the window that opens, click **Options** and select **Display *R*-squared value**. Click **OK**.

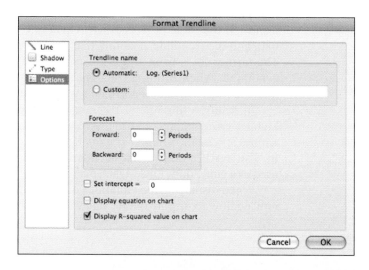

Try different models and compare the R^2 values.

- If one of the R^2 values is considerably smaller than the others, say, more than 0.2 less, then that model very likely can be eliminated.
- If one of the R^2 values is considerably larger than the others, say, more than 0.2 greater, then that model very likely is the correct one.

In any case, you should always consider Guidelines 1 through 3 above to minimize the likelihood of error.

WARNING!

While practicing with trendlines in the preceding chapters, you may have noticed a choice for polynomial models. Only rarely would this be the proper choice, but we mention it here for one specific reason—a polynomial model can always be found that will perfectly fit any data set. In general, if there are N data points, a polynomial of order $N − 1$ can be found that goes exactly through all N points. Excel can only calculate polynomials up to sixth order. For example, a data set with five data points is plotted below. A fourth-order polynomial can be found that perfectly fits the data. Let us consider a simple spring stretching example to illustrate why a perfect fit to the data is not necessarily the correct model.

The graph shows the five data points for spring displacement as a function of force. As force increases, displacement increases, but the points are certainly not in a straight line. Also shown is a fourth-order polynomial model that goes through every point—a perfect fit. This, however, is a terrible model.

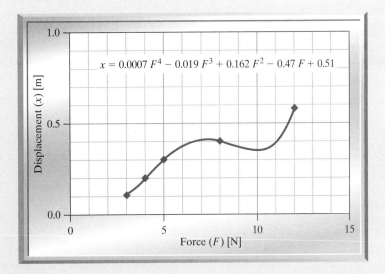

$$x = 0.0007\,F^4 - 0.019\,F^3 + 0.162\,F^2 - 0.47\,F + 0.51$$

Presumably you agree that as force increases, displacement *must* increase as well. The polynomial trendline, however, suggests that as force increases from about 7 to 10 newtons, the displacement *decreases*.

Always ask yourself if the model you have chosen is obviously incorrect, as in this case. We do not use polynomial models in this book, and so discuss them no further.

THE THEORY OF OCCAM'S RAZOR

It is vain to do with more what can be done with less.
or
Entities are not to be multiplied beyond necessity.

—William of Occam

It is probably appropriate to mention Occam's Razor at this point. Those who choose to pursue scientific and technical disciplines should keep the concept of Occam's Razor firmly in mind. **Occam's Razor refers to the concept that the simplest explanation or model to describe a given situation is usually the correct one.** It is named for William of Occam, who lived in the first half of the fourteenth century and was a theologian and philosopher.

● **EXAMPLE 13-2**

The velocity of a ball was recorded as it rolled across a floor after being released from a ramp at various heights. The velocities were then plotted versus the release heights. We want to fit a trendline to the data.

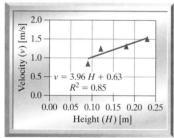

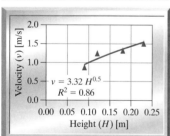

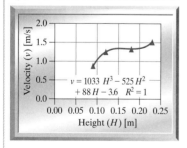

We start with the simplest form, a linear fit, shown on the left. We know that if the ramp is at a height of zero, the ball will not roll down the ramp without any external forces. The linear fit yields an intercept value of 0.6, indicating that the ball will have an initial velocity of 0.6 meter per second when the ramp is horizontal, which we know to be untrue. It seems unlikely experimental variation alone would generate an error this large, so we try another model.

We choose a power fit, shown in the center. With an R^2 value of 0.86, the equation fits the data selection well, but is there a better fit? Using the same data, we try a third-order polynomial to describe the data. The polynomial model, which gives a perfect fit, is shown on the bottom with an R^2 value of 1.

While the polynomial trendline gave the best fit, is this really the correct way to describe the data? Recall that in theory the potential energy of the ball is transformed into kinetic energy according to the conservation of energy law, written in general terms

$$PE_{initial} = KE_{final} \quad \text{or} \quad mgH = \frac{1}{2}mv^2$$

Therefore, the relationship between velocity and height is a relationship of the form

$$v = (2gH)^{1/2} = (2g)^{1/2}H^{1/2}$$

The relation between velocity and height is a power relationship; velocity varies as the square root of the height. The experimental error is responsible for the inaccurate trendline fit. In most instances, the polynomial trendline will give a precise fit but an inaccurate description of the phenomenon. **It is better to have an accurate interpretation of the experimental behavior than a perfect trendline fit!**

Guideline 5: Should We Consider Model Types Not Covered Here?

Many phenomena may be accurately characterized by a linear model, power law model, or exponential model. However, there are innumerable systems for which a different model type must be chosen. Many of these are relatively simple, but some are mind-bogglingly complicated. For example, modeling electromagnetic waves (used for television, cell phones, etc.) or a mass oscillating up and down while hanging from a spring requires the use of trigonometric functions.

You should always keep in mind that the system or phenomenon you are studying may not fit the three common models we have covered in this book.

NOTE ON ADVANCED MATH

Actually, sinusoids (sine or cosine) can be represented by exponential models through a mathematical trick first concocted by Leonhard Euler, so we now refer to it as Euler's identity. The problem is that the exponents are imaginary (some number times the square root of −1).

Euler's identity comes up in the study of calculus, and frequently in the study of electrical or computer engineering, and early in the study of electric circuits. Euler's identity can be expressed in several different forms. The basic identity can be stated as the following equation, where i is the square root of −1.

$$e^{i\pi} = -1$$

Another form often used in electrical engineering is

$$\cos\theta = 0.5(e^{i\theta} + e^{-i\theta})$$

13.2 INTERPRETING LOGARITHMIC GRAPHS

A "regular" plot, shown on a graph with both axes at constant-spaced intervals, is called **rectilinear**. When a linear function is graphed on rectilinear axis, it will appear as a straight line. Often, it is convenient to use a scale on one or both axes that is not linear, where values are not equally spaced but instead "logarithmic," meaning that powers of 10 are equally spaced. Each set of 10 is called a **decade** or **cycle**. A logarithmic scale that ranges either from 10 to 1,000 would be two cycles, 10–100 and 100–1,000. Excel allows you to select a logarithmic scale for the abscissa, the ordinate, or both.

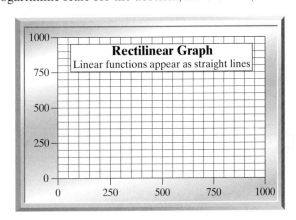

If one scale is logarithmic and the other linear, the plot is called **semilogarithmic** or **semilog**. Note in the figure that the abscissa has its values equally spaced and so is a linear scale. However, the ordinate has powers of 10 equally spaced and thus is a logarithmic scale.

If both scales are logarithmic, the plot is called **full logarithmic** or **log–log**. Note in the figure that both axes have powers of 10 equally spaced.

There are four different combinations of linear and logarithmic axes, each corresponding to one of four specific trendline types that will appear linear on that particular graph type. If the plotted data points are more or less in a straight line when plotted with a specific axis type, the corresponding trendline type is a likely candidate, as discussed earlier.

Once the data are plotted as logarithmic, how do you read data from this graph? This is perhaps best shown through examples.

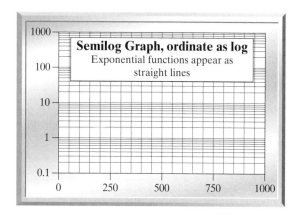

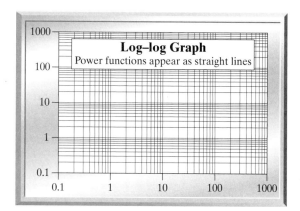

Derivation

Consider a power law model:

$$y = bx^m$$

Now take the logarithm of both sides of the equation.

$$\log y = \log (bx^m) = \log b + \log x^m = \log b + m \log x$$

Using the commutative property of addition, you can write:

$$\log y = m \log x + \log b$$

Since b is a constant, $\log b$ is also a constant. Rename $\log b$ and call it b'. Since x and y are both variables, $\log x$ and $\log y$ are also variables. Call them x' and y', respectively.

Using the new names for the transformed variables and the constant b:

$$y' = mx' + b'$$

This is a linear model! Thus, if the data set can be described by a power law model and you plot the logarithms of both variables (instead of the variables themselves), the transformed data points will lie on a straight line. The slope of this line is m, although "slope" has a somewhat different meaning than in a linear model.

● **EXAMPLE 13-3**

An unknown amount of oxygen, kept in a piston type container at a constant temperature, was subjected to increasing pressure (P), in units of atmospheres; as the pressure (P) was increased, the resulting volume (V) was recorded in units of liters. We have found that a log–log plot aligns the data in a straight line. Using the figure, determine the mathematical equation for volume (V) in units of liters, and of a piston filled with an ideal gas subjected to increasing pressure (P) in units of atmospheres.

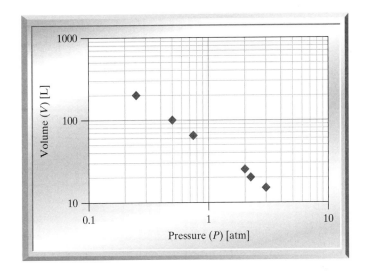

Since the graph appears linear on log–log paper, we can assume a power law relationship exists of the form:

$$V = bP^m$$

For illustration, a line has been sketched between the points for further clarification of function values.

To determine the power of the function (m), we can estimate the number of decades of "rise" (shown as a vertical arrow) divided by decades of "run" (horizontal arrow):

$$\text{Slope} = \frac{\text{Change in decades of volume}}{\text{Change in decades of pressure}}$$

$$= \frac{-1 \text{ decade}}{1 \text{ decade}} = -1$$

To establish the constant value (b), we estimate it as the ordinate value when the abscissa value is 1, shown in the shaded circle. When the pressure is 1 atmosphere, the volume is 50 liters.

The resulting function:

$$V = 50P^{-1}$$

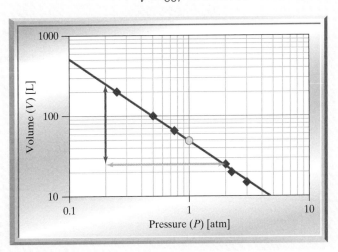

● **EXAMPLE 13-4**   When a body falls, it undergoes a constant acceleration. Using the figure, determine the mathematical equation for distance (d), in units of meters, of a falling object as a function of time (t), in units of seconds.

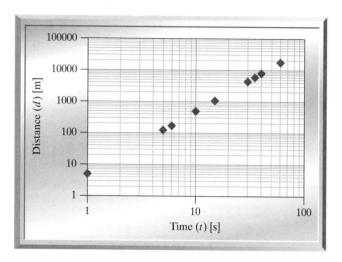

Since the graph appears linear on log–log paper, we can assume a power law relationship exists of the form:

$$d = bt^m$$

For illustration, a line has been sketched between the points for further clarification of function values.

To establish the power of the function (m), we estimate the number of decades of "rise" (shown as vertical arrows) divided by the decades of "run" (horizontal arrow):

$$\text{Slope} = \frac{\text{Change in decades of distance}}{\text{Change in decades of time}} = \frac{2 \text{ decade}}{1 \text{ decade}} = 2$$

To establish the constant value (b), we estimate it as the ordinate value when the abscissa value is 1, shown in the shaded circle. When the time is 1 second, the distance is 5 meters.

The resulting function:

$$d = 5t^2$$

This matches well with the established theory, which states

$$d = \frac{1}{2}gt^2$$

The value of ½ g is approximately 5 m/s².

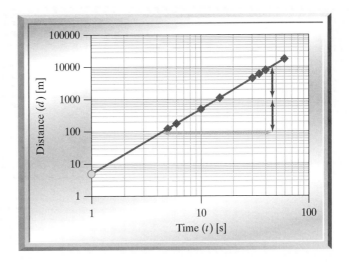

Derivation

Consider an exponential model:

$$y = be^{mx}$$

Now take the logarithm of both sides of the equation.

$$\log y = \log(be^{mx}) = \log b + \log e^{mx} = \log b + (mx)\log e$$

Using the commutative property of addition, you can write:

$$\log y = m(\log e)x + \log b$$

Since b is a constant, $\log b$ is also a constant. Rename $\log b$ and call it b'. Since y is a variable, $\log y$ is also a variable; call it y'.

Using the new names for the transformed variable y and the constant b:

$$y' = m(\log e)x + b'$$

This is a linear model! Thus, if the data set can be described by an exponential law model, and you plot the logarithm of y (instead of y itself) versus x, the transformed data points will lie on a straight line. The slope of this line is $m(\log e)$, but again, "slope" has a somewhat different interpretation. The term $(\log e)$ is a number, approximately equal to 0.4343; the slope is 0.4343 m.

● **EXAMPLE 13-5**   A chemical reaction is being carried out in a reactor; the results are shown graphically in the figure. Determine the mathematical equation that describes the reactor concentration (C), in units of moles per liter, as a function of time spent in the reactor (t), in units of seconds.

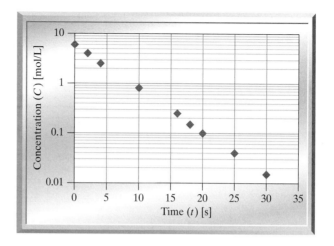

Since the graph appears linear on semilog paper where the ordinate is logarithmic, we can assume an exponential law relationship exists of the form:

$$C = be^{mt}$$

For illustration, a line has been sketched between the points for further clarification of function values.

Since this is an exponential function, to determine the value of m, we must first determine the slope:

$$\text{Slope} = \frac{\text{Change in decades of concentration}}{\text{Change in time}}$$

$$= \frac{-1 \text{ decade}}{21.5 \text{ s} - 10 \text{ s}} = -0.087 \text{ s}^{-1}$$

The value of m is then found from the relationship: slope = $m(\log e)$.

$$m = \frac{\text{slope}}{\log e} = \frac{-0.087 \text{ s}^{-1}}{0.4343} = -0.2 \text{ s}^{-1}$$

When time = 0 seconds, the constant (b) can be read directly and has a value of 6 [mol/L]. The resulting function:

$$C = 6e^{-0.2t}$$

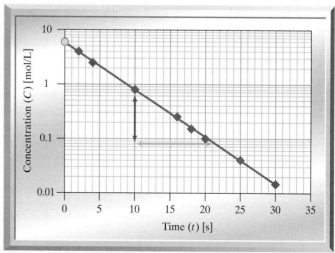

● **EXAMPLE 13-6**   The data shown graphically in the figure describe the discharge of a capacitor through a resistor. Determine the mathematical equation that describes the voltage (V), in units of volts, as a function of time (t), in units of seconds.

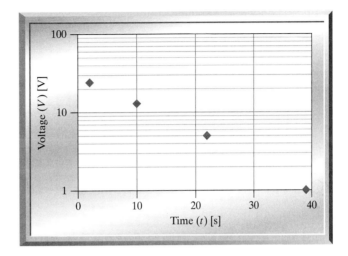

Since the graph appears linear on semilog paper where the ordinate is logarithmic, we can assume an exponential law relationship exists of the form:

$$V = be^{mt}$$

For illustration, a line has been sketched between the points for further clarification of function values.

Since this is an exponential function, to determine the value of m, we must first determine the slope:

$$\text{Slope} = \frac{\text{Change in decades of voltage}}{\text{Change in time}}$$

$$= \frac{-1 \text{ decade}}{36 \text{ s} - 10 \text{ s}} = -0.038 \text{ s}^{-1}$$

The value of m is then found from the relationship: Slope $= m \, (\log e)$.

$$m = \frac{\text{slope}}{\log e} = \frac{-0.038 \text{ s}^{-1}}{0.4343} = -0.89 \text{ s}^{-1}$$

When time = 0 seconds, the constant (b) can be read directly and has a value of 30 volts.

The resulting function:

$$V = 30e^{-0.89t}$$

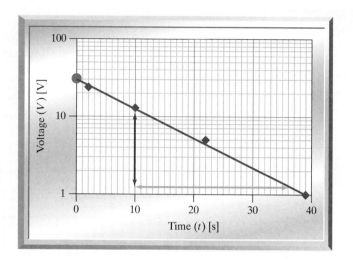

13.3 CONVERTING SCALES TO LOG IN EXCEL

To convert axis to logarithmic:

- Right-click the axis to convert to logarithmic. The **Format Axis** window will appear.
- Click **Axis Options**, then check the box for **Logarithmic scale**.

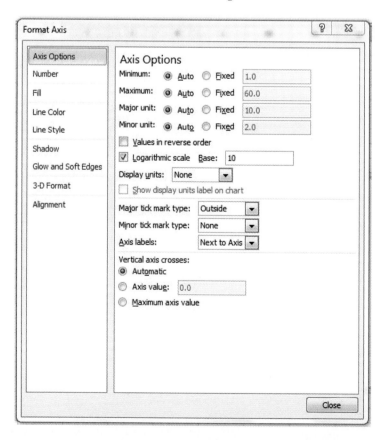

Alternatively:

- Click the chart. In the tool bar, select **Layout > Axes** and choose the axis to convert.
- In the corresponding menu, select **Show Axis with Log Scale**.

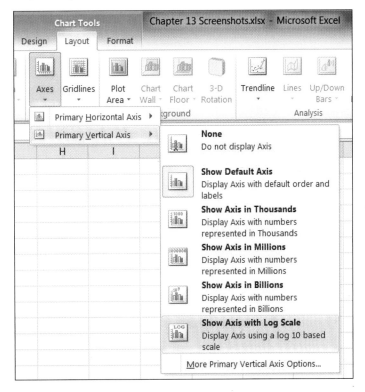

⌘ **Mac OS:** Double-click on the axis you want to convert to logarithmic. The Format Axis window will appear. Click **Scale** in the list on the left side of the window, and then click the checkbox near the bottom that says "Logarithmic scale."

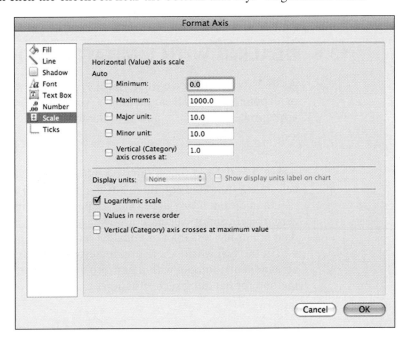

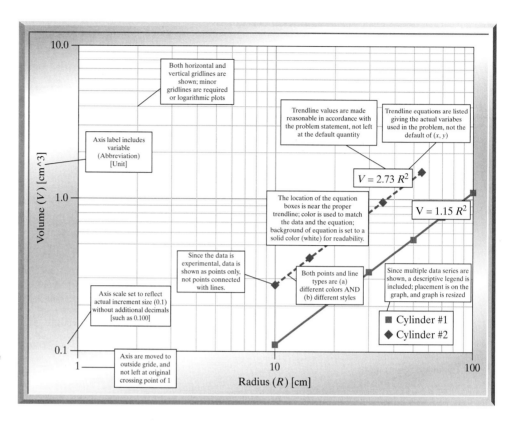

Figure 13-1 Example of a proper plot, showing multiple experimental data sets with trendlines and logarithmic axes.

Above is an example of a properly formatted graph, showing an experimental data series with power trendlines. The axes have been made logarithmic to allow the data series to appear liner.

13.4 DEALING WITH LIMITATIONS OF EXCEL

As we have mentioned earlier, Excel will not correctly calculate a trendline for a power or exponential model containing a vertical offset. In other words, it can calculate appropriate values for b and m in the forms

$$y = bx^m + c \quad \text{or} \quad y = be^{mx} + c$$

only if $c = 0$. Note that if the data inherently has a vertical offset, Excel may actually calculate a trendline equation, but the values of b and m will not be accurate.

In addition, if any data value, dependent or independent, is less than or equal to zero, Excel cannot calculate a power law model. If one or more dependent variable data points are less than or equal to zero, exponential models are unavailable.

In the real world, there are numerous systems best modeled by either a power or an exponential model with a nonzero value of c or with negative values, so we need a method for handling such situations.

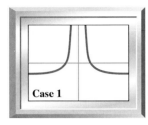

Case 1: Vertical Asymptote

Since a vertical asymptote implies a power model with a negative exponent, there will be a horizontal asymptote as well. If the horizontal asymptote is not the horizontal axis (implying a vertical offset), Excel will calculate the model incorrectly or not at all. The object here is to artificially move the asymptote to the horizontal axis by subtracting the offset value from every data point. If you have a sufficient range of data, you may be able to extract the offset from the data.

For example, if the three data points with the largest values of x (or smallest if the asymptote goes to the left) have corresponding y values of 5.1, 5.03, and 5.01, the offset is likely to be about 5. (This assumes there are other values in the data set with considerably different y values.) You can also try to determine from the physical situation being modeled at what nonzero value the asymptote occurs. In either case, simply subtract the offset value from the vertical component of *every* data point, plot this modified data, and determine a power trendline. Once the trendline equation is displayed, edit it by adding the offset to the power term. (Note that you subtract the value from the data points but add it to the final equation.)

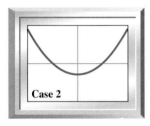

Case 2: No Horizontal Asymptote

Assuming it has been established that the model should not be linear, Case 2 implies a power model with a positive exponent. If you have a data point for $x = 0$, the corresponding y value should be very close to the vertical offset. Also, you may be able to determine the offset value by considering the physical situation. In either case, proceed as in case 1, subtracting the offset from every data point, etc.

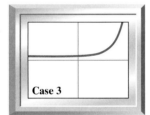

Case 3: Horizontal Asymptote, No Vertical Asymptote

Case 3 implies an exponential model. The object, as in Case 1, is to artificially move the asymptote to the horizontal axis. Also as in Case 1, you may be able to determine the offset by considering the physical system or by looking at the data points with the largest or smallest values. Again, subtract the determined value from every data point, etc.

Case 4: A Few Values with Small Negative Value, Most Positive

In Case 4, the negative values may be a result of measurement inaccuracy. Either delete these points from the data set or change the negative values to a very small positive value.

Case 5: Many or All Data Points Negative

If the independent values are negative, try multiplying every independent value by -1. If this works, then make the calculated value of b negative after the trendline equation is calculated. You may have to apply some of the procedures in the previous cases after negating each data value.

Negative dependent values may simply be a negative offset to the data. If you can determine the asymptote value, ask if essentially all values are greater than the asymptote value. If so, it is probably just an offset. If not, then multiply every dependent data value by -1, and proceed in a manner similar to that described in the preceding paragraph.

NOTE

Sometimes, due to inaccuracy of measurement, one or more of the data points near the asymptote may be negative after the offset value is subtracted, and Excel will be unable to process the data. You can circumvent this either by deleting such data points or by making the vertical component a very small positive value.

● **EXAMPLE 13-7**

The following data were collected in an experiment. We wish to determine an appropriate model for the data.

As the independent variable gets larger, the dependent variable appears to be approaching 10. This is even more apparent when graphed. Subtracting this assumed offset from every data point gives a new column of modified dependent data.

Since we subtracted 10 from every data point, we need to correct the equation by adding 10, giving $y = 14.3e^{-0.5x} + 10$.

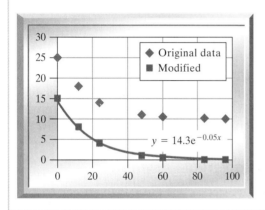

Independent	Dependent	Modified Dependent
0	25.0	15.0
12	18.0	8.0
24	14.0	4.0
48	11.0	1.0
60	10.5	0.5
84	10.2	0.2
96	10.1	0.1

● **EXAMPLE 13-8**

The following data were collected during an experiment. We wish to determine an appropriate model for the data.

You notice that all of the dependent values are negative. Since the value when the independent variable is zero is nonzero, this is likely to be an exponential model, although it could possibly be a power law model with an offset.

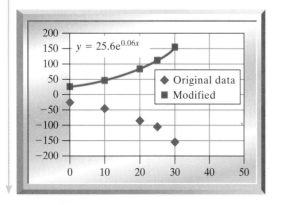

Independent	Dependent	Modified Dependent
0	−25	25
10	−45	45
20	−85	85
25	−106	106
30	−154	154

Following the first assumption (that the model is exponential), multiply every data point by −1, plot the inverted data, and fit an exponential trendline. This seems to fit the data quite well. Finally, negate the trendline equation to obtain the true model, giving $y = -25.6e^{0.06x}$.

If All Else Fails

If you are convinced that the model is exponential or power with an offset but you cannot determine its value, consider making further measurements for larger or smaller values of the independent variable. Particularly in data that has a horizontal asymptote, further measurements may make the value of the asymptote more obvious. In the first chart below, the value of the asymptote is not clear. By extending the measurements in the direction of the asymptote (positive in this case), it is clear that the asymptote has a value of 2. Note also that it becomes much clearer that the data are not linear.

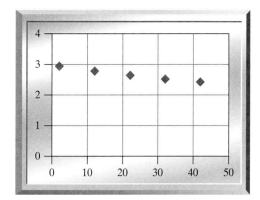
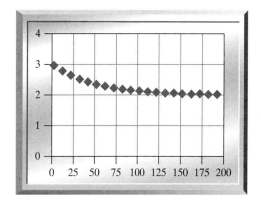

EXAMPLE 13-9

The data shown describe the discharge of a capacitor through a resistor. Before the advent of microprocessors, intermittent windshield wipers in automobiles often used such circuits to create the desired time delay. We wish to determine an appropriate model for the data.

Time (*t*) [s]	2	10	22	39
Voltage (*V*) [V]	24	13	5	1

- *Select the data series and create a linear trendline, being sure to display the equation and the R^2 value.*
- *Without deleting the first trendline, click one of the data points again—be sure you select the points and not the trendline—and again add a trendline, but this time choose a power trend.*
- *Repeat this process for an exponential trend.*

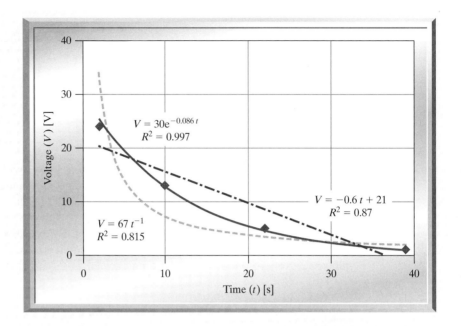

You should now have a chart with three trendlines. Things to note:

- *Neither the linear nor power trendlines are very good compared to the exponential line, and both have an R^2 value less than 0.9. These are probably not the best choice.*
- *The exponential model fits the data very closely and has an R^2 value greater than 0.95; thus, it is probably the best choice.*

As a model check, compare the graph by using logarithmic scales. If the model is exponential, the data should appear linear on a semilogarithmic plot with the ordinate shown as logarithmic.

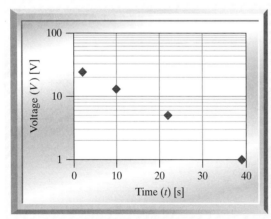
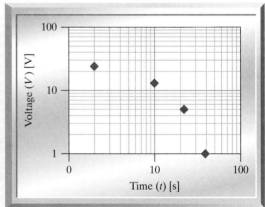

Based on this analysis, you would choose the exponential model. As it turns out, the exponential model is indeed the correct one, being the solution to a differential equation describing the capacitor's behavior. Most students learn about this in second semester physics, and some study it in much more depth in electrical and computer engineering courses.

● EXAMPLE 13-10 These data describe the temperature of antifreeze (ethylene glycol) in the radiator of a parked car. The temperature of the surrounding environment is −20 degrees Fahrenheit. The initial temperature (at $t = 0$) is unknown.

Time (t) [min]	10	18	25	33	41
Temperature (T) [°F]	4.5	1.0	−2.1	−4.6	−6.4

- Determine an appropriate model type for these data.
- Determine the vertical offset of the data.
- Plot the modified data and generate the correct trendline equation to describe this data.

It seems reasonable that the temperature will be asymptotic to the surrounding temperature (−20 degrees Fahrenheit) as time goes on. Also, there is no known mechanism whereby the temperature could possibly go to infinity for any finite value of time, so there is no vertical asymptote. This indicates an exponential model with a negative exponent. Since the asymptote is at −20 degrees Fahrenheit, subtract −20 (i.e., add 20) from every data point before plotting.

Time (t) [min]	10	18	25	33	41
Temperature (T) [°F]	4.5	1.0	2.1	4.6	−6.4
Offset temperature (T_O) [°F]	24.5	21.0	17.9	15.4	13.6

Since you subtracted −20 from every data point, you should add −20 to the trendline equation, giving

$$T = 29.5e^{-0.019t} - 20$$

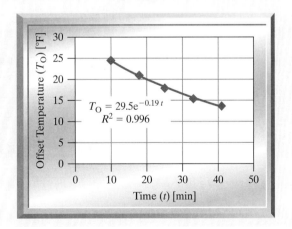

● EXAMPLE 13-11   Assume the car in Example 13.10 is cranked up and driven 50 feet into a garage. The temperature inside the garage is 5 degrees Fahrenheit. These data describe the temperature of antifreeze in the radiator after it is driven into the garage and the motor turned off.

Time (t) [min]	5	13	25	34	51
Temperature (T) [°F]	−13.0	−10.0	−6.8	−4.5	−1.5

The antifreeze temperature will increase asymptotically to 5 degrees Fahrenheit. Since all of the data points are less than this, multiply through by −1 before applying the offset of 5 degrees Fahrenheit. Also, be rather careful here. Since you are dealing with the negatives of the temperatures, you must deal with the negative of the offset as well. Thus, multiply each temperature by −1, then subtract the negative of the asymptote value of 5 degrees Fahrenheit (add 5 to each value). This approach gives the following.

Time (t) [min]	5	13	25	34	51
Temperature (T) [°F]	−13.0	−10.0	−6.8	−4.5	−1.5
Adjusted temperature (T_A) [°F]	18.0	15.0	11.8	9.5	6.5

Since you are working with the negatives of the values, put a minus by the exponential term. Since you subtracted −5 from the data, you might initially think to add −5 to the equation. However, since you used the negatives of the values, add 5 as an offset in the equation since the two negatives cancel. This gives the final equation

$$T = 5 - 20e^{-0.022t}$$

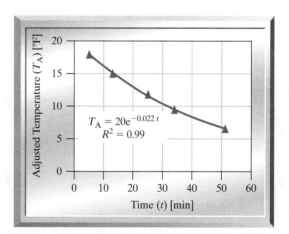

In-Class Activities

ICA 13-1

Capillary action draws liquid up a narrow tube against the force of gravity as a result of surface tension. The height the liquid will move up the tube depends on the radius of the tube. The following data were collected for water in a glass tube in air at sea level. Show the resulting data and trendline, with equation and R^2 value, on the appropriate graph type (rectilinear, semilog, or log–log) to make the data appear linear.

Radius (r) [cm]	0.01	0.05	0.10	0.20	0.40	0.50
Height (h) [cm]	14.0	3.0	1.5	0.8	0.4	0.2

ICA 13-2

Several reactions are carried out in a closed vessel. The following data are taken for the concentration (C) of compounds A, B, and C [grams per liter, g/L] as a function of time (t) [minutes, min], from the start of the reaction. Show the resulting data and trendlines, with equation and R^2 value, on the appropriate graph type (rectilinear, semilog, or log–log) to make the data appear linear.

Time (t) [min]	2	5	8	15	20
Concentration of A (C_A) [g/L]	0.021	0.125	0.330	1.120	2.050
Concentration of B (C_B) [g/L]	0.032	0.202	0.550	1.806	3.405
Concentration of C (C_C [g/L]	0.012	0.080	0.200	0.650	1.305

ICA 13-3

An environmental engineer has obtained a bacteria culture from a municipal water sample and allowed the bacteria to grow. The data are shown below. Show the resulting data and trendline, with equation and R^2 value, on the appropriate graph type (rectilinear, semilog, or log–log) to make the data appear linear.

Time (t) [h]	0	2	3	5	6	7	9	10
Concentration (C) [ppm]	0	21	44	111	153	203	318	385

ICA 13-4

In a turbine, a device used for mixing, the power requirement depends on the size and shape of impeller. In the lab, you have collected the following data. Show the resulting data and trendline, with equation and R^2 value, on the appropriate graph type (xy scatter, semilog, or log–log) to make the data appear linear.

Diameter (D) [ft]	0.5	0.75	1	1.5	2	2.25	2.5	2.75
Power (P) [hp]	0.004	0.04	0.13	0.65	3	8	18	22

ICA 13-5

Being quite interested in obsolete electronics, Angus has purchased several electronic music synthesis modules dating from the early 1970s and is testing them to find out how they work. One module is a voltage-controlled amplifier (VCA) that changes the amplitude (loudness) of an audio signal by changing a control voltage into the VCA. All Angus knows is that the magnitude of the control voltage should be less than 5 volts. He sets the audio input signal to an amplitude of 1 volt, then measures the audio output amplitude for different control voltage values. The table below shows these data. Show the resulting data and trendline, with equation and R^2 value, on the appropriate graph type (*xy* scatter, semilog, or log–log) to make the data appear linear.

Control voltage (V) [V]	−4.0	−2.5	−1.0	0.0	1.0	2.5	4.0
Output amplitude (A) [V]	0.116	0.324	0.567	0.962	1.690	3.320	7.270

ICA 13-6

Referring to the previous ICA, Angus is also testing a voltage-controlled oscillator. In this case, a control voltage (also between −5 and +5 volts) changes the frequency of oscillation in order to generate different notes. The table below shows these measurements. Show the resulting data and trendline, with equation and R^2 value, on the appropriate graph type (*xy* scatter, semilog, or log–log) to make the data appear linear.

Control voltage (V) [V]	−4.0	−2.5	−1.0	0.0	1.0	2.5	4.0
Output frequency (f) [Hz]	28	99	227	539	989	3,110	8,130

ICA 13-7

A growing field of inquiry that poses both great promise and great risk for humans is nanotechnology, the construction of extremely small machines. Over the past couple of decades, the size that a working gear can be made has consistently gotten smaller. The table shows milestones along this path.

Years from 1967	0	5	7	16	25	31	37
Minimum gear size [mm]	0.8	0.4	0.2	0.09	0.007	2E-04	8E-06

(a) Show the resulting data and trendline, with equation and R^2 value, on the appropriate graph type (*xy* scatter, semilog, or log–log) to make the data appear linear. What do you think is the limiting factor on the smallest size that gears can eventually be manufactured?

(b) According to this model, how many years does it take (from any point in time) for the minimum size to be cut in half?

(c) According to the model, during what year will the smallest gear be one-tenth the size of the smallest gear in 2009?

ICA 13-8

If an object is heated, the temperature of the object will increase. The thermal energy (Q) associated with a change in temperature (ΔT) is a function of the mass of the object (m) and the specific heat (C_p), which is the rate at which the object will lose or gain energy. Specific heat is a material property, and values are available in literature. In an experiment, heat is applied to the end of an object, and the temperature change at the other end of the object is recorded. An unknown material is tested in the lab, yielding the following results.

Heat applied (Q) [J]	2	8	10	13	18	27
Temp change (ΔT) [K]	1.5	6.0	7.0	9.0	14.0	22.0

(a) Show the resulting data and trendline, with equation and R^2 value, on the appropriate graph type (*xy* scatter, semilog, or log–log) to make the data appear linear.

(b) If the material was titanium, what mass of sample was tested?

(c) If a 4-gram sample was used, which of the following materials was tested?

Material	Specific Heat Capacity (C_p) [J/(g K)]
Aluminum	0.91
Copper	0.39
Iron	0.44
Lead	0.13
Molybdenum	0.30
Titanium	0.54

ICA 13-9

For the graph shown, determine the equation of the trendline. The trendline will fit one of the following forms. In each case, the constants "m" and "b" may be positive or negative. You must choose the correct trendline equation from choices (A)–(C) below, replacing the variables "y" and "x" with those used in the graph, and determine the value and units of the constants (m and b).

(A) $y = mx + b$
(B) $y = bx^m$
(C) $y = be^{mx}$

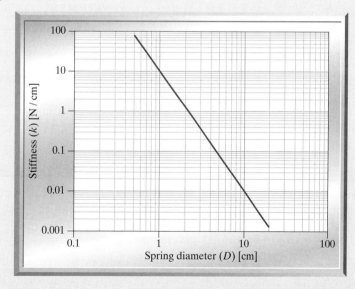

ICA 13-10

For the graph shown, determine the equation of the trendline. The trendline will fit one of the following forms. In each case, the constants "m" and "b" may be positive or negative. You must choose the correct trendline equation from choices (A)–(C) below, replacing the variables "y" and "x" with those used in the graph, and determine the value and units of the constants (m and b).

(A) $y = mx + b$
(B) $y = bx^m$
(C) $y = be^{mx}$

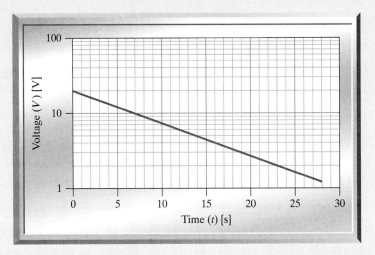

ICA 13-11

For the graph shown, determine the equation of the trendline. The trendline will fit one of the following forms. In each case, the constants "m" and "b" may be positive or negative. You must choose the correct trendline equation from choices (A)–(C) below, replacing the variables "y" and "x" with those used in the graph, and determine the value and units of the constants (m and b).

(A) $y = mx + b$
(B) $y = bx^m$
(C) $y = be^{mx}$

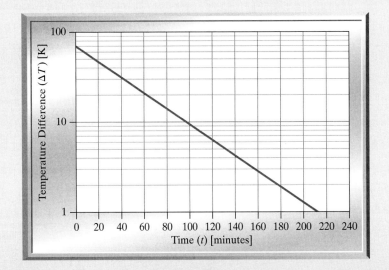

ICA 13-12

For the graph shown, determine the equation of the trendline. The trendline will fit one of the following forms. In each case, the constants "m" and "b" may be positive or negative. You must choose the correct trendline equation from choices (A)–(C) below, replacing the variables "y" and "x" with those used in the graph, and determine the value and units of the constants (m and b).

(A) $y = mx + b$
(B) $y = bx^m$
(C) $y = be^{mx}$

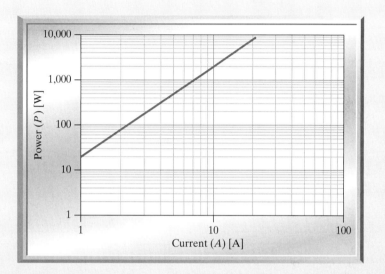

ICA 13-13

As a reminder, Reynolds Number is discussed in Chapter 9.

When discussing the flow of a fluid through a piping system, we say that friction occurs between the fluid and the pipe wall due to viscous drag. The loss of energy due to the friction of fluid against the pipe wall is described by the friction factor. The **Darcy friction factor** (f) was developed by Henry Darcy (1803–1858), a French scientist who made several important contributions to the field of hydraulics. The friction factor depends on several other factors, including flow regime, Reynolds' number, and pipe roughness. The friction factor can be determined in several ways, including from the Moody diagram (shown below).

Olive oil having a specific gravity of 0.914 and a viscosity of 100.8 centipoise is draining by gravity from the bottom of a tank. The drain line from the tank is a 4-inch diameter pipe made of commercial steel (pipe roughness, $\varepsilon = 0.045$ millimeters). The velocity is 11 meters per second. Determine the friction factor for this system, using the following process:

Step 1: Determine the Reynolds number: $Re = \dfrac{\rho v \mathrm{D}}{\mu}$.

Step 2: Determine flow regime.

- If the flow is laminar ($Re \leq 2,000$), proceed to Step 4.
- If the flow is turbulent or transitional ($Re > 2,000$), continue with Step 3.

Step 3: Determine the relative roughness ratio: (ε/D).
Step 4: Determine the *Darcy friction factor* (f) from the diagram.

ICA 13-14

Repeat ICA 13-13 with the following conditions:

Olive oil having a specific gravity of 0.914 and a viscosity of 100.8 centipoise is draining by gravity from the bottom of a tank. The drain line from the tank is a 4-inch diameter pipe made of commercial steel (pipe roughness, $\varepsilon = 0.045$ millimeters). The velocity is 1.5 meters per second. Determine the friction factor for this system.

ICA 13-15

Repeat ICA 13-13 with the following conditions:

Lactic acid, with a specific gravity of 1.249 and dynamic viscosity of 40.33 centipoise, is flowing in a 1½-inch diameter galvanized iron pipe at a velocity of 17 meters per second. Assume the pipe roughness (ε) of galvanized iron is 0.006 inches. Determine the friction factor for this system.

ICA 13-16

Repeat ICA 13-13 with the following conditions:

Lactic acid, with a specific gravity of 1.249 and dynamic viscosity of 40.33 centipoise, is flowing in a 1½-inch diameter galvanized iron pipe at a velocity of 1.5 meters per second. Assume the pipe roughness (ε) of galvanized iron is 0.006 inches. Determine the friction factor for this system.

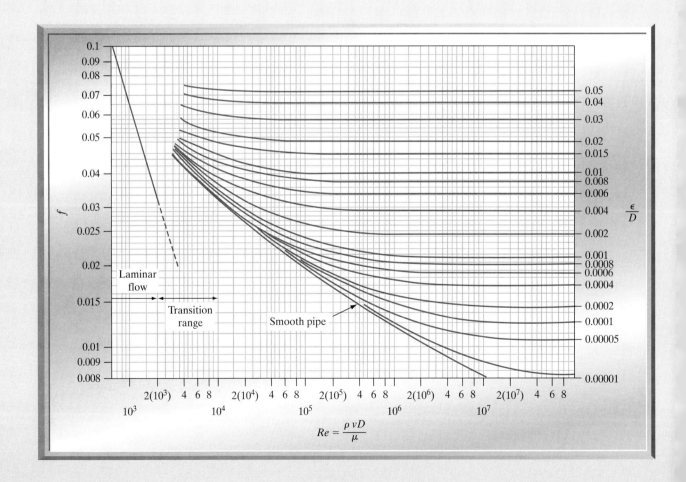

$$Re = \frac{\rho\,vD}{\mu}$$

Chapter 13 REVIEW QUESTIONS

1. An environmental engineer has obtained a bacteria culture from a municipal water sample and has allowed the bacteria to grow.

Time (t) [min]	1	2	4	6	7	9	10
Concentration (C) [ppm]	9	15	32	63	102	220	328

   (a) Plot the data and fit an appropriate trendline model, using the possible choices of linear, power, or exponential.
   (b) Suppose the growth of bacteria is governed by a process described by the following equations: Type A bacteria, $C_A = a + bt$; Type B bacteria, $C_B = ae^{bt}$; and Type C bacteria, $C_C = at^b$. What type of bacteria is the engineer dealing with? Justify your answer using logarithmic plots, and determine the values of a and b for the model chosen.

2. An environmental engineer has obtained a bacteria culture from a municipal water sample and allowed the bacteria to grow.

Time (t) [min]	1	2	4	6	7	9	10
Concentration (C) [ppm]	11.9	17.1	27.0	37.3	42.0	52.3	56.9

   (a) Plot the data and fit an appropriate trendline model, using the possible choices of linear, power, or exponential.
   (b) Suppose the growth of bacteria is governed by a process described by the following equations: Type A bacteria, $C_A = a + bt$; Type B bacteria, $C_B = ae^{bt}$; and Type C bacteria, $C_C = at^b$. What type of bacteria is the engineer dealing with? Justify your answer using logarithmic plots, and determine the values of a and b for the model chosen.

3. An environmental engineer has obtained a bacteria culture from a municipal water sample and allowed the bacteria to grow.

Time (t) [min]	1	2	4	6	7	9	10
Concentration (C) [ppm]	0.5	4.2	32.5	107.5	170.6	346.0	489.8

   (a) Plot the data and fit an appropriate trendline model, using the possible choices of linear, power, or exponential.
   (b) Suppose the growth of bacteria is governed by a process described by the following equations: Type A bacteria, $C_A = a + bt$; Type B bacteria, $C_B = ae^{bt}$; and Type C bacteria, $C_C = at^b$. What type of bacteria is the engineer dealing with? Justify your answer using logarithmic plots, and determine the values of a and b for the model chosen.

4. The viscosity of fluids depends on temperature. For many fluids, the higher the temperature, the lower the viscosity or the easier it becomes to move the fluid. Use the two graphs below to answer the following questions.
   (a) What is the dynamic viscosity and kinematic viscosity of kerosene at 50 degrees Fahrenheit? What is the specific gravity of kerosene?
   (b) What is the dynamic viscosity and kinematic viscosity of glycerin at 100 degrees Fahrenheit? What is the specific gravity of glycerin?

(c) What is the dynamic viscosity and kinematic viscosity of gasoline at 50 degrees Fahrenheit? What is the specific gravity of gasoline?

(d) What is the dynamic viscosity and kinematic viscosity of crude oil at 150 degrees Fahrenheit? What is the specific gravity of crude oil?

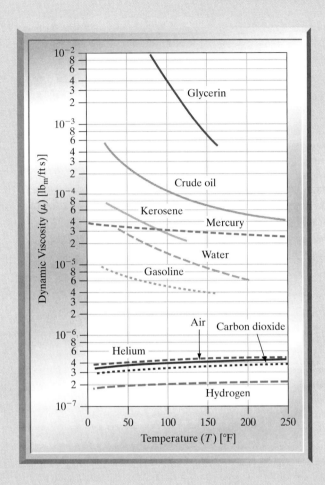

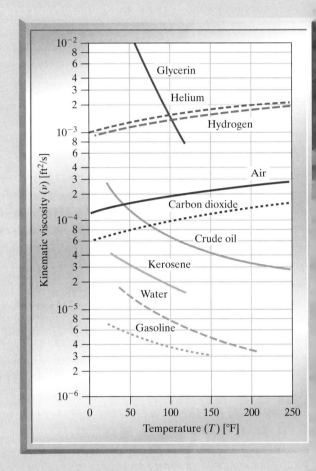

5. Eutrofication is the result of excessive nutrients in a lake or other body of water, usually caused by runoff of nutrients (animal waste, fertilizers, and sewage) from the land, which causes a dense growth of plant life. The decomposition of the plants depletes the supply of oxygen, leading to the death of animal life. Sometimes, these excess nutrients cause an algae bloom—or rapid growth of algae, which normally occur in small concentrations in the water body. To predict if this will be a problem in Lake Betchacan, we have sampled the algae concentration over time for the primary growing season.

For the graph shown, determine the equation of the trendline. The trendline will fit one of the following forms. In each case, the constants "m" and "b" may be positive or negative. You must choose the correct trendline equation from choices (A)–(C) below, replacing the variables y and x with those used in the graph, and determine the value and units of the constants (m and b).

(A) $y = mx + b$

(B) $y = bx^m$

(C) $y = be^{mx}$

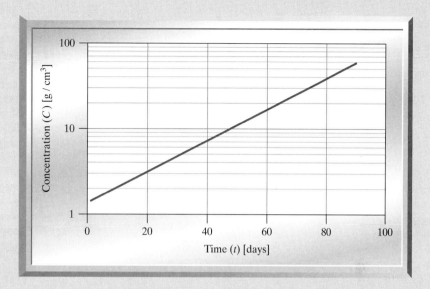

6. The incidence of traffic accidents increases as the volume of traffic on a particular road increases. A civil engineer conducts a traffic study at two locations and creates the following graph. For the graph shown, determine the equation of the trendlines. The trendlines will fit one of the following forms. In each case, the constants m and b may be positive or negative. You must choose the correct trendline equation from choices (A)–(C) below, replacing the variables y and x with those used in the graph, and determine the value and units of the constants (m and b) for each data series shown on the graph.

(A) $y = mx + b$

(B) $y = bx^m$

(C) $y = be^{mx}$

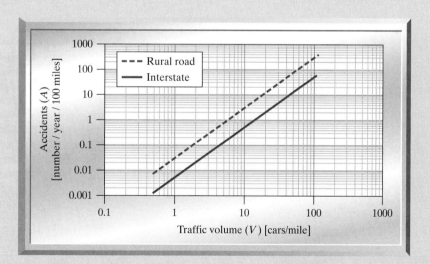

7. Biosystems engineers often need to understand how plant diseases spread in order to formulate effective control strategies. The rate of spread of some diseases is more or less linear, some increase exponentially, and some do not really fit any standard mathematical model.

   Grey leaf spot of corn is a disease (caused by a fungus with the rather imposing name of *Cercospora zeae-maydis*) that causes chlorotic (lacking chlorophyll) lesions and eventually necrotic (dead) lesions on corn leaves, thus reducing total photosynthesis and yield. In extremely severe cases, loss of the entire crop can result.

   During a study of this disease, the number of lesions per corn leaf was counted every 10 days following the initial observation of the disease, which we call day 0. At this time, there was an average of one lesion on every 20 leaves, or 0.05 lesions per leaf. The data collected during the growing season are tabulated.

   (a) Show the resulting data and trendline, with equation and R^2 value, on the appropriate graph type (*xy* scatter, semilog, or log–log) to make the data appear linear.
   (b) According to the model, how many lesions were there per leaf at the start of the survey?
   (c) How many lesions are there per leaf after 97 days?
   (d) If the model continued to be accurate, how many days would be required to reach 250 lesions per leaf?

Day	Lesions per Leaf	Day	Lesions per Leaf
0	0.05	110	4
20	0.10	120	6
30	0.20	140	17
40	0.26	150	20
60	0.60	170	40
80	1.30	190	112
90	2	200	151

8. A **pitot tube** is a device used to measure the velocity of a fluid, typically, the airspeed of an aircraft. The failure of a pitot tube is credited as the cause of Austral Líneas Aéreas flight 2553 crash in October 1997. The pitot tube had frozen, causing the instrument to give a false reading of slowing speed. As a result, the pilots thought the plane was slowing down, so they increased the speed and attempted to maintain their altitude by lowering the wing slats. Actually, they were flying at such a high speed that one of the slats ripped off, causing the plane to nosedive; the plane crashed at a speed of 745 miles per hour.

   In the pitot tube, as the fluid moves, the velocity creates a pressure difference between the ends of a small tube. The tubes are calibrated to relate the pressure measured to a specific velocity. This velocity is a function of the pressure difference (ΔP, in units of pascals) and the density of the fluid (ρ_R in units of kilogram per cubic meter).

$$v = \left(\frac{2}{\rho}\right)^{0.5} P^m$$

Pressure (P) [Pa]	50,000	101,325	202,650	250,000	304,000	350,000	405,000	505,000
Velocity fluid A (v_A) [m/s]	11.25	16.00	23.00	25.00	28.00	30.00	32.00	35.75
Velocity fluid B (v_B) [m/s]	9.00	12.50	18.00	20.00	22.00	24.00	25.00	28.00
Velocity fluid C (v_C) [m/s]	7.50	11.00	15.50	17.00	19.00	20.00	22.00	24.50

Fluid	Specific Gravity
Acetone	0.79
Citric acid	1.67
Glycerin	1.26

(a) Show the resulting data and trendline, with equation and R^2 value, on the appropriate graph type (*xy* scatter, semilog, or log–log) to make the data appear linear.

(b) First determine what the units of the constant in front of the pressure variable must be to yield velocity when multiplied, and then determine how to obtain those units from density. Determine the value and units of the density for each fluid.

(c) From the chart at left, match each fluid (A, B, and C) with the correct name according to the results of the density determined from the trendlines.

9. The drag force on a sphere moving through a fluid is given by

$$F_D = \frac{1}{2}\rho v^2 A_p C_d$$

The variables used are fluid density, ρ; coefficient of drag, C_d (dimensionless); velocity, v; and the drag force, F_D. The area of the object the force acts upon is A_p, and so for spheres is given by the area of a circle.

The Reynolds number in this situation is written as

$$Re = \frac{D_p \rho v}{\mu}$$

where D_p is the diameter of the object the force acts upon and μ is the viscosity of the fluid. At small Reynolds numbers ($Re < 10$), the drag coefficient of a sphere can be determined by a dimensionless relationship known as Stoke's law.

(a) Analyze the data shown to determine two dimensionless parameters: Reynolds number (Re) and coefficient of drag (C_d).

(b) Be sure to *watch the units*! Create a graph of the resulting coefficient of drag (C_d, on the ordinate) versus the Reynolds number (Re, on the abscissa).

(c) Fit a trendline to the data. Linearize the data by creating a rectilinear, semilog, or log–log chart. What is the relationship between C_d and Re?

10. The relationship of the power required by a propeller (shown as the power number, on the ordinate) and the Reynolds number (abscissa) is shown in the graph below. For a propeller, the Reynolds number (Re) is written slightly differently, as

$$Re = \frac{D^2 n \rho}{\mu}$$

where D is the blade diameter [meters] and n is the shaft speed [hertz]. The power number (N_p) is given by the following, where P is the power required [watts].

$$N_p = \frac{P}{\rho n^3 D^5}$$

Use the chart below to answer the following questions:

(a) If the Reynolds number is 120, what is the power number for a system described by Curve A?

(b) If the power number (N_p) is 30, what is the Reynolds number for a system described by Curve B?

(c) If the Reynolds number is 4,000, what is the power (P) required in units of watts at a shaft speed (n) of 0.03 hertz? Assume the system contains acetone, with a kinematic viscosity of 0.419 stokes. The density of acetone is 0.785 grams per cubic centimeter. Use Curve B in the graph to determine your answer. (*Hint:* Use the Reynolds number of the system to first calculate the diameter, then find the power number, and then calculate the power.)

(d) If the power number (N_p) is 5, what is the diameter (D) of the blade in units of centimeters at a shaft speed (n) of 0.02 hertz? Assume the system contains brine, with a kinematic viscosity of 0.0102 stokes. Use Curve A in the graph to determine your answer. (*Hint:* Find the Reynolds number of the system first, and then calculate the diameter.)

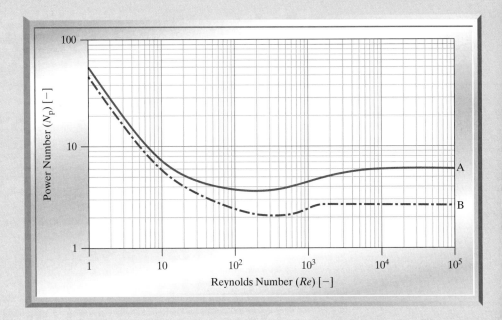

11. When a fluid flows around an object, it creates a force, called the drag force, that pulls on the object. The coefficient of drag (C_d) is a dimensionless number that describes the relationship between the force created and the fluid and object properties, given as

$$C_d = \frac{F_D}{\frac{1}{2}\rho v^2 A_p}$$

where F_D is the drag force, ρ is the fluid density, and v is the velocity of the object relative to the fluid. The area of the object the force acts upon is A_p, and for spheres is given by the area of a circle. The Reynolds number in this situation is written as

$$Re = \frac{D_p \rho v}{\mu}$$

where D_p is the diameter of the object the force acts upon.

(a) If the Reynolds number is 500, what is the coefficient of drag?
(b) If the Reynolds number is 3,000, what is the coefficient of drag?
(c) If the coefficient of drag is 5, what is the Reynolds number?
(d) If the coefficient of drag is 2, what is the Reynolds number?

Ethylene glycol has a dynamic viscosity of 9.13 centipoise and a specific gravity of 1.109.

(e) If the fluid flows around a sphere of diameter 1 centimeter travelling at a velocity of 2.45 centimeters per second, determine the drag force on the particle in units of newtons. (*Hint:* First determine the Reynolds number.)
(f) If a coefficient of drag of 10 is produced, what is the diameter of the particle? Assume the fluid is moving at 1 centimeter per second. (*Hint:* First determine the Reynolds number.)

Methyl ethyl ketone (MEK) has a specific gravity of 0.805 and a dynamic viscosity of 0.0043 grams per centimeter second.

(g) If MEK flows around a sphere of diameter 0.5 centimeter travelling at a velocity of 1.5 centimeters per second, determine the drag force on the particle in units of newtons. (*Hint:* First determine the Reynolds number.)
(h) If a coefficient of drag of 20 is produced, what is the diameter of the particle? Assume the fluid is moving at 15 centimeters per second. (*Hint:* First determine the Reynolds number.)

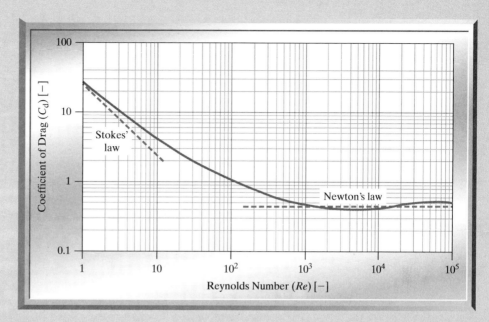

12. The data below was collected during testing of an electromagnetic mass driver. The energy to energize the electromagnets was obtained from a bank of capacitors. The capacitor bank was charged to various voltages, and for each voltage, the exit velocity of the projectile was measured when the mass driver was activated.

NOTE

Due to several complicated nonlinear losses in the system that are far beyond the scope of this course, this is a case of a model in which the exponent does not come out to be an integer or simple fraction, so rounding to two significant figures is appropriate. In fact, this model is only a first approximation— a really accurate model would be considerably more complicated.

Voltage (V) [kV]	9	13	15	18	22	25
Velocity (v_p) [m/s]	430	530	580	650	740	810

(a) Plot this data with voltage as the dependent variable. Determine the most appropriate type of trendline to fit this data. (*Hint:* All three models have high R^2 values.) Consider the physical meaning of the vertical intercept of each of the three possible models: linear, exponential, and power. Which one makes the most sense?
(b) What would the velocity be if the capacitors were charged to 1,000 volts?
(c) What would the velocity be if the capacitors were charged to 100,000 volts?
(d) What voltage would be necessary to accelerate the projectile to 1,000 meters per second?
(e) Assume that the total capacitance is 5 farads. If the capacitors are initially charged to 10,000 volts and are discharged to 2,000 volts during the launch of a projectile, what is the mass of the projectile if the overall conversion of energy stored in the capacitors to kinetic energy in the projectile has an efficiency of 0.2? Recall that the energy stored in a capacitor is given by $E = 0.5CV^2$, where C is capacitance in farads and V is voltage in volts.
(f) Assuming that the capacitors are initially charged to 10,000 volts and are discharged to 2,000 volts during the launch of a projectile, what is the total capacitance in farads if the projectile has a mass of 500 grams and the overall conversion of energy stored in the capacitors to kinetic energy in the projectile has an efficiency of 0.25?
(g) Assume the capacitors are discharged to 20% of their initial voltage when a projectile is launched. Is the energy conversion efficiency a constant, or does it depend on the initial capacitor voltage? If efficiency depends on initial capacitor voltage, determine a mathematical model for efficiency (η) as a function of voltage.

13. As part of an electronic music synthesizer, you need to build a gizmo to convert a linear voltage to an exponentially related current. You build three prototype circuits and make several measurements of voltage and current in each. The collected data is given in the table below.

Circuit A		Circuit B		Circuit C	
Voltage (V_A) [V]	Current (I_A) [mA]	Voltage (V_B)[V]	Current (I_B) [mA]	Voltage (V_C) [V]	Current (I_C) [mA]
−2.7	0.28	−2.7	0.11	0	0.79
−0.4	1.05	−1.5	0.36	0.5	1.59
0	1.74	0	1.34	1.4	5.41
1.2	3.17	0.8	2.37	2.3	20.28
2.9	7.74	2.6	14.53	2.9	41.44

(a) Plot this data and determine an appropriate mathematical model for each circuit. If the model chosen is not linear, adjust the axis to the appropriate log scales to make the data appear linear. Justify your choice of model.

(b) Which circuit is closest to achieving the goal of doubling the current for each 1 volt increase?

(c) Estimate from these data sets the value that should appear in the exponent if the current is to double for each increase of 1 volt.

(d) Calculate the value that should appear in the exponent if the current is to double for each increase of 1 volt. Note that you do NOT need the data above or the chart you created to do this calculation.

14. The Volcanic Explosivity Index (E) is based primarily on the amount of material ejected from a volcano, although other factors play a role as well, such as height of plume in the atmosphere. The table below shows the number of volcanic eruptions (N) over the past 10,000 years having a VEI of between 2 and 7. (There are also VEI values of 0, 1, and 8. There is a level 0 volcano erupting somewhere on the Earth essentially all the time. There are one or more level 1 volcanoes essentially every day. The last known level 8 volcano was about 26,000 years ago.)

Volcanic Explosivity Index (E) [−]	Number of Eruptions (N) [−]
2	3,477
3	868
4	421
5	168
6	51
7	5

(a) Create a graph of this data and determine an appropriate trendline.

(b) How many level 1 volcanoes does the model predict should have occurred in the last 10,000 years?

(c) How many level 8 volcanoes does the model predict should have occurred in the last 10,000 years?

(d) Assuming that the last 10,000 years is representative of much earlier periods in the Earth's history, how many level 8 volcanoes should have occurred in the last 100,000 years?

15. When discussing the flow of a fluid through a piping system, we say that friction occurs between the fluid and the pipe wall due to viscous drag. The loss of energy due to the friction of fluid against the pipe wall is described by the friction factor. The Darcy friction factor (f) was developed by Henry Darcy (1803–1858), a French scientist who made several important contributions to the field of hydraulics. The friction factor depends upon several other factors, including flow regime, Reynolds number, and pipe roughness. The friction factor can be determined in several ways, including the Moody diagram (shown below, discussed in ICA 13-13) and several mathematical approximations presented here.

In the laminar flow range, the *Darcy friction factor* can be determined by the following formula, shown as the linear line on the Moody diagram. (See page 454 for the Moody diagram.)

$$f = \frac{64}{Re}$$

In the turbulent range, the friction factor is a function of the Reynolds number and the roughness of the pipe (ε). For turbulent flow smooth pipes (where the relative roughness ratio (ε/D) is very small), the *Blasius formula* can be used to calculate an approximate value for the Darcy friction factor.

$$f = 0.316(Re)^{-\frac{1}{4}}$$

This simple formula was developed by Paul Richard Heinrich Blasius (1883–1970), a German fluid dynamics engineer. Later, a more accurate but more complex formula was developed in 1939 by C. F. Colebrook. Unlike the Blasius formula, the Colebrook formula directly takes into account the pipe roughness.

The *Colebrook formula* is shown below. Notice that both sides of the equation contain the friction factor, requiring an iterative solution.

$$\frac{1}{\sqrt{f}} = -2\log\left(\frac{\varepsilon/D}{3.7} + \frac{2.51}{Re\sqrt{f}}\right)$$

To begin the iteration, the Colebrook calculation must have an initial value. Use the Blasius approximation as the initial guess to begin the iterative calculation utilizing a conditional statement.

Prepare an Excel worksheet to compute the friction factor.

	A	B	C	D	E	F	G
1							
2	*Enter input values in blue cells*						
3	*Press F9 button to calculate. If in turbulant conditions, press F9 until the Colebrook value does not change.*						
4							
5							
6	**Fluid**	Water		**Pipe Type**	Commercial Steel		
7	Density	Viscosity	Volumetric Flowrate	Diameter	Roughness	Reynolds Number	
8	(ρ) [g/cm3]	(μ) [cP]	(Q) [gpm]	(D) [in]	(ε) [mm]	(Re) [-]	Flow Regime
9	1	1.002	500	6	0.045	263,020	Turbulent
10							
11	Darcy Friction Factor (f) [-]						
12	Laminar	Colebrook					
13		0.0172					

Input Parameters:

■ Fluid: should be chosen from a drop-down list using the material properties listed below. Used with the LOOKUP function to determine:

- Density   (ε) [grams per cubic centimeter]
- Viscosity   (μ) [centipoises]
- Volumetric flow rate   (Q) [gallons per minute]
- Diameter   (D) [inches]

■ Type of pipe: should be chosen from a drop-down list using the properties listed below. Used with the LOOKUP function to determine:

- Pipe roughness   (ε) [millimeters]

Output Parameters:

Be sure to include the appropriate unit conversions.

- Reynolds number.
- Flow regime (laminar, transitional, or turbulent).
- Only the correct Darcy friction factor (one of these two values) should be displayed.
  - For laminar flow, use the equation: $f = 64/Re$.
  - For turbulent flow determined with the Colebrook formula.

Pipe	Equivalent Roughness (ε) [mm]
Cast iron	0.26
Commercial steel	0.045
Concrete	1.5
Drawn tubing	0.0015
Galvanized iron	0.15
Riveted steel	4.5
Smooth pipe	0
Wrought iron	0.045

Compound	Specific Gravity	Viscosity [c_p]
Acetone	0.791	0.3311
Acetic acid	1.049	1.2220
Benzene	0.879	0.6470
Iso-butyl alcohol	0.802	3.9068
Castor oil	0.960	1,027
Ethyl alcohol	0.789	1.1943
Glycerol	1.261	4,220
Lactic acid	1.249	40.3300
Methyl ethyl ketone	0.805	0.4284
Olive oil	0.914	101
Pentane	0.626	0.2395
Phenol	1.071	12.7440
Tetrachloroethylene	1.466	0.8759
Turpentine	0.860	1.4870
Water	1.000	1.0020
m-Xylene	0.864	0.6475

CHAPTER 14
STATISTICS

Probability is associated with assessing the likelihood that an event will or will not occur. For example:

Airplane crash	Earthquake	Nuclear reactor accident
Tornado	Failure of equipment	Terminal cancer
River breaching a levee	Microprocessor failure	Space probe data reception

Statistics are used for design-concept evaluation because they provide quantitative measures to "things" that behave in a random manner. This evaluation helps us make rational decisions about events and manufactured products. Statistics, as well as probability, use numerical evidence to aid decision making in the face of uncertainty. Roles of statistics in engineering include the following:

- Evaluation of new or alternative designs, concepts, and procedures
- Estimation of amount to bid on projects
- Management (human uncertainty, economic uncertainty, and others)
- Determination of degree of acceptable item-to-item variation (quality control)

Often, the best way to analyze an engineering problem is to conduct an experiment. When we take this approach, we face several questions:

- How many tests do we need?
- How confident are we in the results?
- Can we extrapolate the results to other conditions?
- Can we estimate how often the result will lie within a specified range?

There are many other related issues, but addressing all of them requires a separate book. Many readers will take or have taken an entire course in probability or statistics, so for now we just touch on some of the important fundamentals.

- **Repeated tests:** When a test is conducted multiple times, we will not get the same (exact) result each time. For example, use a ruler to measure the length of a particular brand of shoe manufactured by company X. You can produce a table of values, all of which will be nearly, but not exactly, the same. This is because you make slight errors in measurement, so even if every shoe is identical, there will always be some errors in your measurement. Moreover, every shoe is not identical.
- **Differences in a population:** What is the heart rate of all students in a class? Obviously, not everyone will have the same heart rate. We do expect, however, that everyone's rate will lie between, say, 40 and 140 beats per minute. Through measurement, we can

determine this variation. In fact, we may find that the average rate for females and males differs. Statistical procedures help us analyze situations such as this.

■ **Manufacturing errors:** Suppose you are manufacturing a run of widgets and a buyer wants all of them to be exactly alike. Obviously, this is impossible, but you can make them almost alike and then tell the buyer how much variation to expect. If you measure each widget as it comes off the assembly line, there will certainly be some variation.

■ **Design criteria for products:** When you build a house, you would like it to stand safely for some period of time. For example, you might specify that the house be designed to withstand a windstorm that would occur, on average, every 50 years. For that case, you must be able to calculate the wind speeds associated with such a storm. Statistical methods allow you to do this.

14.1 HISTOGRAMS

To illustrate several common statistical concepts, we use data representing the height of several freshman engineering students. Table 14-1 shows the height, to the nearest inch, of each student in a typical class. Table 14-2 shows the same data, summarized by number of students at each height.

Table 14-1 Student height

Student ID	Height (*H*) [in]
A	67
B	73
C	71
D	69
E	68
F	64
G	70
H	72
I	67
J	71
K	70
L	68
M	66
N	71
O	74
P	71
Q	68
R	72
S	67
T	64
U	75
V	74
W	72

Table 14-2 Summary of height data

Height (*H*) [in]	Number of Students
62	0
63	0
64	2
65	0
66	1
67	3
68	3
69	1
70	2
71	4
72	3
73	1
74	2
75	1
76	0
77	0
Total	23

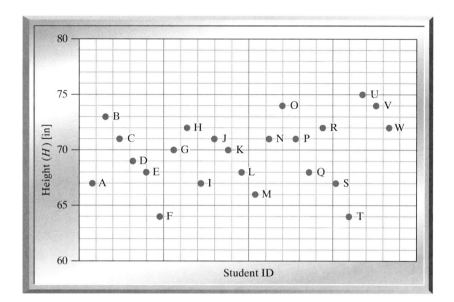

Figure 14-1 Example of student height, shown on scatter plot.

When we graph the values shown in Table 14-1, we end up with a scatter plot with data that is exactly the same: scattered, as shown in Figure 14-1.

Instead of using a scatter plot, we can group the data and plot the group values in a chart similar to a column chart, shown in Figure 14-2. Using the summarized data shown in Table 14-2, we will place two height ranges into a single column or **bin**. The first bin will contain all student-height values less than 62 inches. The next bin will contain student-height values of 62 and 63 inches. The next bin will contain student-height values of 64 and 65 inches, and so on. The abscissa of the graph is the height values; the ordinate is the number of students measured. Graphs of this nature are called **histograms**. By counting the number of blocks, we find the area under the curve represents the total number of samples taken, in this case, the total number of students (23) observed.

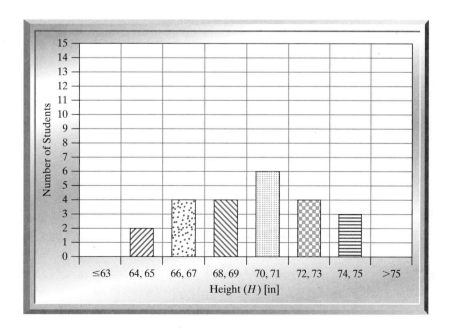

Figure 14-2 Example of student height, shown on a histogram.

Technically, before beginning this example we should have mathematically determined a bin size, rather than arbitrarily grouping the measurements in pairs (62 and 63 in one bin, 64 and 65 in the next bin, etc.). There are several ways to calculate the bin size that will best display the information; below is one method.

DETERMINATION OF BIN SIZE

Step One: Determine the number of bins needed.

Number of Bins = Square root of number of data points, rounded to whole number

Step Two: Determine the range of the data.

Range = $X_{max} - X_{min}$

Step Three: Determine the number of items in each bin.

Bin Size = Range divided by Number of Bins, rounded to whole number

---

Let us apply this to our example.

Step 1: As shown, we have a class of 23 students, so we would need about five bins, since the square root of 23 is about 4.8, which rounds to 5. Four would probably also work fine, as would 6. Remember that this is just a rule of thumb.

Step 2: The shortest person is 64 inches tall and the tallest is 75, so the range is 11 inches.

Step 3: Dividing the range determined in Step 2 by the number of bins determined in Step 1, we get 2.2, or about 2 inches per bin. On the other hand, we might instead decide to have four bins. If we divide the range by 4, we have 2.75 or 3 inches per bin.

Depending on the number of bins, we sometimes get two different, but acceptable, bin sizes. By changing the bin size, we can change the appearance of the data spread, or the data **distribution**.

What happens to the student height data if we alter the bin size? The plot on the preceding page shows a 2-inch bin interval, and Figure 14-3(a) shows a 3-inch interval.

In Figure 14-3(b), we have used a 4-inch bin interval, and while it is not what we obtained from the "rule of thumb" (2- or 3-inch intervals), it is still mathematically correct but not as informative as the other two.

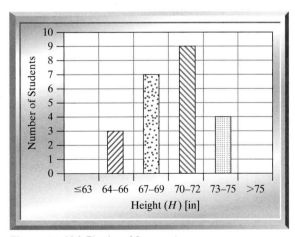

Figure 14-3(a) Bin size of 3.

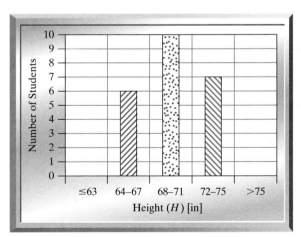

Figure 14-3(b) Bin size of 4.

14.2 STATISTICAL BEHAVIOR

When we have gathered the data and plotted a distribution, the next step is to explain the outcome to others. For convenience, we identify a set of parameters to describe distributions.

One parameter of a distribution is the average value. The **average**, or **mean**, is an estimate of the value most representative of the population. This is often called the central tendency of the data. The computation of the mean ($\overline{X}$) of a data set containing N values is given in the equation below.

NOTE

Average or Mean = typical, expected value of the data set; sensitive to outliers.

Median = value representing the exact middle value of the list; typically unaffected by outliers. Data must be in ascending order to determine!

$$\text{Mean} = \overline{X} = \frac{1}{N}(X_1 + X_2 + \cdots + X_N) = \frac{1}{N}\sum_{j=1}^{N} X_j$$

In other words, the mean is simply the sum of all of the values divided by the total number of values.

The **median**, another measure of central tendency, is the value between the lower half and the upper half of the population. In other words, if all data points are listed in numerical order, the median is the value exactly in the middle of the list. If the number of data points is odd, the median will be the middle value of the population. If the number of data points is even, however, the median will be the average of the two values at the center. A few examples should clarify this.

Set	Data	Mean	Median
1	1, 2, 3, 4, 5, 6, 7	4	4
2	1, 50, 70, 100	55	60
3	5, 10, 20, 40, 80	31	20
4	50, 50, 50, 50, 50, 1,000	208	50

Review the data shown in set 4. It would seem logical if every data point has a value of 50 except one, the average of the data should be about 50; instead, it is 208! This illustrates the sensitivity of the mean to **extreme values**, or **outliers**. Note that the median is unaffected or only slightly affected. It is for this reason that the mean is insufficient to describe the central tendency of all distributions.

Two other terms are useful in describing a distribution: **variance** and **standard deviation**. Both of these terms quantify how widely a set of values is scattered about the mean. To determine the variance (V_x^2), the difference between each point and the mean is determined, and each difference is squared to keep all terms positive. This sum is then divided by one less than the number of data points.

NOTE

Variance = measure of data scatter; has SQUARED UNITS of the original data set.

Standard deviation = square root of the variance; has units of the original data set.

$$\text{Variance} = V_x^2 = \frac{1}{N-1}((\overline{X} - X_1)^2 + (\overline{X} - X_2)^2 + \cdots + (\overline{X} - X_N)^2)$$

$$= \frac{1}{N-1}\sum_{j+1}^{N} (\overline{X} - X_j)^2$$

The standard deviation (SD_x) is found by taking the square root of the variance:

$$\text{Standard deviation} = SD_x = \sqrt{V_x^2}$$

Height (*H*) [in]	Number of Students
62	0
63	0
64	2
65	0
66	1
67	3
68	3
69	1
70	2
71	4
72	3
73	1
74	2
75	1
76	0
77	0
Total	**23**

If we again examine the data found in Table 14-2, we can calculate the mean, median, variance, and standard deviation for our height data.

Calculation of the Mean:

Total number of points (N)
= 23 students

The sum of all heights
= (2 students * 64 inches/student) + (1 * 66) + (3 * 67) + (3 * 68) + (1 * 69)
+ (2 * 70) + (4 * 71) + (3 * 72) + (1 * 73) + (2 * 74) + (1 * 75)
= 1,604 inches

Mean
= 1,604 inches/23 students
= 69.7 inches/student

Calculation of the Median:

Put data in order of value, listing each entry once
64, 64, 66, 67, 67, 67, 68, 68, 68, 69, 70, 70, 71, 71, 71, 71, 72, 72, 72, 73, 74, 74, 75
Find the center value since the total number of students is odd
64, 64, 66, 67, 67, 67, 68, 68, 68, 69, **70**, 70, 71, 71, 71, 71, 72, 72, 72, 73, 74, 74, 75
Median = 70 inches

Calculation of Variance: Note that the variance will have the same units as the variable in question squared, in this case, "inches squared."

$$\text{Variance} = \frac{1}{23-1}((69.7 - 64)^2 + (69.7 - 64)^2 + (69.7 - 66)^2 + \cdots$$

$$+ (69.7 - 75)^2) = 9.5 \text{ in}^2$$

Calculation of Standard Deviation: The standard deviation has the same units as the variable in question, in this case, "inches."

$$\text{Standard deviation} = \sqrt{9.5} + 3.08 \text{ in}$$

● **EXAMPLE 14-1**

Consider the following velocity data, listed in units of feet per second. Determine the mean, median, variance, and standard deviation of the data.

1	28	14	32	35	25	14	28	5
16	42	35	26	5	33	35	16	14

Calculation of the Mean:

Total number of points (N) = 18

Sum of all data $(\Sigma X_j) = (1) + (2 * 5) + (3 * 14) + \cdots + (42) = 404$
Mean   = 404/18 = 22.4 *feet per second*

Calculation of the Median:

Put data in order, listing each entry once.

$$1, 5, 5, 14, 14, 14, 16, 16, 25, 26, 28, 28, 32, 33, 35, 35, 35, 42$$

Find the center two values and average them, since total number of entries is even (18).

$$1, 5, 5, 14, 14, 14, 16, 16, \textbf{25}, \textbf{26}, 28, 28, 32, 33, 35, 35, 35, 42$$

Median = (25 + 26)/2 = 25.5 feet per second

Calculation of Variance:

$$Variance = \frac{1}{18-1}((22.4 - 1)^2 + \cdots + (22.4 - 42)^2) = 147 \text{ (ft/s)}^2$$

Calculation of Standard Deviation:

Standard deviation = $\sqrt{147}$ = 12.1 feet per second

● EXAMPLE 14-2

Consider the following energy data, given in units of joules. Determine the mean, median, variance, and standard deviation of the data.

159	837	618	208	971	571	379	220	31

Calculation of the Mean:
Total number of points (N) = 9
Sum of all data ($\sum X_j$) = 159 + 837 + $\cdots$ + 31 = 3,994

Mean   = 3,994/9 = 443.7 = 444 joules

Calculation of the Median:
Put data in order, listing each entry once.

$$31, 159, 208, 220, 379, 571, 618, 837, 971$$

Find the center value since the total number of students is odd (9; center value at entry 5).

$$31, 159, 208, 220, \textbf{379}, 571, 618, 837, 971$$

Median = 379 joules

Calculation of Variance:

$$Variance = \frac{1}{9-1}[(444 - 31)^2 + \cdots + (444 - 971)^2] = 105,059 \text{ joules}^2$$

Calculation of Standard Deviation:

Standard deviation + $\sqrt{105,059}$ = 324 joules

**COMPREHENSION
CHECK 14-1**

For the following mass data given in units of kilograms, determine the mean, median, variance, and standard deviation.

8	7	9	11	16
12	2	9	10	9

**COMPREHENSION
CHECK 14-2**

For the following temperature data given in units of degrees Celsius [°C], determine the mean, median, variance, and standard deviation.

105	120	110	100	102
103	58	110	100	118

14.3 DISTRIBUTIONS

From the charts on the preceding pages discussing histograms, you can see a similarity in the shape of all three plots. The values start small, increase in size, and then decrease again. In the case of student height, this means that a few people are short, most people have some "average" height, and a few people are tall. This same conclusion is true in many places in our world. For example:

- If we weigh many standard-size watermelons (neither miniature nor giant), we will find that most weigh between 20 and 30 pounds. A few weigh less than 20 pounds, and a few weigh more than 30 pounds.
- As we look through a dictionary, we find that there are many words with between four and six letters. There are a few with one, two, or three letters and a few with more than six, but clearly most have between four and six letters.
- To improve efficiency in the office, we had an expert to monitor the length of phone calls made by the staff. The expert found that most of the time, phone calls lasted between 3 and 5 minutes, but a few were longer than 5 and some others lasted only a minute or 2.

Normal Distributions

We wanted to know how many "flexes" it takes to cause a paper clip to fail, so we asked volunteers to test the bending performance of paper clips by doing the following:

- Unfold the paper clip at the center point so that the resulting wire forms an "S" shape.
- Bend the clip back and forth at the center point until it breaks.
- Record the number of "flexes" required to break the clip.

Using these data, we created Figure 14-4. This is the same as the earlier histogram, but the "boxes" are replaced by a smooth curve through the values.

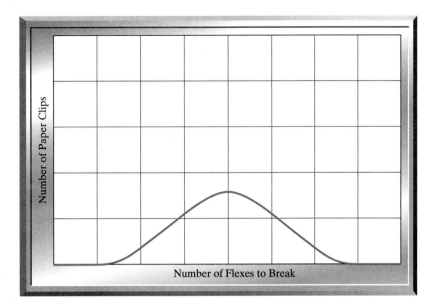

Figure 14-4 Distribution of paper clip failure.

When you are interested in the *shape* of the curve rather than the exact data values, you can replace the bars of the histogram with a smooth curve and rename the graph *distribution*. A distribution is considered *normal* if the following rules hold true. This is known as the **68-95-99.7 rule**, shown in Figure 14-5.

- 68% of values are within one standard deviation (1σ) of the mean (μ).
- 95% of values are within two standard deviations (2σ).
- 99.7% of values are within three standard deviations (3σ).

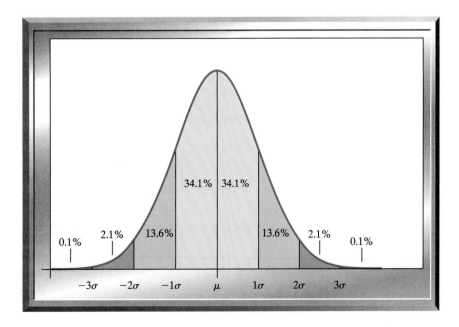

Figure 14-5 "Normal" distribution, showing the 68–95–99.7 rule.

● **EXAMPLE 14-3**   Suppose we ask a class of students how many states they have visited. The results might appear as shown below.

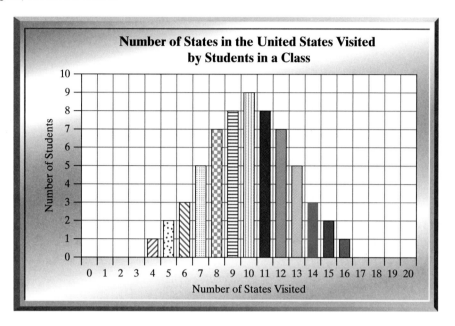

It seems that most have visited between 8 and 12, and that as many have visited more than 10 as have visited fewer than 10. A few have visited as many as 16 states, and all the students have visited at least 4. Let us calculate some values pertinent to this situation.

How many students are there in the class?

To do this we simply add the number of students represented by each bar, or

$$1 + 2 + 3 + 5 + 7 + 8 + 9 + 8 + 7 + 5 + 3 + 2 + 1 = 61$$

What is the cumulative number of state visits?

We answer this by totaling the product of the bar height with the number of states represented by the bar. For example, 5 students have visited 7 states, so those 5 students have visited a total of 5 * 7 = 35 states. Or, 8 students have visited 11 states, so those students have visited a total of 88 states. We calculate

$$1*4 + 2*5 + 3*6 + \cdots + 2*15 + 1*16 = 610$$

What is the average number of states visited by a student?

Once we have the values from our first two answers, this is straightforward division: the total number of visits divided by the total number of students.

$$610/61 = 10$$

Notice that the value 10 is in the center of the distribution. For distributions that are symmetrical (such as this one), the average value is the one in the center, the one represented by the largest number of occurrences.

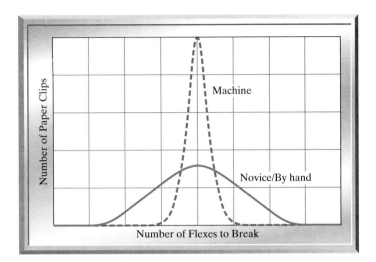

Figure 14-6 Distribution of paper clip failure after a decrease in variance.

Decrease in Variance

Let us examine the shape of the distribution of paper clip failures discussed earlier. How would the distribution change if we brought in a machine that did it "exactly" the same way each time? Both the distribution from the data class and the distribution of the machine are shown below. The same number of clips was tested in each case, so the areas under each curve must be the same. The comparison is shown in Figure 14-6.

This exercise illustrates that distributions that have the same mean (and median) can look very different. In this case, the difference between these two distributions is in their "spread," or their variation about the mean.

Shift in Mean

Redraw the paper clip distribution; then on the same plot, sketch the distribution if each volunteer tested clips that were manufactured by the same manufacturer as before but were stronger and typically required 10 more flexes to fail. The result is shown in Figure 14-7.

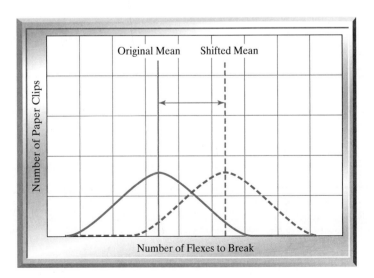

Figure 14-7 Distribution of paper clip failure with a shift.

Skewed Data

It is often easy to place an upper limit on the value of the possible outcome. In these cases, the distribution is no longer symmetric—it is **skewed**. A population is **positively skewed** if the mean has been pulled higher than the median, and **negatively skewed** if the mean has been pulled lower than the median (see Figure 14-8). You have probably heard news reports that use the median to describe a distribution of income in the United States. The median is used in this case because the distribution is positively skewed. This skew is caused by two factors, the presence of extreme values (millionaires) and the range restriction, the latter because income cannot be lower than $0. The extreme values causing the positive skew are not shown on the graph. Most of these would be far off the page to the right.

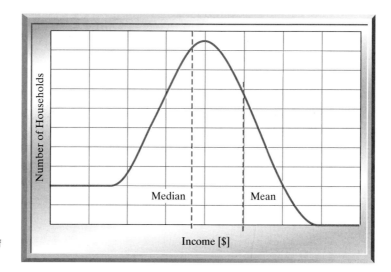

Figure 14-8 Distribution of positively skewed data.

COMPREHENSION CHECK 14-3

For each graph shown below, decide if the mean, variance, or population size has changed.

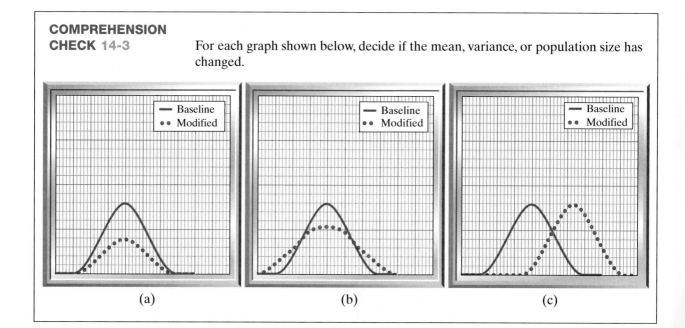

(a)   (b)   (c)

● **EXAMPLE 14-4**   For each scenario, identify one graph from the following that best illustrates how the baseline curve would change under the conditions of that scenario. Each graph shows the usual distribution (labeled baseline) and the way the distribution would be modified from the baseline shape (labeled modified) under certain conditions.

The graphs show SAT composite (verbal + quantitative) scores, for which 400 is generally considered to be the minimum possible score and 1,600 is considered to be the maximum possible score.

(a) The designers of the SAT inadvertently made the test more difficult.

This is shown by Curve (F): Area is the same; mean shifted to left.

(b) The variability of scores is reduced by switching to true/false questions.

This is shown by Curve (D): Area is the same; distribution is narrower.

(c) A population boom increases the number of students seeking college admission.

This is shown by Curve (B): Area increases, distribution stays the same.

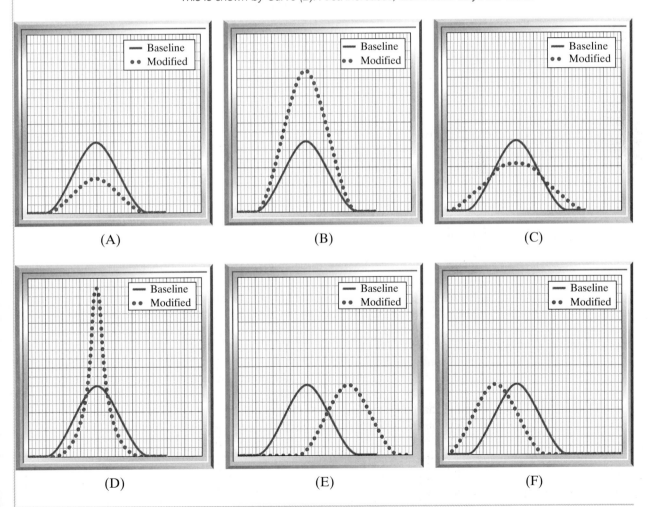

14.4 CUMULATIVE DISTRIBUTION FUNCTIONS

For the earlier student height plot using two heights per bin, we will graph the bin data but now show the values on the ordinate as a fraction rather than a whole number. To do this, we divide the number of students in each bin of the histogram by the total number of students. If we now add the heights of all the bars in the new plot, they should equal 1. This is called a **normalized plot**, shown in Figure 14-9. We "normalized" the values by dividing by the total number of data points. This graph holds no new information; it is simply a rescaling of the histogram we drew earlier.

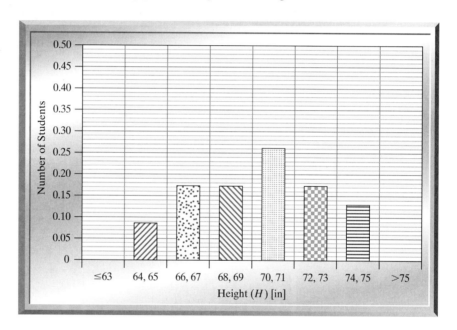

Figure 14-9 Normalized plot of student height, originally shown in Figure 14-2.

This plot can be used as an intermediate step to obtain a final plot called a **cumulative distribution function (CDF)**. We derive this plot by summing the values for each bin on the normalized plot from the first bin up to each individual bin. For example, suppose the values in the first three bins were 0, 0.08, and 0.18. By adding the values, we get new "cumulative" values: bin 1 = 0; bin 2 = 0 + 0.08 + 0.08; and bin 3 = 0 + 0.08 + 0.18 + 0.26. It should be obvious that we can obtain the CDF value for each bin by adding the normalized value of that bin to the CDF value of the bin before it. The CDF values are usually shown as percentages rather than fractions, for example, 50% instead of 0.5.

As we move across the plot, the values should go from 0 to 1. Using the height data in the normalized plot below, we have produced a cumulative distribution shown in Figure 14-10. Sometimes, the CDF is shown as a continuous, curved line rather than a column chart. Both the original histogram and the cumulative distribution plot are useful tools in answering questions about the composition of a population.

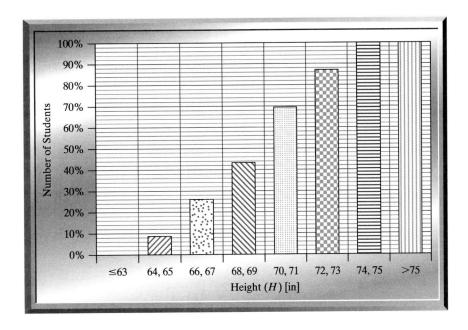

Figure 14-10 CDF of student height, originally shown in Figure 14-2.

● **EXAMPLE 14-5**   Consider the following pressure data, given in units of pascals. Draw the histogram and CDF of the data.

36	9	33	11	23	3
34	39	56	51	39	1
27	25	2	1	53	32
14	41	55	28	29	19
51	15	25	10	35	38

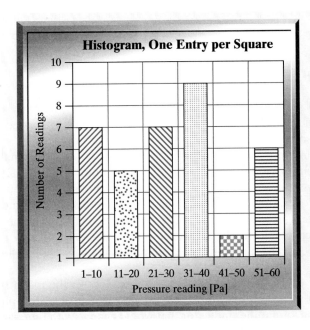

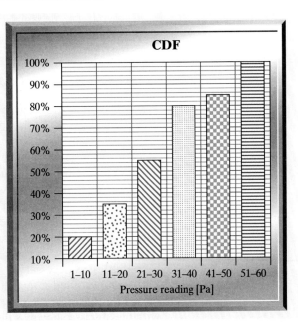

COMPREHENSION CHECK 14-4

Consider the weight of shipping boxes sent down an assembly line, given in units of newtons. Draw the histogram and CDF of the data.

38	103	20	42	16
20	74	63	90	61
114	79	61	50	64

COMPREHENSION CHECK 14-5

Data are presented below for 25 entries. Use the information from the CDF to create the histogram of the data.

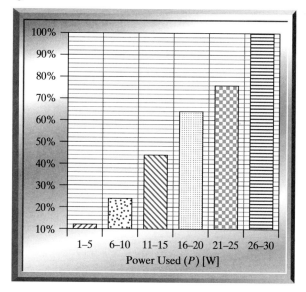

14.5 STATISTICAL PROCESS CONTROL (SPC)

We showed that a histogram such as one shown in Figure 14-11 visually summarizes how a set of values is distributed. Sometimes, however, we are not only interested in the values themselves, but also in how the distribution changes over time.

For example, as a machine in a factory operates, it may slowly (or occasionally quickly) lose proper alignment or calibration due to wear, vibrations, and so on. If a machine was making bolts with a mean length of 1 inch and a standard deviation of 0.01 inch when it first began operating, after it had made 100,000 bolts, the alignment may have drifted so that the mean was only 0.95 inches with a standard deviation of 0.02 inch. This may be unacceptable to the customer purchasing the bolts, so the parameters of the process need to be monitored over time to make sure the machine is readjusted as necessary.

A graph called a **quality chart** is often used to show how close to the mean the results of a process are when measured over time. The graph is usually a scatter plot, with the abscissa shown as time or another indicator that would change with time, such as batch number. Figure 14-12 shows a sample control chart, the mean, and the standard deviation.

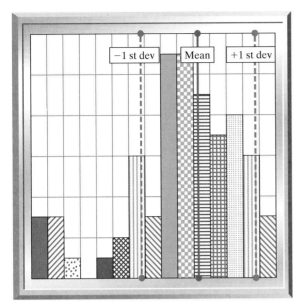

Figure 14-11 Sample histogram.

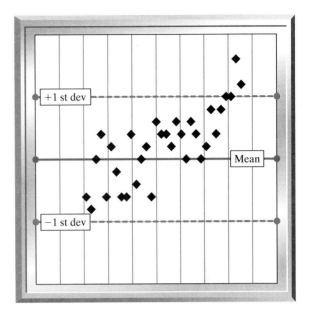

Figure 14-12 Sample quality chart.

When tracking a manufacturing process, engineers are often concerned with whether the process is "in control," or behaving as expected. **Statistical Process Control (SPC)** is a method of monitoring, controlling, and improving a manufacturing process. In some situations, the desired mean and acceptable deviation limits may be preset for a variety of reasons (chemistry, safety, etc.). Often, the upper or lower limits of control are determined by the desired end result.

- The reactor temperature must not rise above 85 degrees Celsius or the reactant will vaporize.
- The injection pressure should be between 50 and 75 kilopascals to ensure that the part is molded properly.
- A bolt must be machined to ±0.02 inches to fit properly in a chair leg.

An engineer will study how the process relates to the control limits and will make adjustments to the process accordingly. To discuss whether a process is in control, we can divide a chart into the following zones, shown in Figure 14-13, to create a control chart: The mean is determined either by the desired end result (the iron content of the product must average 84%) or by the process itself (the reactor temperature should average 70 degrees Fahrenheit for an optimum reaction to occur). The standard deviation is most often determined by experimentation.

- The purity range for this product is ±0.0005%.
- The standard deviation for the reactor temperature must not exceed 5 degrees Fahrenheit or the reaction will create unwanted by-products.
- The standard deviation of the current gain of the transistors being produced must be less than 15.

Eight Ways to Be Out of Control

A variety of conditions can indicate that a process is out of control. First published by Lloyd S. Nelson in the October 1984 issue of the *Journal of Quality Technology*, the **Nelson Rules** are listed below, with examples and graphs. In the graphs, solid points

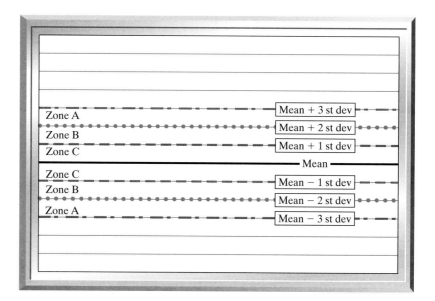

Figure 14-13 Standard deviation ranges.

indicate the rule violations. The actual conditions may vary slightly from company to company, but most take the same standard form. For example, a company may operate with Rule 3 stated as seven or eight points in a row instead of six.

1. **A point falls anywhere beyond Zone A.** The value is more than three standard deviations away from the mean. May occur on either side of the mean.

   Example: The mean temperature of a reactor is 85 degrees Celsius with a standard deviation of 5 degrees Celsius. If the temperature exceeds 100 degrees Celsius, the reactor vessel may explode. If the temperature falls below 70 degrees Celsius, the reaction cannot proceed properly.

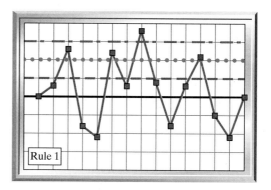

2. **Nine points in a row occur on the same side of the mean.** The actual value seems to be drifting away from the mean.

   Example: The percentage of boron in a semiconductor should be 250 parts per billion. Nine consecutive samples have boron contents less than this value. The machine incorporating the boron into the semiconductor material may need to be cleaned or recalibrated.

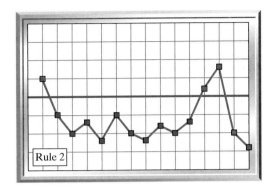

3. **Six points occur with a consistently increasing or decreasing trend.** If this pattern continues, the values will eventually become unacceptable.

    Example: The shaft length of a part is increasing with each successive sample; perhaps the grinding wheel needs to be changed.

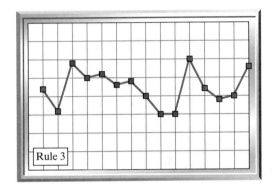

4. **Fourteen points in a row alternate from one side of the mean to the other.** The process is unstable.

    Example: The control system for a crane errs from one side to the other. This may indicate a sensor failure or the need to reprogram the controller.

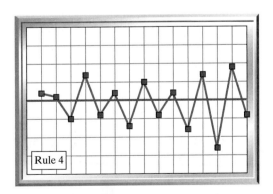

5. **Two out of three points in a row occur in Zone A.** The process is close to the upper limit; take preventive measures now.

Example: A robot that is spot-welding parts in an automobile is coming close to the edge of the material being welded. It probably needs attention.

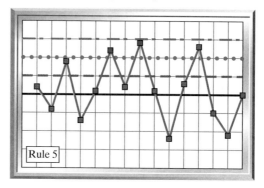

6. **Four out of five points in a row occur in Zone B.** The process is very close to the upper limit; take preventive measures now.

Example: Four out of five customers in the bank teller queue have waited more than one standard deviation to be helped. Perhaps another teller is needed.

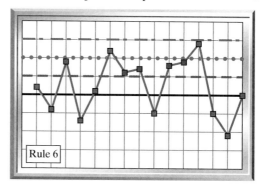

7. **Fifteen points in a row occur in Zone C; points can occur on either side of the mean.** The process is running too perfectly; in many applications the restrictions can be loosened to save money.

Example: The thickness of all washers being manufactured for quarter-inch bolts is within 0.0005 inch of the desired mean. Very few applications require washers with such close tolerances.

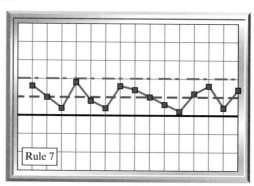

8. **Eight points in a row occur beyond Zone C; points can occur on either side of the mean.** The process does not run close enough to the mean; the parts are never quite on target; may indicate a need for a process adjustment.

Example: The postmark machine in a regional postal distribution center is stamping the envelopes too high or too low; it probably needs attention.

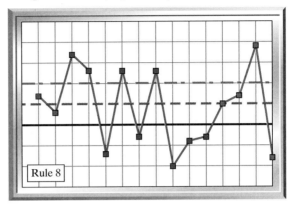

● EXAMPLE 14-6		

The data shown in the table were collected from a manufacturing process that makes bolts.

Assume the process specifies an average bolt length of 10 inches, with a standard deviation of 0.25 inches. Is this process under statistical control?

The control chart for these data is shown below. Rule 1 is violated since a point falls outside of Zone A (part 6); therefore, the process is not in statistical control.

Part	Length (L) [in]
1	10.00
2	10.25
3	10.65
4	9.50
5	9.36
6	9.00
7	10.50
8	10.20
9	9.80
10	10.00
11	10.25
12	10.65
13	9.50
14	9.36
15	9.25
16	10.50
17	10.20
18	9.80
19	10.45
20	10.10

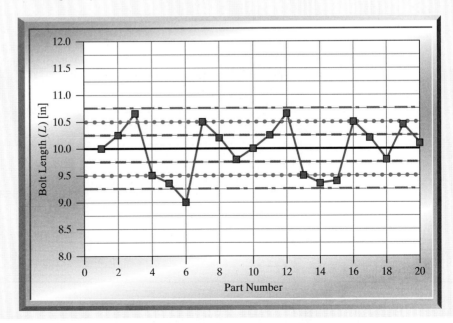

COMPREHENSION CHECK 14-6

The pressure in a water filter is monitored in a chemical plant. The filter should operate at 18 pounds-force per square inch [psi], with a standard deviation of ±2 psi. Analyze the data shown to determine if the filter is behaving as expected (the process is in control; the filter does not require any attention) or if the filter required attention (the process is out of control; the filter should be cleaned). Refer to the Nelson Rules to explain your conclusion.

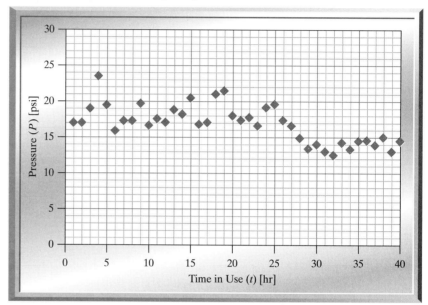

COMPREHENSION CHECK 14.7

The following data were collected from manufacturing machines over the weekend. Using the 8 SPC rules, determine whether each machine is in statistical control. If the machine is not in statistical control, indicate which rule is violated.

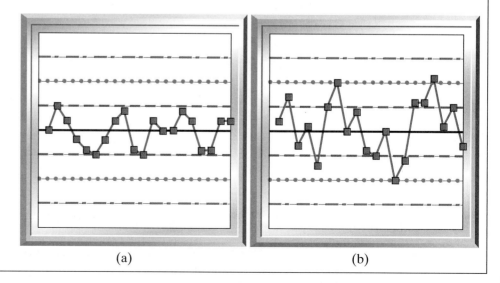

(a)   (b)

14.6 STATISTICS IN EXCEL

To create **histograms** and **CDFs** with Excel, you need to first activate the Analysis ToolPak in Microsoft Excel.

- In Excel, go to the Office button and click **Excel Options**.
- Choose the **Add-Ins** tab on the left menu of the Excel Options window to display all the active add-in applications in Excel. Notice in our list that the Analysis ToolPak is listed as inactive.

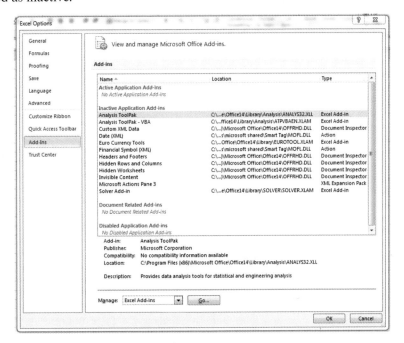

- At the bottom of the Excel Options window, select **Excel Add-Ins** in the **Manage** drop-down menu and click **Go**.
- In the Add-Ins window, check the **Analysis ToolPak** option and click **OK**.
- A prompt might pop up telling you to install the add-in—click **Yes** and finish the installation, using the Office Installer.

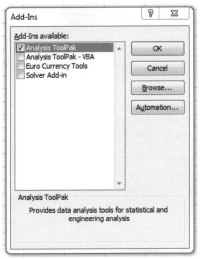

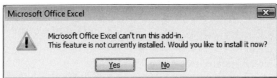

● **EXAMPLE 14-7**

The outline below gives the steps necessary to use the data analysis tool in Excel for basic statistical analysis of a data set. This is presented with an example of the high and low temperatures during the month of October 2006.

- *If necessary, input the data; the data for this example have been provided online. Use Column A to input an identifier for the data point, in this case, the date. Columns B and C will contain the actual high and low temperatures for each day, respectively.*
- *Next, decide on the bin range. This discussion focuses on the high temperatures, but can easily be repeated with the low temperatures.*
  - *A rule of thumb is that the number of bins is approximately equal to the square root of the number of samples. While it is obvious in this example how many total samples are needed, the COUNT function is often very useful. October has 31 days, and the square root of 31 is 5.57; thus, you should choose either 5 or 6 bins.*
  - *Examine your data to determine the range of values. Using the MAX and MIN functions, you can determine that the highest high temperature during October was 86 degrees Fahrenheit and the lowest high temperature was 56 degrees Fahrenheit. Thus, your range is 86 − 56 = 30°F.*
  - *Since 5.57 is closer to 6 than to 5, choose 6 bins. Remember, however, that you might want to try a different number of bins to see if that would result in a clearer representation of the data. With a range of 30 degrees Fahrenheit, 6 bins gives 30°F/6 bins = 5°F per bin.*
- *Type the range of values that will appear in each bin. For example, the first bin will contain temperatures 55, 56, 57, 58, and 59; the second bin will contain temperatures 60–64, and so on.*
- *In the adjacent column, type the corresponding upper value of temperature for each of the bins listed.*

✎ **To create histograms and CDF charts:**

- *Go to **Data** > **Analysis** > **Data Analysis** and under Analysis Tools choose **Histogram**. Click **OK**.*

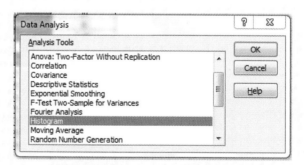

- *In the **Input Range**, click the icon at the right end of the blank box. You can then highlight the range (in this case, $B$6:$B$36). Close the box by clicking the icon at the right-hand end of this small box where the range is shown.*
- *Repeat this procedure for the **Bin Range**, highlighting the cells that contain the upper values.*
- *Next, for the **Output Range**, click the circle and identify a single cell to begin the placement of the output data.*
- *Finally, check the boxes to activate the options of **Cumulative Percentage** calculations and **Chart Output**.*
- *Click **OK**.*

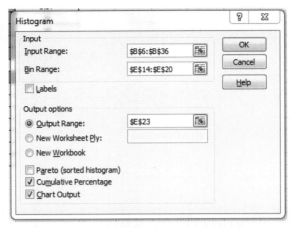

Your worksheet should now look like this:

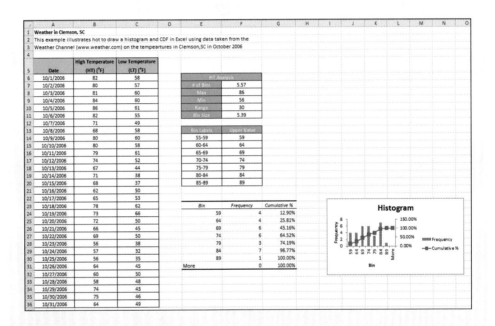

- **Replace the values in the histogram data table for "Bin" with the "Bin Labels"** you entered earlier. This will change the axis labels to the range, rather than the upper value, for each bin.
- **Move the histogram location** to a new worksheet rather than imbedded in the original worksheet to allow the data to be seen clearly. After selecting the chart, use the **Chart > Location** option to select "As new sheet."
- **Modify the histogram to be a proper plot** just as you would with any other chart. The same rules for a "proper plot" apply to a histogram also, so make sure the background is white and alter the series colors, etc., as appropriate. The histogram generated with the directions above is shown below, properly formatted.
- **Change the vertical scale on the left axis** to be a multiple of 2, 5, or 10 to allow the cumulative percentages on the right axis to line up with the gridlines. This is important to do!
- **Change the vertical scale on the right axis** to be a maximum value of 100%. This is important to do. The resulting analysis should appear as follows.

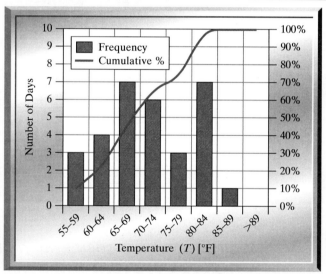

	A	B	C	D	E	F	G
1	Weather in Clemson, SC						
2	This example illustrates hot to draw a histogram and CDF in Excel using data taken from the						
3	Weather Channel (www.weather.com) on the tempeartures in Clemson,SC in October 2006						
4							
5	Date	High Temperature (HT) [°F]	Low Temperature (LT) [°F]				
6	10/1/2006	82	58		HT Analysis		
7	10/2/2006	80	57		# of Bins	5.57	
8	10/3/2006	81	60		Max	86	
9	10/4/2006	84	60		Min	56	
10	10/5/2006	86	61		Range	30	
11	10/6/2006	82	55		Bin Size	5.39	
12	10/7/2006	71	49				
13	10/8/2006	68	58		Bin Labels	Upper Value	
14	10/9/2006	80	60		55-59	59	
15	10/10/2006	80	58		60-64	64	
16	10/11/2006	79	61		65-69	69	
17	10/12/2006	74	52		70-74	74	
18	10/13/2006	67	44		75-79	79	
19	10/14/2006	71	38		80-84	84	
20	10/15/2006	68	37		85-89	89	
21	10/16/2006	62	50				
22	10/17/2006	65	53				
23	10/18/2006	78	62				
24	10/19/2006	73	66		Bin	Frequency	Cumulative %
25	10/20/2006	72	50		55-59	3.00	9.7%
26	10/21/2006	66	45		60-64	4.00	22.6%
27	10/22/2006	69	50		65-69	7.00	45.2%
28	10/23/2006	56	38		70-74	6.00	64.5%
29	10/24/2006	57	32		75-79	3.00	74.2%
30	10/25/2006	56	35		80-84	7.00	96.8%
31	10/26/2006	64	45		85-89	1.00	100.0%
32	10/27/2006	60	50		>89	0.00	100.0%
33	10/28/2006	58	48				
34	10/29/2006	74	43				
35	10/30/2006	75	46				
36	10/31/2006	64	49				

COMPREHENSION CHECK 14-8   Repeat this analysis, using the daily low temperatures during October 2006.

⌘ Statistics on the Mac OS

Unfortunately, as this book goes to press, Microsoft has chosen not to include the histogram tool in Excel 2008 or Excel 2011 for the Mac OS. You have a few options.

- If you have an Intel-based Mac, you can use Excel 2007 or 2010 for Windows. If you do not know how to activate the Windows option on your machine, ask your friendly local Mac guru at your computer center.

- You can use Excel 2004 for the Mac OS, which did include a histogram tool.
- You can create the histogram manually according to the instructions below.

Create columns for the data, bin ranges, and upper value in each bin as described above.

Next, determine the number of data points in each bin. After doing a few, you will find it easy. Use the advanced Excel function, known as an array function, to accomplish this determination. A detailed explanation of array functions is beyond the scope of this book, but if you follow the instructions *carefully*, you should not have any trouble. The specific function to use is called FREQUENCY.

1. In the cell immediately to the right of the topmost "upper bin value" cell (this would be cell G14 in the example above) enter the formula

$$= \text{FREQUENCY (DataRange, UpperBinValueRange)}$$

and press return. In the example above, this would be

$$= \text{FREQUENCY (B6:B36, F14:F20)}$$

2. Click-and-hold the cell into which you entered the formula, then drag straight down to the cell in the row *following* the row containing the last "upper bin value." This would be cell G21 in the example above. Release the mouse button. At this point, you will have a vertical group of cells selected (G14:G21 in the example), the top cell will contain the number of data points in the first bin, and the rest of the selected cells *will be blank*. The formula you entered in the topmost of these cells will appear in the formula bar at the top of the window.
3. Click once in the formula in the formula bar. The top cell of the selected group will be highlighted.
4. Hold down the Command (⌘) key and press Return. The selected cells will now contain the number of data points in each bin immediately to the left. The bottommost selected cell will contain the number of data points larger than the upper bin value in the final bin. In the example, this "extra" cell should contain a 0, since no values are larger than those in the final bin. Note the formulae that appear in these cells are all identical—the cell references are exactly the same. This is normal for an array function.

Use these values to create the histogram.

1. Select the cells containing the bin ranges (E14:E20 in our example), then hold down the command (⌘) key while you select the cells containing the number of data points per bin. In our example, since the "extra" cell at the bottom contains a 0, you need not include it. If this were nonzero, you might want to add a cell at the bottom of the cell ranges that said something like >89. You should now have the two columns for bin ranges and number per bin selected (E14:E20 and G14:G20 in our example).
2. In the toolbar, select **Gallery > Charts > Column**. A row of column chart icons should appear.
3. Click the first icon, which shows pairs of columns. The chart that appears shows the histogram. Be sure to follow all appropriate proper plot rules for completing the histogram.

Finally, generate the CDF. If you have survived this far, you should be able to do this with minimal guidance. Create another column of values next to the column containing the number of data points per bin. In the cell next to the topmost bin cell, enter the number of data points in that bin. In the next cell down, enter a formula that will add the cell above to the cell beside it containing the number of data points in that bin. Replicate this formula down to the last bin. Each cell in the new column should now contain the sum of all data points in all bins to that point.

In-Class Activities

ICA 14-1

This exercise includes the measurement of a distributed quantity and the graphical presentation of the results. You are to determine how many flexes it takes to cause a paper clip to fail.

Test the bending performance of 20 paper clips by doing the following:

- Unfold the paper clip at the center point so that the resulting wire forms an "S" shape.
- Bend the clip back and forth at the center point until it breaks.
- Record the number of flexes required to break the clip.

On a copy of the table below, record the raw data for the paper clips you break. Then, summarize the data for the team by adding up how many clips broke at each number of flexes. Each team member should contribute 20 data points.

(a) Use the data to plot by hand the following: a histogram with an appropriate bin size, the normalized plot, and CDF.

(b) Use the data to plot using Excel the following: a histogram with an appropriate bin size, and CDF.

Paper clip flexing data

Paper Clip	Flexes to Break
1	
2	
3	
4	
5	
6	
7	
8	
9	
10	
11	
12	
13	
14	
15	
16	
17	
18	
19	
20	

Summary of data

No. of Flexes	No. of Clips

ICA 14-2

For the following pressure data, recorded in units of pound-force per square inch, answer the following questions.

1	14	2	15	10
6	3	1	18	

(a) What is the mean of the data?
(b) What is the median of the data?
(c) What is the variance of the data?
(d) What is the standard deviation of the data?

ICA 14-3

The table below lists the number of computer chips rejected for defects during random testing over the course of a week on a manufacturing line. Four samples of 20 parts are pulled each day. Use the following data to generate a histogram and CDF.

1	1	8	0	2	0
0	2	10	1	3	2
0	1	12	0	2	1
1	6	15	0	0	
3	8	1	2	5	

ICA 14-4

One of the NAE Grand Challenges for Engineeering is **Develop Carbon Sequestration Methods**. According to the NAE website: "In pre-industrial times, every million molecules of air contained about 280 molecules of carbon dioxide. Today that proportion exceeds 380 molecules per million, and it continues to climb. Evidence is mounting that carbon dioxide's heat-trapping power has already started to boost average global temperatures. If carbon dioxide levels continue up ward, further warming could have dire consequences, resulting from rising sea levels, agriculture disruptions, and stronger storms (e.g., hurricanes) striking more often."

The Mauna Loa Carbon Dioxide Record is the longest continuous record of atmospheric concentrations of carbon dioxide (CO_2), the chief greenhouse gas responsible for global climate warming. These data are modeled as the Keeling Curve, a graph showing the variation in concentration of atmospheric CO_2 based on measurements taken at the Mauna Loa Observatory in Hawaii under the supervision of Charles David Keeling. It is often called the most important geophysical record on Earth and has been instrumental in showing that mankind is changing the composition of the atmosphere through the combustion of fossil fuels.

The Keeling Curve also shows a cyclic variation in each year corresponding to the seasonal change in the uptake of CO_2 by the world's land vegetation. Most of this vegetation is in the northern hemisphere, where most of the land is located. The level decreases from northern spring onward as new plant growth takes CO_2 out of the atmosphere through photosynthesis and rises again in the northern fall as plants and leaves die off and decay to release the gas back into the atmosphere.

Data and wording for this problem set were obtained from: www.esrl.noaa.gov/gmd/ccgg/trends/. Additional information on the Mauna Loa Observatory can be found at: http://scrippsco2.ucsd.edu/.

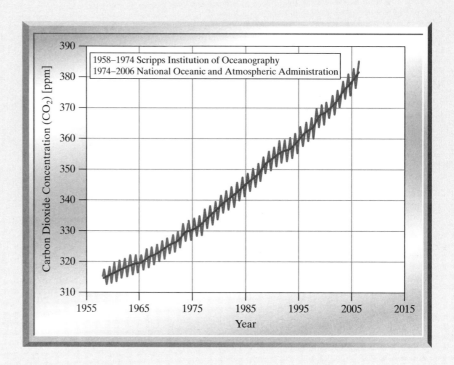

Examine the increase in monthly CO_2 emissions for 2009, taken from the Mauna Loa data set. All values given are in parts per million [ppm] CO_2 as the difference between the December 2008 and the monthly 2009 reading.

1.38	4.64	−0.77	3.23	2.20	0.45
1.87	3.89	−1.16	3.92	0.37	1.73

(a) What is the mean of these data?

(b) What is the median of these data?

(c) The variance of the data set shown here is 3.5 parts per million squared [ppm²]. What is the standard deviation of these data?

(d) The estimated annual growth rates for Mauna Loa are close, but not identical, to the global growth rates. The standard deviation of the differences is 0.26 parts per million per year [ppm/year]. What is the variance?

You use the data from the Mauna Loa observatory to create the following histogram and CDF. These data reflect the observed yearly increase in CO_2 emissions for the past 51 years. The annual mean rate of growth of CO_2 in a given year is the difference in concentration between the end of December and the start of January of that year. If used as an average for the globe, it would represent the sum of all CO_2 added to, and removed from, the atmosphere during the year by human activities and by natural processes.

ICA 14-11

The data below were collected from a manufacturing process for making plastic cylinders. According to the specifications, the cylinder diameter should be 100 inches (the average diameter is 100 inches) and the standard deviation is ± 5 inches.

Graph the data on a control chart. Be sure to clearly indicate the "Zones" of control.

Using the eight SPC rules, determine whether the process is in statistical control. If it is not in statistical control, indicate which rule or rules are violated.

Part	Diameter (D) [in]
1	100
2	106
3	103
4	99
5	90
6	95
7	105
8	107
9	97
10	96
11	89
12	89
13	87
14	92
15	94
16	87
17	96
18	98

Chapter 14 REVIEW QUESTIONS

1. Over 12 school terms, Clemson University recycled the following amount of waste (www.clemson.edu/facilities/recycling). For each set of six academic years (1994 to 1995 through 1999 to 2000 and 2000 to 2001 through 2005 to 2006), determine the mean, median, variance, and standard deviation by hand.

Year	1994–1995	1995–1996	1996–1997	1997–1998	1998–1999	1999–2000
Waste in tons	324.5	204.5	413.7	192.5	280.9	536.9

Year	2000–2001	2001–2002	2002–2003	2003–2004	2004–2005	2005–2006
Waste in tons	671.3	705.9	784.7	681.5	750.7	811.1

2. A technician tested two temperature probes by inserting their probes in boiling water, recording the readings, removing and drying the probes, and repeating the process. The results are shown below, giving temperature reading in degrees Celsius.
   For each probe, determine the mean, median, variance, and standard deviation by hand.

Probe 1	87.5	86.5	88	89.5	87	88.5	89
Probe 2	95.5	100	101.5	97.5	90.5	91.5	103.5

3. Ten students are asked to participate in a golf-putting contest. Each student is given five balls, placed at a distance of 60 feet from the hole. They are to try to get each ball in the hole with a single putt. The number of feet that each ball is from the hole when it comes to rest is measured. For simplicity, the assumption is that no students make their putt.

   (a) On a paper, sketch a histogram of this situation (number of students on the ordinate and distance the putt stops from the hole on the abscissa) using the data given.
   (b) In Excel, create a histogram of this situation (number of students on the ordinate and distance the putt stops from the hole on the abscissa), using the data given.

Distance from hole [ft]	0	1	2	3	4	5	6	7	8	9	10	11	12	13	14
Number of balls	0	1	3	2	3	5	4	6	9	7	5	3	2	0	0

Photos Courtesy of L. Benson

4. Metal plates and screws are used to fix complex bone fractures, such as in cases with multiple bone fragments or a large gap in the bone. There are many plate and screw designs with different features, and surgeons need information on how these designs perform in order to choose the correct designs for their cases. For example, Limited Contact Dynamic Compression (LCDC) plates have spherical screwheads that can pivot within the holes in the plate. This allows the bone fragments to compact when they are loaded, which can promote healing. Locking Compression (LC) plates have screws with threaded heads that actually lock into the holes in the plate when they are tightened down. This makes the plate-screw combination stiffer and stronger than the traditional LCDC design and holds the fragments in place when the bone is loaded.

Biomechanical testing mimics the types of loads that are applied to the plates and screws when they are implanted and measures the amount of bending or deformation that a plate undergoes while loaded. The following table shows results from biomechanical testing for the two types of plates described above, under four different types of loading: compression along the long axis of the bone, torsion (or twisting) along the long axis of the bone, four-point bending in the anterior-posterior (A-P, or forward-backward) direction, and four-point bending in the medial-lateral (M-L, or side-to-side) direction. The plates and screws were fastened to composite resin models of the humerus, with a gap between the bone fragments.

The bone plate test setup is the same for loading in compression and torsion, shown in the figure at left. The bone is clamped at both ends in a mechanical testing machine, and loads are applied along the long axis of the bone. The test setup for the four-point bending tests is shown below.

3.5-mm Locking Compression Plate Stiffness				
Test	Compression	Torsion	A-P Bending	M-L Bending
1	400.18	1.05	419.45	587.94
2	863.95	0.96	427.31	435.99
3	526.52	1.03	408.22	533.99
4	596.58	1.03	388.35	543.57

4.5-mm Limited Contact Dynamic Compression Plate Stiffness				
Test	Compression	Torsion	A-P Bending	M-L Bending
1	556.5	1.08	422.27	868.76
2	441.2	1.39	435.63	716.2
3	545.87	1.35	576.77	794.48
4	487.95	1.38	413.1	1,025.75

(a) For each of the four loading conditions, calculate the mean and standard deviation for the stiffness of both plate designs.

(b) The most common mode of fracturing the humerus is a fall where the person puts an arm out to break the fall. This could load the humerus in the medial-lateral direction. Which of these plate designs would you recommend to withstand medial-lateral loading? Justify your answer.

(c) Test results are not considered "significantly different" if the ranges of one standard deviation overlap. For example, if Product A has a mean of 10 and a standard deviation of 5, and Product B has a mean of 14 and a standard deviation of 2, the products are not significantly different since Product A can range from 5 to 15 and Product B from 12 to 16; the two ranges overlap. Compare the two designs in the anterior-posterior bending and in torsional loading. Can one design be considered superior to the other design in these modes of loading?

5. Polyetheretherketone (PEEK)™ are polymers that are resistant to both organic and aqueous environments; they are used in bearings, piston parts, and pumps. Several tests were conducted to determine the ultimate tensile strengths in units of megapascals [MPa]. The following CDF shows results from 320 points.

(a) What is the frequency value of A on the chart?
(b) What is the frequency value of B on the chart?
(c) What is the frequency value of C on the chart?
(d) What is the frequency value of D on the chart?
(e) What is the frequency value of E on the chart?

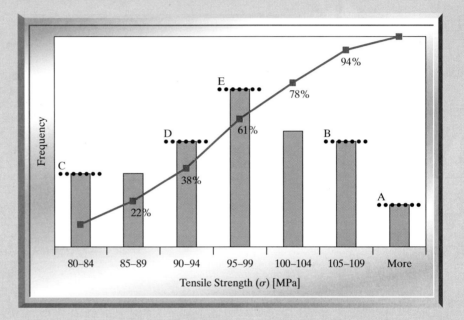

6. A company that fabricates small, custom machines has been asked to generate a machine that throws darts at a dart board as precisely and accurately as possible. To assess the precision and accuracy of each proposed design, the engineers build a model and record the distance from the bullseye of the dart board to the location of each dart thrown—both the straight-line distance (A) and the horizontal (B) and vertical (C) distances are recorded separately with regard to the bullseye, as demonstrated in the figure. The engineers throw 15 darts with their prototype machine and record the three data points for each dart.

Using the data collected for a design, create a histogram and a CDF for the straight-line distance (A), as well as the horizontal (B) and vertical (C) distances and determine which graph or graphs are better for assessing the performance of the design. Justify your answer with a few sentences about why you selected the graph or graphs.

Dart	Straight Distance (A) [in]	Horizontal Distance (B) [in]	Vertical Distance (C) [in]
1	0.703	0.400	−0.578
2	2.740	1.976	1.898
3	1.555	−.484	0.466
4	0.387	−0.387	0.010
5	1.180	−1.147	−0.278
6	1.424	−1.418	0.132
7	0.899	0.835	0.335
8	2.069	0.330	2.043
9	0.547	0.514	0.187
10	3.451	1.566	3.076
11	2.532	2.361	−0.912
12	3.070	1.602	2.619
13	1.386	0.176	1.375
14	1.637	1.458	−0.744
15	0.888	−0.822	0.337

7. You will be given an Excel spreadsheet, titled "Baseball Data," that contains salary data from major league baseball in 2005. Use the data provided to determine the following:

(a) Salary mean, median, and standard deviation for all players.
(b) Salary mean, median, and standard deviation for the Arizona Diamondbacks.
(c) Salary mean, median, and standard deviation for the Atlanta Braves.
(d) Salary mean, median, standard deviation for all pitchers. (*Hint:* Sort first!)
(e) Salary mean, median, standard deviation for all outfielders. (*Hint:* Sort first!)
(f) Which team had the highest average salary?
(g) Which team had the lowest average salary?
(h) Which position had the highest average salary? (*Hint:* Sort first!)
(i) Which position had the lowest average salary? (*Hint:* Sort first!)
(j) Draw a histogram and CDF in Excel for all players.
(k) What percentage of players earned more than $1 million?
(l) What percentage of players earned more than $5 million?

8. This information was taken from the report of the EPA on the U.S. Greenhouse Gas Inventory (http://www.epa.gov).

"Greenhouse gas emission inventories are developed for a variety of reasons. Scientists use inventories of natural and anthropogenic emissions as tools when developing atmospheric models. Policy makers use inventories to develop strategies and policies for emission reductions and to track the progress of those policies. Regulatory agencies and corporations rely on inventories to establish compliance records with allowable emission rates. Businesses, the public, and other interest groups use inventories to better understand the sources and trends in emissions.

The Inventory of U.S. Greenhouse Gas Emissions and Sinks supplies important information about greenhouse gases, quantifies how much of each gas was emitted into the atmosphere, and describes some of the effects of these emissions on the environment.

In nature, carbon is cycled between various atmospheric, oceanic, biotic, and mineral reservoirs. In the atmosphere, carbon mainly exists in its oxidized form as CO_2. CO_2 is released into the atmosphere primarily as a result of the burning of fossil fuels (oil, natural gas, and coal) for power generation and in transportation. It is also emitted through various industrial processes, forest clearing, natural gas flaring, and biomass burning."

The EPA website provides data on emissions. The data found in the file "Carbon Dioxide Emissions Data" were taken from this website for the year 2001 for all 50 states and the District of Columbia.

(a) Use the data on all 50 states (+DC) provided in "Carbon Dioxide Emissions Data" to create a histogram with an appropriate bin size; use Excel.

(b) Determine the mean and median of the data. Indicate each of these values on your graph for part (a).

(c) Which value more accurately describes the data? Indicate your choice (mean or median) and the value of your choice. Justify your answer.

9. The Excel data provided online was collected by Ed Fuller of the NIST Ceramics Division in December 1993. The data represent the polished window strength, measured in units of kilopounds per square inch [ksi], and were used to predict the lifetime and confidence of airplane window design. Use the data set to generate a histogram and CDF in Excel (http://www.itl.nist.gov/div898/handbook/eda/section4/eda4291.htm).

10. Choose *one* of the following options and collect the data required. For the data source you select, do the following:

■ Construct a histogram, including justification of bin size.
■ Determine the mean, median, variance, and standard deviation values.
■ Construct a cumulative distribution function.

(a) On a campus sidewalk, mark two locations 50 feet apart. As people walk along, count how many steps they take to go the 50 feet. Do this for 125 individuals.

(b) Select 250 words at random from a book (fiction). Record the number of letters in each word. Alternatively, you can count and record the words in 250 sentences.

(c) Go to one section of the library, and record the number of pages in 125 books in that same section.

(d) Interview 125 people to determine how far their home is, in miles, from the university.

11. You test several temperature probes by inserting them in boiling ethanol (theoretical boiling point is 78.4 degrees Celsius), recording the readings, removing and drying the probe, and repeating the process 20 times. The distribution curves for the probes are shown below. The solid line "baseline" curve in every graph is the same curve, for a previous probe tested 20 times in boiling ethanol.

(a) Which probe was tested 40 times instead of 20 times?

(b) Which probe has the highest standard deviation?

(c) During the testing of one probe, you suspect your assistant of using formic acid (which boils at 101 degrees Celsius) instead of ethanol. Which probe did your assistant incorrectly test?

(d) Which probe has the lowest standard deviation?

(e) Which probe was tested 10 times instead of 20 times?

(f) During the testing of one probe, you suspect your assistant of using chloroform (which boils at 61 degrees Celsius) instead of ethanol. Which probe did your assistant incorrectly test?

(g) If you could choose between probes C and F, which probe would you choose to use? In a single sentence, describe how you would use the probe you chose to ensure that you found the correct boiling point.

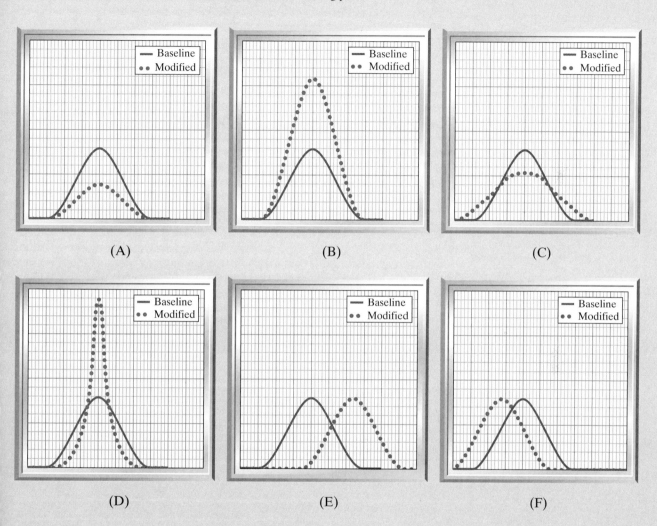

(A)   (B)   (C)

(D)   (E)   (F)

12. You work for a company developing a new hybrid car. Before releasing it to the market, the company has conducted many tests to assess different aspects of the cars' performance. For each characteristic being tested, 500 nonhybrid cars of the same basic design as the new hybrids are tested along with 500 hybrids. The nonhybrid tests are considered the baseline.

    You are given six plots (see previous page) showing the results of testing different aspects of the cars' performance. For each type of test described below, determine which plot is most likely to represent the results of that particular test and explain why. Depending on the characteristic being assessed the abscissa variable will vary.

    (a) This test assesses gas mileage. The hybrid cars are supposed to have better gas mileage than the nonhybrids.

    (b) This test assesses horsepower of the gasoline engine in both hybrid and nonhybrid cars. To ensure engine efficiency, the computer-controlled ignition sequence for the hybrids must have less car-to-car variation than the nonhybrids, thus providing better performance (horsepower).

    (c) This test assesses acceleration time. For the nonhybrid cars, the average 0–60 miles per hour acceleration time is 8 seconds. For the hybrid models, it was determined that only 3 of the 500 cars had times of more than 11 seconds. Since the complex manufacturing process for the new hybrids still has some kinks in it, the hybrids do not perform as consistently as the nonhybrid. Assume the hybrids and nonhybrids had the same average acceleration time.

13. The data below were collected from a manufacturing process involving reactor temperature measured in degrees Celsius. The following values are desired: average = 100 degrees Celsius; standard deviation = ±10 degrees Celsius.

    Graph the data on a control chart. Be sure to clearly indicate the "Zones" of control.

    Using the eight SPC rules, determine whether the process is in statistical control. If it is not in statistical control, indicate which "rule" or "rules" are violated.

Reading No.	Temperature (T) [°C]	Reading No.	Temperature (T) [°C]
1	100	11	103
2	105	12	101
3	106	13	100
4	97	14	98
5	98	15	97
6	95	16	96
7	101	17	104
8	100	18	102
9	96	19	95
10	105	20	101

14. The data below were collected from a manufacturing process involving reactor temperature measured in degrees Celsius. The following values are desired: average = 100 degrees Celsius; standard deviation = ±5 degrees Celsius.

Graph the data on a control chart. Be sure to clearly indicate the "Zones" of control.

Using the eight SPC rules, determine whether the process is in statistical control. If it is not in statistical control, indicate which "rule" or "rules" are violated.

Reading No.	Temperature (T) [°C]	Reading No.	Temperature (T) [°C]
1	101.0	11	97.5
2	103.5	12	100.0
3	98.5	13	95.0
4	100.5	14	97.0
5	96.5	15	103.0
6	102.5	16	103.0
7	105.0	17	105.5
8	100.0	18	100.5
9	102.0	19	102.5
10	98.0	20	98.5

Part 4

PUNCTILIOUS PROGRAMMING

LEARNING OBJECTIVES

The overall learning objectives for this part include:

- Defining the scope of a problem and creating a written or graphical algorithm to solve the problem.
- Performing basic matrix operations.
- Reading and interpreting MATLAB programs written by others.
- Writing a program or function using MATLAB and including I/O, plots, conditionals, and loops.
- Designing formatted output in a MATLAB program.
- Debugging a program to identify different types of errors.

Computers are controlled by software that can be designed in a variety of programming languages. Computer programs are a translation of what you want to accomplish into something the computer can understand, so the term "programming language" is particularly appropriate. Some computer programs are installed permanently or temporarily on computer chips, and others are installed on a variety of other media, such as hard drives or removable media like CD-ROMs.

Computers relentlessly produce a particular result given a particular set of input conditions. It can be frustrating when you make a simple mistake in a computer program—the computer will do exactly what you tell it to do, even if your mistake would be obvious to a person.

The biggest difference between a computer and a person is that you can ask a person open-ended questions—questions like design questions that can have many answers. Computers can only process questions that have a single answer.

This makes the process of programming a computer a bit like trying to ask another person to solve a problem when they are on the other side of a wall and you can communicate only by passing them slips of paper asking questions that can have only one answer and waiting for the person to pass back a slip of paper with the answer on it.

SOME ADVANTAGES OF COMPUTERS

Given our description of how computers work, it may sound to some as if computers are too simple to be useful. The value of programming is linked to a few important characteristics of computers.

- *Calculation speed:* Although computers can only answer analytical questions, they can answer such questions very quickly—in small fractions of a second. Computer programs can therefore ask the computer a lot of questions in a short time, and thus find the answer to more complicated problems by breaking down the complicated question into a series of simple questions.
- *Information storage:* In "Memory: Science Achieves Important New Insights into the Mother of the Muses" (*Newsweek*, September 29, 1986), Sharon Begley estimates that the mind can store an estimated 100 trillion bits of information. The typical computer has a small amount of storage compared to that, but computers are gaining. Where computers have a bigger advantage is that new information can be incorporated in a fraction of the time it takes a human to learn it.
- *Information recall:* Computers have nearly 100% recall of information, limited only by media failures. The human brain can be challenged to recall information in exactly the same form as it was stored.

If a computer always produces the same result every time given the same input conditions, then why does my computer crash sometimes when I am doing something that should work?

HERE IS WHY

The computers you use are simultaneously running a large number of complicated computer programs, including the operating system, background programs, and whatever programs you have started intentionally. Sometimes these programs compete for resources, causing a conflict. Other times, programs are complicated enough that the "input conditions," including the configuration of data in memory and on the hard disk, the time on the system clock, and other factors that change all the time while the computer is running create a combination of circumstances that the programmers never anticipated and so did not include programming code to handle, and the system crashes.

WISE WORDS: WHAT IS YOUR FAVORITE PART OF YOUR CURRENT JOB?

The sheer hour-to-hour diversity of projects and work that I get to see. I learn every day, and that's a blast.

E. Basta, CME

I work with many different companies and get to continually learn new industries, new technologies, and ways companies tackle similar challenges differently. Each engagement typically lasts three months.

M. Ciuca, ME

Finding ways to improve our parts so that our customers can run their hard drives at higher speeds and with greater densities.

E. D'Avignon, CpE

Every day, I get to see a 2-dimensional plan be transformed into something 3D and real. It is exciting to see the progress of a schedule on a daily basis, and exciting to deal with the problems that come about when plans don't exactly work. Coming up with solutions to problems and helping the progress of a building or structure is what keeps me on my toes.

L. Edwards, CE

I love the people I work with. They make work fun. I also love the feeling of seeing a product on the shelf in a store and knowing I was a part of getting it there. I also love the fact that I am never really doing the same thing twice. There is always a new and different challenge, so the job is never boring.

S. Forkner, ChE

Mentoring my engineering staff. Seeing each of them grow and develop professionally is an indescribable feeling.

J. Huggins, ME

I am always using what I learned in engineering school and continually learning new technology. I love that I am consistently stimulated by new challenges, whether technical or legal.

M. Lauer, EnvE

I love working on the next rocket to the Moon; I've enjoyed learning a new discipline and meeting the people involved.

E. Styles, CpE

The flexibility and responsibility. As an engineer, you are already assumed to be "a cut above" everyone else. Management typically gives an engineer a project goal, and the engineer develops it. There is very little hand-holding.

A. Thompson, EE

HOW I LEARNED TO PROGRAM . . .

by David Bowman

Learning how to program a computer is different from learning a new concept or theory. Especially for someone not typically interested in going to school to graduate as a programmer, learning to program a computer might seem like a waste of time. However, computers are so integrated with every aspect of engineering, business, and research that it is critical for any engineer to have some degree of experience with how a computer can be beneficial for solving problems.

As a computer engineer, I realized the value of programming but was more interested in the hardware side of computer engineering. Regardless, I had to learn how to program a computer just like every other student entering the discipline. It is important to recognize that learning to program has very little in common with learning something like calculus. Programming is a skill that is part of the bigger picture of engineering problems, like soldering is an important skill for being able to wire a circuit. Here are a few tips to guide you while you are learning this new skill.

Learning to Write Computer Programs

- **Spend more time planning than typing.** Regardless of what type of program or function you are writing, before you even begin to type any code, you should have a plan. That plan should include a scope definition, written algorithm, flowchart, and a tentative list of any variables, functions, or other built-in controls you might need. When I write a new program from scratch, I like to draw a few pictures on a whiteboard to get an idea of how my program and functions will interact.

- **Do not study. Practice!** Studying programming does not really make sense. Instead, you should practice writing code.
- **Do not get overwhelmed.** Pace yourself! There is almost no chance that you will sit down and write a program from start to finish and have that program work 100% without the need to fix problems in your code. Sometimes the problems are just typos or syntax errors, but sometimes the problems represent a higher misconception about the problem. It is always OK to step back, reevaluate the algorithm, and start again from scratch.
- **Recycling is not just for cans and newspapers.** If you wrote a clever piece of code, remember it—if you ever need to solve a problem that involves all or part of a program you have written in the past, there is no need to reinvent the wheel. If you write a piece of code that you think will be helpful with most programs, turn it into a function and use it when you need it.
- **Documentation is almost more important than the code itself.** As you are writing your code, pay attention to including comments throughout the code. Those comments will prove to be invaluable later as you, your boss, your coworker, or anyone else who is reading your code needs to change the functionality of the program. Your comments should be clear enough so that someone who does not understand programming can pick up your code and comprehend what is going on.
- **Keep it simple, keep it clean.** Only create variables when you absolutely need to in the code. Do not recalculate a value that you stored in a variable earlier in the code. Keep everything properly indented so that you can follow the structure of your program. If your own code is difficult for you to understand, it is going to be difficult for someone else to understand.

Developing a Program from Scratch

Step 1: Clearly define what you want your program to do. Come up with a problem statement that says exactly what your code is supposed to accomplish. In addition, define the scope of the problem.

Step 2: Create an algorithm. Create a written algorithm or flowchart that defines each step in the sequence for solving the problem.

Step 3: Create a blank file with a header and comments. Write a well-documented header for the program; include the problem statement, documentation on all variables needed in the code, a creation and "last modified" date, and the names of each contributor to the program. Populate the file with comments. As a rule of thumb, each step in a written algorithm should be a comment in the code.

Step 4: Write the program. Write the code necessary to solve the problem. If necessary, develop any extra functions needed to be called from the main program.

Step 5: Test the program. Test the program to see if all bugs have been removed. Bugs might include syntax errors, errors in formula calculations, misuse of variables, or other less obvious problems. If the problems are not immediately identified, trace through the code by hand to identify where the problem occurs.

HOW I LEARNED TO PROGRAM . . .

by Matthew Ohland

I was a bit nervous about programming at the beginning—my sister attended engineering school six years ahead of me, and I had seen one of her computer programs—a row of punch cards about two feet long that tracked the population of fish in a pond. My classmates and I were among the first to have the chance to experience personal computers in junior high school—the TRS-80 was the popular model at the time. I got a little programming experience there, but I never had the time (or took the time) to see them as powerful tools—they were more a curiosity. Sensing the onset of the revolution in computing, I took a summer class in computing in the summer between high school graduation and starting college. I believe the three main things I learned were:

1. **Structured programming is better**—it is easier to read programs written without "goto" and "break" statements allowing the flow to jump around like BASIC did.
2. Procedural programming is like other designed systems I was familiar with (like a car), which have well-defined subsystems. This made sense to me, and made it clearer that **computer programs were something that people *designed*.**
3. **People should *never* use their spouse's name for a password**, especially if they are teaching a computer programming class.

While I had learned to break computer programs down into procedures, my programming was pretty linear in format, and I did not really understand the power of computing. In college, I learned FORTRAN in my second semester as part of a statics class. Even then, I resisted using iteration to generate an approximate solution. We had to write a program to determine the maximum stress in a system, which required computing the stress in different orientations iteratively. Since the problem was based on trigonometric principles, it was possible to solve it without iteration, so that is what I did. On the one hand, my program was much shorter and more efficient than anyone else's. On the other hand, I completely missed the point of the assignment, which was to learn how to have a computer guess at a lot of configurations quickly and then establish criteria to describe an optimal solution. Sometimes in the early stages of learning to program, an assignment may seem trivial or contrived, but there is often an underlying point and broader learning outcome to the assignment.

When I program, I need to visualize the whole program. As an experienced programmer, if the program is simple enough, I can see it all in my head, and I just start typing. If the program is longer, or if I am collaborating with someone else, I draw a graphical algorithm showing what information the program needs, what outputs the program is generating, and finally what has to happen in between to connect the two.

I have learned to program in a few languages—Pascal, FORTRAN, MATLAB, and Maple. What is really interesting is that I have found that I can read code written in programming languages that I have never learned. Learning programming has helped me develop logical thinking patterns that seem to have universal applicability—not just in programming, but also more generally in any design process.

WISE WORDS: WOULD YOU CONSIDER YOUR CURRENT POSITION TO BE "PURE ENGINEERING," A "BLEND OF ENGINEERING AND ANOTHER FIELD," OR "ANOTHER FIELD?"

I always feel my work is not "pure engineering," but rather often a blend of engineering, sales, accounting, research, inspection, and maintenance.

E. Basta, CME

I would consider my career in another field from engineering, however, highly reliant on my engineering background. As a management consultant, I have to break down complex problems, develop hypotheses, collect data I believe will prove or disprove the hypotheses, and perform the analysis. My focus area is companies who develop highly engineered products.

M. Ciuca, ME

My position is mostly pure engineering.

E. D'Avignon, CpE

I work in a blend of engineering and business. I spend most of my time working on business-related activities—forecasting, variance reporting, and timing/work decisions—but I also have to work closely with our field engineers and understand our project scopes. I use both my business and engineering knowledge on a daily basis—without each, I would not be able to succeed at my job.

R. Holcomb, IE

It is definitely a blend of engineering and law with a heavy dose of technical writing. It takes the thinking of an engineer or scientist to truly comprehend the inventions and the skill of a writer to convey the inventor's ideas in written and image terms that others will understand (including juries of lay people). It takes the thinking of a lawyer to come up with creative strategies and solutions when faced with a certain set of facts.

M. Lauer, EnvE

My current position is definitely a blend of engineering and at least one other field, but more like five other fields. I definitely use my engineering background in the way I think, the way I analyze data, how I approach problems, and how I integrate seemingly unrelated information together. The project management skills that I learned in engineering are helpful, too.

B. Holloway, ME

Even though my boss calls Hydrology "Voodoo Engineering," it is pure engineering.

J. Meena, CE

A blend of mechanical/aerospace engineering and human factors engineering—and management.

R. Werneth, ME

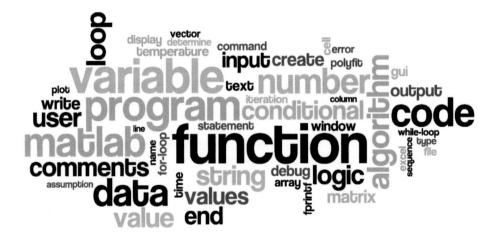

CHAPTER 15
ALGORITHMS

Learning to create effective algorithms is a crucial skill for any aspiring engineer. In general, an **algorithm** is a well-defined sequence of instructions that describe a process. Algorithms can be observed in everyday life through oral directions ("Simon says: raise your right hand"), written recipes ("Bake for 15 minutes at 350 degrees Fahrenheit"), graphical assembly instructions, or other graphical cues. As an engineer, writing any algorithm requires a complete understanding of all the necessary actions and decisions that must occur to complete a task.

When writing an algorithm, you must answer a few questions before attempting to design the process. To even begin thinking of a strategy to describe a process, you must have carefully defined the scope of the problem. The **scope** of an algorithm is the overall perspective and result that the algorithm must include in its design.

For example, if we are required to "sum all numbers between 1 and 5," before thinking about an approach to solve the problem, we must first determine if the scope is properly defined. Does the word "between" imply that 1 and 5 are included in the sum? Do "numbers" include only the integer values? What about the irrational numerical values? Clearly, we observe that we cannot properly define the scope of the charge to add all numbers between 1 and 5.

Likewise, imagine you were charged to design a device that transports people from Atlanta, Georgia, to Los Angeles, California. How many people must the device transport? Does the device need to travel on land? Should it travel by air? Should it travel by water? Does the device require any human interaction?

This section covers two methods of defining a process: with **written algorithms** and with **graphical algorithms**. Both methods require properly identifying the scope of the problem and all of the necessary input and output of the process.

15.1 SCOPE

One of the most difficult steps in designing an algorithm is properly identifying the entire scope of the solution. Like solving a problem on paper involving unit conversions and equations, it is often necessary to state all of the known and unknown variables in order to determine a smart solution to the problem. If information is left out of the problem, it might be necessary to state an assumption in order to proceed with a solution. After all variables and assumptions about the problem have been identified, it is then possible to create a sequence of actions and decisions to solve the problem.

To clearly understand the scope of the problem, we often find it helpful to formally write out the known and unknown information, as well as state any assumptions necessary to solve the problem. In the following examples, notice that as the problem statements become more and more refined, the number of necessary assumptions decreases and eventually disappears.

● **EXAMPLE 15-1**

For the problem statement, list all knowns, unknowns, and assumptions.

Problem: Sum all numbers between 1 and 10.

Known:

■ *The minimum value in the sum will be 1.*
■ *The maximum value in the sum will be 10.*

Unknown:

■ *The sum of the sequence of numbers.*

Assumptions:

■ *We will only include the whole number values (e.g., 1, 2, 3, . . .) in the sum.*
■ *The sum will include the starting value of 1 and the ending value of 10.*

● **EXAMPLE 15-2**

For the problem statement, list all knowns, unknowns, and assumptions.

Problem: Sum all numbers between (and including) 1 and 10.

Known:

■ *The minimum value in the sum will be 1.*
■ *The maximum value in the sum will be 10.*

Unknown:

■ *The sum of the sequence of numbers.*

Assumptions:

■ *We will only include the whole number values (e.g., 1, 2, 3, …) in the sum.*

● **EXAMPLE 15-3**

For the problem statement, list all knowns, unknowns, and assumptions.

Problem: Sum all whole numbers between (and including) 1 and 10.

Known:

■ *The minimum value in the sum will be 1.*
■ *The maximum value in the sum will be 10.*

Unknown:

■ *The sum of the sequence of numbers.*

Assumptions:

■ *[None]*

COMPREHENSION CHECK 15-1

For the problem statement, list all knowns, unknowns, and assumptions.
Problem: Sum all even numbers between (and including) 2 and 20.

COMPREHENSION CHECK 15-2	For the problem statement, list all knowns, unknowns, and assumptions. Problem: Multiply all powers of 5 between (and including) 5 and 50.

15.2 WRITTEN ALGORITHMS

A written algorithm is a narrative set of instructions required to solve a problem. In everyday life, we encounter written algorithms in the form of oral instructions or written recipes. However, it is extremely common for humans to "fill in the blanks" on a poorly written algorithm. Imagine you are handed a strongly guarded family recipe for tacos. One of the steps in the archaic recipe is to "cook beef on low heat until done." To the veteran cook, it is apparent that this step requires cooking the prepared ground beef on a stove-top in a sauce pan for approximately 10 minutes on a burner setting of 2 to 3. To a first-time cook, the step is poorly defined and could result in potentially inedible taco meat.

Engineers and Written Algorithms

As an engineer, to write effective algorithms you must ensure that every step you include in a written algorithm must not be subject to misinterpretation. It is helpful to write an algorithm as if it were to be read by someone completely unfamiliar with the topic. Each step in the written algorithm should be written such that the stepwise scope is properly defined. The **stepwise scope** is all of the known and unknown information at that point in the procedure. If a step in an algorithm contains an assumption, you must formally declare it before proceeding with the next step. By ensuring that the stepwise scope is well defined, you ensure that your algorithm will not be subject to misinterpretation.

All written algorithms should be expressed sequentially. The most effective algorithms are written with many ordered steps, wherein each step contains one piece of information or procedure. While the author of an algorithm may consider each step in an algorithm to be "simple," it might not be trivial to an external interpreter. When writing an algorithm, it is helpful to assume that the reader of your algorithm can only perform small, simple tasks. Assume that your algorithm can be interpreted by a computer. A computer can execute small tasks efficiently and quickly, but unlike a human, a computer cannot fill in the blanks with information you intended the reader to assume.

Decision-making can be expressed in a written algorithm. Assume you are designing a process to determine if the value read from a temperature sensor in a vehicle indicates it is unsafe for operation. To express the decision in a written algorithm, phrase your decisions in questions that have a "Yes" or "No" response.

Format of Written Algorithms

The first step in writing any algorithm is defining the scope of the problem. After you define the scope of the problem, create an *ordered* or *bulleted list* of actions and decisions. Imagine taking an English class and writing a research report on the influence of 19th-century writers on modern-day fiction authors. Before writing the paper, you would create an outline to ensure that your topics have connectivity and flow. Just like the outline of an English paper, an algorithm is best expressed as a sequential list rather than as complete paragraphs of information.

If a decision is required in the algorithm, indent the actions to indicate the action is only associated with the particular condition.

● **EXAMPLE 15-4**

Create a written algorithm to convert a temperature from relative units [°F or °C] to the corresponding absolute units [K or °R].

Known:

- *Temperature in relative units (degrees Celsius or degrees Fahrenheit).*

Unknown:

- *Temperature in absolute units (kelvin or degrees Rankine).*

Assumptions:

- *Since the problem does not explicitly state the temperature to be converted, assume that the interpreter of the algorithm will input the temperature and units.*

Algorithm:

1. *Input the numeric value of the temperature.*
2. *Input the units of the numeric value of the temperature.*
3. *Ask if the input unit is degrees Fahrenheit.*
   (a) *If yes, convert the value to degrees Rankine.*
   (b) *If no, convert the value to kelvin.*
4. *Display the converted value and absolute unit.*
5. *End the process.*

● **EXAMPLE 15-5**

Create a written algorithm to calculate the sum of a sequence of whole numbers, given the upper and lower bounds of the sequence.

Known:

- *Upper bound of whole number sequence.*
- *Lower bound of whole number sequence.*

Unknown:

- *Sum of all whole numbers between the upper and lower bound.*

Assumptions:

- *Since the problem does not explicitly state the upper and lower bounds, assume the interpreter will ask for the values.*
- *Include the boundary values in the summation.*

Algorithm:

1. *Input the lower bound of the sequence.*
2. *Input the upper bound of the sequence.*
3. *If the lower bound is larger than the upper bound,*

    (a) *Warn the user that the input is invalid.*
    (b) *End the process.*

4. *If the upper bound is larger than the lower bound,*

    (a) *Create a variable to keep track of the sum (S).*
    (b) *Create a variable to keep track of the location in the sequence (I).*
    (c) *Set the initial value of S to be zero.*
    (d) *Set the initial value of I to be the lower bound.*
    (e) *If the value of I is less than or equal to the upper bound,*

       (i) *Add I to the current value of S.*
       (ii) *Add one to the current value of I.*
       (iii) *Return to step 4.e. and ask the question again.*

    (f) *If the value of I is greater than the upper bound,*

       (i) *Display the sum of the sequence (S).*

5. *End the process.*

In step 4.e.iii, we required that the interpreter return to an earlier step in the algorithm after changing the values of our variables. This allows us to create a **feedback loop** necessary to calculate the sequence of values. A feedback loop is a return to an earlier location in an algorithm with updated values of variables. It is important to note that if we failed to update the variables, the feedback loop will never terminate. A nonterminating feedback loop is also known as an **infinite feedback loop**.

COMPREHENSION CHECK 15-3

Create a written algorithm to multiply all integer powers of 5, 5^x, for x between (and including) 5 and 50.

15.3 GRAPHICAL ALGORITHMS

To visualize a process, a graphical representation of algorithms is used instead of a written algorithm. A **flowchart** is a graphical representation of a written algorithm that describes the sequence of actions, decisions, and path of a process. Designing a flowchart forces the author of the algorithm to create small steps that can be quickly evaluated by the interpreter and enforces a sequence of all actions and decisions. Flowcharts are used by many different disciplines of engineering to describe different types of processes, so learning to create and interpret flowcharts is a critical skill for a young engineer. In fact, in the United States, any engineers who discover a new innovative algorithm can submit their concept to a patent office by representing the process in terms of a flowchart.

Three different shapes are used in the creation of flowcharts in this book; a number of other widely used operators are encountered across the world. In this book, we describe all actions with rectangles, all decisions with diamonds, and all connections between shapes with directional arrows.

Rules for Creating a Proper Flowchart

■ The flowchart must contain a START rectangle to designate the beginning of a process.
■ All actions must be contained within rectangles.
■ All decisions must be contained within diamonds.
■ All shapes must be connected by a one-way directional arrow.
■ The flowchart must contain an END rectangle to designate the end of a process.

Actions

Actions are any executable steps in an algorithm that do not require a decision to be made. Based on this definition, any defined variables, calculations, and input or output commands would all be contained within action rectangles.

All simple actions are contained within a single rectangle on the flowchart. For each rectangle, two arrows are always associated with the shape, with two exceptions. The inward arrow to the rectangle represents the input to the action. It is assumed that any variables defined in the stepwise scope of an action rectangle are accessible and can be used in the action. The outward arrow from the rectangle represents the output of the action. If any new variables or calculations are performed within the rectangle, those values are passed along to the next shape's stepwise scope.

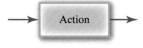

■ Exception One: The START rectangle represents the beginning of the flowchart and does not contain an inward arrow. An oval shape is also commonly used to represent the start of an algorithm.
■ Exception Two: The END rectangle represents the end of the flowchart and does not contain an outward output arrow. An oval shape is also commonly used to represent the end of an algorithm.

Decisions

Decisions are any executable steps in an algorithm that require the answer to a question with "Yes" or "No." All decisions in a flowchart must be represented within a diamond shape.

For each diamond on a flowchart, at least three arrows are always associated with the shape. The inward arrow to the diamond represents the input to the decision. It is assumed that any variables defined in the stepwise scope of a decision diamond are accessible and can be used in the decision. The two outward arrows that exit decision diamonds represent the conditional branch based on the outcome of the question asked within the diamond. If the outcome of the decision is true, the flow of the process will follow the "Yes" branch; otherwise, it will follow the "No" branch. Since no new variables are created in a decision diamond, the stepwise scope that enters the decision diamond is passed on to the next shape of each conditional branch.

● **EXAMPLE 15-6**

Create a flowchart to convert a temperature from relative units [°F or °C] to the corresponding absolute units [K or °R].

Known:

■ *Temperature in relative units (degrees Celsius or degrees Fahrenheit).*

Unknown:

■ *Temperature in absolute units (kelvin or degrees Rankine).*

Assumptions:

■ *Since the problem does not explicitly state the temperature to be converted, the interpreter of the algorithm will ask for the temperature and units.*

Flowchart:

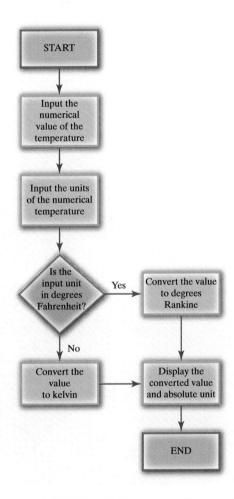

● **EXAMPLE 15-7**

Create a flowchart to calculate the sum of a sequence of whole numbers, given the upper and lower bounds of the sequence.

Known:

■ *Upper bound of whole number sequence.*
■ *Lower bound of whole number sequence.*

Unknown:

- *Sum of all whole numbers between the upper and lower bound.*

Assumptions:

- *Since the problem does not explicitly state the upper and lower bounds, the interpreter will ask for the values. Include the boundary values in the summation.*

Flowchart:

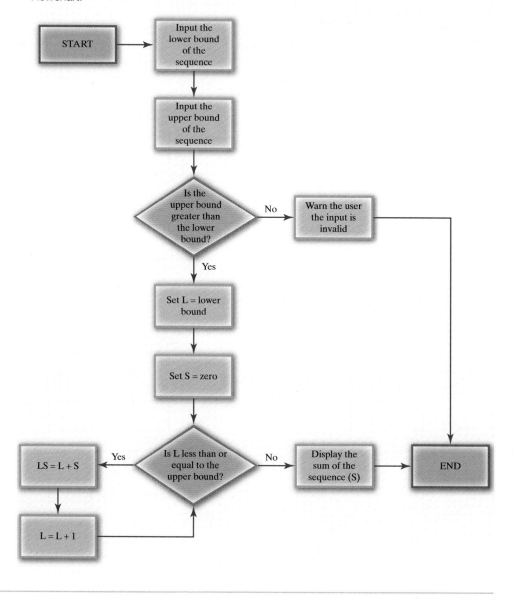

**COMPREHENSION
CHECK 15-4**   Create a graphical algorithm to multiply all integer powers of 5, 5^x, for x between (and including) 5 and 50.

Flowchart Creation in Microsoft Word

Open a new Microsoft Word document. Click the **Insert** ribbon at the top of the Microsoft Word editor window. Click the **Shapes** drop-down menu in the Illustrations box. You will need to use the rectangle and diamond shapes under the Flowchart section as well as the directional arrows under the Lines section.

⌘ **Mac OS:** Access the flowchart shapes by clicking the Object Palette button near the top of the Formatting Palette and then clicking the Shapes button.

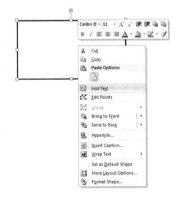

↳ **To add an action:** Click the Rectangle tool from the Shapes menu and click-and-drag into the body of the document to draw a rectangle. Right-click the rectangle and click Add Text to add text on top of the rectangle.

⌘ **Mac OS:** Control-click or two-finger tap to access the **Add Text** menu item.

↳ **To add a decision:** Click the Diamond tool from the Shapes menu and click-and-drag into the body of the document to draw a diamond. Right-click the diamond and click Add Text to add text on top of the rectangle.

↳ **To insert the YES and NO labels:** Use the Text Box option under the Basic Shapes menu. To remove the border, right-click the text box and click Format Text Box. On the Color and Lines tab, click the Color drop-down menu under the Lines section and select No Color. Click OK.

⌘ **Mac OS:** In the main menu, click **Insert > Text Box**. The default is probably "no border," but the border can be modified in the formatting palette.

↳ **To add an arrow:** Click the single direction arrow from the Shapes menu and starting from the source click-and-drag to the destination.

15.4 ALGORITHM BEST PRACTICES

If you have never composed a written or graphical algorithm before, the remaining part of this section details specifics on how to begin planning and writing algorithms from scratch. This section does not intend to be a definitive resource on algorithm development, but it may provide guidance if you are struggling to break down a process into small, achievable steps.

Actions

In every action within an algorithm, there must be a key verb that defines the purpose of that step within an algorithm. The remaining subsections discuss different types of actions and list some of the common verbs associated with that category of action.

Establishing Variables and Constants

After defining the scope of a problem, it might become obvious that there are intermediate calculations or assumed constants that must be contained throughout the process. Along with the explicitly defined known values, these intermediate and constant values are referred to as variables. Algorithmic variables are different from the mathematic definition of a variable because algorithmic variables are treated more like containers to store known values and results of calculations rather than being some unknown entity in a mathematical expression. They are called variables because the stored value can be written, overwritten, and used by other actions or decisions in the algorithm.

Example	Action
We assume the acceleration due to gravity is 9.8 meters per second squared.	**Set** variable g to be 9.8.

Other Verbs					
Set	Define	Assign	Write	Store	Designate
Label	Name	Cast	Insert	Save	Initialize

User Interaction

It is often necessary to write algorithms that can be executed with prompts for input from the person using the algorithm, provide feedback on results, or display any error messages generated in the algorithm.

User Input:

Example	Action
We want the user of the algorithm to provide the amount of water in gallons.	**Input** the amount of water in gallons, save in variable W.

Other Verbs					
Input	Ask	Load	Request	Query	Prompt

User Output:

Example	Action
We want the algorithm to inform the user that the amount of water can't be negative.	**Display** error message to user "Warning: amount of water can't be negative!"

Other Verbs				
Output	Display	Reveal	Write	Warn

Calculations and Conversions

When algorithms involve calculating a value using an equation, it is helpful to write out the full equation and identify which variables in the algorithm correspond to the variables in the expression. For unit conversions, it is not necessary to write out the conversion factors since those are published standards that are readily available to anyone executing your algorithm. When using conversion, it is best to list them individually so they are easily recognizable to the user. For example, when converting from feet to centimeters, the expression $L = L/3.28 * 100$ is easily recognized as the conversion from feet to meters, and then from meters to centimeters. It is harder to recognize the conversion of $L = L * 30.48$. Furthermore, it is easy to make a calculation error; it is easier to allow the program to calculate for you. When dealing with unit conversions, it is ideal to save the converted value back into the original variable to reduce the number of variables you need to keep track of in your algorithm.

Calculations:

Example	Action
We want to calculate the thermal energy of a substance using the expression $Q = m\, C_p\, \Delta T$, where m is the mass, C_p is the specific heat, and ΔT is the change in temperature.	**Compute** the thermal energy: $Q = m\, C_p\, \Delta T$ All variables should appear in the variable list.

Other Verbs					
Calculate	Adjust	Count	Measure	Add	Multiply
Subtract	Divide	Compute	Increment	Decrement	

Conversions:

Example	Action
We want to convert a variable t from minutes to seconds and save the result back in the variable t.	**Convert** t from minutes to seconds, save in t.

Other Verbs				
Convert	Change	Alter	Revise	Switch

Referencing Other Algorithms

When developing a large program that involves a number of built-in or user-defined functions, it is often necessary to reference those functions within an algorithm. As a rule of thumb, each custom function you create should have its own separate algorithm. If you have separate algorithms for a program and the different functions referenced in the code, it makes the algorithms simpler to understand and easier to debug. When calling a function within an algorithm, it is critical to list the variables passed to the function and variables captured by the function. The variables passed to and captured from a function call do not need to be named the same as they are in the function since the function and the code calling the function have different scopes. In general, the most common verb used with functions is "call." If you know the name you plan to use for your function, list it; otherwise, this can be set later.

Example	Action
We want to use a user-defined function named Poltocar that converts coordinates from polar to Cartesian. We will pass in the variable Z as the radius and the variable T as the angle. We will capture the x-coordinate in the variable X and the y-coordinate in the variable Y.	**Call** Poltocar In: Z, T Out: X, Y

Decisions

All decisions made in algorithms must be constructed as binary decisions. Typical decisions in algorithms involve comparing variables, examining the contents of a variable, or examining the dimensionality of a variable. Any decisions that require some amount of computation in the decision should be split so that the calculations occur in actions before reaching the decision block.

Error Checking and Prevention

Including error checking in an algorithm allows for the creation of robust solutions to problems that will not lead to incorrect or unstable answers. In general, a check for an error will either terminate the algorithm or lead to some action that will allow the algorithm to continue; however, in both cases it should notify the user that an error has occurred (see Figure 15-1). If you want your algorithm to re-prompt the user for input to assure that proper data are contained in a variable before proceeding into the remainder of the algorithm, see section on "Error Prevention." The remainder of this section discusses three different types of errors that may occur in an algorithm, but this is only a starting point. The amount and type of errors that can occur in an algorithm are infinite, so it is up to the designer of the algorithm to decide how and when error checking should occur. In most programming languages, the structure that enables error checking is the if-statement; however, some languages like MATLAB include "try" and "catch" blocks that will allow programmers to wrap a calculation inside a try block, and if it causes an error, execute the commands contained within the catch block.

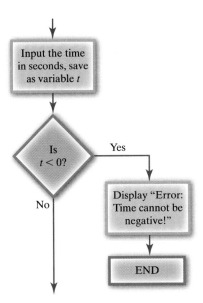

Figure 15-1 An example of error checking embedded in a flowchart.

Division by Zero and Infinite Values

If your algorithm contains a calculation where a combination of one or more variables in the computed expression could lead to a division by zero, it is smart to include a check to see if the result is zero. Some languages like MATLAB will happily compute an expression with a zero divisor and return the result as "INF"—a special MATLAB reserved word representing infinity.

Invalid Dimensions of Variables

When an algorithm assumes that one or more variables contain matrices or vectors, any calculations on those variables must follow the same mathematical rules associated with the matrix operation. For example, if an algorithm requires two matrices to be added together, it would be wise to include a check to see if the two matrices have the same number of rows and columns before attempting to add them together. This will prevent algorithms from crashing due to an invalid computation. In addition, this will prevent issues related to accessing elements of a matrix that do not exist.

Invalid Range of Values

Since variables typically represent some measured or computed value, restrictions on those variables that apply in real life may not be directly enforced in your algorithm. For example, if your algorithm prompts a user to input a quantity that cannot be negative (like length, volume, time, etc.), it is smart to check if the value in the variable is reasonable. Likewise, if your algorithm should not generate complex values (e.g., $3 + 2i$), your algorithm will need to check to see if the result of a computation would generate a complex value instead of the desired real value.

Error Prevention

Error prevention looks for the same type of errors that are detected in error checking, but error prevention will allow your algorithm to prompt or re-prompt the user to correct the erroneous variables. For example, if your algorithm asks the user to type in a

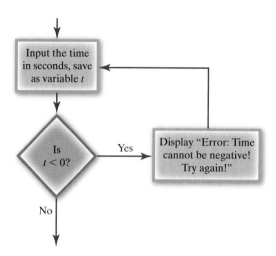

Figure 15-2 An example of error prevention embedded in a flowchart.

volume and the user erroneously types a negative value, your algorithm could detect the incorrect value and go back to the input statement to force the user to type the value again (and again, and again . . .) until the user types a value within the acceptable range (see Figure 15-2). In most programming languages, the structure that enables error prevention is the while-loop.

Iteration

Some algorithms require repeated calculations that typically involve the use of a sequence of values or some operation on a vector or matrix stored in a variable. Such algorithms are considered to be iterative because they require iteration on a counter variable to keep track of when to terminate. For example, assume we have a vector, **V**, which contains positive and negative values in random order. If we want to create two new vectors, **VN** and **VP**, that contain the negative and positive values of V respectively, we will need to iterate through each element of the vector **V**, make a decision about each value, and store it in the corresponding vector. To do this, we would need to create a counter variable, or sometimes called an index variable, that will keep track of the number of times we have repeated a calculation or decision. If we create a counter variable, X, and initialize it to be the number 1, X will actually serve two purposes. In addition to keeping track of the number of times we have repeatedly made decisions and stored new values into **VN** and **VP**, it will also serve as the index variable into the **V** vector so that we can access element V(1), V(2), and so on until we reach the last element in **V**. Figure 15-3 demonstrates this scenario as a flowchart, including the iterated counter variable X.

Algorithms that will require iteration typically have a scenario where you have to repeat some decision or calculation "for each" or "for every" element or value within a sequence or vector. Since there is no "for each" or "for every" building block within an algorithm, this type of structure must be constructed out of the initialization of a counter variable (e.g., Set X to be 1), a decision that involves the value of a counter variable (e.g., If X is less than or equal to the number of elements in **V**), and some action block that increments the counter variable (e.g., Set X equal to the current value of X plus 1). After the counter increment action block, the algorithm can loop or refer back to the decision made on the increment variable. In most programming languages, the structure that enables iteration in an algorithm is the for-loop.

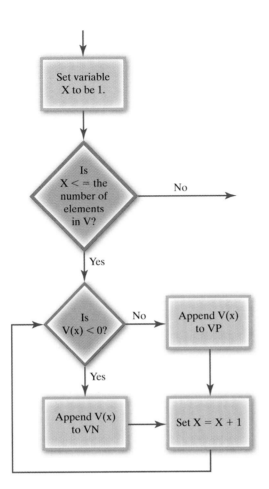

Figure 15-3 An example of iteration embedded in a flowchart.

Testing Your Algorithm

The last step in writing an algorithm is developing test cases that will reveal whether or not your algorithm behaves as expected. The key to writing test cases is figuring out how many test cases are necessary to confirm whether or not your code works. In general, there should be at least one or two test cases that will demonstrate the proper behavior of the algorithm given good input values. Not only should a test case include a list of all of the inputs used to generate the output, but you should also compute the expected output of the algorithm by hand in order to verify that the algorithm works. In addition, there should be one test case that will verify that all of the error checking/prevention built in to the algorithm works properly. Likewise, any decisions that lead to different states within your algorithm should have a test case to verify that the logic you designed is arranged properly in order to generate the desired output.

For example, assume we have designed an algorithm that will calculate power given energy and time. In our algorithm, we included two error checks—the first check to see if energy is greater than 50 joules and the second check to see if time is greater than 0 seconds. If either of these conditions are not true, the algorithm will display an error message "Error: Incorrect input value" and terminate. In addition, we added logic to check to see if the energy in the system is greater than 500 joules and added 5 seconds to the time variable; else we left the time variable alone.

A test case for this scenario would look like this:

Input	Output
$E = 0$ J, $T = 30$ seconds	Error: Incorrect input value
$E = 55$ J, $T = -3$ seconds	Error: Incorrect input value
$E = 400$ J, $T = 5$ seconds	$P = 80$ W
$E = 550$ J, $T = 10$ seconds	$P = 36.7$ W

To help you develop good algorithms, we have included an algorithm template to help you document all of the variables, procedures, and test cases necessary to create correct and verifiable algorithms. Examples of how this template can be used are provided below and on select problems in the remaining chapters.

● **EXAMPLE 15-8**   Create an algorithm to determine the volume of a cylinder, given the radius and height.

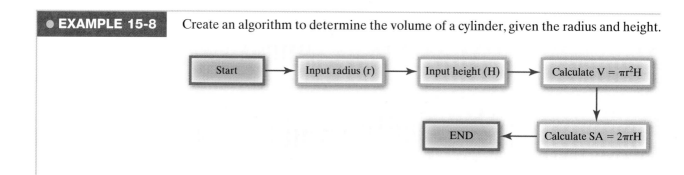

In-Class Activities

ICA 15-1

Create an algorithm (written and/or flowchart as specified by your instructor) that describes the steps necessary to create a jelly sandwich. You may assume that you are starting with a loaf of bread, jar of jelly, a knife, and a plate on the table in front of you. When you are finished, hand your algorithm to the instructor and wait for further instruction.

ICA 15-2

Create an algorithm (written and/or flowchart as specified by your instructor) that describes the steps necessary to create a paper airplane. You may assume that you are starting with a single sheet of 8½ × 11 inch paper. When you are finished, hand your algorithm to the instructor and wait for further instruction.

ICA 15-3

Your instructor will provide you with a picture of a structure created using K'Nex™ pieces. Create an algorithm (written and/or flowchart as specified by your instructor) to recreate the picture. When you are finished, hand your algorithm to the instructor and wait for further instruction.

ICA 15-4

Create an algorithm (written and/or flowchart as specified by your instructor) to determine the factorial value of an input integer between 1 and 10. Add a statement to check for input range.

ICA 15-5

Create an algorithm (written and/or flowchart as specified by your instructor) to cook your favorite meal, including a starting materials list.

ICA 15-6

Create an algorithm (written and/or flowchart as specified by your instructor) to walk from your home or dorm to your engineering classroom without using any street names.

ICA 15-7

Create an algorithm (written and/or flowchart as specified by your instructor) that will repeatedly ask the user to draw a card from a deck of cards until the user pulls out the queen of hearts.

ICA 15-8

Create an algorithm (written and/or flowchart as specified by your instructor) to model the decision of a driver after looking at the color of a traffic light. You may assume that the traffic light contains three bulbs (red, yellow, green) to represent stop, slow down, and go.

ICA 15-9

Given the flowchart below, for what range of values of pressure (in atmospheres) will the light be each of the colors Yellow, Blue, and Violet?

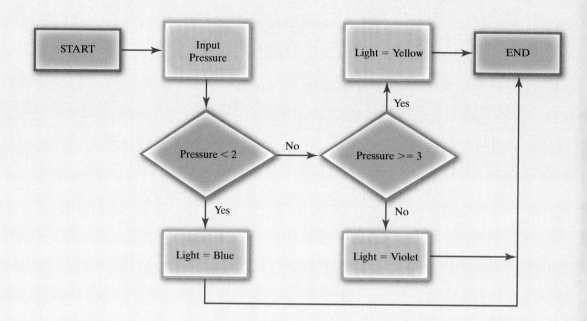

ICA 15-10

A reaction vessel is equipped with both temperature and pressure sensors. Draw a flowchart for the function described below.

- Read the values of temperature and pressure.
- If the temperature exceeds the maximum safe temperature (TMax), the status should be listed as "unsafe," and a variable T should be set to 1. Otherwise, T should equal 0.
- If the pressure exceeds the maximum safe pressure (PMax), the status should be listed as "unsafe," and a variable P should be set to 1. Otherwise, P should equal 0.
- If both temperature and pressure are less than their corresponding maxima, the status should be listed as "safe" and both variables T and P should be set to 0.
- Repeat the entire procedure.

Chapter 15 REVIEW QUESTIONS

1. In the game of chess, each piece can move a different way across the board. A diagram of a chessboard at the beginning of a game is shown for reference. Write an algorithm to show how the following pieces can move. The pieces can begin at any location on the board (not necessarily in their starting position) and will proceed forward as far as possible (within the limitations of their piece or until the end of the board is reached or until it encounters another piece).

   (a) Pawn
   (b) Bishop
   (c) Knight

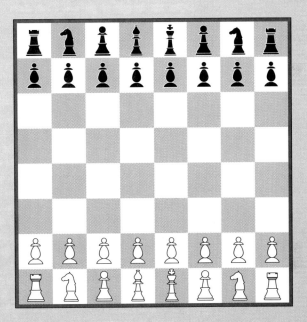

2. You want to calculate your grade point ratio (GPR). Create an algorithm to input your courses for this semester, the semester hours for each course, the letter grade you expect to earn, and number of grade points earned for each course (such as A = 4). Calculate the average GPR for the semester using the following formula: *summation of the product of the fall semester hours and number of points earned divided by the summation of fall semester hours*. After you have written out the algorithm, convert your narrative algorithm to a flowchart.

3. Construct a flowchart to determine the class rank (freshman, sophomore, junior, or senior) of students enrolled in this course. Output the total number of students in each rank.

4. Construct a flowchart to make a hamburger, offering the user a choice of four toppings (lettuce, tomato, onions, cheese), three condiments (mustard, mayonnaise, ketchup), and three ways to cook the burger (medium, medium well, well done).

5. Construct a flowchart to solve the following problem. A rod on the surface of Jupiter's moon Callisto has a volume of 0.3 cubic meters. Determine the weight of the rod in units of pounds-force. The specific gravity is 4.7. Gravitational acceleration on Callisto is 1.25 meters per second squared.

6. Construct a flowchart to solve the following problem. The specific gravity of gold is 19.3. Determine the length of one side of a 0.4 kilogram cube of solid gold, in units of inches.

7. Humans can see electromagnetic radiation when the wavelength is within the spectrum of visible light. Create a written algorithm and a flowchart to determine if a given wavelength [nanometer, nm] is one of the six spectral colors listed in the chart below. Your algorithm should provide a warning if the provided wavelength is not within the visible spectrum.

Color	Wavelength Interval
Red	~ 700–635 nm
Orange	~ 635–590 nm
Yellow	~ 590–560 nm
Green	~ 560–490 nm
Blue	~ 490–450 nm
Violet	~ 450–400 nm

8. Create a flowchart that represents the following written algorithm.

   ■ Input the height of person 1 [in units of feet] as P1
   ■ Input the height of person 2 [in units of feet] as P2
   ■ If person 1 is taller than person 2

     • Display "Person 1 is taller"

   ■ Otherwise, if person 2 is taller than person 1

     • Display "Person 2 is taller"
     • Otherwise, display "They are the same height"

9. Create an algorithm (written and/or flowchart as specified by your instructor) to determine whether a given altitude [meters] is in the troposphere, lower stratosphere, or the upper stratosphere. In addition, your algorithm should calculate and report the resulting temperature in units of degrees Celsius [°C] and pressure in units of kilopascals [kPa]. Refer to atmosphere model provided by NASA: http://www.grc.nasa.gov/WWW/K-12/airplane/atmosmet.html.

10. Create an algorithm (written and/or flowchart as specified by your instructor) to determine whether a given Mach number is subsonic, transonic, supersonic, or hyper sonic. Assume the user will enter the speed of the object in meters per second, and the program will calculate the Mach number and determine the appropriate Mach category. As output, the program will display the Mach rating. Refer to the NASA page on Mach number: http://www.grc.nasa.gov/WWW/K-12/airplane/mach.html.

11. Create an algorithm (written and/or flowchart as specified by your instructor) to solve the following problem. An unmanned X-43A scramjet test vehicle has achieved a maximum speed of Mach number 9.68 in a test flight over the Pacific Ocean. Mach number is defined as the speed of an object divided by the speed of sound. Assume the speed of sound is 343 meters per second. Determine the speed in units of miles per hour.

12. Assume you have to design an algorithm for the temperature controls of an oven. Create a flowchart to implement such an algorithm. Assume all temperatures are in degrees Fahrenheit. The temperature control will be set to the desired oven temperature, which we will call **SetTemp** for convenience.

    When the oven is first turned on, the heaters should be turned on until the oven temperature is greater than or equal to **SetTemp** +5°, at which point the heaters turn off. Note that if the oven had been used recently and had not cooled down, it is possible that the temperature already exceeds this limit. When the oven temperature falls below **SetTemp** −5°, the heaters should be turned back on.

    When the oven is turned off, the heaters should immediately be turned off and the algorithm ends. For example, if the algorithm is in a loop with the heaters on, waiting for the upper temperature limit to be reached, it should *not* continue to heat the oven to the upper limit if the oven is turned off.

    The algorithm should be designed in such a way that if the user changes the set temperature, the controls will immediately react to that change. Note the comment above about the oven being turned off—this is similar.

CHAPTER 16
PROGRAMS AND FUNCTIONS

Programming is the process of expressing an algorithm in a language that a computer can interpret. To correctly automate a process on a computer, a programmer must be able to correctly speak the language both grammatically and semantically. In this text, we communicate with a computer by using the MATLAB programming language.

16.1 PROGRAMMING BASICS

Certain principles apply to programming, regardless of what language is being used. Entire textbooks have been written on this subject, but a few important concepts are addressed here.

Notes About Programming in General

RULE OF THUMB

For each rectangle and diamond in a flow-chart, there should be a few comments in the source code where that action occurs.

- **Programming style:** Just like technical reports should follow a particular format to ensure they can be understood by someone reading them, all programs should have a few common elements of programming style, particularly the inclusion of comments that help identify the source of the program and what it does. For any program, a proper header should be written describing the scope of the problem, including a problem statement and definition of all input and output variables used in the program. Properly commenting the source code is also critical for ensuring that someone else can follow your work on the program. *As a rule of thumb, for each rectangle and diamond in a flowchart, there should be a few comments in the source code where that action occurs.*

- **Program testing:** When you write a program, testing it is a critical step in making sure that it does what you expect. Lots of things can go wrong.

  - *Syntax errors*: These violate the spelling and grammar rules of the programming language. Compilers (programs that interpret your computer program) generally identify their location and nature.

  - *Runtime errors*: These occur when an inappropriate expression is evaluated during program execution. Such errors may occur only under certain circumstances. Examples of runtime errors are division by zero, the logarithm of zero, or the wrong kind of input, such as a user's entering letters when numbers are expected. Computer programs should anticipate runtime errors and alert users to the error (rather than having the program unexpectedly terminate).

  - *Formula coding errors*: It is not uncommon to enter a formula incorrectly in a form that does not create either a syntax or a runtime error. Evaluating even a few simple calculations by hand and comparing them to the values calculated by the program will usually reveal this kind of error.

- *Formula derivation errors*: This difficult error to discover in programming occurs when the computer is programmed correctly but the programming is based on a faulty conceptualization. This kind of error is difficult to figure out because programmers generally assume that a program that executes without generating any errors is correctly written. This situation underscores the importance of checking a solution strategy before starting to write a computer program.

NOTE

Why are units important? Read the "Lessons of the Mars Climate Orbiter" from Section 7.4.

- **Keeping track of units:** Some computer programs can be designed to have the user keep track of units as long as they use consistent units. Other computer programs require input data to be in a particular set of units. Regardless of which approach you use, the way in which units are managed must be clear to the user of the program and in the program's comments.

- **Precision:** All computers have finite precision, the maximum number of significant digits they can manage. Some computer programs further restrict the precision. The issue of finite precision affects the possibility of two numbers being exactly equal. As a result, it is always necessary to consider how close two numbers must be in order to be considered equal rather than test for equality.

- **Numerical versus symbolic solutions:** In algebra, we learn that the equal sign means that whatever is on the left side of the equation identically equals whatever is on the right. This is known as symbolic manipulation. Some computer programs (e.g., Maple and MathCAD) specialize in manipulating symbolic expressions. Although MATLAB can manipulate symbolic expressions, it is primarily used for its numerical capabilities. Numerical computer packages and computer programming languages in general use the equal sign to assign the right-hand value to the variable on the left-hand side. The expression on the right of the "=" is evaluated, and then placed in the variable on the left.

- **Documentation:** In any programming language, there is usually a special character or pair of characters that tell the program interpreter to ignore everything to the right of the special character to allow the programmer to write human readable notes within the code of the program. These notes are commonly referred to as "comments" within the code. A great deal of controversy surrounds how much documentation is necessary in any program, but this text proposes and follows the following guideline to properly document a program:

  For every numbered item in a written algorithm, you should include that item as a comment in your code.

In addition, at the top of each program, it is extremely helpful to at least write a problem statement and document any variables used within the program. In a function, it is helpful to cluster your description of variables into three groups: input variables, output variables, and function variables.

Notes Particular to MATLAB

Before proceeding with the material on MATLAB, it would be wise to review the material on matrices in the appendix materials. Since variables in MATLAB are inherently matrices, it is critical that you understand not only some basic matrix concepts, but also the specific notation used in this text (as well as MATLAB) when referring to matrices. Certain details specific to MATLAB are important to note early on.

■ **The MATLAB interface:** The MATLAB program window has a number of subwindows. We are not concerned with the Launch Pad or Command History windows, but two windows are of interest.

NOTE

clc = clear the
    Command Window

• The *Command Window* can be used as a powerful but awkward calculator. The ▣ button at the top-right side of the Command Window menu bar will "undock" the Command Window from the MATLAB program window. The ▣ button will close the Command Window. To restore the Command Window, select **Desktop > Command Window**. To clear any text from the Command Window, enter the special reserved MATLAB expression clc in the Command Window.

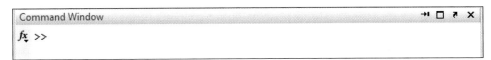

FILE TYPES

Word = .docx

Excel = .xlsx

Powerpoint = .pptx

Matlab = .m

• Although you can enter MATLAB statements in the Command Window, the *Editor/Debugger window* is the interface of choice for developing MATLAB programs. The editor stores your commands in M-files. This permanent record of your commands allows you to make small changes to either the program or the inputs and to re-execute a complicated set of commands. The editor also assists in the formatting of MATLAB programs, using both indentation and color to distinguish program elements.

■ **Naming programs, functions, and variables in MATLAB:** In MATLAB, program names, function names, and variable names must all follow certain rules to prevent unintended side effects:

• All names must consist only of letters, numbers, and the underscore character (displayed by holding "Shift" and typing a hyphen).
• Names can *only* begin with alphabetic letters (no numbers or special characters).
• Names cannot be longer than 63 characters.
• Names in MATLAB are case sensitive: Frog and frog are different.
• Names cannot have the same name as any other identifier (program name, function name, built-in function, built-in constant, reserved word, etc.), because MATLAB's order of execution will prevent the program from running in the intended order.

COMPREHENSION CHECK 16-1

Which of the following are valid MATLAB variable names?

(a) my name	**(d)** m	**(g)** my_var	**(j)** for
(b) length	**(e)** m6	**(h)** @clemson	**(k)** HELLO[]
(c) MyLength	**(f)** 4m	**(i)** my.variable	

NOTE

Order of Execution:

■ Variable

■ Built-in function

■ Program or function
    in current directory

■ Program or function
    in current path

■ **Order of execution:** Order of execution is the sequence MATLAB goes through when you type a name in either the Command Window or in a program or function. If you type Frog in the Command Window.

1. MATLAB checks if Frog is a variable, and displays its value if it is.
2. MATLAB checks if Frog is one of MATLAB's built-in functions, and executes that function if it is.
3. MATLAB checks if Frog is a program or function in its current directory, and executes the program or function if it is.

4. MATLAB checks if `Frog` is a program anywhere in its path, and executes the program or function if it is. The location where you store your MATLAB programs or functions must be included in MATLAB's path. You can check where MATLAB is looking for programs and functions by typing path in the Command Window. Any M-files in the directories listed will be visible to MATLAB. If you try to execute an M-file outside any of the directories listed, MATLAB will ask if you want to switch current directories or add that directory to the list of paths MATLAB can see.

- **doc, help, and lookfor:** In addition to a standard Windows help utility, the first line of MATLAB's help files can be searched for a keyword with the `lookfor` command. To display the entire contents of the help documentation available for a function, type `help` or `doc` followed by a space and the function name. It is recommended that `doc`, `help`, and `lookfor` be used *only* in the Command Window rather than those calls being incorporated in an M-file.

- **The Workspace:** MATLAB's Workspace is the place where all of the variables are visible to the user or where programs are located. In the main interface of MATLAB, all variables created by the user directly in the Command Window or created as a result of executing a program are stored and displayed in the Workspace tab. To clear all the variables from the workspace, type the command `clear` into the Command Window. To remove a specific variable from the workspace, type the command `clear` followed by a space and the name of the variable.

- One important fact to note about variables is that the variables in the workspace are only available within MATLAB's Command Window and to any running program. Functions, explained later, create a private workspace and cannot access any of the variables available in MATLAB's main workspace. The only way to access any variables in MATLAB's main workspace is to "pass" the variable to the program as function input. The phenomenon of having "main workspace" variables inaccessible by functions is because functions exist within a different scope from the programs. The variables created within the "private workspace" of a function are commonly referred to as **local variables**, since they are destroyed after the function executes and do not appear in the "main workspace" unless they are passed back as function output. Special constants like "pi" exist within the scope of programs. Variables that are visible within any scope are commonly referred to as **global variables**.

In the diagram below, consider the two inner boxes to be programs (`Prgm1`) or functions (`Func1`) and the outer box represents all of the other variables in MATLAB's Workspace. Since we do not include any commands to clear the variables from the workspace at the top of the program `Prgm1`, the program is able to "see" and use all of the variables within MATLAB's workspace. In addition, it will create a new variable in MATLAB's workspace (`c`) after we execute program Prgm1.

In contrast, the function cannot "see" or use any of the variables in MATLAB's Workspace and creates its own separate workspace to execute the code inside the function. In order to run the function, the programmer must provide the values of variables `e` and `g` by "passing" the variables into the function, the code calculates the value of `d`, and returns `d` as function output. Assume that we store the output of the function in the variable `h` when the function returns to the MATLAB workspace. After the function has executed, the separate workspace is destroyed, along with variables `d` and `k`, which were never stored in MATLAB's main workspace.

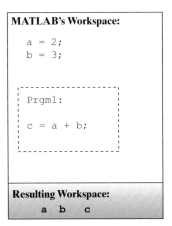

MATLAB's Workspace:

```
a = 2;
b = 3;
```

```
Prgm1:

c = a + b;
```

Resulting Workspace:
a   b   c

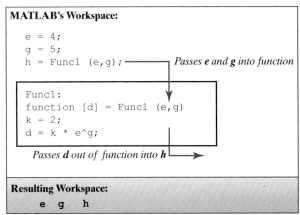

MATLAB's Workspace:

```
e = 4;
g = 5;
h = Func1 (e,g);          Passes e and g into function
```

```
Func1:
function [d] = Func1 (e,g)
k = 2;
d = k * e^g;
```

*Passes **d** out of function into **h***

Resulting Workspace:
e   g   h

- **what and who:** To display a list of all MATLAB programs in the current directory, type `what` into the Command window. To display all variables currently in MATLAB's workspace, type `who` in the Command Window. Remember that all variables automatically clear after you exit MATLAB.
- **Special constants:** In addition to the names of special functions, MATLAB reserves certain names as special constants, such as `pi`. To search for a special constant, use the `lookfor` command.
- **Comments:** For documenting or commenting any code in MATLAB, a special character is used to "comment" out text: the percent symbol (%). The comment character can appear anywhere in a MATLAB program or function, but preferably either at the beginning of a line or middle of a line of MATLAB code. Everything after the percent sign to the end of the line is ignored by MATLAB, so you as a programmer can "comment out" any broken lines of code during code testing in addition to providing comments. Program or function file headers will usually have 6–10 lines of comments at the top of every program and more throughout the rest of the program or function.

16.2 PROGRAMS

Programs are a set of instructions given to a computer to perform a specific task. In this section, we create MATLAB program files, or **M-Files,** to automate our processes.

EXAMPLE 16-1

We want to create a MATLAB program to convert Cartesian coordinates (x, y) to polar coordinates (r, θ). Before we can even attempt to write the program, we need to recall some information about the coordinate systems.

We know the following:

- In the Cartesian coordinate system, the coordinates (x, y) represent how "far away" a particular data point is from the origin $(0, 0)$ by discussing the horizontal and vertical distance with respect to the origin. The x-value represents the horizontal distance and the y-value represents the vertical distance.
- In the polar coordinate system, the coordinates (r, θ) represent the exact distance from the point to the origin (r) as well as the angle of elevation (θ) with respect to the origin $(0, 0)$ in units of radians.

- The conversion from polar coordinates to Cartesian coordinates is

$$x = r \cos \theta$$
$$y = r \sin \theta$$

where θ the angle of elevation in radians.

- The conversion from Cartesian coordinates to polar coordinates is

$$r = \sqrt{x^2 + y^2} \qquad \theta = \tan^{-1}\left(\frac{y}{x}\right)$$

We have the equations necessary to write a program to convert Cartesian coordinates to polar coordinates, and vice versa.

Now, we need to ask MATLAB for more information on doing things like calculating a square root and calculating an inverse tangent of a value.

In MATLAB's Command Window, we search for the built-in functions to calculate a square root and an inverse tangent. After we type `lookfor square root` *into the Command Window, we notice that MATLAB gives us an error message. With the* `lookfor` *command, MATLAB is expecting us to provide only one keyword, so we decide to try our search again, this time including only the word "square."*

Because MATLAB is searching through every built-in function, we need a way to tell MATLAB to stop, because when the function `sqrt` *appears, it seems to be the exact function we need. To terminate a MATLAB command early, we press* CTRL + C *(i.e., the CTRL key and the "C" key at the same time) on our keyboard, and the prompt will return to the Command Window.* CTRL + C *can also be used to stop a program caught in an infinite loop or that otherwise wandered away and got lost.*

To ask for more information on the function `sqrt`, *we type* `help sqrt` *to double-check that the function is going to perform our intended calculation.*

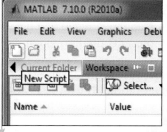

Repeating this process to find the inverse tangent function, we determine that `atan` *is the correct function necessary to calculate the inverse tangent. We note, from the help documentation, that the* `atan` *function will return the inverse tangent in radians.*

To create a program, we click the New M-File button on MATLAB's main screen. After creating a new blank M-file, we need to write out our problem statement and start keeping track of any variables we create within the program.

Next, we start writing our code. In a MATLAB program, it is good practice to start the program by clearing all of the variables from the workspace (clear) and clearing the Command Window (clc). That way, we guarantee that if we run our program on a different computer or run it after restarting MATLAB, our program will create all the variables we need to solve the problem and not rely on the workspace having variables predefined before code execution.

Now we are ready to start our program. Let us look at the step-by-step written algorithm necessary to solve this problem:

1. *Create a variable to contain the x coordinate as a variable named "x."*
2. *Create a variable to contain the y coordinate as a variable named "y."*
3. *Calculate the radius by taking the square root of the sum of the squares of x and y and store the result in a new variable called "r."*
4. *Calculate the angle by taking the inverse tangent of the result of dividing y by x and store the result in a new variable called "Theta."*

In the written algorithm, the first thing we have to do is to manually assign the values of x and y. For this example, we assign x to contain the value of 5 and y to contain the value of 10, but in later chapters, we show how the user of a program can enter values.

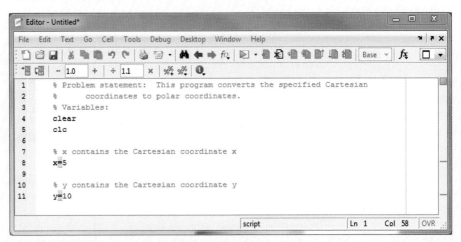

Note that MATLAB highlights the = symbol in the editor. When we move the cursor over the assignment operator (=), a pop-up message tells us that we can suppress the output of the assignment expression by using a semicolon. For now, we omit the semicolon and observe MATLAB's behavior when we leave the code unsuppressed.

Next, we type the equations to calculate the radius and the angle, using the built-in MATLAB functions we discovered with lookup and help.

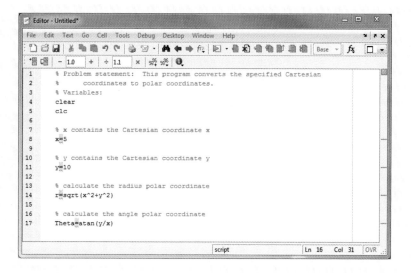

Next, we save the program in MATLAB's path with a file name that follows the program-naming convention. We choose the name "CartesianToPolar" since that name nicely describes the purpose of the program.

To save a program, we click the Save icon in the Editor window.

*Alternatively, we can go to **File > Save** in the Editor menu. When the **Save File As** window appears, we type the name of the program or function and click **Save**.*

⌘ ***Mac OS:*** *MATLAB is one of the few programs that does not modify the main menu at the top of the screen. Instead, its main menu is at the top of the MATLAB window, just as in the Windows implementation. This is because MATLAB is actually running under UNIX®.*

Running the Program

There are a number of different ways to run a MATLAB program located within MATLAB's path.

Method 1: *In the main MATLAB window (not the Editor), type the name of the program in the Command Window and press **Enter**.*

Method 2: *In the main MATLAB window (not the Editor), right-click the program you want to run while in the Current Directory box and select Run.*

Method 3: *With the program loaded in the Editor window, go to the Debug menu and click the option Run file name.*

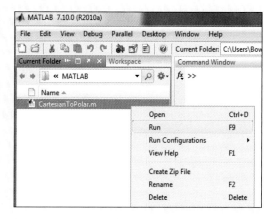

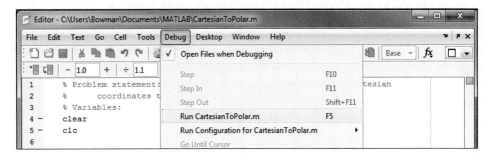

Method 4: With the program loaded in the Editor window, click the Save and Run button in the toolbar.

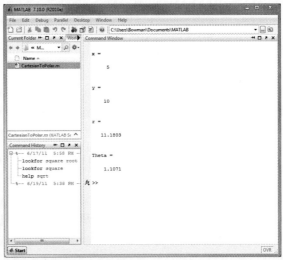

We see that MATLAB spits out the results of every assignment and calculation performed in the program. Let us assume we want to show only the output from the calculations of r and Theta, and not the output of the assignment of x and y. To do this, we add a semicolon (;) at the end of each line of code we do not want showing up in the output of the program.

Now when we save and run our modified program, we will see only the output from the calculations of r and Theta.

Programming Features in Our Coverage of MATLAB

MATLAB has many advanced programming features that make sophisticated numerical analysis, including simulation, possible. In this text, you will learn basic programming structures to solve engineering analysis problems.

- **Arrays:** These offer the opportunity to store data in their natural structure: one-dimensional arrays for vectors, two-dimensional arrays for matrices and paired data, and so on. Arrays are also more efficient for data storage.
- **Input and output:** By gathering input, the program can solve a different problem each time the program is executed. Students can also produce formatted output to communicate with the user.
- **Conditional statements:** These allow a program to take different paths for various user input, intermediate calculations, or final results. Conditional statements allow the programmer to ask questions in order to make decisions.
- **Looping:** Repetition is critical in breaking down complicated problems into many simple calculations.
- **Plotting:** Graphs communicate data relationships in a variety of forms.

Data and Program Structures

MATLAB actually views all data structures as arrays—we discuss the importance and the implications of this later. For now, consider the following data and program structures and how they are used.

> **Numbers:** Numbers are also referred to as scalars, as opposed to vectors. There are many number types—integers and floating-point numbers of different levels of precision. Double-precision floating-point numbers (16 digits stored, 4 of which are displayed) are the default. The display format can be manipulated independently of what kind of number is stored—for example, integers can be displayed to a specified number of decimal, or floating-point numbers can be displayed in scientific notation.

COMPREHENSION CHECK 16-2

Type the following expressions into the MATLAB Command Window and observe how MATLAB stores the variables in the workspace:

```
A = 1     B = 1.2     C = 1e-10     D = 7i+2     E = 2;
```

What effect does the semicolon have on the expression?

> **Text:** Strings of text, or character strings (including letters, numbers, and other special characters such as a question mark) can also be stored. Text strings must be placed in quotes to avoid being mistaken for variable or function names (or causing errors where the strings have no definition as a variable or function). For example, `Name = 'Jonas Doe'` will store the text in quotes in the variable `Name`. Note that single quotes are used.

COMPREHENSION CHECK 16-3

Type the following expressions into the MATLAB Command Window and observe how MATLAB stores the variables in the workspace:

```
A = 'Hello'          B = 'Hello World!'     C = 'It''s 3:00 PM'
```

What effect did two single quotes in the expression have on the output?

Arrays: Also referred to as matrices, arrays are used to store blocks of numbers or text. When a row is entered, spaces or commas separate columns. For example, the row vector [1 2 3] can be entered as it appears or as [1, 2, 3]. Similarly, use a semicolon or the Enter key to separate columns. For example, you can enter the column vector

$$\begin{bmatrix} 1 \\ 2 \\ 3 \end{bmatrix} \text{ as [1; 2; 3] or [1<ENTER>2<ENTER>3].}$$

COMPREHENSION CHECK 16-4

Type the following expressions into MATLAB and observe how MATLAB stores the variables in the workspace:

A = [1 2 3]   B = [1,2,3]   C = [2;4;6]   D = [2:6]

E = [1 2 3; 4 5 6]   F = [1,2,3;4,5,6]   G = [1:3;4,5,6]   H = [1:3;4:6]

- What does the colon operator do if it is used inside square brackets?
- What does the comma do if it is used inside square brackets?
- What is the difference between a blank space and the comma inside square brackets?

Programs: A program is a series of executable MATLAB statements in a special text file with the suffix ".m" instead of ".txt" or ".docx" or ".xlsx," and so forth. Files with the ".m" suffix are commonly referred to in the text as "M-files." The statements in an M-file are executed as if each line were typed into the MATLAB Command Window.

Variables: Special placeholders, called variables, inside MATLAB contain information like text and numbers. MATLAB has built-in variables such as pi, so take care to properly name any variable you use in MATLAB. Also note that you should *not* place a value into such variables (e.g., do not write pi = 3.14).

Built-in Functions: MATLAB has many built-in functions, some of which you will recognize immediately: sqrt, abs, log, sin, cos, tan, etc. These common functions each have a single argument and a single returned value and are called in the traditional manner: sqrt(d), abs(t), sin(x). Some built-in functions act on multiple arguments, such as plot(x, y) —discussed in detail later—and some return multiple output variables.

16.3 MATRIX OPERATIONS

Review the sections on matrices in the appendix, particularly if some of the following read like an extraterrestrial language to you.

One of the most powerful features of MATLAB is its ability to directly manipulate matrices. Many things that MATLAB can do with a single statement would require numerous lines of code, nested loops, etc., in most other computer languages. A few of these matrix operations are shown below.

- *Assigning* part of an array to another array
  - `a = b(2, :)` puts the second row of b into a.
  - `c = d(:, 3:5)` puts the third through fifth columns of d into c.
- *Transposing* matrices—converting rows to columns, and vice versa (the "glitch" (`'`) or single quote operator means transpose)
  - `[1; 2; 3]' = [1 2 3]`
  - `[0 8 6; 3 5 7]' = [0 3; 8 5; 6 7]`
- *Incrementing* a vector or an array (increasing all terms by a scalar)
  - If `TempC` is an array of Celsius temperatures, `TempK = TempC + 273.15` converts all array elements to units of kelvins.
- *Scaling* a vector or an array (multiplying all terms by a scalar)
  - If `LengthInches` is an array of lengths in inches, `LengthCentimeters = LengthInches * 2.54` converts all array elements to units of centimeters.
- *Adding* and *subtracting* arrays
  - `[1 2 3; 4 5 6] - [8 7 6; 4 2 0] = [-7 -5 -3; 0 3 6]`
  - If `A` is a vector distance traveled on day 1 and `B` is a vector distance traveled on day 2, then the vector addition `A + B` is the total vector distance traveled.
- *Repeating* a basic operation on every element of the array
  - `sqrt(A)` takes the square root of all elements of array A.
- *Operating term-by-term*, as when array operators (preceded by a period, as in `.*`) cause elements of one array to operate on the corresponding elements in another array
  - `[1 3; 5 8] .* [0 1; 4 2] = [0 3; 20 16]`
  - If `distance` is a vector of distances traveled by a bicyclist in each leg of a cycling event and `speed` is a vector of the speed the cyclist travels for each leg, then `time = distance/speed` computes a vector that contains the time it takes the cyclist to complete each leg.
  - If `accounts` is a vector of the current balance of a number of accounts and `rates` is a vector of the interest rates paid to each of those accounts, then `interest = accounts . * rates` calculates the interest accrued to each account.
  - If `radius` is a vector of circle radii, then `area = pi * radius .^ 2` is a vector of the circle areas.

> **NOTE**
>
> Some MATLAB operations do *not* perform element-by-element operations if the dot is omitted, but instead perform matrix operations. For example, A * B multiplies array A by array B using matrix multiplication. This is very different from the element-by-element multiply of A.*B.

Creating Matrices Efficiently

Simple matrices can be constructed in MATLAB programs by listing numbers inside square brackets [ ] and assigning the matrix into a variable. Assume we want to place the following 2×3 (two rows and three columns) matrix into a variable:

$$\begin{bmatrix} 3 & 6 & 4 \\ 2 & 9 & 7 \end{bmatrix}$$

To do this, we need to understand the syntax of separating the rows and columns inside this matrix. To separate columns, we can either separate the numbers in the first row using spaces or commas. To separate the rows, we need to use the semicolon operator to designate the end of a row. For example:

(a) $M = \begin{bmatrix} 3 & 6 & 4; & 2 & 9 & 7 \end{bmatrix};$

(b) $N = \begin{bmatrix} 3, & 6, & 4; & 2, & 9, & 7 \end{bmatrix};$

The commands (a) and (b) shown above are equivalent and will save the 2×3 matrix into different, but equivalent matrices M and N. Note that MATLAB will construct this matrix from top to bottom and left to right according to the order you list the values in the construction of the matrix.

In addition to the square bracket notation, we can actually access the individual elements of a matrix using special row, column notation to create matrices:

(c) P(1, 1) = 3;
    P(1, 2) = 6;
    P(1, 3) = 4;
    P(2, 1) = 2;
    P(2, 2) = 9;
    P(2, 3) = 7;

Note that in (c), the matrix we created is the same matrix mentioned earlier, so it is also equivalent to matrices M and N. The order of the numbers inside the parenthesis is row number, followed by the column number, so when we look at the line of code P(2,3) = 7, MATLAB takes the number 7 and stores it in the 2nd row, 3rd column of the variable P.

A special case of a matrix with either 1 row or 1 column is called a vector. It is often necessary to create a sequence of numbers, so MATLAB includes special notation through the use of the colon operator to create linear sequences of numbers. For example, if we wanted to create a matrix that contained one row of every number between 2 and 200 incrementing by 2 $(2, 4, \ldots, 198, 200)$, we can use the colon operator to quickly create that matrix:

(d) Q = [2:2:200]

Note that the first number in the syntax is the starting value, the second number is increment value, and the last number is the largest possible value that should appear in the sequence.

To demonstrate the impact of the last number in this syntax, assume we want to create a matrix that contained one row of every number between 1 and 20, incremented by 3 $(1, 4, 7, \ldots)$, we can use the same notation to generate this vector:

(e) R = [1:3:20];

Note that the sequence created by MATLAB (1, 4, 7, 10, 13, 16, and 19) does not actually contain the number 20, but it also does not contain the next highest number (22) either because that value is greater than 20.

It is also possible to use a single colon operator to create a vector of values incrementing by 1. For example, if we wanted to generate a vector that starts at the number 2.5 and increments by 1 until it gets to 8.5, we could type into MATLAB:

(f) W = [2.5:8.5];

Advanced Addressing and Matrix Construction

We can use MATLAB's colon operator to extract portions of matrices. Consider matrix M = [1 2 3; 4 5 6; 7 8 9], or

$$M = \begin{bmatrix} 1 & 2 & 3 \\ 4 & 5 & 6 \\ 7 & 8 & 9 \end{bmatrix}$$

If we wanted to pull out rows 2 and 3 from this matrix and store it into a new matrix, M2, we could type:

(g) M2 = M(2:3, 1:3);

The syntax of this special use of the colon operator allows you to pull out specific rows and columns of a predefined matrix. In (g), the first argument to the matrix address (2:3) tells MATLAB to extract rows 2 through 3 and the second argument (1:3) tells MATLAB to extract columns 1 through 3.

Since we wanted to extract all three columns of the matrix data into this new variable M2, we could have used a special use of the colon operator to extract all of the columns without specifying the column numbers:

(h) M3 = M(2:3, :);

Note that M2 and M3 are equivalent matrices.

We could also use the single colon operator to extract all rows of a matrix. Assume we want to build a new matrix, M4, which contains just the first and second columns of matrix M:

(i) M4 = M(:, 1:2);

We can also combine existing matrices in MATLAB's workspace to build new matrix structures by concatenating one or more matrix into a single variable. Consider matrices A = [1, 2; 3, 4] and B = [5, 6; 7, 8]:

$$A = \begin{bmatrix} 1 & 2 \\ 3 & 4 \end{bmatrix} \qquad B = \begin{bmatrix} 5 & 6 \\ 7 & 0 \end{bmatrix}$$

If we wanted to create a new matrix, C, where the top half of this new matrix should be A and the bottom half should be matrix B, we can create this quickly in MATLAB by using the square bracket notation:

(j) C = [A; B];

The resulting matrix C is a 4×2 matrix with the elements of A and B comprising the rows and columns.

Likewise, we could create a new matrix, D, where the left portion of the matrix contains matrix A and the right portion of the matrix contains matrix B:

(k) D = [A, B];

LAW OF ARGUMENTS

In computer programming, and often in math as well, the word "argument" does *not* refer to an altercation or vehement disagreement. In this context, it means the information that is given to a function as input to be processed. If you use the function sqrt to find the square root of 49, you would write sqrt(49) : 49 is the argument of sqrt—the value upon which the sqrt function performs its calculations. Functions may have zero arguments (e.g., rand will calculate a single random number between 0 and 1—no input is required), one argument (e.g., sqrt(4)), or two or more arguments (e.g., power(a,b) raises a to the power b).

16.4 FUNCTIONS

Functional programming allows the designer of an algorithm to recycle a segment of code to reduce the amount of redundant code. Decreasing code redundancy allows for cleaner source code that is easier to read. **Functions** are a special class of programs that require the user to pass in input variables and capture output variables.

User-defined functions are similar to programs in that they contain a series of executable MATLAB statements in M-files that exist independently of any program we create in MATLAB. To allow any MATLAB program to call a user-defined function, we must follow a few guidelines to ensure that our function will run properly.

Function Creation Guidelines

- The first line in a function must be a function definition line:

```
function [output_variables] = function_name(input_variables)
```

  - `output_variables`: A comma-separated list of output variables in square brackets
  - `input_variables`: A comma-separated list of input variables in parentheses
  - `function_name`: Function names cannot be the same as a built-in function, user-created variables, or any reserved word in MATLAB. Avoid creating variables with the same name as a user-defined function because you will not be able to call your function.
- **The name of the function must be the same as the file name**. For example, if the function name is `sphereVolume`, the function must be saved in an M-file named `sphereVolume.m`, where `.m` is just the extension of the M-file. In addition to the aforementioned function-naming rules, the name of the function must also follow all program-naming rules, including omitting spaces in the file name.
- To run a function, you must call it from MATLAB's Command Window by passing in input variables or properly calling it from another program or function. Running the function using the "Run" commands in the Editor/Debugger window will report errors to the Command Window.
- Do *not* run the `clear` command after the function header because it will clear the input variables. The `clear` and `clc` commands should only appear at the top of programs, not functions!

Function Structure and Use

The following examples demonstrate the structure of different functions and how MATLAB programs can use the functions to reduce code redundancy.

● **EXAMPLE 16-2**

Assume we are required to create a function, areaCircle, to calculate the area of a circle given the radius. The function should accept one input (radius) and return one output (area). Furthermore, we are told to suppress all calculations within the function.

Source Code

```
% Problem Statement: This function calculates the area of a
  circle
% Input:
%   r—Radius [any length unit]
% Output:
%   A—Area [units of input^2]
function [A]=areaCircle(r)
% Calculation of the area of a circle
A=pi*r^2;
```

Typical Usage (can be typed in Command Window or can be a separate program):

You can call the function by typing the name of the function and passing in a value for the radius. Note that the result is stored in a variable called "ans" when you do not specify a variable to capture the output.

```
>> areaCircle(3)
ans=
     28.2743
```

You can call the function by typing the name of the function and passing in a variable.

```
>> Rad=4;
>> CircleA=areaCircle(Rad)
CircleA=
   50.2655
```

COMPREHENSION CHECK 16-5

What is the output when you "pass in" the value 10 to the function in Example 16.2?

Assume that you want to store the result of the previous function call in a variable called Dogs. What is the command you would type in MATLAB?

● **EXAMPLE 16-3**

Assume we are required to create a function, volumeCylinder, to calculate the volume of a cylinder given the radius. The function should accept two inputs (radius, height) and return one output (volume). For this example, you may assume that the user will input the radius and height in the same units. Any calculations in the functions must be suppressed.

Algorithm

Known/input:	Unknown/output:	Assumptions:
r = radius [cm] H = height [cm]	V = volume [cm^3]	

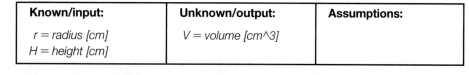

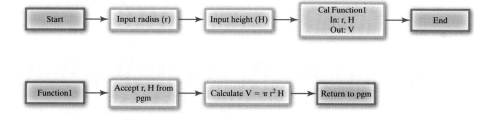

Source Code

```
% Problem Statement: Calculate the volume of a cylinder.
% Input:
%    r—Radius [any length unit]
%    H—Height [same length unit as radius]
% Output:
%    V—Volume [units of input^3]
function [V]= volumeCylinder(r, H)
% Calculation of the volume of a cylinder
V=pi*H*r^2;
```

Typical Usage

```
>> Vol1 = volumeCylinder(1,2)
Vol1=
   6.2832

>> Vol2= volumeCylinder(2,1)
Vol2=
   12.5664
```

Note that the order of arguments passed into the function matters. The order of variables expected by the function is based on the order they are listed in the function header. Therefore, the first input argument is the radius, followed by the height.

● EXAMPLE 16-4

Assume we are required to create a function, `circleCalculations`, to calculate the area and perimeter of a circle given the radius. The function should accept one input (radius) and return two outputs (area, perimeter). Any calculations in the functions must be suppressed.

Source Code

```
% Problem Statement: This function calculates the area and
  perimeter of a circle.
% Input:
%   r—Radius [any length unit]
% Output:
%   A—Area [units of input^2]
%   P—Perimeter [units of input]
function [A, P]=circleCalculations(r)
% Calculation of the area of a circle
A=pi*r^2;
% Calculation of the perimeter of a circle
P=2*pi*r;
```

Typical Usage

```
>> [Area,Perimeter]=circleCalculations(5)
Area=
   78.5398
Perimeter=
   31.4159
```

● EXAMPLE 16-5

Assume we are required to create a function, `cylinderCalculations`, to calculate the volume and lateral surface area of a cylinder given the radius and height. The function should accept two inputs (radius, height) and return two outputs (volume, surface area). For this example, you may assume that the user will input the radius and height in the same units. Any calculations in the functions must be suppressed.

Source Code

```
% Problem: Function calculates volume and surface area of
  cylinder.
% Input:
%   r—Radius [any length unit]
%   H—Height [same length unit as radius]
% Output:
%   V—Volume [units of input^3]
%   SA—Surface area [units of input^2]
function [V, SA]=cylinderCalculations(r, H)
% Calculation of the volume of cylinder
V=pi*H*r^2;
% Calculation of the surface area of a cylinder
SA=2*pi*r*H;
```

Typical Usage

```
>> [Volume,Surface]=cylinderCalculations(3,2)
Volume=
    56.5487
Surface=
    37.6991
```

16.5   DEBUGGING MATLAB CODE

If a program has syntax errors, MATLAB (and other compilers) will give you useful feedback regarding the type and location of the error to help you fix, or **debug**, your program. If this information is not sufficient for you to understand the problem, check with others who might have seen this error previously—some errors are particularly common. If you still cannot identify the error, you might need to use a more formal process to study the error. For other kinds of errors, this formal process is essential for diagnosing the problems. When you first write a program, you must test the output of the program with a set of inputs for which you know result. Such test cases would include:

- Simple cases for which you can quickly compute the expected output;
- Cases provided by an instructor or textbook with a published solution;
- Cases for which results have already been produced by a previously tested program that does the same thing;
- Test cases that are customarily used to test programs.

When the output of a program is different from what you expect, the debugging process begins. It is common to debug shorter programs simply by reading them and writing a few notes. Longer programs may require the use of the MATLAB Debugger, which is available with the MATLAB Editor. Whether you are using the MATLAB Debugger or are debugging "by hand" or in the Command Window, the same techniques apply.

Preparing for Debugging

When using the Editor/Debugger, open the program for debugging. If it is open, make sure changes are saved—MATLAB will run the saved version without including recent changes. If you are debugging from a printed program and output, make sure that the printed output came from the version you are reading. In preparing to debug a program, it is critical to be able to reproduce the conditions that caused the bug in the first place. If the bug occurs in processing data from a large data set, you may have to split the data set to find the specific data that triggered the bug.

Setting Breakpoints

When using the Editor/Debugger, establish **breakpoints** that allow you to check your agreement with the program at various stages. A breakpoint stops the program at the specified location to allow you to examine the contents of variables, etc., before the program continues. This will help you find the specific location when the program does

NOTE

Why do we say that computers (and programs) have bugs? Grace Murray Hopper is quoted in the April 16, 1984, issue of *Time Magazine* as saying, "From then on, when anything went wrong with a computer, we said it had bugs in it"—referring to when a 2-inch-long moth was removed from an experimental computer at Harvard in 1947.

NOTE

A breakpoint stops the program at the specified location to allow you to examine the content before the program continues.

something you were not expecting. When debugging shorter programs, write down everything the program does or make all program results display in the Command Window by removing semicolons that were used to suppress output to the Command Window.

MATLAB has three types of breakpoints: standard (set by location in the program), conditional (triggered at a specified location by specified conditions), and error (triggered by a particular error). We focus on standard breakpoints. Breakpoints are normally shown in red. If they are shown in gray, either the file has not been saved since changes were made to it or there is a syntax error in the file. All breakpoints remain in a file until you clear (remove) them or until they are automatically cleared.

Clear a breakpoint after determining that a line of code is not causing a problem. You can conduct limited testing without setting a breakpoint by using the Go Until Cursor menu selection. When `clear name` or `clear all` appears as a statement in an M-file that you are debugging, it clears the breakpoints.

Stepping Through a Program

When using the Editor/Debugger, you must start by running the program. You can choose to step through the program one line at a time or have it continue until a breakpoint is encountered. A green arrow will point to the program line where execution will begin if you elect to continue the program. When stepping through a program, you can choose to step through functions line by line or to simply return their results immediately.

Examining Values

When the program is paused at a breakpoint or when you are stepping through the program a line at a time, you can view the value of any variable to see whether a line of code has produced the expected result. If the result is as expected, continue running or step to the next line. If the result is not as expected, then that line, or a previous line, contains an error. You can examine the value of a variable in the Workspace window in the main MATLAB program window or by holding the mouse over the variable in the Editor/Debugger. It is also possible (again, when the program is paused) to select either a variable or an expression and select "Evaluate Selection" from the context menu.

Correcting Problems and Ending Debugging

If a problem is difficult to diagnose, one method to help define or correct problems is to change the value of a variable in a paused program to see if continuing with the new value produces expected results. A new value can be assigned through the Command Window, Workspace browser, or Array Editor. Do not make changes to an M-file while MATLAB is in debug mode—it is best to exit debug mode before editing an M-file. If you edit an M-file while in debug mode, you can get unexpected results when you run the file.

In-Class Activities

ICA 16-1

Do the following scripted exercise. The >> prompt indicates what you should type into the MATLAB Command Window. Note that some of these commands may result in error messages.

>> clc	Press enter (this clears the screen ONLY)
>> 30/10	Press enter (this serves just as a calculator)
>> r=30/10	Press enter and note the output
>> r	Press enter (the variable "r" has now been assigned a value)
>> s=10*r	Press enter (now "s" is defined as well)
>> v=sin(s)	Press enter (s is defined, it can be used to compute v) Why is not sin(30) = 0.5?
>> r	Press enter (r still has its original value)
>> clear r	Press enter
>> r	Press enter (r is undefined; "clear r" deleted the variable r)
>> s	Press enter (we still have a value for s)
>> r=s;	Press enter (semicolon suppresses the display)
>> r	Press enter (the value of r was assigned even though display was suppressed)
>> clear	Press enter (this clears *all* variables)
>> help elfun	Gives elementary function list
>> help sin	Provides info about sine function
>> lookfor sine	The lookfor command will help you find things when you do not know what they might be called. Wait until MATLAB finds a few examples and then hold down the CTRL key and type the letter C to stop this. The program continues to search the documentation for the word "sine" unless we stop it.

ICA 16-2

What is the value of W after this MATLAB program is finished?

```
X = 2;
Y = 4;
Z = 2;
W = (Y + Z/X + Z^1/2)
```

ICA 16-3

Which of the following are not valid program/function filenames? Circle all that apply.

(A) 2b_solved.m
(B) calc_circum.m
(C) graph-data.m
(D) help4me.m
(E) MATLAB is fun.m
(F) matrix*matrix.m
(G) Mult2#s.m
(H) pi.m
(I) ReadFile.m
(J) SuperCaliFragiListicExpiAliDocious.m

ICA 16-4

Do the following scripted exercise. The >> prompt indicates what you should type into the MATLAB Command Window. Note that some of these commands may result in error messages.

>> [10 20] * 30	% Multiply a row vector containing 10 and 20 by 30
>> m=[10,20]*30	% Multiply a row vector containing 10 and 20 by 30, store in variable m
>> m1=[10;20]*30	% Multiply a column vector containing 10 and 20 by 30, store in variable m1
>> n=m+1	% Add 1 to every element in variable m, store in variable n
>> p=m1+10	% Add 10 to every element in variable m1, store in variable p
>> r=m+n	% Add row vector in variable m to column vector in variable n, store in r. Note that even though this is dimensionally inconsistent, this works in MATLAB because it will allow any vectors (row or column) to be added together as long as they contain the same number of elements
>> v=[2:2:6]	% Create a row vector of numbers between 2 and 6, increasing by 2, store in v
>> u=[1:0.5:3]	% Create a row vector of numbers between 1 and 3, increasing by 0.5, store in u
>> j=u+v	% Add vector v to vector u—why will this not work?
>> a=sin(u)	% Take the sin of each element in the vector u
>> b=sqrt(u)	% Take the square root of each element in the vector u
>> c=u^2	% Square vector u—why will this not work?
>> d=u.^2	% Square each element of vector u
>> e=[10,20;30,40]	% Create a matrix of numbers 10 and 20 in the first row, 30 and 40 in the second row, store in the variable e
>> f=[1,2,3;4,5]	% Create a matrix of numbers 1, 2, and 3 in the first row, 4 and 5 in the second row. Why will this not work?

ICA 16-5

Complete the following table by writing the MATLAB code necessary to create or calculate the following matrix computations given the algorithmic step in the "Action" column. If a calculation is not possible, explain why in the "Result" column. You may assume that the variables created in question 1 are available for the remaining computations in questions 1 through 10.

	Action	MATLAB Code	Result Computed by Hand	No. of Rows	No. of Cols
(a)	Save matrix $\begin{bmatrix} 3 & 8 \\ 5 & 7 \end{bmatrix}$ in variable W		N/A		
(b)	Save matrix $\begin{bmatrix} 4 & 8 & 1 \\ 5 & 1 & 6 \end{bmatrix}$ in variable X		N/A		
(c)	Save matrix $\begin{bmatrix} 2 \\ 3 \end{bmatrix}$ in variable Y		N/A		
(d)	Save matrix $\begin{bmatrix} 4 & 2 \\ 2 & 9 \end{bmatrix}$ in variable Z		N/A		
(e)	Add W + Z, save in variable A				
(f)	Subtract Y − Z, save in variable B				
(g)	Multiply W * X, save in variable C				
(h)	Multiply X * Z save in variable D				
(i)	Transpose X, save in variable E				
(j)	Multiply W * Z term-by-term, save in variable F				
(k)	Square Z, save in variable G				
(l)	Square X, save in variable H				
(m)	Add 20 to each element in X, save in variable I				

ICA 16-6

Write a program to store the following matrices into MATLAB and perform the following calculations. All variable assignments should be suppressed, but you should leave the calculations unsuppressed. If the variable assignment or calculation is not possible or causes an error message on the screen, write MATLAB comments explaining the problem.

Input variables:

$$A = \begin{bmatrix} 4 & 6 \\ -3 & 8 \end{bmatrix} \quad B = \begin{bmatrix} 3 & 5 \\ 5 & 5 \end{bmatrix} \quad C = \begin{bmatrix} 0 & 1 \end{bmatrix} \quad D = \begin{bmatrix} 1 \\ 1 \end{bmatrix} \quad E = \begin{bmatrix} 3 & 6 & 5 \\ 7 & 1 & 5 \end{bmatrix} \quad F = \begin{bmatrix} 2 & 8 \\ 4 & 3 \\ 1 & 5 \end{bmatrix}$$

$$G = \begin{bmatrix} 7 & 8 & 3 \\ 4 & 8 & 2 \\ 5 & 9 & 5 \end{bmatrix} \quad H = \begin{bmatrix} 1 & 2 & \ldots & 299 & 300 \end{bmatrix} \quad I = \begin{bmatrix} 200 \\ 198 \\ \vdots \\ -298 \\ 300 \end{bmatrix}$$

$$J = \begin{bmatrix} 20 & 40 & \ldots & 180 & 200 \\ 10 & 20 & \ldots & 90 & 100 \end{bmatrix} \quad K = \begin{bmatrix} 5 & 50 \\ 10 & 45 \\ \vdots & \vdots \\ 45 & -45 \\ 50 & -50 \end{bmatrix}$$

Calculations:

The result of the following calculations should each be stored into variable names of your choosing. Each calculation should be stored in a different variable. Here, note the "T" indicated the matrix should be transposed.

(a) A * B

(b) A + B

(c) A + C

(d) C − E

(d) E * F

(f) F * E

(g) G^2

(h) F^T

(i) C * D

(j) H + 15

(k) I * 30

(l) $E^T + 30$

(m) $E^2 * 50$

(n) $B^2 * 50$

ICA 16-7

Write a MATLAB program to evaluate the following mathematical expression. The equation should utilize variables for a, b and c. Test the program with $a = 1$, $b = 2$ and $c = 3$.

$$x = \frac{b + \sqrt{b - 4a}}{c^7}$$

ICA 16-8

Write a MATLAB program that will implement the following mathematical expression. The equation should utilize a variable for x. Test the program with $x = 30$.

$$A = \frac{x^2 \cos(2x + 1)}{(6x) \log(x)}$$

ICA 16-9

Write a MATLAB program that will implement the following mathematical expression. The equation should utilize variables for x, μ, and σ. Test the program with $x = 0$, $\mu = 0$, and $\sigma = 1$.

$$P = \frac{1}{\sigma\sqrt{2\pi}} e^{-(x-\mu)^2/(2\sigma^2)}$$

This program calculates a *Gaussian normal distribution*.

ICA 16-10

Write a function that, given any two 2 × 2 arrays (A and B) as arguments, returns the product matrix C = AB. The function is not expected to work properly for any other matrix dimensions. You must calculate each term in C individually from terms in A and B—you may not use MATLAB's ability to multiply matrices directly. Use the following values to test the program: A = [1 1; 2 2] and B = [3 3; 4 4]

ICA 16-11

Write a function that, given any two 2 × 2 arrays (A and B) as arguments, returns the sum of the two matrices, or C = A + B. The function is not expected to work properly for any other matrix dimensions. You must calculate each term in C individually from terms in A and B—you may not use MATLAB's ability to add matrices directly. Use the following values to test the program: A = [1 1; 2 2] and B = [3 3; 4 4]

ICA 16-12

A member of your team gives you the following MATLAB program. Your job is to create the functions used in the program.

```
A = 10; % area [cm^2]
H = 30; % height [cm]
V = 50; % volume [cm^3]
% calculates the radius [cm] of the circle
R1 = RadiusCircle(A);
% calculates the radius [cm] of a cone
R2 = RadiusCone(H,V);
```

ICA 16-13

Debug the MATLAB programs/functions provided for you with the online materials for this ICA:

```
atm to Pa.m
ft to m.m
ICA #7.m
```

These files must be corrected to eliminate any syntax, runtime, formula coding, and formula derivation errors. You may assume that the header comments at the top of the program and functions are correct. In addition to the corrected files, you must submit an algorithm template, showing the main program and both functions.

ICA 16-14

Debug the MATLAB programs/functions provided for you with the online materials for this ICA:

```
SG to rho.m
N_2_lbf.m
ICA S-D.m
```

These files must be corrected to eliminate any syntax, runtime, formula coding, and formula derivation errors. You may assume that the header comments at the top of the program and functions are correct. In addition to the corrected files, you must submit an algorithm template, showing the main program and both functions.

ICA 16-15

Consider the following MATLAB program and function, stored in MATLAB's Current Directory:

MyRadFunction.m

```
clear clc
function [Out] = RadF (In1, In2, In3)
Out = (In1+In2)/2 + (In2+In3)/4;
```

MyRadProgram.m

```
clear clc

InVar1=1;  InVar2=3;  InVar3=-1;

M=MyRadFunction(invar1), MyRadFunction(invar2),
myradfunction(invar3);
```

Fix the program and function to eliminate all of the error messages. Note that for the variables provided in the program as InVar1, InVar2, and InVar3, the numerical result stored in M should be 2.5.

ICA 16-16

A novice MATLAB user created the following code with poor choice of variable names. Fill in the blanks provided to comment this code to determine the purpose of the code.

Main Program	Comments
`T = [30, 45, 120, 150];`	**(a)** % T =
`Z = 2;`	**(b)** % Z =
	% The purpose of the function DTOR is ...
`[W] = DTOR (T)`	**(c)** %
	(d) % W =
	% The purpose of the function PCAR is ...
	(e) %
`[X, Y] = PCAR (Z, W)`	**(f)** % X =
	(g) % Y =

```
function [A] = DTOR (T)
A = T*2*pi/360;
```

```
function (P,Q) = PCAR (M, N)
P = M * cos (N);
Q = M * sin (N);
```

(h) The output of this code, when run, is
_____

Chapter 16 REVIEW QUESTIONS

1. The specific gravity of gold is 19.3. Write a MATLAB program that will determine the length of one side of a 0.4 kilogram cube of solid gold, in units of inches.

2. An unmanned X-43A scramjet test vehicle has achieved a maximum speed of Mach number 9.68 in a test flight over the Pacific Ocean. Mach number is defined as the speed of an object divided by the speed of sound. Assuming the speed of sound is 343 meters per second, write a MATLAB program to determine the record speed in units of miles per hour.

3. A rod on the surface of Jupiter's moon Callisto has a volume of 0.3 cubic meters. Write a MATLAB program that will determine the weight of the rod in units of pounds-force. The specific gravity is 4.7. Gravitational acceleration on Callisto is 1.25 meters per second squared.

4. The maximum radius a falling liquid drop can have without breaking apart is given by the equation:

$$R = \sqrt{\frac{\sigma}{g\rho}}$$

where σ is the liquid surface tension, g is the acceleration due to gravity, and ρ is the density of the liquid. Write a MATLAB program to determine the surface tension in units of joules per meter squared. As a test case, consider the fluid is acetone at 20 degrees Celsius, the maximum radius of a drop is 1 centimeter and the specific gravity of the liquid is 0.785.

5. Write a function that implements the quadratic equation. Given three inputs (a, b, and c), calculate the roots (r1 and r2) of the quadratic formula. Recall the quadratic equation:

$$r = \frac{-b \pm \sqrt{b^2 - 4ac}}{2a}$$

6. Write a function that implements the Pythagorean theorem. Recall that the theorem states that the length of the hypotenuse (z) can be calculated by the sum of the squares of the adjacent sides (x and y), or

$$z = \sqrt{x^2 + y^2}$$

7. In the starting file provided on online, there are data sets from two different data collection sessions. In the first data collection session, a lab technician collected three different measurements and recorded mass in grams, height in feet, and time in minutes. However, in data collection session two, a different lab technician collected four different measurements and recorded mass in pounds-mass, height in centimeters, and time in hours.

    Your job is to write a function that will calculate the potential energy in joules and power in watts for each data condition (7 total). In addition, you will need to write a program to call the function using the datasets provided in the starting file. Your function should only consider variables in SI units, so you will need to convert the vectors in the program before passing them in to your function. Note that all conversions must be done in MATLAB code—you may not hard code any values you calculate by hand.

MATLAB
examples

8. As part of a team investigating the effect of mass on the oscillation frequency of a spring, you obtain data from three different lab technicians, provided for you in the starter file online.

The data consists of frequency data on three different springs recording the amount of time it takes each spring with different masses attached to oscillate a certain number of times. In this experiment, each technician recorded the time it took for the spring to oscillate 25 times, stored in the variable N in the starter file.

As part of the analysis, you need to write a program containing the experimental data provided. The data should be converted to make all units consistent. The data should then be passed into a function.

The function should accept three inputs: a vector containing mass measurements [grams], a vector containing time measurements [seconds], and a variable containing the number of oscillations observed in the experiment.

The function should return two matrices. The first matrix should contain the mass measurements [kilograms] in the first column and the period [seconds], or the length of time required for one oscillation for each mass in the second column. The second matrix should contain a calculation of the force applied [newtons] by each mass in the first column and the frequency, the number of oscillations per second, or the inverse of the period [hertz] for each mass measured in the second column.

9. We have made many measurements of coffee cooling in a ceramic coffee cup. We realize that as the coffee cools, it gradually reaches room temperature. Consequently, we report the value of the coffee temperature in degrees above room temperature (so after a long time, the temperature rise will be equal to 0). Also, we realize that the hotter the coffee is initially (above room temperature), the longer it will take to cool. The values presented here are in degrees Fahrenheit.

Temperature Rise (T) [°F]						
Initial Temp Rise (T_0) [°F]	Cooling Time Elapsed (t) [min]					
	0	10	20	30	40	50
20	20	13	9	6	4	3
40	40	27	18	12	8	5
60	60	40	27	18	12	8
80	80	54	36	24	16	11

Write a MATLAB function that will perform a single interpolation given four numbers as input arguments and return the interpolated value as the only function output.

Write a MATLAB program that will calculate the following scenarios. Store each part in a different variable (e.g., part **(a)** should be stored in a variable named PartA, part **(b)** should be stored in a variable named PartB, etc.).

(a) What is the temperature (rise) of the cup of coffee after 37 minutes if the initial rise of temperature is 40 degrees Fahrenheit?

(b) If the coffee cools for 30 minutes and has risen 14 degrees Fahrenheit at that time, what was the initial temperature rise?

(c) Find the temperature rise of the coffee at 17 minutes if the initial rise is 53 degrees Fahrenheit.

(d) What is the temperature (rise) of the cup of coffee after 23 minutes if the initial rise in temperature is 80 degrees Fahrenheit?

(e) If the coffee cools for 10 minutes and has risen 32 degrees Fahrenheit at that time, what was the initial temperature rise?

(f) Find the temperature rise of the coffee at 34 minutes if the initial rise is 45 degrees Fahrenheit.

10. In a factory, various metal pieces are forged and then plunged into a cool liquid to quickly cool the metal. The types of metals produced, as well as their specific heat capacity, are listed in the table below.

Material	Specific Heat [J/(g °C)]
Aluminum	0.897
Cadmium	0.231
Iron	0.450
Tungsten	0.134

The metal pieces vary in mass, and are produced at a temperature of 300 degrees Celsius. The ideal process lowers the temperature of the material to 50 degrees Celsius. The liquid used to cool the metal is glycerol. The properties of glycerol are listed below.

Material Property	Value [Units]
Specific heat	2.4 J/(g °C)
Specific gravity	1.261
Initial temperature	25°C

There are data sets from four different data collection sessions. In the first data collection session, a lab technician collected seven different measurements and recorded mass in grams of aluminum rods. The second data set contains cadmium rods; the third data set contains iron rods, and the fourth data set contains tungsten rods.

Mass of Object [g]			
Aluminum	Cadmium	Iron	Tungsten
2,000	3,000	2,500	4,800
2,500	4,000	3,500	6,400
3,000	6,500	4,500	10,400
4,000	8,000	5,000	12,800
5,500	10,000	5,500	16,000
7,500	11,000	7,500	17,600
8,000	15,000	9,000	24,000

Your job is to write two functions: (1) to calculate the thermal energy in joules that must be removed for each rod to cool it from 300 to 50 degrees Celsius for each mass, and (2) to determine the volume of fluid needed in gallons to properly cool the rod for each mass. The result from the second function should be a matrix with the first column being the mass of the rod, and the second column the volume of fluid needed.

In addition, you will need to write a program to call the functions using the datasets provided in the final table. You will call each function four times, once for each material. Your function should only consider variables in SI units, so you will need to convert the vectors in the program before passing them to your function as necessary. Note that all conversions must be done in MATLAB code—you may not hard code any values you calculate by hand.

CHAPTER 17
INPUT/OUTPUT IN MATLAB

When writing a program, it is wise to make the program understandable and inter-active. This chapter describes the mechanisms in MATLAB that allow the executor of a program to input values during runtime. In addition, this chapter describes a few different mechanisms for displaying clean, formatted output to the executor that can be read and interpreted as if it were prepared by hand.

17.1 INPUT

If all programs required the programmer to enter the input quantities at the beginning of the program, everyone would need to be a computer programmer to use the pro-gram. Instead, MATLAB and other programming languages have ways of collecting input from the program's user at runtime. This section focuses on program-directed user input, using three different approaches.

Numerical Input Typed by the User

The `input` function allows the user to input numerical data into MATLAB's Command Window. The user is prompted to input values at the point the input state-ment occurs in the code.

```
Variable = input ('String')
```

- `Variable`: The variable where the input value will be stored
- `'String'`: The text that will prompt the user to type a value

It is important to note the difference between user input and functional input. User input prompts the user in runtime to type in a value. Functional input are values passed in as arguments to a function before a function is executed. Traditionally, all user input occurs in programs instead of functions.

● EXAMPLE 17-1

Imagine you are writing a program to calculate the speed of a traveling rocket. Prompt the user to input the distance the rocket has traveled.

Before writing an input statement, it is critical to remember that MATLAB cannot handle units, so it is up to the programmer to perform any unit conversions necessary in the algo-rithm. Imagine that the user wants to report the speed of the rocket in miles per hour. It would be wise to have the user input the distance traveled in miles and the travel time in hours.

```
Distance = input('How far has the rocket traveled [miles]?')
Time = input('How long has the rocket been in the air? [hours]')
Speed = Distance/Time
```

In the preceding code segment, the functions were executed without a semicolon being at the end of the input call. If you insert a semicolon at the end of the input call, output of the stored variables to the screen is suppressed, which allows for cleaner display on the Command Window.

```
% Variable output suppressed
Distance = input('How far has the rocket traveled [miles]?');
Time = input('How long has the rocket been in the air [hours]?');
```

COMPREHENSION CHECK 17-1

- Write an input statement to ask for the user's height in inches.
- Write an input statement to ask the user to enter the temperature in degrees Fahrenheit.

Text Input Typed by the User

The input function also allows the user to input text data into MATLAB's Command Window. The programmer must include an additional argument ('s') to the input function to tell MATLAB to interpret the user input as text. The term 's' is used to indicate a string.

```
Variable = input ('String','s')
```

- `Variable`: The variable where the input value will be stored
- `'String'`: The text that will prompt the user to type a value
- `'s'`: Input type is text, not numeric

● EXAMPLE 17-2

Imagine you are required to ask for the user's name and for the name of the month he/she was born. Suppress all extraneous output.

```
Name = input('Type your full name and press enter: ','s');
DataMonth = input('What month were you born? ','s');
```

COMPREHENSION CHECK 17-2

- Write an input statement to ask the user for the color of his/her eyes.
- Write an input statement to ask the user to type the current month.

Menu-Driven Input

In the preceding example, the user was required to type the name of the month he/she were born, but there is no good way to predict the way the user will actually type the name of the month into MATLAB. For example, the user may abbreviate, represent the number numerically, or misspell the name of the month completely. MATLAB has a built-in input function (menu) that allows the programmer to specify a list of options for user selection. The menu function will display a graphical prompt of a question and responses.

```
Variable = menu ('String', 'opt1','opt2',…)
```

- `Variable`: The variable where the ordinal value will be stored
- `'String'`: The text that will prompt the user to select an item in the menu
- `'optN'`: The options that appear as buttons in the menu. The first option must always be the menu title

The first argument of the menu function is always the question to display and the following arguments are the possible responses. The menu function does not store the actual text response, but rather the ordinal number of the response in the list. An example will help clarify this idea.

● **EXAMPLE 17-3**

Suppose we want to ask the user to input a favorite color. However, we want to force the user to choose from red, green, or blue. Suppress all extraneous output.

```
Color = menu('Favorite Color','Red','Green','Blue');
```

This command will create the menu shown. The value returned upon selection is an integer. For example, if the user were to select "Red" in the menu, the value stored in `Color` *would be 1 since "Red" is the first option available in the list.*

COMPREHENSION CHECK 17-3

- Write a menu statement to ask the user to select the current month from a list.
- Given the menu statement below, what is stored in variable `C` when the user clicks "Red"?

```
C = menu('My favorite color:','Green','Blue','Yellow',
'Black','Red')
```

17.2 OUTPUT

As critical as input is to enhancing the versatility of a program, output is equally important—this is how the user gets feedback on the computer's solution. Furthermore, as discussed earlier, program output of intermediate calculations is an important diagnostic tool. Once a program is working correctly, you should eliminate superfluous program output by using the semicolon, as demonstrated earlier. The program user will also find it helpful if the remaining output is neatly formatted.

Using `disp` for User Output

One of the basic methods for displaying the contents of any variable is through the use of the disp function.

```
disp (Variable)
```

- `Variable`: The variable to be displayed on the screen

Given some variable x defined in MATLAB's Workspace, `disp(x)` displays the value of x to the screen. However, the `disp` command cannot display a value and text on the same line. Furthermore, `disp` relies on using MATLAB's default numerical display format for displaying any numbers, so it does not provide a capability for changing the form in which the number is displayed.

Formatting Output with `fprintf`

The formatted print command, `fprintf`, gives extensive control over the output format, including spacing.

```
fprintf ('String',Var1, Var2,...)
```

- `'String'`: The formatted output string
- `VarN`: The variables referenced in the formatted output string

The `fprintf` command can intersperse precisely formatted numbers within text, guarantee table alignment, align the decimal points of a column of numbers, etc. The `fprintf` command uses control and format codes to achieve this level of precision, as shown in Tables 17-1 and 17-2.

Table 17-1 Output control with `fprintf`

Control Character	Use	Example
\n	Inserts a new line	`fprintf('Hello\n');`
\t	Inserts a tab	`fprintf('A\tB\tC\n');`
\\	Inserts a backslash	`fprintf('Hello\\World\n');`
'' (two single quotes)	Inserts a single quote mark	`fprintf('Bob''s car\n');`
%%	Inserts a percent symbol	`fprintf('25%%');`

The biggest benefit of using `fprintf` is the ability to plug values contained in variables into formatted sentences. In addition to the formatting controls shown above, MATLAB contains a number of special controls to represent different types of variables that might be displayed within a sentence. Inserting a variable control character within a string tells MATLAB to plug in a variable at that location in the string.

Table 17-2 Variable (numeric) and TextVariable (text) control with `fprintf`

Control Character	Use	Example
%0.Nf	Inserts a fixed point value with N decimal places	`fprintf('0.1%f',Variable);`
%e or %E	Inserts a number in exponential notation	`fprintf('%e',Variable);` `fprintf('%E',Variable);`
%s	Inserts text	`fprintf('%s',TextVariable);`

● **EXAMPLE 17-4**

Assume that the following variables are defined in MATLAB's Workspace:

```
Age=20;
Average=79.939;
Food='pizza';
```

   Display the variables in formatted `fprintf` statements. At the end of each formatted output statement, insert a new line.

Since age is a decimal value, it is best to use the %.0f control character:

```
fprintf('My age is %.0f\n',Age);
```

To display the average with two decimal places:

```
fprintf('The average is %.2f\n',Average);
```

Note that MATLAB rounds values when using `%f`.

To display a string within a sentence:

```
fprintf('My favorite food is %s',Food);
```

● **EXAMPLE 17-5**

Assume the variables in Example 17-4 are still defined in the workspace. Create a single formatted output statement that displays each sentence with a line break between each sentence:

```
fprintf('My age is %.0f\n The average is %.2f\nMy favorite
food is %s\n',Age,Average,Food);
```

In MATLAB, the previous line would be typed on a single line. Note that the order of the arguments to the function corresponds to the order of the variables in the string.

COMPREHENSION CHECK 17-4

Assume that the variable M is stored in the workspace with the value 0.3539.

- What would be the control code used to display M with two decimal places?
- What would be the control code used to display M with three decimal places?
- What would be the output of M if a control code is used to display it to two decimal places?
- What would be the output of M if a control code is used to display it to three decimal places?

COMPREHENSION CHECK 17-5

Consider the following segment of code. What appears in the Command Window if this is executed?

```
clear
clc
Scores=[2010,29,7;2009,34,17;2008,14,31];
fprintf('The Past Three Years of the Clemson-USC Football
Rivalry:\n\n');
fprintf('Year\tUSC\t\tClemson\n');
fprintf('%.0f\t%.0f\t\t%.0f\n',Scores(1,1),
Scores(1,2),Scores(1,3));
fprintf('%.0f\t%.0f\t\t%.0f\n',Scores(2,1),
Scores(2,2),Scores(2,3));
fprintf('%.0f\t%.0f\t\t%.0f\n',Scores(3,1),
Scores(3,2),Scores(3,3));
```

Error Messages

You might have noticed that when you run MATLAB code with syntax errors, the Command Window will display errors in red letters to report the error message rather than the standard black text font. In MATLAB, you can actually create your own custom error messages.

Error messages in MATLAB begin with three question marks in red letters:

```
??? Undefined function or variable 'A'.
```

However, if you would like to build in your own error checking in MATLAB, you can actually program custom error messages that may be more helpful than the default messages displayed by MATLAB using the error function.

EXAMPLE 17-6

We are given the weight (in variable W) in newtons of an object as well as the mass (in variable M) in kilograms and want to determine the gravity in meters per second squared. To solve this, we will use the equation $g = W/M$, but we want to wrap that expression inside a conditional just in case the user provides a zero mass, which does not make sense in our situation, but will not cause an error in MATLAB—if you divide a number by 0 in MATLAB, the value "Inf" is calculated, which represents infinity.

```
>> W=530;
>> M=0;
>> g=W/M
g=
   Inf
```

Instead, we could write the following segment of code to force MATLAB to throw an error message.

```
if M==0
    error('Error: The mass cannot be zero!');
else
    g=W/M;
end
fprintf('The gravity is %.1f m/s^2\n',g);
```

When you run this code given the W and M above, the message below will display in the Command Window and your code will stop executing.

```
??? Error: The mass cannot be zero!
```

Warning Messages

If you want to provide a warning to the user in the same red letters used in an error message, but do not want MATLAB to stop running, you can use special syntax with the fprintf function.

● **EXAMPLE 17-7**

Just like in the previous example, we are given the weight (in variable *W*) in newtons of an object as well as the mass (in variable *M*) in kilograms and want to determine the gravity in meters per second squared. We will again use the equation $g = W/M$, but this time if the user sets the mass variable to be zero, we want to display a warning message to the user and re-define the mass to be 3 kilograms to compute gravity. The code highlighted below in the `fprintf` statement (the number 2, followed by the string) will tell MATLAB to display the text as a warning, but continue executing the code.

```
W=530;
M=0;
if M==0
    fprintf(2,'Warning: The mass cannot be zero! Using 3 kg
    for the mass\n');
    M=3;
    g=W/M;
else
    g=W/M;
end
fprintf('The gravity is %.1f m/s^2\n',g);
```

When you run this code, the warning message will display in the Command Window and your code will continue executing.

```
Warning: The mass cannot be zero! Using 3 kg for the mass
The gravity is 176.7 m/s^2
```

17.3 PLOTTING

MATLAB has many plotting options, not all of which concern us in this book. The simplest way to plot in MATLAB is the `plot` command.

Plotting Variables

If the vectors time and distance contain paired data, `plot(time,distance)` will plot the time data on the abscissa and the distance data on the ordinate.

In addition to time and distance, assume a second distance vector, `distance2`, contains additional distance recordings at the same time values in time. To plot both distance and distance2 on the same graph, the plot command will contain both the distance and distance2 vectors: `plot(time,distance,time,distance2)`.

`plot (A,B,C,…)`

- `A`: The vertical axis values
- `B`: The horizontal axis values
- `C`: Specification of the line/symbol type/color
- `A,B,C`: Can be repeated for multiple data series (`plot (A1, B1, C1, A2, B2, C2)`)

Creating Proper Plots

A number of other functions allow the programmer to automatically insert information onto a graph generated in MATLAB. Many plot options are defined by separate commands, discussed below. These can be applied when the plot is created or afterwards, as long as the plot is unchanged.

- **xlabel:** Insert a label on the abscissa.

  `xlabel('Time (t) [s]')`

- **ylabel:** Insert a label on the ordinate.

  `ylabel('Height (H) [m]')`

- **title:** Insert a title.

  `title('Flight Across The United States')`

- **legend:** Insert a legend on the graph. The order of elements provided in the legend is based on the order they are passed into plot or fplot.

  `legend('Airplane 1','Airplane 2')`

- **grid:** Allows the grid to be turned on or off.

  `grid on`

- **axis:** Allows the setting of the minimum and maximum values on the abscissa and ordinate (in that order).

  `axis([0 50000 0 90000])`

Alternately, the functions `xlim` and `ylim` can be used to define the [minimum, maximum] limits on the abscissa and ordinate, respectively.

The following examples use several special graphing commands. For more information on graphing properties, refer to the MATLAB Graphing Properties, Linespec, and Special Character reference tables contained in the endpages of your textbook.

● EXAMPLE 17-8

The following code, when executed, creates the graph shown of multiple experimental data series.

```
clear
clc
H1=[10, 15, 25, 35, 55];    H2=[10, 30, 50, 70, 100];
P1=[0.27,0.41,0.68,0.95,1.5];
P2=[0.11,0.33,0.54,0.76,1.09];
plot(H1,P1,'o',H2,P2,'+');
axis([0 120 0 1.75]);
title('Correct Format of an Experimental Data, Multiple Data
Series Plot');
```

```
legend('Mass = 100 kg','Mass = 250 kg');
xlabel('Height (H) [m]');
ylabel('Power (P) [hp]');
grid
```

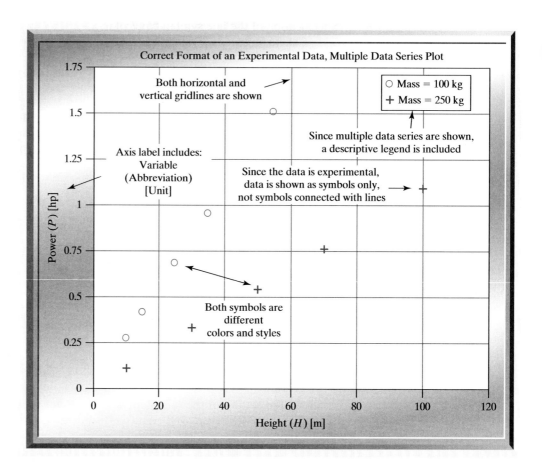

Plotting Theoretical Expressions

A handy feature is that MATLAB can plot a function by using `fplot`.

The way the function `fplot` works is to create a function of the variable (x), which takes the range [a b] in the following format:

```
fplot('function(x)', [a b])
```

It can only use a SINGLE VARIABLE for the function variable, such as an **x** or a **t** or an **m**. So for example, try to run this plot:

```
fplot ('100*x + 52 / x + 11', [10 30])
```

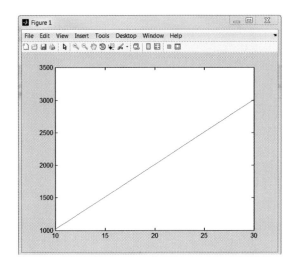

Then try this:

```
fplot ('[100*m + 52, 300*m + 100]', [10 30])
```

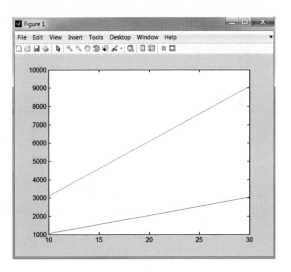

● **EXAMPLE 17-9**   The following code, when executed, creates the graph shown of a single theoretical data series.

```
clear
clc
fplot('0.0273*x+0.05',[0 60]);
title('Correct Format of a Theoretical Data, Single Data Series
Plot');
xlabel('Mass (m) [g]');
ylabel('Density (\rho) [g/cm^3]');
grid
```

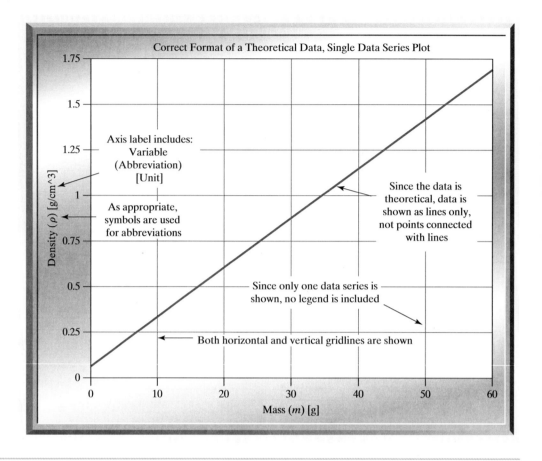

Correct Format of a Theoretical Data, Single Data Series Plot

Axis label includes:
Variable
(Abbreviation)
[Unit]

As appropriate, symbols are used for abbreviations

Since the data is theoretical, data is shown as lines only, not points connected with lines

Since only one data series is shown, no legend is included

Both horizontal and vertical gridlines are shown

Density (ρ) [g/cm^3]

Mass (m) [g]

● **EXAMPLE 17-10**   The following code, when executed, creates the graph shown of multiple experimental data series on logarithmic axes.

```
clear
clc
R1=[10,30,50,70,100];
V1=[0.11,0.33,0.54,0.76,1.09];
R2=[10,15,25,35,55];
V2=[0.27,0.41,0.68,0.95,1.5];
loglog(R1,V1,'s',R2,V2,'o');
grid
xlim([1 100]);
ylim([0.1 10]);
title('Correct Format of a Logarithmic, Experimental Data,
Multiple Data Series Plot');
xlabel('Radius (R) [cm]');
ylabel('Volume (V) [cm^3]');
legend('Cylinder #1','Cylinder #2')
```

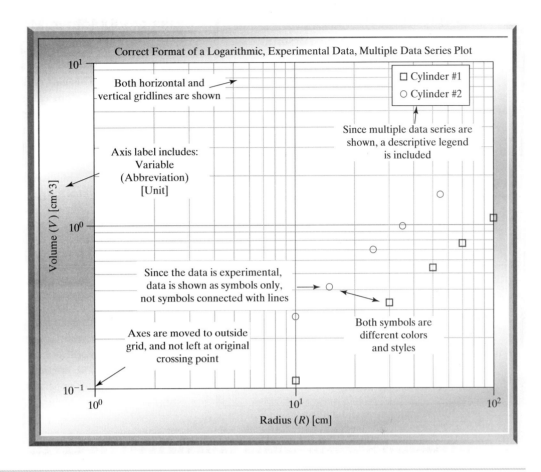

Correct Format of a Logarithmic, Experimental Data, Multiple Data Series Plot

Creating Figures in MATLAB

When MATLAB creates a new plot, it automatically draws the plot to a "Figure" window. If a Figure window is already open, MATLAB will replace the current plot with the new plot within this window. To create multiple plot windows, type the command figure to create a new window before drawing your next plot. Add all titles, x- and y-axis labels, etc., before you initiate the figure command, because those functions execute to the newest figure. If you need to reference a previous figure, you can type figure(N), where N is the number of the figure. By default, MATLAB names the first figure created Figure 1 and increases figure number by 1 as new figures are created.

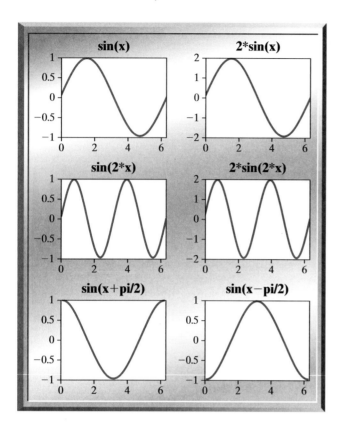

Using Subplots in MATLAB

In addition to generating single plots, MATLAB can generate multiple plots, called subplots, in a single window. This can make it easier to compare the results of different cases or to study a process from a variety of perspectives simultaneously. In this case, the `fplot` command has been used to study the effect of changing the input parameters to the general sine function $A \sin(Bx + C)$. Notice that even plots in MATLAB are arranged in matrices. The command to generate the grouping of plots is shown below. Note that parameter A is amplitude, B is frequency, and C represents a phase shift.

The `subplot` command requires three numbers in parentheses. The first is the number of rows of graphs to appear in the window. The second is the number of columns of graphs to appear in the window. The third is the position of the graph about to be inserted (following a comma or on a new line). The graph positions are numbered starting in the top-left corner and going across each row. The following program was used to create the subplot figure shown:

```
subplot(3,2,1),fplot('sin(x)',[0 2*pi])          plots sin(x) in top left
subplot(3,2,2),fplot('2*sin(x)',[0 2*pi])        plots 2 sin(x) in top right
subplot(3,2,3),fplot('sin(2*x)',[0 2*pi])        plots sin(2x) in middle left
subplot(3,2,4),fplot('2*sin(2*x)',[0 2*pi])      plots 2 sin(2x) in middle right
subplot(3,2,5),fplot('sin(x+pi/2)',[0 2*pi])     plots sin(x + π/2) in bottom left
subplot(3,2,6),fplot('sin(x-pi/2)',[0 2*pi])     plots sin(x − π/2) in bottom right
```

17.4 POLYFIT

Trendline tools in software like Microsoft Excel will allow you to add trends to data sets on graphs to extract that information, but in a MATLAB program, it might be desired to automatically calculate the slope and y-intercept without the use of graphical tools. The `polyfit` function will allow a programmer to pass in two equally sized vectors and determine the polynomial of order n. Consider the generic form of a polynomial:

$$p(x) = c1\ xn + c2\ xn - 1 + \cdots + cn\ x + cn + 1$$

In the equation above, the coefficients $(c1, c2, \ldots, cn+1)$ will be what the `polyfit` function will return, given the order of the polynomial (n) in the input.

The syntax of `polyfit` is:

```
C=polyfit(X,Y,n)
```

where C is a vector of the resulting coefficients, X is a vector of values that correspond to the abscissa, Y is a vector (of equal length to X) of values that correspond to the ordinate, and n is the order of the polynomial. For a linear relationship, the value 1 would be passed in for n.

Linear Relationships

As you might recall, a linear relationship takes the form $y = mx + b$, where m is the slope of the line and b is the y-intercept. The syntax of `polyfit` for a linear relationship is:

```
C=polyfit(X,Y,1)
```

● EXAMPLE 17-11

You are part of a firm designing nanoscale speedometers to measure speeds of small moving creatures like centipedes. To test your sensor, you gather a centipede and you measure the following data, given-in MATLAB notation. Use the data to create a graph and a mathematical model.

```
T=[0, 20, 40, 60, 80, 100, 120];      % time (t) [s]
V=[0, 105, 197, 310, 390, 502, 599];  % velocity (v) [mm]
```

To calculate the speed using the polyfit function:

```
C=polyfit(T,V,1)
```

Note the resulting vectors in C ([4.9714, 2.1429]) *are the coefficients of the polynomial form, or*

$$v = 4.9714t + 2.1429 \quad \text{or} \quad v = 5t + 2$$

Therefore, the speed of the centipede is approximately 5 mm/s. Assume you used the built-in round *function and saved the reasonable values of the slope and y-intercept in the variables m and b as integers:*

```
m=round(C(1));
b=round(C(2));
```

To add this trendline to your data, you will need to create a theoretical data set in MATLAB:

```
T2=[0:5:120];
V2=m*T2+b;
```

To plot both the data as red diamond symbols and the trendline as a straight, red solid line, use the plot *function:*

```
plot(T,V,'dr',T2,V2,'-r');
```

Finally, in order to add a trendline equation to your graph, you will need to construct a properly formatted string in a variable. In order to do this, you will need to use the num2str *function to convert the numbers stored in variables b and m into strings and concatenate all of the strings into a single string variable; in this case,* TE:

```
TE=['v = ',num2str(m),' t + ',num2str(b)];
```

To add this expression on your graph near the trendline, you will need to provide coordinates (determined experimentally) for where the equation should be using the `text` *function:*

```
text(45,450,TE)
```

which take the form `text(x coordinate, y coordinate, variable)`.
After applying additional MATLAB commands to make this plot a proper plot, your plot should look something like the figure shown.

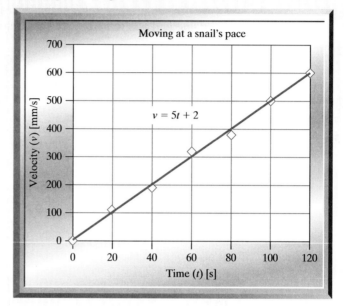

Power Relationships

As you might recall, a power relationship takes the form: $y = bx^m$. Unfortunately, MATLAB does not have a nice, clean function like `polyfit` for power functions, so some simple math must be performed in order to use the `polyfit` function with a power relationship. If we take the log of both sides of the general power relationship equation, we get:

$$\log(y) = m \log(x) + \log(b)$$

This expression is similar to an order-1 polynomial, so we can use `polyfit` to create a linear fit between $\log(x)$ and $\log(y)$ to calculate the value of m and $\log(b)$. Note that this calculation requires the use of the common (base 10) logarithm, not the natural logarithm. In MATLAB, the base-10 logarithm function is `log10` and the natural logarithm function is `log`.

Assuming our abscissa and ordinate values are stored in variables X and Y respectively:

```
C = polyfit(log10(X),log10(Y),1)
```

The `C(1)` value will contain m, but b can be calculated by exploiting the rules of logarithms, $b = 10\char`\^C(2)$.

EXAMPLE 17-12

Joule's first law, also known as the Joule effect, relates the heat generated to current flowing in a conductor. It is named for James Prescott Joule, the same person for whom the unit of Joule is named. The Joule effect states that the electric power (P) can be calculated as $P = I^2R$, where R is the resistance and I is the electrical current. The following data are collected in MATLAB notation. Use the data to create a graph and a mathematical model.

```
I=[0.50,1.25,1.50,2.05,2.25,3.00,3.20,3.50];    % Current (I) [A]
P=[1.2,7.5,11.25,20,25,45,50,65];               % Power (P) [W]
```

To determine the value of the resistance in this relationship:

```
C=polyfit(log10(I),log10(P),1)
```

Note the resulting vectors in C `([2.0302,  0.6850])` *are the coefficients of the polynomial form, or*

$$\log(P) = 2.0302 \log(I) + \log(0.6850) \quad \text{or} \quad P = 10^{0.6850} I^{2.0302}$$

```
m=C(1);
b=10^C(2);
```

To add this trendline to your data, you will need to create a theoretical data set in MATLAB:

```
I2=[0.5:0.5:3.5];
P2=b*I2^m;
```

To plot both the data as blue square symbols and the trendline as a straight, blue solid line, use the plot function:

```
plot(I,P,'sb',I2,P2,'-b');
```

Finally, in order to add a trendline equation to your graph, you will need to construct a properly formatted string in a variable. In order to do this, you will need to use the num-2str *function to convert the numbers stored in variables b and m into strings and concatenate all of the strings into a single string variable; in this case,* TE:

```
TE = ['P = ',num2str(b,'%3.1f'),'* I\^',num2str(m,'%1.0f')];
text(1,55,TE)
```

Note this expression is slightly different from the one given for linear equations. Here, we do not format the numbers prior to the num2str *function. Instead, we format the numbers inside the function, using the notation:*

```
num2str(variable,'%#.df')
```

where # is the number of total digits to display, d is the number of decimal places to display, and f is the code for fixed point value, just as in the fprintf *statement. Other* fprintf *codes (e or E) may also be used.*

The graph that results from this code appears as follows. Here, the exponent does not appear raised about the variable but instead is inline with the other text: P = 4.8 * I ^ 2.

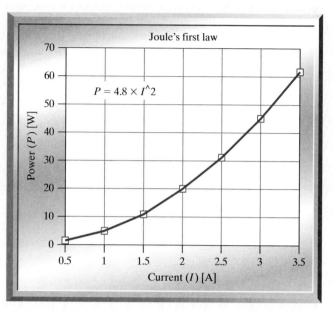

To raise the exponent in superscript format, we need a slightly different placement of `num2str` *in our trendline code:*

```
TE=['P = ',num2str(b,'%3.1f'),'* I^{',num2str(m,'%1.0f'),'}'];
```

This will place the number 2 in superscript format: $P = 4.8 * I^2$.

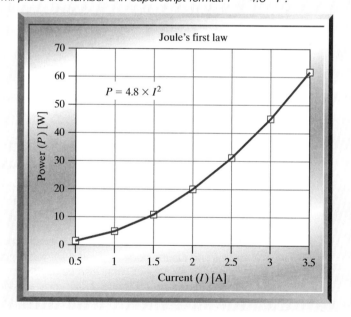

Exponential Relationships

As you might recall, an exponential relationship takes the form: $y = be^{mx}$. Once again, MATLAB does not have a nice, clean function like `polyfit` for exponential functions, so some math must be performed again in order to use the `polyfit` function with an exponential relationship. If we take the natural log of both sides of the general exponential relationship equation, we get:

$$\ln(y) = mx + \ln(b)$$

This expression is similar to an order-1 polynomial, so we can use `polyfit` to create a linear fit between x and $\ln(y)$ to calculate the value of m and $\ln(b)$. Recall that the natural logarithm function in MATLAB is `log`. Assuming our abscissa and ordinate values are stored in variables X and Y, respectively:

```
C=polyfit(X,log(Y),1)
```

The `C(1)` value will contain m. The value of b is calculated by exploiting the rules of natural logarithms:

```
b=exp(C(2))
```

● EXAMPLE 17-13

A reaction is carried out in a closed vessel. The following data are taken for the concentration (C) of the organic material as a function of time (t) [min], from the start of the reaction. A proposed mechanism to predict the concentration at any given time is $C = C_0 e^{-kt}$, where k is the reaction rate constant and C_0 is the initial concentration of the species at time zero. The following data are collected in MATLAB notation. Use the data to create a graph and a mathematical model.

```
T=[36,65,100,160];          % Time (t) [min]
C=[0.145,0.120,0.100,0.080];  % Concentration (C) [g / L]
```

To determine the value of the resistance in this relationship:

```
        P=polyfit(T,log(C),1)
```

Note the resulting vector in

```
    P ([-0.0047, -1.7938])
```

are the coefficients of the polynomial form, or

$$\ln(C) = -0.0047t + \ln(-1.7938)$$

or

$$C = e^{-0.0047t-1.7938} \quad or \quad C = 0.17\,e^{-0.05t}$$

```
        m=P(1);
        b=exp(P(2));
```

To add this trendline to your data, you will need to create a theoretical data set in MATLAB:

```
        T2=[20:10:160];
        C2=b*exp(-m*T2);
        plot(T,C,'*g',T2,C2,'-.g');
        TE=['C  =  ',num2str(b,'%3.2f'),' e^{',num2str(m,'%4.3f'),
        't}'];
        text(43,0.085,TE);
```

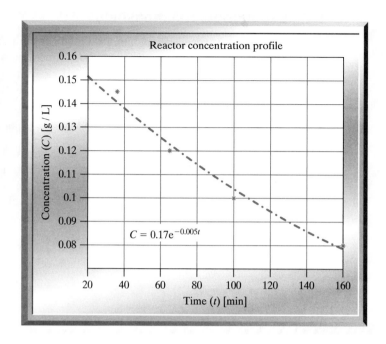

17.5 STATISTICS

Literally hundreds of functions are built into MATLAB. A few were introduced earlier. Here, we expand our use of the functions shown in Table 17-3.

Table 17-3 Common MATLAB statistical functions

MATLAB Function	Definition
ceil(X)	Rounds each element of X up to the next largest integer.
fix(X)	Rounds each element of X to the neighboring integer closest to 0.
floor(X)	Rounds each element of X down to the next smallest integer.
length(X)	If X is a vector, length(X) returns the number of elements in X. If X is a matrix, length(X) returns either the number of rows in X or the number of columns in X, whichever is larger.
max(X) and min(X)	Finds the maximum or minimum value of X. If X is a matrix, returns the maximum or minimum value of the elements of each column in X.
mean(X)	Finds the mean or average of the elements of X. If X is a matrix, mean(X) returns the mean of the elements in each column of X.
median(X)	Finds the median value of X. If X is a matrix, median(X) returns the median value of the elements of each column in X.
round(X)	Rounds each element of X to the nearest integer.
size(X)	Returns a vector of the number of rows and columns of X.
std(X)	Finds the standard deviation value of the elements of X. If X is a matrix, std(X) returns the standard deviation of each column of X.
var(X)	Finds the variance value of X. If X is a matrix, var(X) returns the variance of the elements of each column of X.

● **EXAMPLE 17-14**

You are studying the number of fatal accidents that occur during different times of the day. Using MATLAB and the data shown, determine the following parameters. The data represent the number of accidents between midnight and 6 A.M. for nine consecutive weeks.

Week	Number of Fatal Accidents
A	190
B	202
C	179
D	211
E	160
F	185
G	172
H	205
I	177

Given the accident data:

```
>> accidents=[190 202 179 211 160 185 172 205 177];
```

Mean:

```
>> mean_accidents=ceil(mean(accidents))
mean_accidents =
   187
```

Median:

```
>> median_accidents=median(accidents)
median_accidents =
   185
```

Variance:

```
>> variance_accidents=var(accidents)
variance_accidents =
   281.9444
```

Standard deviation:

```
>> stdev_accidents=std(accidents)
stdev_accidents =
   16.7912
```

Using MATLAB for Statistical Analysis

The outline below gives the steps necessary to use MATLAB for basic statistical analysis of a data set. This is presented as an example of the high and low temperatures during the month of October 2006. The data are given in the table on the following page and in a starting MATLAB file in the online materials. This discussion focuses on the high temperatures, but would hold for the low temperatures as well.

1. Input the data. The first column is simply an identifier (in this case, the date). The second and third columns contain the actual raw data of high and low temperatures for each day, respectively. This step has already been completed in the provided file.
2. Decide on the bin range. A rule of thumb is that the number of bins needed is approximately equal to the square root of the number of samples. While it is obvious in this example how many total samples are needed, the length function is often very useful.

```
>> number_bins=round(sqrt(length(high)))
number_bins=
   6
```

Date	High Temperature [°F]	Low Temperature [°F]
10/1/2006	82	58
10/2/2006	80	57
10/3/2006	81	60
10/4/2006	84	60
10/5/2006	86	61
10/6/2006	82	55
10/7/2006	71	49
10/8/2006	68	58
10/9/2006	80	60
10/10/2006	80	58

3. Examine your data to determine the range of values. Using the max and min functions, you can determine the highest high temperature and the lowest high temperature. Use these values to determine the range and number of points in each bin.

```
>> max_high=max(high)
max_high =
   86
>> min_high=min(high)
min_high =
   56
>> range = max_high - min_high
range =
   30
>> bin_size = range/number_bins
bin size =
   5
```

4. Determine the range of values that will appear in each bin. For example, the first bin will contain temperatures 55–59; the second bin will contain temperatures 60–64. Create a vector with all of the center values of the ranges for each bin and a cell array with all of the bin range labels.

```
>> center_value=[57 62 67 72 77 82 87];
>> bin_range=['55-59'; '60-64'; '65-69';
'70-74'; '75-79'; '80-84'; '85-89'];
```

Using the `hist` Function

In this section, we illustrate three histograms created with the `hist` function.

- **One argument specified (data set):** By default, `hist` separates the data into 10 equally spaced bins and displays the histogram in a figure. Note that the image to the left is not a proper plot.

```
>> hist(high)
```

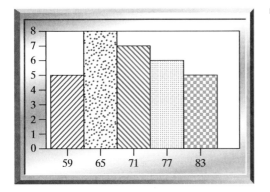

- **Two arguments specified (data set, number of bins):** hist separates the data into the specified number of bins and displays the histogram in a figure. MATLAB divides the range between the minimum and maximum values of the data set into the number of bins specified by the user. Note that MATLAB does not line up the bins on powers of 5, 10, 100, etc., in order to create a reasonable graphing axis. Note that the image to the left is not a proper plot.

```
>> hist(high, 5)
```

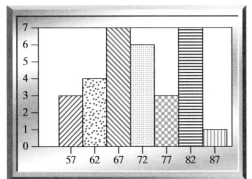

- **Two arguments specified (data set, center value of bins):** hist separates the data into bins specified by the center value of each bin provided in the vector passed to the function and displays the histogram in a figure. Note that the image to the left is not a proper plot.

```
>> hist(high, center_value)
```

In our discussion of the hist function, we have not addressed what data the hist function will return if we assign it to a variable. For example, if we save the result of hist using the centers of each bin and save the result of the function call to the variable N, we see the following result.

```
>> N=hist(high, center_value)
N=
   3   4   7   6   3   7   1
```

It is clear that the data stored in N is the same information conveyed by the histogram, but instead of a graphical representation, we now have integer values of the number of temperatures contained in each bin.

Creating a Cumulative Distribution Function

The Cumulative Distribution Function (CDF) is created in MATLAB by the following procedure.

1. Create the cumulative sum of the histogram data using the cumsum built-in function.

```
>> HTSum=cumsum(N)
HTSum=
   3   7   14   20   23   30   31
```

2. Normalize the cumulative sum.
To normalize our data, we divide by the total number of elements represented in the histogram (in this example, we are dealing with 31 temperatures). In general, we

solve this by dividing by the maximum value of the cumulative sum and multiplying the result by 100 to create a CDF within the range of 0%–100%.

```
>> CDF=HTSum/max(HTSum)*100
   CDF=
   9.6774  22.5806  45.1613  64.5161  74.1935  96.7742
   100.0000
```

Creating CDF Graphs

To create a bar graph of our CDF, use the `bar` function and the functions available for creating proper titles and axis labels.

```
>> bar(CDF)
>> set(gca,'XTickLabel',bin_range);
>> xlabel('Temperature (T) [deg F]')
>> ylabel('CDF Value [%]')
>> title('CDF of Daily High Temperature in Clemson, SC in October, 2006')
```

To create a line graph of the CDF, use the `plot` function and the functions available for creating proper titles and axis labels.

```
>> plot(CDF)
>> set(gca,'XTickLabel',bin_range);
>> xlabel('Temperature (T) [deg F]')
>> ylabel('CDF Value [%]')
>> title('CDF of Daily High Temperature in Clemson, SC in October, 2006')
```

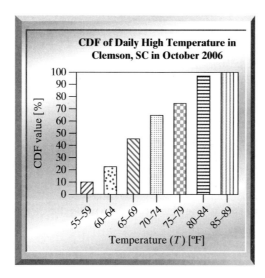
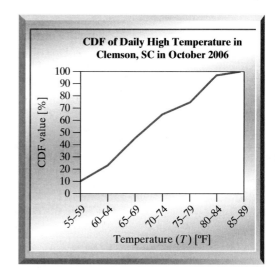

COMPREHENSION CHECK 17-6   Repeat this analysis, using the daily low temperatures during October 2006.

In-Class Activities

ICA 17-1

Write a MATLAB statement that results in the input request shown in bold. The >> shows where your statement is typed, and the | shows where the cursor waits for input. The display must be correctly positioned. Each has a space before the cursor (shown as |). The input variable name and the variable type are shown at the right.

(a) >>
   Enter the length of the bolt in inches: | (bolt, a number)

(b) >>
   Enter the company's name: | (Company, text)

(c) >>

(In this statement, the window displayed will appear for the user to choose a color, and a number representing the color selected will be stored in the variable LineColor.)

ICA 17-2

Write a program that will allow the user to type a liquid evaporation rate in units of kilograms per minute and display the value in units of pounds-mass per second, slugs per hour, and grams per second. Display the result in the Command Window: "The evaporation rate is ____ pounds-mass per second, ____ slugs per hour, or ____ grams per second."

ICA 17-3

In order to calculate the pressure in a flask, write a program that allows the user to type the volume of the flask in liters, the amount of an ideal gas in units of moles, and the temperature of the gas. Display the result in the Command Window: "The pressure is ___ atmospheres."

ICA 17-4

To calculate the efficiency of an electric motor used to raise an object, write a program that will allow the user to type the power in watts, the mass of the object in kilograms, the height the object will be raised in meters, and the time it took to raise the object in seconds. Display the calculated value in the Command Window: "The motor is ____% efficient."

ICA 17-5

Write a MATLAB program that will allow a user to type the specific heat of a value in calories per gram degree Celsius. Display the converted value in units of British thermal units per pound-mass degree Fahrenheit in the Command Window: "The specific heat is ____ BTU / (lb_m*deg F)."

ICA 17-6

For the following questions, $z = 100/810$. Write the MATLAB output that would result from each statement.

(a) `>> disp(z)`
(b) `>> fprintf('%f',z)`
(c) `>> fprintf('%e',z)`
(d) `>> fprintf('%E',z)`

ICA 17-7

For the following questions, $z = 100/810$. Write the MATLAB output statement that displays z in the format shown in bold. In each case, the >> shows where your statement is typed. The display must be correctly positioned.

(a) `>>`
0.123457   (with no blank line after)
(b) `>>`
0.1   (with one blank line after)
(c) `>>`
0.123
(d) `>>`
1.235e-001
(e) `>>`
The value of z is 0.123.   (this must all appear on one line)
(f) `>>`
z is 0.123, so 10 z is 1.235.   (this must all appear on one line)

(g) Check the output of the statement below and indicate how you would modify it so that the decimal points of the two numbers line up vertically with each other. You may only modify the two occurrences of `%.3f` in these statements—you may not use spaces to get these to line up.

```
fprintf('\nThe value of z is %.3f\n',z), fprintf('1000z is equal
to %.3f\n\n',1000*z)
```

(h) Compare the output of the statements below. Why does adding 1 to this number change the last two digits from 78 to 90?

```
fprintf('\n%.17f\n\n',z)
fprintf('\n%.17f\n\n',z+1)
```

ICA 17-8

The tiles on the space shuttle are constructed to withstand a temperature of 1,950 kelvin. Write a MATLAB program that will display the temperature converted to degrees Celsius, degrees Fahrenheit, and degrees Rankine. Each conversion should appear on a new line and each value should be displayed with two decimal values.

ICA 17-9

The specific gravity of acetic acid (vinegar) is 1.049. Write a MATLAB program that will display the density of acetic acid in units of pounds-mass per cubic foot, grams per cubic centimeter, kilograms per cubic meter, and slugs per liter. Each conversion should appear on a new line and each value should be displayed with one decimal value.

ICA 17-10

Joule's first law, also known as the **Joule effect**, relates the heat generated to current flowing in a conductor. It is named for James Prescott Joule, the same person for whom the unit of joule is named. The Joule effect states that electric power can be calculated as $P = I^2R$, where R is the resistance in ohms [Ω] and I is the electrical current in amperes [A]. Create a proper plot of the experimental data.

Current (*I*) [A]	0.50	1.25	1.50	2.25	3.00	3.20	3.50
Power (*P*) [W]	1.20	7.50	11.25	25.00	45.00	50.00	65.00

ICA 17-11

There is a large push in the United States currently to convert incandescent light bulbs to compact fluorescent lights (CFLs). The lumen [lm] is the SI unit of luminous flux, a measure of the perceived power of light. Luminous flux is adjusted to reflect the varying sensitivity of the human eye to different wavelengths of light. To test the power usage, you run an experiment and measure the following data. Create a proper plot of these data, with electrical consumption (EC) on the ordinate and luminous flux (LF) on the abscissa.

Luminous Flux [lm]	Electrical Consumption [W]	
	Incandescent 120 Volt	Compact Fluorescent
80	16	
200		4
400	38	8
600	55	
750	68	13
1,250		18
1,400	105	19

ICA 17-12

Your instructor will select several of the following functions. Plot using subplots. The independent variable (angle) should vary from 0 to 360 degrees.

(a) $\sin(\theta)$
(b) $2\sin(\theta)$
(c) $-2\sin(\theta)$
(d) $\sin(2\theta)$
(e) $\sin(3\theta)$
(f) $\sin(\theta) + 2$
(g) $\sin(\theta) - 3$
(h) $\sin(\theta + 90)$
(i) $\sin(\theta - 45)$
(j) $2\sin(\theta) + 2$
(k) $\sin(2\theta) + 1$
(l) $3\sin(2\theta) - 2$

ICA 17-13

Your instructor will select several of the following functions. Plot using subplots. The independent variable (angle) should vary from 0 to 360 degrees.

(a) $\cos(\theta)$
(b) $2\cos(\theta)$
(c) $-2\cos(\theta)$
(d) $\cos(2\theta)$
(e) $\cos(3\theta)$
(f) $\cos(\theta) + 2$
(g) $\cos(\theta) - 3$
(h) $\cos(\theta + 90)$
(i) $\cos(\theta - 90)$
(j) $2\cos(\theta) + 2$
(k) $\cos(2\theta) + 1$
(l) $3\cos(2\theta) - 2$

ICA 17-14

You want to create a graph showing the relationship of an ideal gas between pressure (P) and temperature (T). The ideal gas law relates the pressure, volume, amount, and temperature of an ideal gas using the relationship: $PV = nRT$. The ideal gas constant, R, is 0.08207 atmosphere liter per mole kelvin. Assume the tank has a volume of 12 liters and is filled with nitrogen (formula, N_2; molecular weight, 28 grams per mole). Allow the initial temperature to be 270 kelvin at a pressure of 2.5 atmospheres. Create a graph, showing the temperature on the abscissa from 270 to 350 kelvin.

ICA 17-15

The decay of a radioactive isotope can be modeled with the following equation, where C_0 is the initial amount of the element at time zero and k is the decay rate of the isotope. Create a graph of the decay of Isotope A [$k = 1.48$ hours]. Allow time to vary on the abscissa from 0 to 5 hours with an initial concentration of 10 grams of Isotope A.

$$C = C_0 e^{-t/k}$$

ICA 17-16

Today, most traffic lights have a delayed green, meaning there is a short time delay between one light turning red and the light on the cross-street turning green. An industrial engineer has noticed that more people seem to run red lights that use delayed green. She conducts a study to determine the effect of delayed green on driver behavior. The following data were collected at several test intersections with different green delay times. These data represent only those drivers who continue through the intersection when the light turns red *before* they reach the limit line, defined as the line behind which a driver is supposed to stop. The data show the "violation time," defined as the average time between the light turning red and the vehicle crossing the limit line, as a function of how long the delayed green has been installed at that intersection.

(a) Graph the violation time (V, on the ordinate) and the time after installation (t, on the abscissa) for all three intersections on a single graph.
(b) Use polyfit to determine linear relationships for each data set.

Time After Installation (t) [Month]	Violation Time (V) [s]		
	Intersection 1	Intersection 2	Intersection 3
	1-Second Delay	2-Second Delay	4-Second Delay
2	0.05	0.1	0.5
5	0.1	0.5	1.5
8	0.3	1	2.5
11	0.4	1.3	3.1

ICA 17-17

The resistance of a typical carbon film resistor will decrease by about 0.05% of its stated value for each degree Celsius increase in temperature. Silicon is very sensitive to temperature, decreasing its resistance by about 7% for each degree Celsius increase in temperature. This can be a serious problem in modern electronics and computers since silicon is the primary material from which many electronic devices are fabricated.

(a) Create a proper plot to compare a carbon film resistor with a resistor fabricated from specially doped silicon ("doped" means impurities such as phosphorus or boron have been added to the silicon).

(b) For relatively small temperature differences from the reference temperature, this process is essentially linear. Use `polyfit` to determine linear relationships for each data set.

Temperature (T) [°C]	Carbon Film	Doped Silicon
15	10.050	10.15
20	10.048	9.85
25	10.045	9.48

ICA 17-18

Cadmium sulfide (CdS) is a semiconducting material with a pronounced sensitivity to light—as more light strikes it, its resistance goes down. In real devices, the resistance of a given device may vary over four orders of magnitude or more. An experiment was set up with a single light source in an otherwise dark room. The resistance of three different CdS photoresistors was measured when they were at various distances from the light source. The farther they were from the source, the dimmer the illumination on the photoresistor.

(a) Create a proper plot of the data.

(b) Use `polyfit` to determine power relationships for each data set.

Distance from Light (d) [m]	Resistance (R) [V]		
	A	B	C
1	79	150	460
3	400	840	2,500
6	1,100	2,500	6,900
10	2,500	4,900	15,000

ICA 17-19

Your supervisor has assigned you the task of designing a set of measuring spoons with a "futuristic" shape. After considerable effort, you have come up with two geometric shapes that you believe are really interesting.

You make prototypes of five spoons for each shape with different depths and measure the volume each will hold. The table below shows the data you collected.

Depth (d) [cm]	Volume of Shape A (VA) [mL]	Volume of Shape B (VB) [mL]
0.5	1	1.2
0.9	2.5	3.3
1.3	4	6.4
1.4	5	7.7
1.7	7	11

(a) Create a proper plot of the data.
(b) Use polyfit to determine power relationships for each data set.
(c) Use your models to determine the depths of a set of measuring spoons comprising the following volumes for each of the two designs.

Volume Needed (V) [tsp or tbsp]	Depth of Design A (dA) [cm]	Depth of Design B (dB) [cm]
¼ tsp		
½ tsp		
¾ tsp		
1 tsp		
1 tbsp		

ICA 17-20

If the voltage across a semiconductor diode is held constant, the current through it will vary with changes in temperature. This effect can be (and has been) used as the basis for temperature sensors. Three different diodes were tested: a constant voltage (0.65 volts) was held across each diode while the current through each was measured at various temperatures. The following data were obtained.

(a) Create a proper plot of the data.
(b) Use polyfit to determine exponential relationships for each data set.

Temperature (T) [K]	Current (I_D) [mA]		
	Diode A	Diode B	Diode C
275	852	2,086	264
281	523	1,506	179
294	194	779	81
309	69	390	35
315	47	301	26

ICA 17-21

The table below lists the number of defective computer chips rejected during random testing over the course of a week on a manufacturing line. Use the following data to generate a histogram and CDF.

1	1	8	0	2	0
0	2	10	1	3	2
0	1	12	0	2	1
1	6	15	0	0	
3	8	1	2	5	

ICA 17-22

This exercise includes the measurement of a distributed quantity and the graphical presentation of the results. We are interested in determining how many "flexes" it takes to cause a paper clip to fail.

Test the bending performance of 20 paper clips by doing the following:

1. Unfold the paper clip at the center point so the resulting wire forms an "S" shape.
2. Bend the clip back and forth at the center point until it breaks.
3. Record the number of "flexes" required to break the clip.

Use the table below to record the raw data for the paper clips you break. Summarize the data for the team by adding up how many clips broke at each number of flexes. Each team member should contribute 20 data points. Use the data to plot a histogram and cumulative distribution function.

Paper Clip Flexing Data		Summary of Data	
Paper Clip	Flexes to Break	No. of Flexes	No. of Clips
1			
2			
3			
4			
5			
6			
7			
8			
9			
10			
11			
12			
13			
14			
15			
16			
17			
18			
19			
20			

1. You are part of an engineering firm on contract by the U.S. Department of Energy's Energy Efficiency and Renewable Energy task force to develop a program to help consumers measure the efficiency of their home appliances. Your job is to write a program that measures the efficiency of stove-top burners. Before using your program, the consumer will place a pan of room temperature water on their stove (with 1 gallon of water), record the initial room temperature in units of degrees Fahrenheit, turn on the burner, and wait for it to boil. When the water begins to boil, they will record the time in units of minutes it takes for the water to boil. Finally, they will look up the power for the burner provided by the manufacturer.

   The output of your program should look like the output displayed below; where the highlighted values are example responses typed by the user into your program. Note that your code should line up the energy and power calculations, as shown below. In addition, your code must display the efficiency as a percentage with one decimal place and must include a percent symbol.

   ```
   Household Appliance Efficiency Calculator: Stove

   Type the initial room temperature of the water [deg F]: 68
   Type the time it takes the water to boil [min]: 21
   Type the brand name and model of your stove: Krispy 32-Z
   Type the power of the stove-top burner [W]: 1200


   Energy required:        1267909 J
   Power used by burner:    1006 W

   Burner efficiency for a Krispy 32-Z stove: 83.9%
   ```

   Here are a few more examples to use as test cases for your code.

Stove Model	Room Temp [°F]	Time to Boil [min]	Rated Burner Power [W]
Krispy 32-Z	68	21	1,200
MegaCook 3000	71	25	1,300
SmolderChef 20F	72	21	1,500
Blaze 1400-T	68	26	1,400
CharBake 5	69	18	1,350

2. We want to conduct an analysis for a wooden baseball bat manufacturer who is interested in diversifying their product line by producing white ash, maple, hickory, and bamboo baseball bats. The manufacturer is set up to be able to produce bats of a single wood type, so they are interested in examining the potential profits with different material types for their bats.

   (a) To help the manufacturer look at the different scenarios, write a program that asks the user of the program to type the following variables:

   - Material cost to produce a single white ash bat (in dollars)
   - Material cost to produce a single maple bat (in dollars)
   - Material cost to produce a single hickory bat (in dollars)
   - Selling price of a white ash bat (in dollars)
   - Selling price of a maple bat (in dollars)
   - Selling price of a hickory bat (in dollars)
   - The total number of bats the manufacturer can produce per week. Assume the manufacturer can produce the same number of bats each week regardless of the material.
   - The number of weeks the manufacturer plans to run its bat production machinery

The labor cost for white ash and maple bats is $2.50 per bat; however, due to the hardness associated with hickory bats, the labor cost is $5.00 per bat. The total variable cost to produce a bat should be considered as the labor cost plus the material cost; energy cost will be considered negligible for this process.

Your program should display the total revenue generated for the scenario, as shown below:

```
Producing ## bats a week for ## weeks will generate:
White ash bat revenue: $####.##
Maple bat revenue: $####.##
Hickory bat revenue: $####.##
Total number of bats produced: #.##e+###
```

Note that the revenue should be displayed with two decimal places and the total number of bats produced should be displayed in exponential notation with two decimal places. In addition, each revenue line should display with a single tab at the front of each sentence, as shown above. The "#" character in the output displayed above will be actual numbers in the program you create.

(b) You should modify the program from part (a) to allow the manufacturer to account for any energy cost associated with each type of bat, which will be typed by the user of the program:

- Fixed energy cost for white ash bats (in dollars)
- Fixed energy cost for maple bats (in dollars)
- Fixed energy cost for hickory bats (in dollars)

(c) You should generate four proper plots with the number of bats produced on the abscissa and the total revenue/cost on the ordinate for each material type.

(d) You should generate a plot that displays only the profit of each bat type with respect to the total number of bats produced in the specified situation.

3. The Fibonacci sequence is an integer sequence calculated by adding previous numbers together to calculate the next value. This is represented mathematically by saying that $F_n = F_{n-1} + F_{n-2}$ (where F_n is the nth value in the sequence, F) or:

<u>0</u>	<u>1</u>	0 + 1= 1	1 + 1= **2**	1 + 2= **3**	2 + 3= **5**	3 + 5= **8**	5 + 8= **13**	8 + 13= **21**	$F_{n-1} + F_{n-2}$ = **... F_n**

Note this sequence starts with the underlined values $(0, 1)$ and calculates the remaining values in the sequence based on the sum of the previous two values.

Professor Bowman found this sequence to be extremely insufficient and created the Bowman sequence, which is an integer sequence calculated by adding the previous three numbers together (instead of two like in the Fibonacci sequence) to calculate the next value. This is represented mathematically by saying that $F_n = F_{n-1} + F_{n-2} + F_{n-3}$ (where F_n is the nth value in the sequence, F) or:

<u>0</u>	<u>1</u>	<u>2</u>	0 + 1 + 2= **3**	1 + 2 + 3= **6**	2 + 3 + 6= **11**	6 + 11 + 20= **37**	$F_{n-3} + F_{n-2} + F_{n-1}$= **... F_n**

Note this sequence starts with the underlined values $(0, 1, 2)$ and calculates the remaining values in the sequence based on the sum of the previous three values.

Write a MATLAB function that implements the Bowman sequence that accepts one input argument, the length of the desired Bowman sequence to generate, and returns one output variable, the Bowman sequence stored inside of an array. This function should also check to see if the number passed in to the function is a valid Bowman sequence length (think about what might constitute valid sequence lengths!). If the input is invalid, your function should display an error message and the output variable should contain only one number: –1. Otherwise if the input is valid, your function should calculate the Bowman sequence and display each value in the sequence in the Command Window.

Sample Output:

```
>> B=FUNCTIONNAME(8);
   Bowman sequence:
   0    1    2    3    6    11    20    37
   The variable B will contain:
   B = 0    1    2    3    6    11    20    37
```

4. Assume that a three-element row vector V already exists. Write a single MATLAB statement that will print the contents of V diagonally from top left to bottom right.

```
Example:  V=[2,74,928]
   Sample Output:
   2
        74
             928
```

5. Write a program that asks a user to input (one at a time) the four numbers to fill a 2×2 matrix. For each number entered, make sure the user knows which number is being entered. Display the completed matrix in the Command Window so the user can check the entries.

6. When one attempts to stop a car, both the reaction time of the driver and the braking time must be considered. Plot the following data using `plot`. This plot should be a proper plot.

Vehicle Speed	Distance (d) [m]	
(v) [mph]	Reaction (d_r)	Braking (d_b)
20	6	6
30	9	14
40	12	24
50	15	38
60	18	55
70	21	75

7. If an object is heated, the temperature of the body will increase. The energy (Q) associated with a change in temperature (ΔT) is a function of the mass of the object (m) and the specific heat (C_p), which is the rate at which the object will lose or gain energy. Specific heat is a material property, and values are available in literature. In an experiment, heat is applied to the end of an object, and the temperature change at the other end of the object is recorded. This leads to the theoretical relationship shown. An unknown material is tested in the lab, yielding the following results:

$$\Delta T = \frac{Q}{mC_p}$$

Heat applied (Q) [J]	12	17	25	40	50	58
Temp change (ΔT) [K]	1.50	2.00	3.25	5.00	6.25	7.00

(a) Graph the experimental temperature change (ΔT, ordinate) versus the heat applied (Q, abscissa) using `plot`. This plot should be a proper plot.

(b) Use `polyfit` to determine a linear relationship for the data set.

(c) Graph the theoretical model that represents the temperature change (ΔT, ordinate) versus the heat applied (Q, abscissa) using `fplot`. This plot should be a proper plot. Consider the mass (m) to be 5 kilograms and the specific heat (C_p) to be copper at 0.39 joules per gram kelvin.

8. Capillary action draws liquid up a narrow tube against the force of gravity as a result of surface tension. The height the liquid moves up the tube depends on the radius of the tube. The following data were collected for water in a glass tube in air at sea level.

Radius (r) [cm]	0.01	0.05	0.10	0.20	0.40	0.50
Height (H) [cm]	14.0	3.0	1.5	0.8	0.4	0.2

(a) Graph the height (H, ordinate) versus the radius (r, abscissa) assuming the data are experimental. This plot should be a proper plot.

(b) Use `polyfit` to determine a power relationship for the data set.

9. In a turbine, a device used for mixing, the power required depends on the size and shape of the impeller. In the lab, we have collected the following data:

Diameter (D) [ft]	0.5	0.75	1	1.5	2	2.25	2.5	2.75
Power (P) [hp]	0	0.04	0.13	0.65	3	8	18	22

(a) Graph the power (P, ordinate) versus the diameter (D, abscissa) assuming the data are experimental. This plot should be a proper plot.

(b) Use `polyfit` to determine a power relationship for the data set.

10. A pitot tube is a device that measures the velocity of a fluid, typically the airspeed of an aircraft. The failure of the pitot tube was credited as the cause of Austral Lineas Aéreas flight 2553's crash in October 1997. The pitot tube had frozen, causing the instrument to give a false reading of slowing speed. As a result, the pilots thought the plane was slowing down, so they increased the speed and attempted to maintain their altitude by lowering the wing slats. Actually, they were flying at such a high speed that one of the slats ripped off, causing the plane to nosedive; the plane crashed at a speed of 745 miles per hour [mph].

In the pitot tube, as the fluid moves, the velocity creates a pressure difference between the ends of a small tube. The tubes are calibrated to relate the pressure measured to a specific velocity, using the speed as function of the pressure difference (P, in units of pascals) and the density of the fluid (ρ, in units of kilograms per cubic meter).

$$v = \left(\frac{2}{\rho}\right)^{0.5} P^m$$

(a) Graph the velocity (v, ordinate) versus the pressure (P, abscissa) assuming the data are experimental. This plot should be a proper plot.

(b) Use `polyfit` to determine the power relationships for the data sets.

(c) Use the `polyfit` results to determine the density of each fluid.

Pressure (P) [Pa]	50,000	101,325	202,650	250,000	304,000	350,000	405,000	505,000
Velocity fluid A (v) [m/s]	11.25	16.00	23.00	25.00	28.00	30.00	32.00	35.75
Velocity fluid B (v) [m/s]	9.00	12.50	18.00	20.00	22.00	24.00	25.00	28.00
Velocity fluid C (v) [m/s]	7.50	11.00	15.50	17.00	19.00	20.00	22.00	24.50

11. A growing field of inquiry that both holds great promise and poses great risk for humans is nanotechnology, the construction of extremely small machines. Over the past couple of decades, the size that a working gear can be made has consistently gotten smaller. The table shows milestones along this path.

Years from 1967	0	5	7	16	25	31	37
Minimum gear size [mm]	0.8	0.4	0.2	0.09	0.007	2E-04	8E-06

   (a) Graph the gear size (ordinate) versus the number of years from 1967 (abscissa) assuming the data are experimental. This plot should be a proper plot.

   (b) Use `polyfit` to determine the exponential relationship for the data set.

12. You will be given an Excel spreadsheet that contains salary data from a major league baseball in 2005, titled "Baseball Data." Use the data provided to determine:

   (a) Salary mean, median, and standard deviation for all players.

   (b) Salary mean, median, and standard deviation for the Arizona Diamondbacks.

   (c) Salary mean, median, and standard deviation for the Atlanta Braves.

   (d) Salary mean, median, standard deviation for all pitchers.

   (e) Salary mean, median, standard deviation for all outfielders.

   (f) Which team had the highest average salary?

   (g) Which team had the lowest average salary?

   (h) Which position had the highest average salary?

   (i) Which position had the lowest average salary?

   (j) Draw a histogram and CDF in Excel for all players.

   (k) What percentage of players earned more than $1 million?

   (l) What percentage of players earned more than $5 million?

13. The following set of data was collected by Ed Fuller of the NIST Ceramics Division in December 1993. The data represent the polished window strength, measured in units of ksi [kilopounds per square inch], and were used to predict the lifetime and confidence of airplane window design. Use the data set to generate a histogram and CDF (http://www.itl.nist.gov/div898/handbook/eda/section4/eda4291.htm).

14. The following information was taken from the report of the EPA on the U.S. Greenhouse Gas Inventory.

"Greenhouse gas emission inventories are developed for a variety of reasons. Scientists use inventories of natural and anthropogenic emissions as tools when developing atmospheric models. Policy makers use inventories to develop strategies and policies for emission reductions and to track the progress of those policies. Regulatory agencies and corporations rely on inventories to establish compliance records with allowable emission rates. Businesses, the public, and other interest groups use inventories to better understand the sources and trends in emissions.

The Inventory of U.S. Greenhouse Gas Emissions and Sinks supplies important information about greenhouse gases, quantifies how much of each gas was emitted into the atmosphere, and describes some of the effects of these emissions on the environment.

In nature, carbon is cycled between various atmospheric, oceanic, biotic, and mineral reservoirs. In the atmosphere, carbon mainly exists in its oxidized form as (CO_2). CO_2 is released into the atmosphere primarily as a result of the burning of fossil fuels (oil, natural gas, and coal) for power generation and in transportation. It is also emitted through various industrial processes, forest clearing, natural gas flaring, and biomass burning."

The EPA website provides data on emissions. The data found in the file "Carbon Dioxide Emissions Data" were taken from this website for the year 2001 for all 50 states and the District of Columbia.

(a) Use the data on all 51 states (including the District of Columbia) provided in the file "Carbon Dioxide Emissions Data" to create a histogram with an appropriate bin size.

(b) Determine the mean and median of the data. Indicate each of these values on your graph for part **(a)**.

(c) Which value more accurately describes the data? Indicate your choice (mean or median) and the value of your choice. Justify your answer in words.

15. Choose *one* of the following options and collect the data required.

(A) On a campus sidewalk, mark two locations 50 feet apart. As people walk along, count how many steps they take to go the 50 feet. Do this for 125 individuals.

(B) Select 250 words at random from a book (fiction). Record the number of letters in each word. Alternatively, you can count the words in 250 sentences and record these.

(C) Go to one section of the library, and record the number of pages in 125 books, all of which are in the same section.

(D) Interview 125 people and determine how far away (in miles) their home is from the university.

For the data source you select, you are to do the following:

(a) Construct a histogram, including justification of bin size.

(b) Determine the mean, median, variance, and standard deviation values.

(c) Construct a cumulative distribution function.

CHAPTER 18
LOGIC AND CONDITIONALS

Outside the realm of computing, logic exists as a driving force for decision making. Logic transforms a list of arguments into outcomes based on a decision. Some examples of everyday decision making are as follows:

- If the traffic light is red, stop. If the traffic light is yellow, slow down. If the traffic light is green, go.

  Argument: three traffic bulbs

  Decision: is bulb lit?

  Outcomes: stop, go, slow

- If the milk has passed the expiration date, throw it out; otherwise, keep the milk.

  Argument: expiration date

  Decision: before or after?

  Outcomes: garbage, keep

To bring decision making into our perspective on problem solving, we need to first understand how computers make decisions. **Boolean logic** exists to assist in the decision-making process, where each argument has a binary result and the overall outcome exhibits binary behavior. **Binary behavior**, depending on the application, is any sort of behavior that results in two possible outcomes.

18.1 TRUTH TABLES

A tool called a **truth table** will help us determine the final outcome of our logical expression. To construct a truth table, we need to know in advance the logical expressions to be evaluated, as well as a list of all arguments used within the evaluations.

For example, assume that we are given the following statement:

If I do not follow the speed limit OR I do not use proper turn signals, I am a danger on the highway.

This statement can be broken down into the following logic sequence:

$$D = NOT(SL) \ OR \ NOT(TS)$$

where D represents "danger," SL represents "following the speed limit," and TS represents "properly using turn signals."

In MATLAB, we would write this logical statement as

$$\sim\text{SL} \;||\; \sim\text{TS}$$

To determine all possible outcomes of the logical sequence, we can construct a truth table:

Input Variables		Output Function
SL	**TS**	**$D = \sim SL \parallel \sim TS$**
0	0	1
0	1	1
1	0	1
1	1	0

That is, when I do not follow the speed limit and do not use turn signals, I am considered a danger on the highway. Likewise, when I either follow the speed limit or use turn signals, but never both at the same time, I am considered a danger on the highway. The only case where I am not considered a danger on the highway is where I both follow the speed limit and use proper turn signals.

The anatomy of the truth table is shown below.

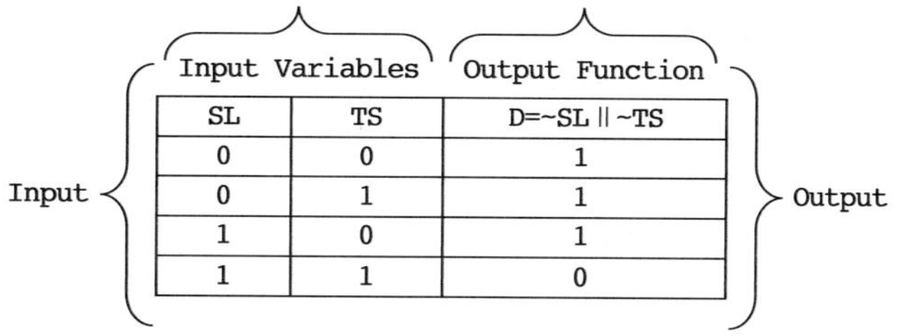

Generally, "true" is assigned the value of 1, and "false" is assigned the value of 0. If you later learn more about Boolean logic, you will discover that sometimes the opposite choice is made: "true" = 0 and "false" = 1. This is sometimes referred to as "negative logic." We assume "positive logic" throughout this text.

● **EXAMPLE 18-1**

Imagine we have two variables that represent Cartesian coordinates, x and y. We want to determine whether these Cartesian coordinates are located inside, outside, or on the perimeter of the unit circle. To determine the location of the coordinate, we calculate the distance from the center of the unit circle $(0, 0)$ to our coordinate with a distance equation:

$$d = \sqrt{x^2 + y^2}$$

How to determine whether a coordinate is in, out, or on the circle:

- If $d > 1$, the coordinate is outside the unit circle.
- If $d < 1$, the coordinate is inside the unit circle.
- Otherwise, the coordinate is on the perimeter of the unit circle.

To make the decisions above, we could have formally prepared a truth table, as follows:

d > 1	d < 1	Outcome
0	0	On
0	1	In
1	0	Out
1	1	–

From this truth table, we recognize that we will need more than one decision to determine the location of the x–y coordinate. Note the use of a dash to indicate a condition that cannot occur.

● **EXAMPLE 18-2**   Imagine we have two variables that represent Cartesian coordinates, *x* and *y*. We want to determine whether these Cartesian coordinates are located in quadrant 1, 2, 3, or 4 of an *x–y* scatter plot.

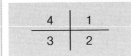

To make the decisions above, we could have formally prepared a truth table, as follows:

x > 0	y > 0	Outcome
0	0	3
0	1	2
1	0	4
1	1	1

From this truth table, we recognize that we will need more than one decision to determine the location of the x–y coordinate.

COMPREHENSION CHECK 18-1

- Draw a truth table that represents the decisions made at a stop light (RED, YELLOW, GREEN).
- Draw a truth table that represents the decisions made to determine if water is frozen (<0 degrees Celsius), steam (>100 degrees Celsius), or liquid (otherwise).
- Draw a truth table that represents the decisions made to determine if a grade is an A, B, C, D, or F on a 10-point scale.

18.2 BINARY NUMBERS

To understand how an input sequence for variables is created, we must first understand how binary numbers work. In computing, integers are often represented as binary numbers. Binary numbers are composed strictly of 1s and 0s. The number system most familiar to people is the decimal system, which is composed of 10 different symbols representing different integers.

Table 18-1 Number system symbols?

Binary (2)	Decimal (10)
0, 1	0, 1, 2, 3, 4, 5, 6, 7, 8, 9

Using the different number systems shown in Table 18-1, we can represent the same numerical value by using a combination of the symbols available in each number system. For binary numbers, we are working with base 2 numbers, in contrast to working with base 10 numbers in the decimal system. For example, let us assume we want to communicate the value 13 in binary and the decimal system.

To do this, we use combinations of different powers of the number system base. Using a combination of 10^X values, we can communicate number 13 in the decimal system by saying

$$1*10^1 + 3*10^0 = 13$$

Table 18-2 Decimal system (base 10)

10^4	10^3	10^2	10^1	10^0
0	0	0	1	3

Note that the coefficients used in front of the 10^X terms (see Table 18-2) are the exact symbols necessary for communicating the required value.

In the binary system, we can communicate the value of 13 by using a combination of values 2^Y

$$1*2^3 + 1*2^2 + 0*2^1 + 1*2^0 = 1101$$

which is the value of 13 in binary.

Table 18-3 Binary system (base 2)

2^4	2^3	2^2	2^1	2^0
0	1	1	0	1

Also note that in the binary example, the $0*2^1$ term was retained to properly extract the sequence of binary digits for the representation of the number in binary, as shown in Table 18-3. This is similar to the fact that 409 has 0 tens, being expressed as

$$4*10^2 + 0*10^1 + 9*10^0 = 409$$

When entering input values in the truth table, the correct input sequence starts from the binary value of 0 and runs up to the binary value of 2^N, where N is the number of input variables. For example, if $N = 3$, the input sequence runs from 000 to 111. The complete sequence is 000, 001, 010, 011, 100, 101, 110, 111.

● **EXAMPLE 18-3**

Convert the following numbers to binary numbers:

■ 19   ■ 165   ■ 1,060

To begin converting any number to its binary value, it is easiest to begin by determining the number of characters necessary to represent the binary value. To do this, we need to look at the calculated values of the first 11 positions in any binary value:

$$2^0 = 1 \qquad 2^4 = 16 \qquad 2^8 = 256$$
$$2^1 = 2 \qquad 2^5 = 32 \qquad 2^9 = 512$$
$$2^2 = 4 \qquad 2^6 = 64 \qquad 2^{10} = 1,024$$
$$2^3 = 8 \qquad 2^7 = 128 \qquad 2^{11} = 2,048$$

To convert 19 to a binary value, we will create a five-digit binary value because 19 is between 16 (2^4) and 32 (2^5).

$$19 = 16 + 2 + 1$$

2^4	2^3	2^2	2^1	2^0
1	0	0	1	1

To convert 165 to a binary value, we will create an eight-digit binary value because 165 is between 128 (2^7) and 256 (2^8).

$$165 = 128 + 32 + 4 + 1$$

2^7	2^6	2^5	2^4	2^3	2^2	2^1	2^0
1	0	1	0	0	1	0	1

To convert 1,060 to a binary value, we will create an 11-digit binary value because 1,060 is between 1,024 (2^{10}) and 2,048 (2^{11}).

$$1,060 = 1,024 + 32 + 4$$

2^{10}	2^9	2^8	2^7	2^6	2^5	2^4	2^3	2^2	2^1	2^0
1	0	0	0	0	1	0	0	1	0	0

Therefore,

Base 10	Base 2
19	10011
165	10100101
1,060	10000100100

18.3 LOGIC AND RELATIONAL OPERATORS IN MATLAB

To connect all of the Boolean arguments to make a logical decision, we turn to a few operators with which we can relate our arguments to determine a final outcome.

NOTE

And = &&
Or = ||
Not = ~

■ **And:** The AND logical operator connects two Boolean arguments and returns the result as true if and only if *both* Boolean arguments have the value of TRUE. In MATLAB, we use the two ampersands (&&) symbol (SHIFT+7 on a keyboard) to represent the AND logical operator.

- **Or:** The OR logical operator connects two Boolean arguments and returns the result as true if *either one* (or both) of the Boolean arguments has the value of TRUE. In MATLAB, we use two pipes (||) symbol (SHIFT+\ on a keyboard) to represent the OR logical operator.
- **Not:** The NOT logical operator inverts the value of a single Boolean argument or the result of another Boolean operation. In MATLAB, we use the tilde (~) symbol (SHIFT+` on a keyboard) to represent the NOT logical operator.

Operator	Meaning
>	Greater than
<	Less than
>=	Greater than or equal to
<=	Less than or equal to
==	Equal to
~=	Not equal to

In order to determine the relationship between two values (numbers, variables, etc.), we have a few operators with which we can compare two variables to determine whether or not the comparison is true or false.

These relational operators are usually placed between two different variables or mathematical expressions to determine the relationship between the two values. This expression of variable–operator–variable is typically called a **relational expression**. Relational expressions can be combined by logical operators to create a **logical expression**. If no logical operator is required in a particular decision, then the single relational expression can be the logical expression.

● **EXAMPLE 18-4**   Express the mathematical inequality in MATLAB code.

Mathematical Inequality	Relational Expression
$4 \leq X < 5$	$4 < = X$ && $X < 5$
$10 < X \leq 20$	$10 < X$ && $X < = 20$
$30 \leq X \leq 100$	$30 < = X$ && $X < = 100$

In this example and others, MATLAB cannot operate with multiple relational operators in a single expression—a common situation in mathematical inequalities. Instead, we must separate out the two relations and combine the expressions with the "and" symbol.

COMPREHENSION CHECK 18-2

What are the relational expressions for the following mathematical inequalities?
- $5 \leq t < 10$
- $-30 < M \leq 20$
- $Y \leq 100$

18.4 CONDITIONAL STATEMENTS IN MATLAB

Conditional statements are commands by which users give decision-making ability to the computer. Specifically, the user asks the computer a question framed in conditional statements, and the computer selects a path forward (it provides an answer—either numerical or as a text comment) based on the answer to the question. Sample questions are given below:

- If the water velocity is fast enough, switch to an equation for turbulent flow!
- If the temperature is high enough, reduce the allowable stress on this steel beam!
- If the pressure reading rises above the red line, issue a warning!
- If your grade is high enough on the test, state: You Passed!

In these examples, the comma indicates the separation of the condition and the action that is to be taken if the condition is true. The exclamation point marks the end of the statement. Just as in language, more complex conditional statements can be crafted with the use of "else" and similar words. In these statements, the use of a semicolon introduces a new conditional clause, known as a nested conditional statement. For example:

- If the collected data indicate the process is in control, continue taking data; otherwise, alert the operator.
- If the water temperature is at or less than 10 degrees Celsius, turn on the heater; or else if the water temperature is at or greater than 80 degrees Celsius, turn on the chiller; otherwise, take no action.

Single Conditional Statements

In MATLAB, a single conditional statement involves the `if` and `end` commands. If two distinct actions must occur as a result of a condition, the programmer can use the `else` command to separate the two actions. The basic structure of a single `if` statement is as follows:

```
if Logical_Expression
        % Actions if true
else
        % Actions if false
end
```

Note that the *Logical_Expression*, as shown in the structure of a single `if` statement, is a logical expression as described previously.

● **EXAMPLE 18-5**

Write a MATLAB statement to represent the following conditional statement.

If the water velocity is less than 10 meters per second, switch to an equation for laminar flow to determine the friction factor of the piping system.

English (Pseudocode)	MATLAB Code
If the water velocity is slow enough,	`if  v < 10`
switch to an equation for laminar flow	`f = 64/Re`
(period)	`end`

In this example, the comma indicates the separation of the condition and the action that is to be taken if the condition is true. The period marks the end of the statement.

● **EXAMPLE 18-6**

Write a MATLAB statement to represent the following conditional statement.
If the speed of the vehicle is greater than 65 miles per hour, display "Speeding" to the Command Window, otherwise; display "OK."

English (Pseudocode)	MATLAB Code

If the speed is greater than 65,
    display "Speeding"
	`if Speed > 65`
	    `fprintf('Speeding')`

Otherwise,
    display "OK"
	`else`
	    `fprintf('OK')`

(period)	`end`

In this example, the comma indicates the separation of the condition and the action that is to be taken if the condition is true. The semicolon after the word "otherwise" marks the beginning of the false action. The period marks the end of the statement.

COMPREHENSION CHECK 18-3

Why is not there a logical expression after the word "`else`" in Example 18-6?

● **EXAMPLE 18-7**

Write a MATLAB statement to represent the following conditional statement.
If the water temperature is at or less than zero degrees Celsius, it is solid; else if the water temperature is at or greater than 100 degrees Celsius, it is vapor; else, it is liquid.

English (Pseudocode)	MATLAB Code

If the temperature is 0 degrees Celsius or less,
    form is solid
	`if Tw <= 0`
	    `form = 'solid'`

else if the temperature is 100 degrees Celsius or greater,
    form is vapor
	`elseif Tw >= 100`
	    `form = 'vapor'`

else
    form is liquid
	`else`
	    `form = 'liquid'`

(period)	`end`

Compound Conditional Statements

Even more complicated questions are possible with the use of logical operators AND, OR, and NOT.

- If the hot plate is on AND the beaker has no water in it, sound an alarm.
- If the stock price falls below $3.00 per share OR if the company's price/earnings ratio goes above $25, sell all shares of the stock.
- If the temperature in the oven falls below 350 degrees Fahrenheit and the internal temperature of the roast is not 250 degrees Fahrenheit, turn the heating element on.

Format of Conditional Statements

Since conditional statements ask questions about data relationships, the first step in programming a conditional statement is defining all the possible relationships. In addition to less than ($<$), greater than ($>$), and equal to ($==$), there are natural extensions of these—less than or equal to ($<=$), greater than or equal to ($>=$), and not equal to ($\sim=$). Note that the conditional test for equality ($==$) is different from the assignment operator ($=$). The logical operator notation for AND ($\&\&$), OR ($||$), and NOT ($\sim$) allows compact logical statements that might look like this:

```
if ChargeLevel==discharged && BatteryStatus~=defective
        DeviceCheck(alternator)
end
```

`if` statements can be nested, implying the AND logic. Sometimes this makes the programming less complex:

```
if ChargeLevel==discharged
        DeviceCheck(battery)
        if BatteryStatus==defective
                DeviceReplace=battery
        else
                DeviceCheck(alternator)
                if AlternatorStatus==defective
                        DeviceReplace=alternator
                end
        end
end
```

COMPREHENSION CHECK 18-4   Rewrite the above nested-`if` statement with the `elseif` control.

COMPREHENSION CHECK 18-5   Rewrite the above nested-`if` statement with logical operators (AND, OR, NOT).

Equivalent Forms of Logic

Conditional statements are similar to traditional language in another way—there are many ways to express the same concept. The order, hierarchy (nesting), and choice of operators are flexible enough to allow the logic statement to be expressed in the way that makes the most sense. All three of the following logic statements are equivalent for integer values of NumTeamMembers, but all three read differently in English:

Conditional Statement	English Translation
| ```matlab
if NumTeamMembers>=4

  if NumTeamMembers<=5

    fprintf('Team size OK.')

  end

end
``` | If the number of team members is 4 or more and if the number of team members is 5 or fewer, the team size is okay. |
| ```matlab
if NumTeamMembers>=4&&NumTeamMembers<=5

  fprintf('Team size OK.')

end
``` | If the number of team members is 4 or more and 5 or fewer, the team size is okay. |
| ```matlab
if NumTeamMembers==4||NumTeamMembers==5

  fprintf('Team size OK.')

end
``` | If the number of team members is 4 or 5 the team size is okay. |
| ```matlab
if NumTeamMembers<4||NumTeamMembers~=5

  fprintf('Adjust team size')

end
``` | If the number of team members is fewer than 4 or more than 5, adjust the team size. |
| ```matlab
if NumTeamMembers~=4&&NumTeamMembers~=5

  fprintf('Adjust team size')

end
``` | If the number of team members is not 4 and not 5, adjust the team size. |
```

In-Class Activities

ICA 18-1

Choose from the list below to answer the questions that follow.

(A) ++   (B) &&   (C) %%   (D) || (E) !!   (F) ~   (G) #

(a) Which of the above operators represents AND?
(b) Which of the above operators represents OR?
(c) Which of the above operators represents NOT?

ICA 18-2

Choose from the list below to answer the questions that follow.

(A) =   (B) !   (C) ~   (D) >   (E) >>   (F) ==   (G) !!   (H) ~~

(I) >=   (J) =>   (K) <>   (L) ><   (M) !=   (N) ~=   (O) !>   (P) ~>

(a) Which of the above operators represents EQUAL TO?
(b) Which of the above operators represents NOT EQUAL TO?
(c) Which of the above operators represents GREATER THAN OR EQUAL TO?

ICA 18-3

A menu is generated using the following code:

```
Status=menu('ClassStanding','Freshman','Sophomore','Junior','Senior');
```

Write a segment of code that will classify the person making a menu selection as a new student (freshman) or a continuing student (sophomore, junior, or senior). The code should display one of the following messages:

```
You are a new student.
You are a continuing student.
```

ICA 18-4

Answer the following questions.
(a) For what integer values of D will the code between if and end execute?

```
if D >= -1 && 3 > D
        code
end
```

(b) For what integer values of G will the code between if and end execute?

```
if G < 2 && -5 <= G
        code
end
```

(c) For what values of F will the code between if and end execute?

```
if F >= 100 || F <= 5
        code
end
```

(d) For what values of R will the code between if and end execute?

```
if R < 75 || R >= 95
        code
end
```

(e) What combinations of values for B and Q will cause a warning to be displayed?

```
if B>100 || (B<20 && Q~=0)
        fprintf('WARNING! Parameters out of bounds.\n')
end
```

(f) What combinations of values for W and M will cause a warning to be displayed?

```
if W<50 || (W>120 && M~=0)
        fprintf('WARNING! Parameters out of bounds.\n')
end
```

ICA 18-5

Write a program that displays a menu of traffic light colors. Depending on which button is pressed, the user should then be told to continue, slow, or stop.

ICA 18-6

Write a program that prompts the user for both the distance and the time at which a vehicle starts and finishes. The input values should be stored in variables StartDist [m], StartTime [s], FinishDist [m], and FinishTime [s]. If the finish time is not greater than the starting time, an appropriate error message should be generated and the program should terminate. Otherwise, the program should return the phrase "Backing up at _____ m/s," "Not moving," or "Moving forward at _____ m/s" as appropriate including numbers where there are blanks.

ICA 18-7

Write a program that asks the user to enter the length of the hypotenuse, opposite, and adjacent sides of a right triangle with respect to an angle (θ). After the user has entered all three values, a menu should pop up asking if the user wants to calculate $\sin(\theta)$, $\cos(\theta)$, $\tan(\theta)$, $\cot(\theta)$, $\csc(\theta)$, or $\sec(\theta)$. The program should display a final message, such as "For a triangle of sides h, o and a, $\sin(\theta) = 0.3$," where h, o and a are replaced by the actual lengths entered by the user. If the user enters lengths that do not work with a right triangle, an error message should appear and the program should terminate.

ICA 18-8

This graph shows the change due if an item is purchased with $1. Regions I and II represent distinctly different situations.

Write a program that asks the user to enter the item cost and the change received. The program should produce an appropriate output message depending on whether the change given is on the line, in Region I, or in Region II. In Region I or II, the message should include the extra amount due to the seller, or the extra amount of change due to the buyer. Assume the purchase price is always $1 or less.

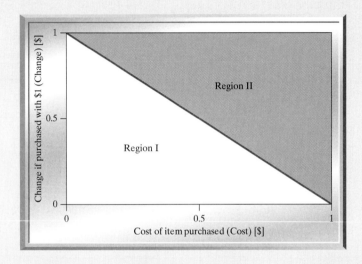

ICA 18-9

What will be displayed by the following code in each of the cases listed below?

```
if flag==1 && alt>=30000
        fprintf('Normal operation at %.0f feet.\n',alt)
elseif flag==0 || alt==0
        fprintf('On Ground')
elseif flag==2 && alt < 30000
        fprintf('Currently at %.0f feet and climbing\n',alt)
elseif flag==3
        fprintf('Currently at %.0f feet and descending\n',alt)
else
        fprintf('Status transitional')
end
```

(a) flag = 1; alt = 25000
(b) flag = 0; alt = 7500
(c) flag = 2; alt = 10000
(d) flag = 1; alt = 37000
(e) flag = 3; alt = 35000

ICA 18-10

We go to a state-of-the-art amusement park. All the rides in this amusement park contain biometric sensors that measure data about potential riders while they are standing in line. Assume the sensors can detect a rider's age, height, weight, heart problems, and possible pregnancy. Help the engineers write the conditional statement for each ride at the park based on their safety specifications.

Variable Definitions:	
A	% Age of the potential rider (as an integer)
H	% Height of the potential rider (as an integer)
HC	% Heart Condition status ('yes' if the person has a heart condition, 'no' otherwise)
P	% Pregnancy status (1 if pregnant, 0 otherwise)

(a) The Spinning Beast: All riders must be 17 years or older and more than 62 inches tall and must not be pregnant or have a heart condition.

```
if _____

    fprintf ('Sorry, you cannot ride this ride');

end _____
```

(b) The Lame Train: All riders must be 8 years or younger and must not be taller than 40 inches.

```
if _____

    fprintf ('Sorry, you cannot ride this ride');

end
```

(c) The MATLAB House of Horror: All riders must be 17 years or older and must not have a heart condition.

```
if _____

    fprintf ('Welcome to the MATLAB House of Horror!');

end
```

(d) The Neck Snapper: All riders must be 16 years or older and more than 65 inches tall and must not be pregnant or have a heart condition.

```
if _____

    fprintf ('Welcome to The Neck Snapper!');

end
```

(e) The Bouncy Bunny: All riders must be between the ages of 3 and 6 (including those 3 and 6 years old)

```
if _____

    fprintf ('This ride is made just for you!');

end
```

ICA 18-11

What is stored in variable A after the following code segment is executed? If the program returns an error, indicate why.

(a)
```
A=5;
if A/2>2
  A = A*-1;
else
  A = A/2;
end
```

(b)
```
x=2; y=4; A=0;
if x+y<4
  A = A+2;
elseif x+y<6
  A = A+4;
end
```

(c)
```
T=6; A=-1;
if T/3>2
  A = A*-1;
else T/3<=2
  A = A+2;
end
```

(d)
```
x=2;y=0;A=10;
if x<=0&&y<=0
  A = A+2;
elseif x>=0&&y<0
  A = A/2;
elseif x>=2&&y<=0
  A = A*2;
end
```

ICA 18-12

A phase diagram for cobalt and nickel is shown below. If it is assumed the lines shown are linear, the mixture has the following characteristics:

- Above 1,500 degrees Celsius, it is a liquid (L).
- Below 1,500 degrees Celsius, there are two possible phases: face-centered cubic (FCC) phase and hexagonal close-packed (HCP) phase.

Write a program to determine the phase. The program should ask the user to enter the weight percent nickel in a variable w and the temperature in a variable T. Call the phases "HCP," "FCC," "L," "FCC + L" (on the line), and "HCP + FCC" (on the line) for simplicity. Store the phase as text in a variable called Phase. The equation of the line dividing the HCP and FCC phases must be found in the program using polyfit. The program should produce a formatted output statement to the command window, similar to "For w weight percent nickel and a temperature of T degrees Celsius, the phase is PHASE," where w, T and Phase are replaced by the actual values.

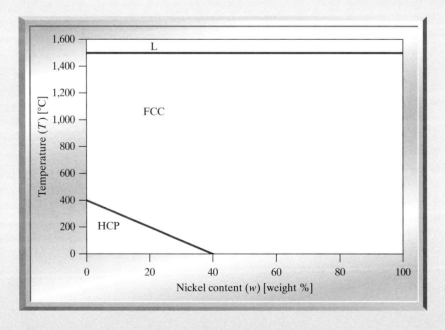

ICA 18-13

A phase diagram for carbon and platinum is shown. If it is assumed the lines shown are linear, the mixture has the following characteristics:

- Below 1,700 degrees Celsius, it is a mixture of solid platinum (Pt) and graphite.
- Above 1,700 degrees Celsius, there are two possible phases: a Liquid (L) phase and a Liquid (L) + Graphite phase. The endpoints of the division line between these two phases are labeled on the diagram.

Write a program to determine the phase. The program should ask the user to enter the weight percent carbon in a variable w and the temperature in a variable T. Call the phases "Pt + G," "L," "L + G," "Pt + L + LG" (on the line), and "L + L + G" (on the line) for simplicity. Store the phase as text in a variable called Phase. The equation of the line dividing the L and L + Graphite phases must be found in the program using polyfit. The program should produce a formatted output statement to the command window, similar to "For w weight percent carbon and a temperature of T degrees Celsius, the phase is PHASE," where w, T and Phase are replaced by the actual values.

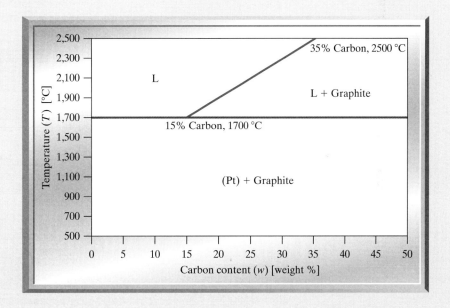

Chapter 18 REVIEW QUESTIONS

1. Create a program to determine whether a user-specified altitude [meters] is in the troposphere, lower stratosphere, or upper stratosphere. In addition, your program should calculate and report the resulting temperature in units of degrees Celsius [°C] and pressure in units of kilopascals [kPa]. Refer to the atmosphere model provided by NASA: http://www.grc.nasa.gov/WWW/K-12/airplane/atmosmet.html.

2. Create a program to determine whether a given Mach number is subsonic, transonic, supersonic, or hypersonic. Assume the user will enter the speed of the object, and the program will determine the Mach number. As output, the program will display the Mach rating. Refer to the NASA page on Mach number: http://www.grc.nasa.gov/WWW/K-12/airplane/mach.html.

3. Humans can see electromagnetic radiation when the wavelength is within the spectrum of visible light. Create a program to determine if a user-specified wavelength [nanometer, nm] is one of the six spectral colors listed in the chart below. Your program should provide a warning if the provided wavelength is not within the visible spectrum.

Color	Wavelength Interval
Red	~ 700–635 nm
Orange	~ 635–590 nm
Yellow	~ 590–560 nm
Green	~ 560–490 nm
Blue	~ 490–450 nm
Violet	~ 450–400 nm

4. Assume a variable R contains a single number. Write a short piece of MATLAB code that will:

   (a) Generate the message: The square root of XXXX is YYYY, if R is nonnegative.
   (b) If R is negative, the code should generate the message: R is negative. The square root of XXXX is YYYYi.

   The value of R should be substituted for XXXX, and the value of the square root of the magnitude of R should be substituted for YYYY. Both numbers should be displayed with three decimal places.

5. Write a program that gathers two data pairs in a 2×2 matrix and then asks the user to input another value of x for which a value of y will be interpolated or extrapolated. Report the value of y in the correct sentence below (with numbers in the blanks).

   Given x = _____, interpolation finds that y = _____ or Given x = _____, extrapolation finds that y = _____.

   For example, if the matrix [0 10, 10 20] is entered and the user chooses 5, the program would return:

   Given x = 5, interpolation finds that y = 15.

6. For the protection of both the operator of a zero-turn radius mower and the mower itself, several safety interlocks must be implemented. These interlocks and the variables that will represent them are listed below.

Interlock Description	Variable Used	State
Brake Switch	Brake	True if brake on
Operator Seat Switch	Seat	True if operator seated
Blade Power Switch	Blades	True if blades turning
Left Guide Bar Neutral Switch	LeftNeutral	True if in neutral
Right Guide Bar Neutral Switch	RightNeutral	True if in neutral
Ignition Switch	Ignition	True if in run position
Motor Power Interlock	Motor	True if motor enabled

For the motor to be enabled (thus capable of running), all the following conditions must be true.

■ The ignition switch must be set to "Run."
■ If the operator is not properly seated, both guide levers must be in the locked neutral position.
■ If either guide lever is not in the locked neutral position, the brake must be off.
■ If the blades are powered, the operator must be properly seated.

(a) Write a function that will accept all of the above variables except Motor, decide whether the motor should be enabled or not, and place true or false in Motor as the returned variable.

(b) Assume the operator has been mowing and desires to dismount from the mower with the engine running. What sequence of actions to the controls (brake, blade power, guide bars, ignition switch) must be taken before the operator dismounts and in what order if the order matters. Note that the operator does not necessarily need to do something with all of the controls.

7. Most resistors are so small that the actual value would be difficult to read if printed on the resistor. Instead, colored bands denote the value of resistance in ohms. Anyone involved in constructing electronic circuits must become familiar with the color code, and with practice, one can tell at a glance what value a specific set of colors means. For the novice, however, trying to read color codes can be a bit challenging.

You are to design a program that, when a user enters a resistance value, the program will display (as text) the color bands in the order they will appear on the resistor.

The resistance will be entered in two parts: the first two digits, and a power of 10 by which those digits will be multiplied. The user should be able to select each part from a menu. You should assume the ONLY values that can be selected by the user are the ones listed in Tables 18-4 and 18-5.

As examples, a resistance of 4,700 ohms has first digit 4 (yellow), second digit 7 (violet), and 2 zeros following (red). A resistance of 56 ohms would be 5 (green), 6 (blue), and 0 zeros (black); 1,000,000 ohms is 1 (brown), 0 (black), and 5 zeros (green). There are numerous explanations of the color code on the web if you want further information or examples.

Table 18-4 Standard resistor values

10	12	15	18	22	27	33	39	47	56	68	82

Table 18-5 Standard multipliers

1	10	100	1000	10000	100000	1000000

Table 18-6 Color codes

0	1	2	3	4	5	6	7	8	9
Black	Brown	Red	Orange	Yellow	Green	Blue	Violet	Grey	White

8. A phase diagram (from http://www.eng.ox.ac.uk/~ftgamk/engall_tu.pdf) for copper–nickel shows that three phases are possible: a solid alpha phase (a solid solution), a liquid solution, and a solid–liquid combination.

   Write a program that takes as input a mass percent of nickel and a temperature and gives as output a message indicating what phases are present. Assume the region borders are linear, and determine a method for dealing with any points that lie directly on the lines. The equation of the line dividing the phases must be found in the program using `polyfit`. The program should produce a formatted output statement to the command window, similar to "For w weight percent nickel and a temperature of T degrees Celsius, the phase is PHASE," where w, T and Phase are replaced by the actual values.

   The program should also produce the graph as shown here with all phase division lines, and indicate the point entered by the user with a symbol.

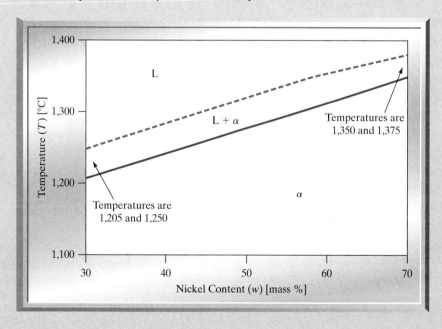

9. This generic phase diagram, based on the temperature and composition of elements A and B, is taken from http://www.soton.ac.uk/~pasr1/build.htm, where a description of how phase diagrams are constructed can also be found. The alpha and beta phases represent solid solutions of B in A and of A in B, respectively.

   The eutectic line represents the temperature below which the alloy will become completely solid if it is not in either the alpha or beta region. Below that line, the alloy is a solid mixture of alpha and beta. Above the eutectic line, the mixture is at least partially liquid, with partially solidified lumps of alpha or beta in the labeled regions.

   Assume the following:

- The melting point of pure A is 700 degrees Celsius.
- The melting point of pure B is 800 degrees Celsius.
- The eutectic line is at 300 degrees Celsius, and the eutectic point occurs when the composition is 50% B.
- The ends of the eutectic line are at 15% A and 85% B.

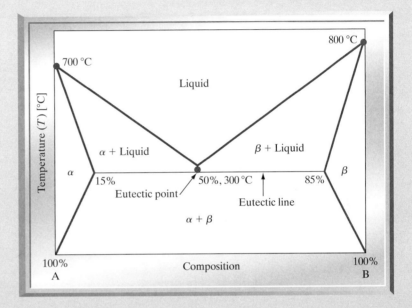

   Using the simplified phase diagram, which substitutes straight lines for the curved lines typical of phase diagrams, write a MATLAB program that takes as input the mass percent of B and the temperature and returns the phases that may exist under those conditions. The equation of the line dividing the phases must be found in the program using `polyfit`. The program should produce a formatted output statement to the command window, similar to "For the composition of x% A, y% B and a temperature of T degrees Celsius, the phase is PHASE," where x, y, T and Phase are replaced by the actual values.

   The program should also produce the graph as shown here with all phase division lines, and indicate the point entered by the user with a symbol.

   Your MATLAB program should include special notes if the provided conditions are on the eutectic line or at the eutectic point.

10. Soil texture can be classified with a moist soil sample and the questions in the table below from the Soil Texture Key found at http://www.pasture4horses.com/soils/hand-texturing.php/. Each answer results in either a soil classification or directions to continue. Write a program that implements this decision tree. You may find it useful to download the key itself from the above address. You may also want to compare your results to an online implementation of the key found at the same website. Your program should include the use of menu windows to gather user responses.

Question	Yes	No
1. Does the soil feel or sound noticeably sandy?	Go to Q2	Go to Q6
2. Does the soil lack all cohesion?	SAND	Go to Q3
3. Is it difficult to roll the soil into a ball?	LOAMY SAND	Go to Q4
4. Does the soil feel smooth and silky as well as sandy?	SANDY SILT LOAM	Go to Q5
5. Does the soil mould to form a strong ball that smears without taking a polish?	SANDY CLAY LOAM	SANDY LOAM
6. Does the soil mould to form an easily deformed ball and feel smooth and silky?	SILT LOAM	Go to Q7
7. Does the soil mould to form a strong ball that smears without taking a polish?	Go to Q8	Go to Q10
8. Is the soil also sandy?	SANDY CLAY LOAM	Go to Q9
9. Is the soil also smooth and silky?	SILTY CLAY LOAM	CLAY LOAM
10. Does the soil mould like plasticine, polish, and feel very sticky when wetter?	Go to Q11	UNKNOWN SOIL
11. Is the soil also sandy?	SANDY CLAY	Go to Q12
12. Is the soil also smooth and buttery?	SILTY CLAY	CLAY

11. We have the menus shown below in a MATLAB program. Write a program to generate these menus and display the choice of the user in a sentence like: "You selected a car with automatic transmission." If the user clicks "Other," an input statement should allow them to type a different vehicle type (as a string). Assume the result of clicking a vehicle in the menu is saved in a variable called `vehicle` and the transmission type is in a variable called `tType` in the workspace.

12. We have the menus shown below in a MATLAB program. Write a program to generate these menus and display the choice of the user in a sentence like: "You selected size 8 Nike shoes." If the user clicks "Other," an input statement should allow them to type a different shoe size (as a number). Assume the result of clicking a brand in the menu is saved in a variable called `brand` and the size is in a variable called `shoeSize` in the workspace.

13. The variable grade can have any real number values from 0 to 100. Ask the user to enter a grade in numerical form. Write an `if-elseif-else` statement that displays the letter grade (any format) corresponding to a numerical grade in an appropriately formatted output statement.

A typical range of grades:

A: $90 \leq$ grade
B: $80 \leq$ grade < 90
C: $70 \leq$ grade < 80
D: $60 \leq$ grade < 70
F: grade < 60

14. Your boss hands you the following segment of MATLAB code that implements a safety control system in an automobile. Recently, the engineer who wrote the code below was fired for their inability to write efficient code. The control system below takes 5 minutes for the car to execute because it was not written using nested-`if` statements, and worse, it does not work with multiple sensors because the former employee could not figure out how to make use of nested-`if` statements! Your boss has instructed you to take the former employee's code and fix it, or you might meet the same fate as the original programmer!

```
% Variable Definition:
% Input:
% Ignition: 0 if engine off, 1 if key in ignition, 2 if engine on
% SeatBelt: 1 if buckled, 0 if unbuckled, -1 if sensor broken
% HeadLamps: 1 if all bulbs ok, 0 if 1 bulb out, -1 if 2 or
more out
% TailLamps: 1 if all bulbs ok, 0 if 1 bulb out, -1 if 2 or
more out
% Light: 1 if sky is bright, 0 if sky is dark (need lamps)
% Output:
% SafetyStatus: 1 if safe, 0 if warn/caution, -1 if unsafe

function[SafetyStatus]=Vehicle(Ignition,SeatBelt,HeadLamps,
TailLamps,Light)
```

```
if Ignition==1 || Ignition == 0
     SafetyStatus = 1;
end

if Ignition==2 && SeatBelt == 1
     SafetyStatus = 1;
end

if Ignition==2 && SeatBelt == 0
     SafetyStatus = 0;
end

if Ignition==2 && SeatBelt == -1
     SafetyStatus = -1;
end

if Ignition==2 && SeatBelt == 1 && HeadLamps == 1
     SafetyStatus = 1;
end

if Ignition==2 && SeatBelt == 1 && HeadLamps == 0
     SafetyStatus = 0;
end

if Ignition==2 && SeatBelt == 1 && HeadLamps == -1
     SafetyStatus = 0;
end

if Ignition==2 && SeatBelt == 1 && HeadLamps == -1 && Light==0
     SafetyStatus = -1;
end

if Ignition==2 && SeatBelt == 1 && TailLamps == 1
     SafetyStatus = 1;
end

if Ignition==2 && SeatBelt == 1 && TailLamps == 0
     SafetyStatus = 0;
end

if Ignition==2 && SeatBelt == 1 && TailLamps == -1
     SafetyStatus = 0;
end

if Ignition==2 && SeatBelt == 1 && TailLamps == -1 && Light == 0
     SafetyStatus = -1;
end
```

15. You are part of an engineering firm on contract by the U.S. Department of Energy's Energy Efficiency and Renewable Energy task force to develop a program to help laboratory technicians measure the efficiency of their lab equipment. Your job is to write a program that measures the efficiency of hot plates.

The program will begin by suggesting four possible fluids for the technician to choose from using a menu: acetic acid, citric acid, glycerol, and olive oil. The technician is then prompted to enter the initial room temperature in units of degrees Fahrenheit, the brand name and model of the hot plate, and the theoretical power for the hot plate provided by the manufacturer.

The program will then call a function, which will determine the following:

■ All fluid properties [Specific Gravity, Specific Heat] should be contained in your function, not in the main program.

- The technician will use 2 liters of fluid.
- The technician should take 2 data points, one at 2 minutes and one at 5 minutes during the heating process. The technician will then begin to heat the fluid. The program will prompt the technician to enter each data point at the time interval. The technician will record the temperature of the fluid [°F] for the two data measurements.
- Once the final data point has been entered, the function will calculate the energy required to heat the fluid, in joules, and the power used to heat the fluid, in watts
- The function will return to the program the time interval, the temperature readings, and the power used.
- The program will then calculate the efficiency of the hot plate, in percentage.

The output of your program should look like the output displayed below; where the highlighted blue values are example responses provided by the user (typed or by pressing a button depending on the requirements mentioned above) into your program, and the highlighted yellow values are the calculated values that will change based upon your starting properties. Note the DATA SHOWN AS RESULTS AND ON THE GRAPH ARE EXAMPLES ONLY AND MAY NOT REFLECT ACTUAL CALCULATIONS! The code should line up the output calculations. In addition, your code must display the efficiency rounded to the nearest integer and must include a percent symbol.

Finally, the program should produce a graph of the two experimental data points relating time and temperature and a trendline with a formatted equation found using polyfit.

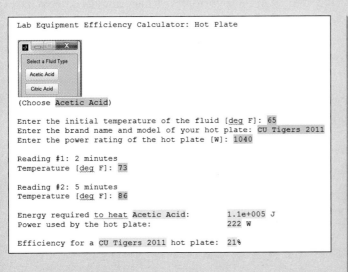

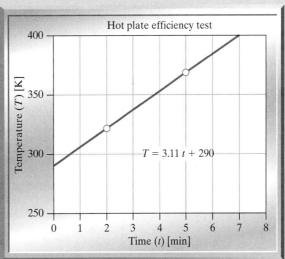

Use the following fluid properties in your program:

Fluid	Specific Gravity [—]	Boiling Point [°C]	Specific Heat [J/(g K)]
Acetic acid	1.049	118	2.18
Citric acid	1.665	153	4
Glycerol	1.261	290	2.4
Olive oil	0.915	300	1.97

CHAPTER 19
LOOPING STRUCTURES

The key to making an algorithm compact is to exploit its variables by means of a loop. A loop is a programming structure that allows a segment of code to execute a fixed number of times or until a condition is true. This chapter describes two looping structures: the `for` loop and the `while` loop.

19.1 for LOOPS

You should have noticed that a number of the algorithms we have already studied have parts that are repetitive. Sometimes, we want to execute an algorithm a certain number of times:

- For every pressure sensor, record the pressure reading.
- For each employee, check the date of the employee's last safety training.

It may be useful to sample less than every possible point, as in:

- For every fifth data point, enter its value into an array for plotting.

The ability to count backwards is also useful:

- For every second counting down from 10 to 0 seconds, announce the time remaining until launch.

The syntax of a `for` loop is illustrated below:

```
for counter = start : step : finish
      % executable statements
   end
```

This programming structure functions as follows: execute the part of the program that is between the `for` statement and the `end` statement for every value of counter from `start` to `finish` counting by `step`. The counter variable can be used in the loop, and most often is to calculate other variables, to index an array, to set function parameters, etc.

A sample `for` loop is shown below. This loop will display on the screen the numbers from 1 to 5, each on a separate line.

```
for k=1:1:5
      fprintf('%.0f\n',k)
   end
```

Understanding the `for` Loop

Most program statements do *exactly* the same type of thing every time, although they may be manipulating different information. The `for` loop is slightly different. A `for` statement can be arrived at during program execution in one of two ways:

1. It can be executed immediately following a statement that was outside the loop, typically a statement immediately above it.
2. It can be executed upon returning from the `end` statement that marks the end of the loop after the loop has executed one or more times.

What the `for` statement does in these two cases is different and is crucial to understanding its operation.

1. When the `for` statement is arrived at from another statement *outside* the loop, it does the following:
   - Places the start value into the counter variable.
   - Decides whether or not to do the instructions inside the `for` loop.

2. When the `for` statement is arrived at from its corresponding `end` statement below, it does the following:
   - Adds the step value to the counter variable.
   - Decides whether or not to do the instructions inside the `for` loop.

The second step is the same in both cases and requires a bit more explanation. To determine whether or not to execute the instructions inside the loop (those between the `for` and its corresponding `end`), the `for` statement asks the question "Has the counter gone beyond the finish value?"

   - If the answer is no, execute the statements inside the loop.
   - If the answer is yes, skip over the loop and continue with the first statement (if any) immediately following the corresponding `end` statement.

NOTE

Since the loop can count either up or down, we use the word "beyond."

If the loop is counting up (step > 0), then we could ask "Is the counter greater than the finish value?"

If the loop is counting down (step < 0), we could use "less than" instead. Several examples are given below.

● **EXAMPLE 19-1**

Write the command to execute the following loop the desired amount of times.

(a) The statements in the loop will be executed four times, once each with i = 1, 2, 3, and 4:

```
for i=1:1:4
        % loop statements
end
```

(b) The statements in the loop will be executed two times, once each with time = 1 and 3. When time is incremented to 5, it is "beyond" the finish value:

```
for time=1:2:4
        % loop statements
end
```

(c) The statements in the loop will be executed three times, once each with gleep = 4, 2, and 0:

```
for gleep=4:-2:0
        % loop statements
end
```

(d) The statements in the loop will be executed three times, once each with raft = 13, 7, and 1. When time is decremented to -5, it is "beyond" the finish value:

```
for raft=13:-6:-4
        % loop statements
end
```

(e) The statements in the loop will not be executed. k is "beyond" the finish value when the loop is first entered:

```
for k=6:1:4
        % loop statements
end
```

(f) The statements in the loop will be executed five times, once each with wolf = 5, 12, 19, 26, and 33:

```
S1=5;
S2=7;
for wolf=S1:S2:S1^2+S2+1
        % loop statements
end
```

Arithmetic Sequences

Changing the `for` loop control parameters allows `for` more intuitive design of loop function. Some of the sequences above can be achieved by changing the `start`, `step`, or `finish` values in the above sample loop. Specifically, any **arithmetic sequences**— those that require only changing k by adding or subtracting a number from it—can be achieved by changing the loop controls.

Referring back to the first sample `for` loop, let us examine how the same sequences can be generated using either the loop controls or calculations inside the loop itself. The first example is repeated here for convenience.

```
for k=1:1:5
        fprintf('%.0f\n',k)
end
```

This generates the sequence k = 1, 2, 3, 4, 5.

In each of the following examples, a sequence of five numbers is given, with two solutions shown to generate that sequence. The first code segment in each example uses the `for` loop to count from 1 to 5, and the values desired are calculated inside the loop. The second segment of code accomplishes the same goal by manipulating the `start`, `step`, and `finish` values instead.

● **EXAMPLE 19-2**   We desire the sequence to be 3, 6, 9, 12, 15.

```
for k=1:1:5
    fprintf('%.0f\n',3*k)
end
```

Given that k goes through the sequence k = 1, 2, 3, 4, 5, the desired sequence could be calculated as 3k.

```
for k=3:3:15
    fprintf('%.0f\n',k)
end
```

This illustrates the idea that multiplication and division of a sequence by a constant to generate a new sequence affects all three loop control parameters: start, step, *and* finish.

● **EXAMPLE 19-3**   We desire the sequence to be 3, 4, 5, 6, 7.

```
for k=1:1:5
    fprintf('%.0f\n',k+2)
end
```

This can be calculated as k + 2

```
for k=3:1:7
    fprintf('%.0f\n',k)
end
```

Adding and subtracting a constant to create a new sequence affects only the start *and* finish *values.*

● **EXAMPLE 19-4**   We desire the sequence to be 4, 6, 8, 10, 12.

Note that the standard order of operations applies, and the multiplication must be done first.

```
for k=1:1:5
    fprintf('%.0f\n',2*k+2)
end
```

This can be calculated as 2k + 2.

```
for k=4:2:12
    fprintf('%.0f\n',k)
end
```

Note that all three loop parameters are affected: start *and* finish *are subject to both multiplication and addition, but* step *is only affected by multiplication.*

<table>
<tr><td>**COMPREHENSION CHECK** 19-1</td><td>Write a for loop to display every even number from 2 to 20 on the screen.</td></tr>
</table>

<table>
<tr><td>**COMPREHENSION CHECK** 19-2</td><td>Write a for loop to display every multiple of 5 from 5 to 50 on the screen.</td></tr>
</table>

<table>
<tr><td>**COMPREHENSION CHECK** 19-3</td><td>Write a for loop to display every odd number from 13 to −11 on the screen.</td></tr>
</table>

Using Variable Names to Clarify Loop Function

Using variables with meaningful names as loop controls makes it easier for the programmer and those trying to interpret the program to keep track of what the loop is doing. The following loop keeps track of the position of an object moving at a constant speed. By defining the step distance as Speed*TimeStep thus an increment in position, the number of steps in the loop is flexible as well. Note that these variables would need to be defined earlier in the program.

```
for Position=StartPosition:Speed*TimeStep:FinalPosition
        fprintf('%.0f\n',Position)
end
```

NOTE

The **length** function determines the number of elements in a vector.

>> A = [30,40,50];

>> LA = length(A);

Variable LA contains 3 since there are 3 elements in the vector.

The **size** function determines the number of rows and columns in a matrix.

>> B = [10,20,30; 40,50,60];

>> LB = size(B);

LB contains a vector [2,3] where the first element (2) is the number of rows and the second element (3) is the number of columns.

Using a for Loop in Variable Recursion

An important use of loops is for **variable recursion**—passing information from one loop execution to the next. A simple form of recursion is to keep a running total. Some of the sequences discussed in Comprehension Check 19-4 require recursion. The loop below totals the sales for all the vendors at a football game. The loop assumes that the array Sales contains the total sales for each vendor. The number of vendors is computed when the loop begins, using the length command, and the total sales are accumulated in TotalSales. Note that TotalSales must be initialized before the loop is entered, since it appears on the right-hand side of the equation:

```
TotalSales = 0;
for Vendor=1:1:length(Sales)
        TotalSales = TotalSales + Sales(Vendor);
end
```

Manipulating a for Loop Counter

Many kinds of computations can be performed in a loop. The loop above will output the value of k to the screen each time through the loop. Other sequences can be achieved by having an expression other than k in the executable part of the loop.

COMPREHENSION CHECK 19-4

Consider the following table of values. Determine the formulae to represent columns A through J. In some cases, the formulae in rows 1–10 will be similar. In other cases, the entry in the first row will be a number rather than a formula.

Create a `for` loop to display the values; use tabs in a formatted `fprintf` statement to create each column.

k	A	B	C	D	E	F	G	H	I	J
1	2	1	1	1	7.25	10	1	2	1	1.00
2	4	3	4	3	7.50	9	−1	4	2	0.50
3	6	5	9	6	7.75	8	1	8	6	0.33
4	8	7	16	10	8.00	7	−1	16	24	0.25
5	10	9	25	15	8.25	6	1	32	120	0.20
6	12	11	36	21	8.50	5	−1	64	720	0.17
7	14	13	49	28	8.75	4	1	128	5020	0.14
8	16	15	64	36	9.00	3	−1	256	40320	0.13
9	18	17	81	45	9.25	2	1	512	362880	0.11
10	20	19	100	55	9.50	1	−1	1024	3628800	0.10

Using the Counter Variable as an Array Index

The example above used the counter variable `Vendor` to access each element of the `Sales` array in sequence to add them together. If we wanted to add only the odd-numbered elements, the only change necessary would be to change the step variable to 2. If we wanted every fourth element of the array beginning with the fourth element, the `for` statement would become

```
for Vendor=4:4:length(Sales)
```

and so forth.

To access elements in a two-dimensional (2-D) array, we could use a single index since MATLAB will handle this, but we would have to be very careful to make certain which row and column is being accessed. For example, if A is a 3 × 2 matrix, A(4) = A(1,2). Since this can be quite confusing, it is usually better to use the double index notation when accessing values in a 2-D array. One method to accomplish this would be to use nested `for` loops, that is, a `for` loop inside another `for` loop.

Let us modify the football sales example above as an example.

Assume we have a 2-D array named `Sales` where each row represents a specific vendor and each column represents a different category of item, such as drinks, hot dogs, hats, etc. Now if we want the total sales, we need to step through every element in every column (or every element in every row) and add them all up:

NOTE

Matrices are described dimensionally as the number of rows "by" the number of columns. Given a matrix as input, the size function saves the number of rows into the first output variable and the number of columns into the second output variable.

```
% Initialize the sum of all sales to 0
TotalSales = 0;
% determine the number of vendors (rows) and items (columns)
[NumbVendors, NumbItems]=size(Sales)
for Vendor=1:1:NumbVendors        % this will index the rows
      for Item=1:1:NumbItems      % this will index the columns
            TotalSales = TotalSales + Sales(Vendor, Item);
      end
end
```

Note that the inner `for` loop (`Item` loop) goes through all items (columns) before exiting, at which point the outer `for` loop increments to the next vendor (row) and the inner loop resets to the first column and steps through all items again, but for a different vendor.

If there were four vendors (rows) and three items (columns), the order in which the entries in the `Sales` array would be added to `TotalSales` would be:

(1,1); (1,2); (1,3); (2,1); (2,2); (2,3); (3,1); (3,2); (3,3); (4,1); (4,2); (4,3);

19.2 `while` LOOPS

Once a `for` loop has started, it executes a specified number of times, regardless of what happens in the loop. At other times, we want to continue executing a loop until a particular condition is satisfied. This requires the use of a `while` loop, which executes until the specified condition is false. For example:

■ While the calculation error is unacceptable, refine the calculations to improve accuracy.

```
while error >= 0.01
      % require less than 1% error
      % perform another iteration of the calculation
      % recalculate error
end
```

The conditional part of the `while` statement has the same syntax as that of the `if` statement.

Converting "Until" Logic for Use in a `while` Statement

In some cases, it may make more sense to use the word *until* in phrasing a conditional loop, but MATLAB does not have such a structure. As a result, it is sometimes necessary to rephrase our conditions to fit the `while` structure.

```
while Logical_Expression
      % executable statements
end
```

"Until" Logic Condition	while Logic Translation

until cows == home
    party
end
	while cows ~ = home
	    party
	end

until homework == done
        TVpower = off
end
	while homework ~ = done
	        TVpower = off
	end

Initializing while Loop Conditions

In for loop constructions, the loop initializes and keeps track of the loop counter. In a while loop, it is necessary to make sure that variables are properly initialized.

```
while error >= 0.01
        % require less than 1% error
        % perform another iteration of the calculation
        % recalculate error
    end
```

If error is not initialized, a syntax error results since the error term does not exist when MATLAB tries to compare it to 0.01. If error is incorrectly initialized to a value less than 0.01, the loop never executes.

● **EXAMPLE 19-5**

We want to input a number from the user that is between (and including) 5 and 10, but reject all other numbers. All extraneous output should be suppressed.

```
X=0;
while X < 5 || X > 10
    X=input('Enter a number between 5 and 10');
end
```

COMPREHENSION CHECK 19-5

Write a while loop that requires the user to input a number until a nonnegative number is entered.

● **EXAMPLE 19-6**

We have a number stored in a variable T that we want to repeatedly divide by 10 until T is smaller than 3×10^{-6}. After determining the first value of T that meets the condition, it should be displayed in a formatted fprintf statement where the value of T is displayed in exponential notation. All extraneous output should be suppressed.

```
T=3993;
while T > 3E-6
    T=T/10;
end
fprintf('%E\n',T);
```

● **EXAMPLE 19-7**

Given a vector of positive integers, V, we want to create two new vectors: vector Even will contain all of the even values of V; vector Odd will contain all of the odd values of V. To begin solving this problem, we need to seek out a MATLAB function that will help us to determine whether or not a given value is even or odd. The built-in function rem accepts two input arguments (Z = rem(X,Y)) such that Z is the remainder after dividing X by Y. For example, if we type rem(3,2), the function would return 1 because $\frac{3}{2} = 1$ with a remainder of 1.

```
clear,clc;
V=[2, 3, 91, 87, 5, 8];
Even=[];
Odd=[]; % initially, Even and Odd are both empty vectors
fprintf('V ='),disp(V);
while length(V)>0
        if rem(V(1),2)==0      % is the first element of V even?
            Even=[Even V(1)]; % add it to the end of Even
        else                   % else first element of V is odd
            Odd=[Odd V(1)];    % add it to the end of Odd
        end
        V(1)=[];               % delete the first element of V
end
fprintf('Even ='),disp(Even);
fprintf('Odd ='),disp(Odd);
```

The output of this code segment would be:

V	=	2	3	91	87	5	8
E	=	2	8				
O	=	3	91	87	5		

19.3 APPLICATION OF LOOPS: CELL ARRAYS AND CELL MATRICES

So far in our discussion of arrays and matrices, all elements of those structures have been numeric since we have discussed them only in a mathematical context. MATLAB provides a similar structure called **cell arrays** or **cell matrices** that allow us to store both numeric values and non-numeric values such as text within a single structure stored in a variable. The syntax for cell arrays and cell matrices uses braces { } instead of parentheses ( ) in standard arrays and matrices.

Pay particular attention to the syntax in this section! In MATLAB:

Square brackets [ ] are used to create numerical arrays and matrices.
Parentheses ( ) are used to indicate the index or address of numerical arrays and matrices.
Braces { } are used to indicate the index or address of cell arrays and matrices.

● **EXAMPLE 19-8**

Assume we want to build a cell array to contain various types of information about different cartoon characters in a cell array named C. In particular, we want to save their first name, last name, date of birth, and the number of different voice actors who have voiced the character. For Mickey Mouse, this cell array would look like:

```
C{1}='Mickey';
C{2}='Mouse';
C{3}=[11,18,1928];
C{4}=4;
```

The first and second elements of the cell array C are both of type text to represent the character's first and last name. The third element of the cell array is a numeric array that contains the month, day, and year of "birth" for the character, as recognized by the cartoon authors. Finally, the fourth element of the array is a number that indicates how many different voice actors have voiced the character.

To access the elements of the C array, we would use a similar notation:

```
fprintf('%s %s ',C{1},C{2});
fprintf('was born on %.0f/%.0f/%.0f.\n',C{3}(1),C{3}(2),C{3}(3));
fprintf('%s %s ',C{1},C{2});
fprintf('has had %.0f different voices.\n',C{4});
```

Note that the syntax used for cell arrays is nearly identical to numeric arrays, except for the use of braces instead of parentheses. For the third element (the array), we used standard notation to reference the third element of the cell array (C{3}), but also used parentheses at the end of that expression to access particular elements of the array. For the month of birth, we typed C{3}(1) since the month was stored in the first element of the numeric array, which was stored in the third element of the cell array C. If we ran the code segments above, we should generate the following output:

```
Mickey Mouse was born on 11/18/1928.
Mickey Mouse has had 4 different voices.
```

● **EXAMPLE 19-9**

Assume we want to expand the cell array we built in Example 19-8 to allow for multiple characters to be stored in a cell matrix. We want to build a cell matrix to contain the same information about different cartoon characters in a cell matrix named C2. For Mickey Mouse and Donald Duck, this cell array would look like:

```
C2{1,1}='Mickey';
C2{1,2}='Mouse';
C2{1,3}=[11,18,1928];
C2{1,4}=4;

C2{2,1}='Donald';
C2{2,2}='Duck';
C2{2,3}=[6,9,1934];
C2{2,4}=2;
```

Note that the syntax for creating a cell matrix is similar to creating a numeric matrix, including the syntax for addressing rows and columns within the matrix. Many of

MATLAB's built-in functions commonly used with arrays and matrices like `size` and `length` also work with cell arrays and cell matrices. Consider the following segment of code, which could be used to display the contents of our cell matrix `C2`.

```
[numRows,numCols]=size(C2);
for i=1:1:numRows
    fprintf('%s %s ',C2{i,1},C2{i,2});
    fprintf('was born on %.0f/%.0f/%.0f.\n',C2{i,3}(1),C2{i,3}(2),
    C2{i,3}(3));
    fprintf('%s %s ',C2{i,1},C2{i,2});
    fprintf('has had %.0f different voices.\n',C2{i,4});
end
```

If we ran the code segment above, we would generate the following output:

```
Mickey Mouse was born on 11/18/1928.
Mickey Mouse has had 4 different voices.
Donald Duck was born on 6/9/1934.
Donald Duck has had 2 different voices.
```

Accessing elements of cell arrays and matrices is syntactically similar to standard numeric arrays and matrices; however, deleting elements from cell arrays and matrices behaves differently. In particular, the use of parentheses versus braces when attempting to delete elements from cell arrays or matrices have dramatically different results.

● **EXAMPLE 19-10**

Assume we have the variable States, as defined below:

```
States={'SC','NC','NC','VA'}
```

If we wanted to delete the third element of the States variable to create a cell array with only three states (SC, NC, VA), we will still need to use the empty array operator [ ] to delete the value. However, it is important to recognize the syntax and the difference in the following comments:

```
States{3}=[]
```

Result: `States = {'SC','NC','','VA'}`. *Note this array has four elements, with the third element being an empty string.*

```
States(3)=[]
```

Result: `States = {'SC','NC','VA'}`. *Note this array has only three elements.*

The difference in these two commands is subtle; however, it is important to recognize the difference between removing an element from a cell array and making an element of a cell array blank.

Unlike `length` and `size`, not, all numerical array functions (like `max`, `min`, or `sort`) will behave the same way using cell arrays. In particular, `max` and `min` will not work on the entirety of a cell array, even if every element of the cell array is a number.

In other words, if we have a cell array, A, defined to be A={1,3,2} and we type M=max(A) into MATLAB, the code will crash because the max function is not defined for cell inputs. However, if we modify A to be A={ [1,3,2], [5,9,3] }, it is perfectly acceptable to use MATLAB to find the maximum value of either the first (or second) numerical array stored in A, or M=max(A{1}). Similarly, sort will only work on cell arrays that only contain strings, and will crash if a cell array used as input into sort contains any non-string data types. Assuming we define a variable C to be C={'mouse','aardvark','dog'}, executing D=sort(C) would create cell array D={'aardvark','dog','mouse'}.

Consider the following code segment:

```
UserInfo={{'Last Name','First Name'},{'Username','Password'},
'Age','Birthday';{'Bear','Yogi'},{'YBEAR','PBasket'},25,
[05,07];{'Doo','Scooby'},{'YABBA','DabbaDoo'},35,[04,21];
{'Runner','Road'},{'RRUNNER','MeepMeep'},40,[03,31]};
```

Cell arrays are used to store mixed variable types inside of a single variable. The cell array UserInfo contains a header row (1×2 cell, 1×2 cell, text, text) and 3 rows of data (1×2 cell, 1×2 cell, number, vector). The 1×2 cells each contain text (first name and last name in the first column; username and password in the second column). If you want to extract or store anything from/in the cell array, you must use braces { }—not parentheses ( ) or square brackets [ ]—when indexing the cell array.

Consider the following expressions that utilize the UserInfo variable.

```
A=UserInfo(1,1)
```

This returns a 1×1 cell that contains a 1×2 cell —{{'Last Name','First Name'}}. This is wrong because we expected the code to return a 1×2 cell. Do not use parentheses because it does not return what you would expect it to return—however, it will not crash, display an error message, or display a warning. If you try to use a cell as a number or text in something like an fprintf statement, it will return an error message similar to this:

```
fprintf('%s',A)
??? Error using ==> fprintf
Function is not defined for 'cell' inputs.
B=UserInfo{1,1}
```

Returns 1×2 cell—contains {'Last Name','First Name'}

```
C=A{1,2}
```

This code will crash because there is no 1, 2 element in the variable A.

```
D=B{1,2}
```

Stores 'First Name' in variable D.

Tip: Whenever you use parentheses on a cell array, it will always return a cell that contains the content at the location you requested. This is wrong, but it will not crash.

Tip: Whenever you use braces on a cell array, it will return the actual content of the specified location in the cell array.

```
E=UserInfo{1,1}{1}
```

Stores 'Last Name' in variable E.

```
F=B{1}
```

Equivalent code to the expression for E, but using cell array we created in B above.

```
G=UserInfo{2,3}
```

Stores the number 25 in variable G.

```
H=UserInfo{3,4}
```

Stores a 1×2 vector [4 21] in variable H.

```
J=UserInfo{3,4}(1)
```

Stores the number 4 in variable J. This is how to reference a number stored in a vector in a cell array.

```
K=['Road','Runner']
```

Creates "RoadRunner" as text in variable K. You can store numbers in a regular matrix, but you cannot store text in a regular matrix. You can use matrices to concatenate strings.

```
L={{'Clemson','SC'}, {'Columbia','SC'}};
```

The code for L correctly creates a 1×2 cell matrix.

```
M=[{'Clemson', 'SC'}, {'Columbia','SC'}];
```

The code for M does not create a 1×2 cell matrix.

If you are building a cell array, like we did when we built UserInfo, the structure must be contained within braces, not parentheses or square brackets—this will lead to unexpected results.

19.4 APPLICATION OF LOOPS: EXCEL I/O

> **WARNING!**
>
> Due to incompatibilities between Mac OS and Microsoft Excel, you must run MATLAB in Microsoft Windows in order to use the built-in Excel I/O (input/output) functions discussed in this document. Likewise, Microsoft Office must be installed to use the Excel I/O functions.

In this section, you will learn to read data from Microsoft Excel workbooks into MATLAB and write information from MATLAB to an Excel Workbook. File input and output is ubiquitous in computer programming if you want to give your programs and functions the ability to save computed information for use at a later time. We discuss file input and output using Microsoft Excel Workbooks because it is a common file format used in many different disciplines.

Reading Microsoft Excel Workbooks

For importing data from Excel files, there are two built-in functions in MATLAB that will allow us to read in all of the information we need from our Excel files; they are `xlsfinfo` and `xlsread`.

xlsfinfo

The `xlsfinfo` function in MATLAB allows us to extract information about Microsoft Excel files stored in MATLAB's current directory. The `xlsfinfo` function will return three outputs:

        [FileType,Sheets,ExcelFormat]=xlsfinfo(fileName)

where:

- `FileType` is a variable containing a string that describes the type of the file. If we call `xlsfinfo` on a file that is not a Microsoft Excel workbook, this string will be empty (`''`); otherwise, 'Microsoft Excel Spreadsheet' will be saved in the `FileType` variable. This is not particularly helpful information, but does give us a mechanism for determining whether or not the file we are trying to open is a valid Microsoft Excel workbook.
- `Sheets` is a cell array variable containing the names of each Worksheet within the Excel Workbook. Each sheet name is stored as text within the cell array.
- `ExcelFormat` is a variable containing a string that describes the version of the Microsoft Excel Workbook. In general, this information is not useful, but if the string is blank, it is an indicator that Microsoft Excel is not installed on your computer.
- `fileName` is the name of the Microsoft Excel file in which we are interested. The file name must be passed into `xlsfinfo` as a string and must contain the file extension of the Microsoft Excel Workbook as part of the file name. The file extension of Microsoft Excel Workbooks is usually .xls or .xlsx.

The following examples will use the starter file titled "ClemsonWeather.xlsx" found in the online materials. If you open the file, you will notice that there are three worksheets within the workbook, each containing day-by-day weather information for Clemson, SC, from March 20, 2010, all the way back to January 1, 2008.

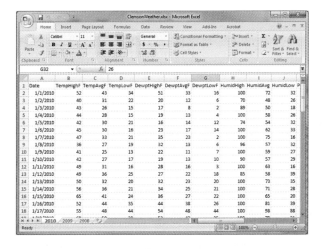

For each day, we have the following values recorded in each worksheet:

- High-, average-, and low-temperature readings measured in degrees Fahrenheit (TempHighF, TempAvgF, TempLowF).

- High-, average-, and low-dew-point temperatures measured in degrees Fahrenheit (DewptHighF, DewptAvgF, DewptLowF).
- High-, average-, and low-humidity percentages (HumidHigh, HumidAvg, HumidLow).
- Maximum and minimum atmospheric pressure measured in inches of mercury [in. Hg] (PressureMaxIn, PressureMinIn).
- Maximum and average wind speed measured in miles per hour (WindMaxMPH, WindAvgMPH).
- Maximum gust speed measured in miles per hour (GustMaxMPH).
- Total daily precipitation measured in inches (PrecipSumIn).

The weather information provided in this worksheet is from Clemson University's Entomology Department data feed on Weather Underground: http://www.wunder ground.com/weatherstation/WXDailyHistory.asp?ID=KSCCLEMS1

● EXAMPLE 19-11

Use xlsfinfo on the ClemsonWeather.xlsx Microsoft Excel Workbook to extract information about the file.

```
>>[f,sn,e]=xlsfinfo('ClemsonWeather.xlsx')
f=
Microsoft Excel Spreadsheet
sn=
    '2010'        '2009'        '2008'
e=
    xlOpenXMLWorkbook
```

As expected, variables f and e contain strings that tell us that the file we asked about ("ClemsonWeather.xlsx") is indeed a Microsoft Excel workbook. Variable sn contains a cell array of the different sheet names we have in our Excel workbook ('2010', '2009', and '2008'). Note that even though the worksheet names are numeric values, MATLAB still stores the names of the worksheets as text values in the cell array, thus these cannot be used in mathematical equations.

---

xlsread

The xlsread function allows us to read in data from Microsoft Excel workbooks. There are a number of different variations on how xlsread can be used, but we will only learn about a few of the variants involving different combinations of arguments and returned values. The default behavior of xlsread:

```
D=xlsread(fileName)
```

will read all of the numeric data from the default (the left-most worksheet) worksheet in the Excel file passed in as a string in the fileName variable and store all of the values in a numeric matrix, D.

● EXAMPLE 19-12

Use xlsread's default behavior on ClemsonWeather.xlsx.

```
WeatherData=xlsread('ClemsonWeather.xlsx');
```

It is particularly important to remember to include the semicolon to suppress output when using the xlsread function because reading data from Excel files generates a large amount of unnecessary output. If we look at the dimension of the WeatherData variable we just created, we notice that by default, MATLAB used the weather data for 2010 to populate our matrix since the dimension of that matrix is 79 × 15. If we open `WeatherData` *in MATLAB's Variable Editor (double click the variable name in the Workspace), we notice that* `xlsread` *imported only the numeric values from the worksheet, starting with cell B2 in the "2010" worksheet. All other values in the original worksheet, including text and dates, were not imported by default.*

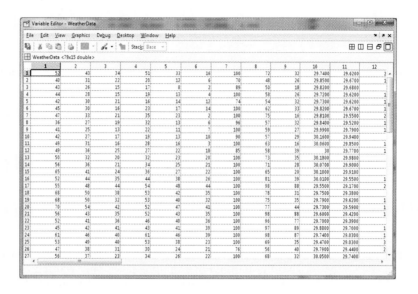

If we want to tell `xlsread` to import values from a particular sheet in our Excel Workbook, we can use the following syntax:

```
D = xlsread(fileName,Sheet)
```

where

- `fileName` is the name of the Excel Workbook including the file extension, as a string.
- `Sheet` is a string containing the name of the Excel Worksheet we want to import into our variable `D`.

● **EXAMPLE 19-13**   Use `xlsread` to import all of the weather data from 2008 from ClemsonWeather.xlsx.

```
WD08=xlsread('ClemsonWeather.xlsx','2008');
```

Notice that the dimension of `WD08` *is 366 × 15, which is what we expected since 2008 was a leap year. If we open* `WD08` *in MATLAB's Variable Editor, we can verify that our matrix contains all of the numeric values in the "2008" worksheet, beginning at cell B2.*

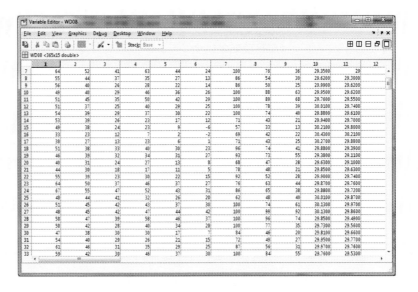

If we only want to import a particular range of an Excel file, we can add in another argument into `xlsread` that will allow us to type in ranges of values in Microsoft Excel notation. For example, if we only wanted to import all of the cells between B2 and B5, we could provide a range B2:B5 just like we would type into Microsoft Excel. The revised syntax for `xlsread`:

```
D = xlsread(fileName,Sheet,Range)
```

where

- `fileName` is the name of the Excel Workbook including the file extension, as a string.
- `Sheet` is a string containing the name of the Excel Worksheet we want to import into our variable `D`.
- `Range` is a string containing the cell range written in Microsoft Excel notation.

● **EXAMPLE 19-14**

Use `xlsread` to import the first 10 high temperatures from 2009 from ClemsonWeather.xlsx. If we open ClemsonWeather.xlsx in Microsoft Excel, we notice that the first 10 values occur between B2 and B11.

```
WD09=xlsread('ClemsonWeather.xlsx','2009','B2:B11');
```

MATLAB creates a 10 × 1 matrix containing a column of values representing the first 10 high temperatures recorded in 2009.

If we need to extract any of the text information from the Excel Workbook, we need to modify our syntax to capture two outputs from `xlsread`. To extract nonnumeric values from an Excel Workbook:

```
[D,T] = xlsread(fileName,Sheet)
```

where

- `fileName` is the name of the Excel Workbook including the file extension, as a string.
- `Sheet` is a string containing the name of the Excel Worksheet we want to import.
- `D` is a matrix of all of the numeric values in Sheet.
- `T` is a cell matrix of all nonnumeric (text, dates, etc.) values in the worksheet. If a cell contains a numeric value, that value will be stored as an empty string (`''`) in `T`. In other words, matrix `D` and `T` will not necessarily have the same dimensions. For more information, refer to Examples 19-15 and 19-16.

It is worth noting that even though we are using the syntax requiring a sheet name to save text values, any of the previously covered notations will allow you to capture text values in an Excel worksheet.

● EXAMPLE 19-15

Use `xlsread` to import the first 10 high temperatures and the corresponding text values for the dates from 2009 from ClemsonWeather.xlsx. If we open ClemsonWeather.xlsx in Microsoft Excel, we notice that the first 10 values occur between B2 and B11 and the dates are listed between A2 and A11, so the range we need to import is A2:A11.

```
[WD09,WT09]=xlsread('ClemsonWeather.xlsx','2009','A2:B11');
```

MATLAB creates two 10 × 1 matrices containing a column of values representing the first 10 high temperatures recorded in 2009 in WD09 and the corresponding text dates in a cell matrix WT09. In this special case, since the entire second column in WT09 was blank (all values were numeric), MATLAB automatically removed the second column so that WT09 is a 10 × 1 matrix.

● EXAMPLE 19-16

Use `xlsread` to import the numeric and nonnumeric values in the 2009 sheet in ClemsonWeather.xlsx.

```
[D09,T09]=xlsread('ClemsonWeather.xlsx','2009');
```

If we look at the dimensions of D09 and T09, we note that D09 is a 365 × 15 matrix, but T09 is a 366 × 16 cell matrix. This is due to the fact that T09 contains the header row with all of the column labels, as well as the entire first column containing the corresponding dates. If we examine T09 in the Variable Editor, we see that the only values in the cell matrix are the first row and first column and all of the other values in the cell matrix are empty strings (''). In light of these phenomena, special care must be taken when attempting to write MATLAB code that attempts to relate cell matrices and numeric data matrices imported using xlsread.

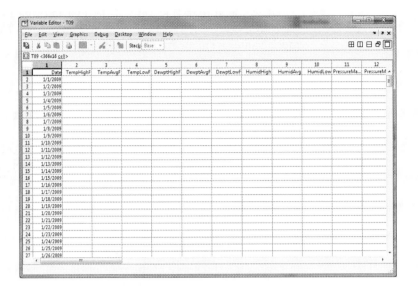

Writing Microsoft Excel Workbooks

MATLAB provides one built-in function, `xlswrite`, that enables writing Microsoft Excel Workbooks in MATLAB programs and functions.

xlswrite

Just like with `xlsread`, `xlswrite` accepts different syntax for different writing scenarios. For the sake of brevity, we will only cover the standard use of `xlswrite`, but refer to MATLAB's `help` and `doc` documentation for more information. `xlswrite` returns no output and only requires input arguments:

```
xlswrite(fileName,Matrix,Sheet,Cell)
```

where

- `fileName` is a string containing the file name you want to write to. If the Excel file does not exist, MATLAB will create a blank workbook with three blank worksheets ('Sheet1,' 'Sheet2,' and 'Sheet3') and display a warning "Warning: Added specified worksheet." This is common and is not considered an error message.
- `Matrix` is either a numeric or cell matrix containing the data you want to write to Excel file.
- `Sheet` is a string that contains the sheet name where you want to write your data.
- `Cell` is a string that designates the top-left corner of where MATLAB should start writing the data in `Matrix` in the Excel workbook. If `Cell` is not specified, MATLAB will default to writing with the top-corner set to cell A1.

● **EXAMPLE 19-17**   Use `xlsread` to import the numeric and nonnumeric values in the 2009 sheet in ClemsonWeather.xlsx, convert all of the average high temperatures to degrees Celsius, and write the output to a new Excel file, newWeather.xlsx.

```
[D09,T09]=xlsread('ClemsonWeather.xlsx','2009');
T09{1,2}='TempHighC';
D09(:,1)=(D09(:,1)-32)/9*5;
xlswrite('newWeather.xlsx',T09,'2009');
xlswrite('newWeather.xlsx',D09,'2009','B2');
```

Note that we are able to use the `xlswrite` function twice: first to write the headers and dates to the Excel file, followed by writing the matrix containing the converted high temperatures.

19.5 APPLICATION OF LOOPS: GUI

The purpose of this section is to familiarize you with creating graphical user interfaces (GUIs—often pronounced "gooey") in MATLAB and provide some insight into user-centered design to give you the tools you need to build a robust but clear interaction platform without the much "heavy" work on behalf of the program user. Designing user interfaces is different from designing algorithms to implement as programs or functions because there is an additional layer of complexity necessary to obscure some of the programming or language-specific requirements of your code. Graphical interface design is especially important when delivering a final "usable" product to a client or customer (the user) who may not necessarily have knowledge of the programming language you used to implement your solution.

Different graphical interface systems have different sets of best practices. For example, Microsoft Windows applications tend to have boxes in the upper right corner that allow users to click buttons to minimize, maximize, or close a program, whereas Mac OS X applications contain three circles in the upper left corner of windows that perform similar functions. Graphical interfaces also exist in many mobile operating systems like iOS or Android that allow handheld devices like iPhones and iPads to interact using push buttons or touch gestures to complete tasks. The remainder of this document focuses on WIMP interaction—"window, icon, menu, pointing device" like that found on programs run in Microsoft Windows (like MATLAB itself—see Figure 19-1) or Mac OS X, but similar controls and development ideologies exist for mobile applications.

Graphical User Interface Development Environment

In MATLAB, the tool used for designing a user interface is the Graphical User Interface Development Environment (GUIDE). GUIDE can be launched in the main MATLAB window by selecting File > New > GUI or by typing **guide** into the Command Window. The GUIDE Quick Start window (Figure 19-2) gives you the options to either create a new GUI or open an existing GUI created with GUIDE. By default, there are a few built-in templates that give examples of how interactions are built using MATLAB commands with the GUI, but in general, most GUIs in MATLAB are built by selecting Blank GUI, the default option, and clicking the OK button.

Figure 19-1 MATLAB itself is built with a GUI that allows you to click different toolbar buttons, menu items, launch new windows like the Editor Window, or type commands using the keyboard into the Command Window.

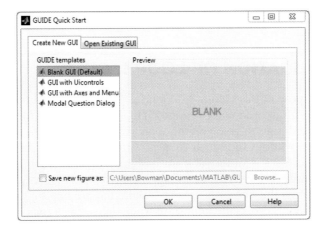

Figure 19-2 GUIDE Quick Start window.

The resulting window (Figure 19-3) is the graphical layout editor where you can arrange different user interaction controls that will allow, for example, the user of your program to click buttons, select options from check boxes or drop down menus, type values into text boxes, or an assortment of other options. This graphical layout is saved as a MATLAB Figure file with the file extension ".fig." For any GUI built in MATLAB

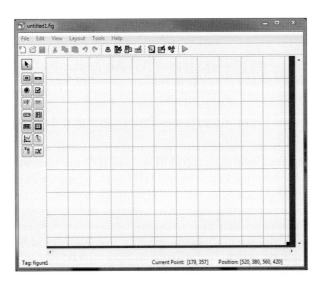

Figure 19-3 Graphical layout editor in GUIDE.

using GUIDE, there will be both an M-File and a Figure file—both named the same and following the standard set of MATLAB file naming rules.

The graphical layout editor contains a set of buttons on the left side of the window that allows you to insert different interface controls on your GUI. To the right of the button set, the canvas of the GUI allows you to lay out exactly where the different interface controls should appear on the window of your application. Finally, the status bar at the bottom of the window contains different information depending on which user interface control is selected on the canvas. By default, the "figure" control is selected with the tag name "figure1"—this will be the main area where other controls will be placed in each window. In addition, as you move your mouse over the canvas, the "Current Point" value in the status bar updates with the current XY coordinate of your mouse within the GUI. That position will be helpful in lining up interface controls on your GUI so that it looks clean and consistent. In Table 19-1, the different user interface controls are described along with typical examples of when these controls might be used in a program window.

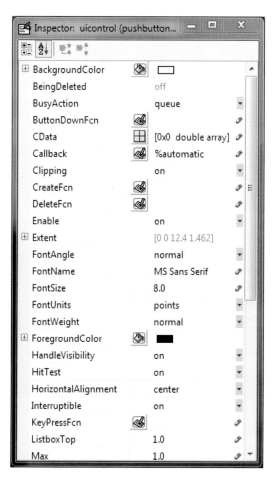

Figure 19-4 Property Inspector window for a push button control.

Interacting with User Interface Controls in GUIDE

Before laying out a program interface using GUIDE, it is critical to understand a few key ideas with user interface controls—specifically, how to change or extract values or properties of controls that are set automatically by your program. To demonstrate this, if you add a push button to the canvas in GUIDE, the tag (see the bottom left of the status bar in GUIDE when the push button is selected) of the push button is automatically set as "pushbutton1" by GUIDE. With this control, and all other controls, "tag" is a user interface control property that uniquely identifies that specific control on a GUI, much like how MATLAB would not allow you to have multiple variables with the same name. For each user interface control, there are a number of different properties that are automatically set by GUIDE, and understanding how those properties behave will allow for greater customization of the user interface.

The Property Inspector window allows the programmer to manually change the values in GUIDE of different user interface controls. In Figure 19-4, the Property Inspector window shows the different properties for a push button user interface control that can be changed by typing different values if the property box contains a pencil icon or selecting different values on properties that contain a pop-up menu of different property values. For example, in this window, several properties like `FontName`, `FontSize`, `FontAngle`, and `FontWeight` will allow for customization of the font on the push button. Setting or getting the values of GUI controls can also be done in code using the `get` and `set` functions.

Table 19-1 User interface controls in MATLAB

Control	Name/Description	Sample Uses	Not Used For . . .
[OK]	**Push Button** Launches a calculation/action or opens a new dialog window.	Submit/calculate action, quit program, close window, load/open data file, cancel operation.	Setting or unsetting a parameter—use a checkbox or toggle button instead.
[slider]	**Slider** Slides a value between a minimum and maximum value by a step size.	Increasing or decreasing a variable value, incrementing or iterating through an expression.	Setting a value outside of the minimum or maximum value of the range—use a text box instead (or in addition to the slider).
[radio]	**Radio Button** Allows the user to select a single value from a mutually exclusive list of options. These are generally used inside of button group controls.	Selecting gender (male or female), water phase (solid, liquid, or gas), or other options from small groups.	Groups that have more than 5 or 6 different options (unit systems, when all SI units are included)—use a pop-up menu instead.
[✓]	**Check Box** Allows the user to turn on (or off) a single value. In general, check boxes are presented as a group of check boxes.	Selecting font style: bold, italic, underline. The user may want to use 1, 2, or all 3 options, or select none.	Allowing conflicting selections— a person can't be both male and female! Use a radio button instead.
[EDIT]	**Edit Text** Allows the user to manually type data into the program. Also called "entry fields" in some GUI environments.	Typing text (first name, last name, phone number) or numbers (zip code, weight, height, tire pressure).	Typing an element from a closed set of options—you don't want the user to type "Male"—let them select it using a radio button.
[TXT]	**Static Text** Allows the program to display or update text in a control that is not editable by the user. Also called "labels" or "protected fields" in some GUI environments.	Display program header labels, text entry field labels, slider values, calculated values, small error messages, or other information of interest to the program user.	Editing or typing text—use an edit text control instead.
[popup]	**Pop-up Menu** Allows the user to select a single option from a list instead of typing values in an edit text box.	Selecting state of birth from a list of 50 states, selecting a car manufacturer from a list, selecting the car model from a list.	Pop-up menus don't allow the user to select more than one option (e.g. Jeep and Dodge)— use a listbox control instead.
[listbox]	**Listbox** Allows the user to select multiple options from a list instead of typing multiple values into multiple text boxes.	Select multiple beam lengths, pipe sizes, data sets for a plot, students in an engineering class, or other scenarios where you want to choose more than one option.	If you want to restrict the selection to only one option, use a pop-up menu instead.

Table 19-1 (Continued)

	Toggle Button Allows the user to turn on (or off) a value. Toggle buttons can be presented in groups, but can also stand alone. For exclusive selection of toggle buttons, use the toggle controls inside a button group.	Turning on or off debug output. Selecting font style: bold, italic, underline. The user may want to use 1, 2, or all 3 options, or select none. Can be used like a check box.	Can only be used in binary states (on or off)—for more than 2 states, use a control like a radio button or pop-up menu.
	Table Allows the program to display tabular data in a visually attractive and interactive (e.g. scrollable) control.	Displaying a list of spring stiffnesses for multiple springs. Displaying a matrix of values calculated by a program.	Not useful for displaying single text strings or values—use a static text control instead.
	Axes Allows the program to display a plot or other data that can be displayed on axes in MATLAB (like an image) on the program window.	Displaying a plot of data, displaying an image of a card from a deck of cards, displaying the frequency response of an audio signal.	For multiple graphs on a GUI, you might want to consider using a single axis control and a subplot to allow MATLAB to automatically handle graph alignment.
	Panel Allows the programmer to group similar user interface controls to make the window visually easier to navigate.	Grouping all of the input controls in an "input" panel, grouping all of the output controls in an "output" panel.	Does not enable exclusivity within controls like radio or toggle buttons—use a button group if you require exclusive selection of options.
	Button Group Allows the programmer to group radio and/or toggle buttons to allow the user to make a mutually exclusive (single) selection from a group of multiple controls.	Selecting gender (male or female), water phase (solid, liquid, or gas), or other options from small groups.	Groups that have more than 5 or 6 different options (unit systems, when all SI units are included)—use a pop-up menu instead.
	ActiveX Control Allows the programmer to embed a control from a different program installed on the computer.	Embedding an Adobe PDF document in a GUI, embedding a QuickTime movie player.	Cross-platform programs— many ActiveX controls are operating system specific, so a GUI built-in Windows may not run in Mac OS X or Linux versions of MATLAB.

Interacting With User Interface Controls Programmatically: `get` *and* `set`

Before discussing the use of the `get` and `set` function to change or store property values in user interface controls, recall the use of the built-in graphing tool `fplot`, which we use for displaying functions in a MATLAB Figure window. In this code segment, we are plotting the `sin` function from $-\pi$ to π:

```
fplot('sin(x)',[-pi,pi]);
xlabel('X-Axis Text','FontSize',12,'FontWeight','bold');
ylabel('Y-Axis Text','FontSize',12,'FontWeight','bold');
title('Title Text','FontSize',12,'FontWeight','bold',
'Color','blue');
grid
```

In this code segment, we are using the extra arguments in the `xlabel`, `ylabel`, and `title` functions to change properties like `FontSize`, `FontWeight`, and `Color` by listing the property names and values as extra arguments in the function call. All plotting functions like `fplot` or `plot` actually create GUIs that are built in to MATLAB that have specialized functions like `xlabel`, `ylabel`, and `title` that are simply just short-cut tools that manually set the text values of the title and axis labels on plots. Since these plot windows are GUIs, the title and axis labels are actually just user interface controls like the static text labels described in Table 19-1, so these controls can actually be manually set or referenced creating a handle to the user interface control. A handle is actually just a reference that knows which control we want to modify that will allow us to access the different properties of the user interface.

To set user interface control properties using the `set` function, we need to pass in three arguments:

```
set(h,'PropertyName','PropertyValue');
```

where h is the variable that contains the handle reference, `'PropertyName'` is the name of the property (like `'FontWeight'` in the code above), and `'PropertyValue'` is the value to set (like `'bold'` in the code segment above).

The equivalent code to the code segment above written using `set` function calls:

```
fplot('sin(x)',[-pi,pi]);
xl=xlabel('X-Axis Text');
yl=ylabel('Y-Axis Text');
t=title('Title Text');
set(xl,'FontSize',12);
set(yl,'FontSize',12);
set(xl,'FontWeight','bold');
set(yl,'FontWeight','bold');
set(t,'FontSize',12);
set(t,'FontWeight','bold');
set(t,'Color','blue');
grid
```

Note that in the block of MATLAB code above, the handles to the user interface controls are created by capturing the handle references in different variables (`xl`, `yl`, and `t`) as outputs of the `xlabel`, `ylabel`, and `title` functions. Plotting functions like `xlabel` and `ylabel` were designed to be able to modify property values using extra function input arguments to reduce the number of lines of code when writing common code like creating graphs, so that is the preferred method for setting properties with those functions. The use of the set function in this fashion is only presented here as an example to help explain how this function works later when presented in the context of GUIs.

The get function works similarly to the set function by referencing user interface controls using a handle:

```
V=get(h,'PropertyName')
```

where h is the handle reference to the user interface control, 'PropertyName' is the property in the control, and V is a variable that will contain the property value for 'PropertyName'.

Consider the following segment of MATLAB code that uses the fplot function to again plot sin(x) from $-\pi$ to π and displays a title on the graph.

```
fplot('sin(x)',[-pi,pi]);
t=title('Title Text');
fs=get(t,'FontSize');
fw=get(t,'FontWeight');
fc=get(t,'Color');
st=get(t,'String');
```

In this code segment, we use the get function to read the default property values for the FontSize, FontWeight, Color, and String property values into the variables fs, fw, fc, and st. Note that fw and st are both text variables, fs is a number, and fc is a vector that contains the amount of red, green, and blue in the text font face color.

● EXAMPLE 19-18

In this example, the canvas (Figure 19-5) contains a simple GUI with a few static and edit text boxes and a push button that will allow us to multiply the numbers typed into the boxes on the GUI and display them in the interface.

To begin, the controls were laid out on the GUI canvas as shown below. A list of the properties changes for the different user interface controls are documented below as well.

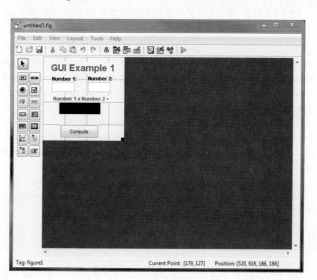

Figure 19-5 GUI for Example 19-18.

Tag: labelGUIHeader
   (static text)
String: GUI
   Example 1
ForegroundColor: blue
FontSize: 16
FontWeight: bold

Tag: labelNumber1
   (static text)
FontWeight: bold
String: Number 1:

Tag: labelNumber2
   (static text)
FontWeight: bold
String: Number 2:

Tag: label
   Multiplication
   (static text)
FontWeight: bold
String: Number 1 ✕
   Number 2
ForegroundColor: red

Tag: num2 (edit text)
String: (blank)

Tag: labelResult (static text)
FontSize: 14
FontWeight: bold
ForegroundColor: white
BackgroundColor: black
String: (blank)

Tag: num1 (edit
text)
String: (blank)

Tag: button
   Compute (push
   button)
String: Compute

In addition to the changes above, the canvas was resized so that the GUI window does not contain a bunch of unnecessary unutilized space. After the GUI is properly arranged and renamed on the canvas, click the "Save and Run" button on GUIDE and the dialog shown in ▷.

Figure 19-6 will pop-up on the screen. Click Yes on the question dialog and type the name for the GUI Figure .fig file: GUIExample1.

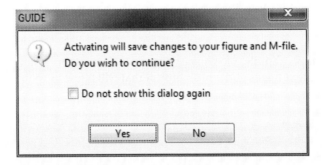

Figure 19-6 GUIDE save changes dialog.

Next, two windows will pop up—the MATLAB Figure (Figure 19-7) generated by GUIDE, as well as an M-File (Figure 19-8) containing over 100 lines of MATLAB code and comments necessary to interact with the GUI. All of the functions within the M-File are actually subfunctions within the M-File, which means they can be called anywhere within this M-File, but you cannot call any of these functions from a different M-File. Subfunctions can only be written in M-Files that are implemented as a function, so the main GUI code is actually written as a function. Note that the M-File is named GUIExample1.m, which is the same file name as the MATLAB Figure we typed earlier. It is important that you do not rename this file since the Figure and M-File contain special references to "GUIExample1" throughout the code.

Now that the MATLAB Figure and M-File for the GUI exist, edit the M-File to insert some of the back-end code to implement the multiplication. To do this, we need to modify the callback function for the push button. Callback functions are codes that run as soon as some interaction occurs on the user interface. Examples of interaction might include clicking, double clicking, selecting an option from a menu, clicking a check box, or even events as simple as moving the mouse over a user interface control. To edit the callback function (as seen in Figure 19-9) for the push button, right click the Compute button in GUIDE and select View Callbacks > Callback from the context menu.

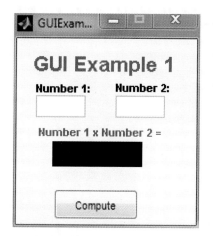

Figure 19-7 MATLAB Figure generated by GUIDE in Example 19-18.

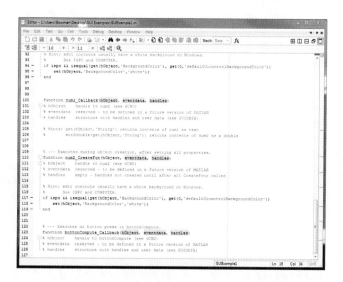

Figure 19-8 MATLAB M-File generated by GUIDE in Example 19-18.

At this point, the M-File Editor should pop up and display the automatically generated code that handles the event that fires when the user clicks on the Compute button in your GUI:

```
% --- Executes on button press in buttonCompute.
function buttonCompute_Callback(hObject, eventdata, handles)
% hObject handle to buttonCompute (see GCBO)
% eventdata reserved - to be defined in a future version of
MATLAB
% handles structure with handles and user data (see GUIDATA)
```

From here, we need to use the get and set functions to pull down the values in the edit text boxes, multiply the values together, and display the result on the GUI. To do this, the function header generated by MATLAB contains a function input called handles that is a structure variable that contains handle references to all of the user interface controls in the GUI. We can access the different user interface controls by typing handles.tagName, where tagName is the actual tag name embedded in different control properties. In this example, if we wanted to access the contents of the num1 edit text box, we could type the following:

```
Num1Text=get(handles.num1,'String');
```

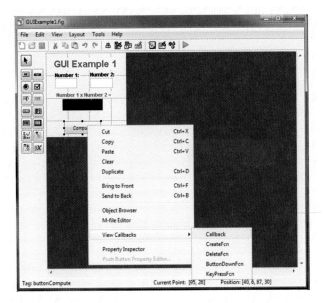

Figure 19-9 Accessing the push button callback function.

When we access some of the properties from user interface controls, be aware that some of the variable types might not be what you expect. In this case, the Num1Text *variable is actually a string of text instead of a number. Before we do our computation, we must take the strings we* get *from the two edit text boxes and convert them into numbers before we can multiply them. Likewise, the* String *property in the static text field that will contain the computed result expects the property to be set as a string, so we must convert the computed value into a string before we use the* set *function with the* String *property. The final code in the* buttonCompute *callback function should look something like:*

```
% --- Executes on button press in buttonCompute.
function buttonCompute_Callback(hObject, eventdata, handles)
% hObject handle to buttonCompute (see GCBO)
% eventdata reserved - to be defined in a future version of
MATLAB
% handles structure with handles and user data (see GUIDATA)
Num1Text=get(handles.num1,'String');
Num2Text=get(handles.num2,'String');
M=str2double(Num1Text) * str2double(Num2Text);
set(handles.labelResult,'String',num2str(M,'%.0f'));
```

Callback and Related Functions for Different Controls

Like we saw in Example 19-18, many of the different user interface controls use callback functions or similar functions that will run when the user clicks on a control or somehow interacts with your GUI. These callback functions will only run the code within the function when the control is activated and will run again the next time the control is clicked. This idea is referred to as event-driven programming, where code will only execute when the user interacts with a control in an interface. User interaction with a control is referred to as an event, and when a callback function is executed, it is often said that the event for that control has fired.

In Table 19-2, the list of the important functions on different user interface controls is given with important properties that contain information about the text, values, or other data that might be necessary to use in the callback function. Note that there are other callback functions that may not be listed in the table that will fire on certain events like pressing a certain key on the keyboard, but those are not explicitly mentioned because they behave similarly to the standard callback functions.

Table 19-2 User interface controls

Control	Callback	Important Properties	Fires When . . .
Push Button	Callback	**'String'** – text on the button	Button is pressed
Slider	Callback	**'Value'** – current value of slider, number **'Min'** – minimum slider value, number **'Max'** – maximum slider value, number	Slider value changes
Radio Button	Callback	**'Value'** – current value of radio, number **'Min'** – minimum radio value (not selected), number **'Max'** – maximum radio value (is selected), number	Radio button is pressed
Check Box	Callback	**'Value'** – current value of radio, number **'Min'** – minimum radio value (not selected), number **'Max'** – maximum radio value (is selected), number	Check box is clicked (checked or unchecked)
Edit Text	Callback	**'String'** – text in the edit text box	After editing text, user clicks outside of the textbox
Static Text	N/A	**'String'** – text in the static text box	N/A
Pop-up Menu	Callback	**'String'** – the pop-up menu contents, as a cell array of text values **'Value'** – the index into the cell array of the selected item in the pop-up menu	Selection in the pop-up menu is changed
Listbox	Callback	**'String'** – the listbox contents, as a cell array of text values **'Value'** – the index or indices into the cell array of the selected item in the pop-up menu, either a number or a vector depending on the number of selected items	Selection in the listbox is changed

(Continued)

Table 19-2 (Continued)

Control	Callback	Important Properties	Fires When . . .
Toggle Button	`Callback`	**'Value'** – current value of the toggle button, number **'Min'** – minimum toggle button value (not pressed), number **'Max'** – maximum toggle button value (pressed), number	The user toggles the toggle button
Table	`N/A`	**'Data'** – the numerical values in the table **'ColumnName'** – the names of the columns, as a cell array **'RowName'** – the names of the rows, as a cell array	N/A
Axes	`N/A`	The standard plot functions will work with the axes control. If the axes handle name is axes1, the plot commands need to be preceded by `axes(handles.axes1)`	N/A
Panel	`N/A`	The panel control is generally used for layout purposes on the GUI, but does contain some functions that will fire if the GUI is resized	N/A
Button Group	`Selection-` `ChangeFcn`	The tag name of the selected button can be extracted by calling `get(eventdata.NewValue,'Tag')`	The user changes the selected object in the button group

Built-In Dialogs and Interaction Interfaces

There are a number of other built-in dialog and interaction interfaces that can complement your GUI by allowing all interaction to appear in the user interface rather than occasionally displaying or typing values in the Command Window. These functions include:

Built-In Dialog Boxes

- `dialog`: Displays text in a clickable textbox as a pop-up dialog.
- `msgbox`: Displays text in a clickable textbox as a pop-up dialog with customizable icons and dialog title.
- `errordlg`: Displays text in a clickable textbox as a pop-up dialog as an error message.
- `helpdlg`: Displays text in a clickable textbox as a pop up dialog as a help message.
- `inputdlg`: Prompts the user to type in a value into an input dialog.
- `listdlg`: Prompts the user to select an option from a listbox in a pop-up dialog.
- `questdlg`: Prompts the user to answer a question (yes/no) in a question dialog.

- `warndlg`: Displays text in a clickable textbox as a pop-up dialog as a warning message.
- `printdlg`: Displays the print dialog to allow the user to print your GUI.

Built-In Interaction Dialogs

- `uigetdir`: Allows the user to select a directory that contains files.
- `uigetfile`: Allows the user to select a file (e.g., select an Excel file to open with `xlsread`).
- `uiputfile`: Allows the user to save a file (e.g., select where and what to name an Excel file you are writing with `xlswrite`).
- `uisetcolor`: Allows the user to pick a color from a color list.
- `uisetfont`: Allows the user to change font properties.

For more information on any of these functions, refer to the MATLAB `help` and `doc` pages for each function.

For more examples of GUI, look at the following .zip files available in the online materials:

- SliderAndPlotExample
- UIGetFileAndPlotExample
- UITableExample
- CardGameDemo

In-Class Activities

ICA 19-1

Write a program containing a `for` loop to compute the factorial of any positive integer. Include an `if . . . else . . .` statement that displays the phrase, "Factorial of #A is #B" (where #A is the number for which the factorial is being computed, and #B is the computed factorial) if the user entered value is positive.

(a) If the user enters a negative value, the program should tell the user of their error and terminate.

(b) If the user enters a negative value, the program should tell the user of their error and prompt the user to enter another value.

(c) After the program has displayed the output phrase for the entered value, the program should ask the user if they wish to enter another value. If yes, the program should repeat. If not, the program should terminate.

ICA 19-2

A materials engineer is testing the uniformity of a large sheet of a new high-strength transparent material by measuring the refractive index, η (a measure of how fast light travels in the material), in every square centimeter of the sheet. The measured values are placed in a 2-D array named `Refract`, each element of which corresponds to the locations on the sheet, for analysis.

Write a program that will generate two outputs.

- A variable named `TotalHigh` that equals the total number of measurements with a refractive index greater than that of common window glass ($\eta = 1.5$).
- An array named `GridLoc` containing two columns and `TotalHigh` rows. Each row in `GridLoc` will contain the row and column indices of the locations with $\eta > 1.5$.

ICA 19-3

A symmetric matrix is a square matrix, A, that is equivalent to its transpose, $A = A^T$. The entities of a symmetric matrix are symmetric with respect to the main diagonal (top left to bottom right). An example of a symmetric matrix is shown below. Write a MATLAB function called `IsSymmetric` that determines if the input matrix A is a symmetric matrix.

(a) If the user enters a matrix that is not able to be transposed, the program should inform the user of their error and terminate.

(b) If the user enters a matrix that is not able to be transposed, the program should tell the user of their error and prompt the user to enter another matrix.

(c) After the program has displayed the output phrase stating if the matrix is symmetric, the program should ask the user if they wish to enter another matrix. If yes, the program should repeat. If not, the program should terminate.

$$\begin{bmatrix} 1 & 2 & 3 \\ 2 & 4 & -8 \\ 3 & -8 & 9 \end{bmatrix}$$

ICA 19-4

Write a function for sorting numerical values in a vector of data. It should receive an array as an argument and return an array with the same values sorted without using any MATLAB built-in functions. It should ask the user if the data is to be sorted in ascending or descending order.

ICA 19-5

Write a function that receives as arguments two matrices A and B (in that order) and returns the product C = AB. The program must use `for` loops and matrix indices to compute each term in C, and cannot use MATLAB's ability to multiply either matrices or vectors. If the matrices A and B cannot be multiplied because their inner matrix dimensions do not agree, you must give an error message. The function should include:

- Determination of argument matrix dimensions using the length command; this determines if there is an error and establishes the loop limits.
- Error reported if inner matrix dimensions do not agree.
- `for` loops addressing each term in C.
- Correct vector product calculations of each term in C.

ICA 19-6

It is sometimes necessary to "smooth" a data set in order to simplify certain analyses. Write a program that asks the user to enter a vector of any length and returns a smoothed vector to the command window using formatted output. The program should determine the smooth vector by:

- Each array term computed as the average of the five terms surrounding the original term. For example, for the average of element 5 of a 10 element vector, this would include the average of elements 3, 4, 5, 6 and 7.
- First and last terms computed as special cases, counting the original term three times in the average (and the neighboring points once each). For example, the first points would be determine by the sum of three times the first element, the second element, and the third element, divided by five.
- Second and next-to-last terms computed as special cases by counting the original term and the first or last term 1.5 times each in the average (and neighboring points once each). For example, the second point would be determined by the sum of 1.5 times the first element, 1.5 times the second element, the third element and the fourth element, divided by five.

As an example, if your smoothing function is called on the vector [1 2 3 4 5 6 7 8 9 10], you should get the resulting vector

[1.6 2.3 3.0 4.0 5.0 6.0 7.0 8.0 8.7 9.4]

(a) If the user enters a vector shorter than 5 elements, the program should tell the user of their error and terminate.

(b) If the user enters a vector shorter than 5 elements, the program should tell the user of their error and prompt the user to enter another value.

(c) After the program has displayed the output phrase for the entered value, the program should ask the user if they wish to enter another vector. If yes, the program should repeat. If not, the program should terminate.

ICA 19-7

Assume you are given two matrices M1 and M2, both with the same dimensions. Using nested `for` loops, write a function named `MatComp` that will return a matrix D with the same dimensions as M1 and M2 with each element containing either 1, 0, or −1 based on the contents of M1 and M2 as described below.

For all r and c within the matrix size:

- If M1(r,c) = M2(r,c), then D(r,c) = 0.
- If M1(r,c) − M2(r,c) > 0, then D(r,c) = 1.
- If M1(r,c) − M2(r,c) < 0, then D(r,c) = −1.
- If the matrices passed to the function are not the same size, an error message should appear on the screen and an empty matrix be returned for D (D = [ ] will establish D as an empty matrix).

Example: M1 = [2 -5 9; -4 0 3] M2 = [1 -7 12; -3 9 3] D = [1 1 -1; -1 -1 0]

ICA 19-8

Write a program for a simple guessing game. The program should first prompt the user for a number between 1 and 100. The program should then generate a single random number `TargetNum` between 0 and N. This can be done with `TargetNum = N*rand(1)`.

Next, the program should prompt the user to guess the number. After each guess, the program should inform the user whether the guess is too high or too low. When the user guesses the value of `TargetNum` to be within 1% (`TargetNum ±0.01*N`), the program should print the following, where XXXX represents the appropriate values. You will need to consider how many significant digits you should print.

```
TargetNum = XXXX. Your guess of XXXX is within XXXX%.
This required XXXX guesses.
Would you like to play again?
```

If the user responds with "N," the program should terminate.

If the user responds with "Y," the program should prompt the user for a new value of N, generate a new random number, etc.

ICA 19-9

The question below refers to the `ClemsonWeather.xlsx` file provided online.

(a) Write a MATLAB program to read in the Excel data. Use formatted `fprintf` statements to display the following table in the command window, filling in the missing information

Year	Ave. High Temp [deg C]	Ave. Low Temp [deg C]	Total Precip. [in]
2008			
2009			
2010			

(b) Create a program that will create a new Excel file that only contains the weather data for days when the average temperature was below 32 degrees Fahrenheit and the total precipitation for the day was greater than 0 inches.

(c) A heat wave occurs when the daily maximum temperature during five consecutive days or more exceeds the average maximum temperature by 5 Celsius degrees (9 Fahrenheit degrees) or more.

For this problem, we will assume a heat wave begins when five consecutive days exhibit an average daily temperature greater than or equal to 80 degrees Fahrenheit. Create a new Excel file that contains a list of "heat wave" days. For example, if August 3, 4, 5, 6, 7, and 8 all have higher temperatures exceeding or equal to 80 degrees Fahrenheit, then the weather data for August 7th and August 8th would go on the list.

ICA 19-10

Note that some of the commands may crash—the purpose of this assignment is to get familiar with the differences between cell arrays and regular vectors in MATLAB.

Assume the following command is typed into MATLAB:

```
StudentInfo={'John','Doe','ENGR 141','099',[100,95,89,70]};
```

If the following commands are typed into MATLAB, what will appear on the screen as the output or error message?

(a) `fprintf('My name is %s\n',StudentInfo(1))`
(b) `fprintf('My name is %s\n',StudentInfo{1})`
(c) `fprintf('My grades are %d\n', StudentInfo{5})`
(d) `fprintf('My grades are %d\n', StudentInfo{5}*5)`
(e) `fprintf('My first ICA grade is %d\n', StudentInfo{5}(1))`
(f) `A=length(StudentInfo)`
(g) `A=length(StudentInfo(5))`
(h) `A=length(StudentInfo{5})`
(i) `C=min(StudentInfo{5})`
(j) `Student=StudentInfo{1:2}`
(k) `Student=StudentInfo(1:2)`
(l) `StudentInfo(2,1)=5`
(m) `StudentInfo{2,1}=5`

ICA 19-11

Assume the following command is typed into MATLAB:

```
Teams={{'Rays',96},{'Reds',91},{'Phillies',97}};
```

If the following commands are typed into MATLAB, what will appear on the screen as the output or error message?

(a) `LT=length(Teams)`
(b) `LT2=length(Teams{1})`
(c) `T=Teams{1}{1}`
(d) `W=Teams{1}{2}`
(e) Fill in the blanks in the following code needed to produce the output shown.

```
fprintf('2010 Baseball Season\n\n');
fprintf('Team\tWins\n\n');
for i=1:1:length(_____)
  fprintf('%s\t%.0f\n', _____, _____)
end
```

1. Write a function for finding the maximum value in a vector of data. It should receive an array as an argument and return the maximum of the values without using MAX or other MATLAB built-in functions.

2. A matrix named mach contains three columns of data concerning the energy output of several machines. The first column contains an ID code for a specific machine, the second column contains the total amount of energy produced by that machine in calories, and the third column contains the amount of time required by that machine to produce the energy listed in column 2 in hours.

   Write a function named MPower to accept as input the matrix mach and return a new matrix named P containing two columns and the same number of rows as mach. The first column should contain the machine ID codes, and the second column should contain the average power in units of watts generated by each machine.

3. As early as 650 BC, mathematicians had been composing magic squares, a sequence of n numbers arranged in a square such that all rows, columns, and diagonals sum to the same constant. Used in China, India, and Arab countries for centuries, artist Albrecht Dürer's engraving Melencolia I (year: 1514) is considered the first time a magic square appears in European art. Each row, column, and diagonal of Dürer's magic square sums to 34. In addition, each quadrant, the center four squares, and the corner squares all sum to 34. An example of a "magic square" is displayed below.

16	3	2	13
5	10	11	8
9	6	7	12
4	15	14	1

Write a program to prove a series of numbers is indeed a 4×4 magic square. Your program should:

- Ask the user to enter their proposed magic square in a single input statement (e.g., [1 2 3 4; 5 6 7 8; 9 10 11 12; 13 14 15 16]—note that this is a 4×4 matrix, but NOT a magic square)

- Check for an arrangement of 4×4

  (a) If the matrix is not a 4×4, warn the user and exit the program
  (b) If the matrix is not a 4×4, warn the user and ask the user to try again until they type a 4×4 matrix

- Check that all values are positive; ** for-loop or nested for-loop required

  (a) If one or more of the values in the matrix are negative, warn the user and exit the program
  (b) If one or more of the values in the matrix are negative, warn the user and ask the user to try again until they type a magic square without negative values

- Determine the type of magic square by determining if it meets the following requirements:

  1. If each row and column sums to the same value, the magic square is classified as "semimagic;"
  2. If, in addition to criterion 1, each diagonal sums to the same value, the magic square is classified as "normal;"
  3. If, in addition to #1 and #2, the largest value in the magic square is equal to 16, the magic square is classified as "perfect;"

4. If your code determines that the matrix is properly formatted (4 × 4, all positive), but isn't a magic square, your code should display a warning that the matrix isn't a magic square and should run again until the user types a properly formatted, magic square that at least meets criteria #1.

Once the user types a properly formatted magic square, the program should output the magic constant, or sum of the elements in each row and column, and the magic square classification. Format your magic square classification similar to the format shown below. You may choose to format your table differently, but each classification should contain a "yes" or "no" next to each magic square category.

The magic constant for your magic square is 24.
The classification for your magic square:

Semi-magic	Normal	Perfect
yes	yes	yes

A few test cases for you to consider:

- Albrecht Dürer magic square: [16, 3, 2, 13; 5, 10, 11, 8; 9, 6, 7, 12; 4, 15, 14, 1];
- Chautisa Yantra magic square: [7, 12, 1, 14; 2, 13, 8, 11; 16, 3, 10, 5; 9, 6, 15, 4];
- Sangrada Familia church, Barcelona magic square: [1, 14, 14, 4; 11, 7, 6, 9; 8, 10, 10, 5; 13, 2, 3, 15];
- Random magic square: [80, 15, 10, 65; 25, 50, 55, 40; 45, 30, 35, 60; 20, 75, 70, 5];
- Steve Wozniak's magic square: [8, 11, 22, 1; 21, 2, 7, 12; 3, 24, 9, 6; 10, 5, 4, 23].

4. A mummy picks up a calculator and starts adding odd whole numbers together, in order: $1 + 3 + 5 + \ldots$ etc. What will be the last number the mummy will add that will make the sum on his calculator greater than 10,000? Your task is to write the MATLAB code necessary to solve this problem for the mummy or he will eat your brain.

5. Write a function called `Balloon` that will accept a single variable named S. The function should replace S with the square of S and repeat this process until S is either greater than 10^{15} or less than 10^{-15}. The function should return a value Q containing the number of times S was squared during this process.

Examples:
- If $S = 100$, $Q = 3$ (S equals 10^{16} after the third square).
- If $S = 0.1$, $Q = 4$ (S equals 10^{-16} after the fourth square).
- If $S = 3$, $Q = 5$ (S equals 1.853×10^{15} after the fifth square).

6. You have written three functions for three different games. The names of the games (and the functions that implement them) are Dunko, Bouncer, and Munchies. Each function will accept an integer between one and three indicating the level of difficulty (1 = easy, 3 = hard), and when play is complete will return a text string, either "won" or "lost," to the program that executed the function.

Write a program that will use a menu to ask the user if he or she wants to play a game. If not, the program should terminate. If so, the program should generate another menu to allow the user to select one of the three games by name.

After selecting a game, the program should display another menu asking the user for the desired level of difficulty (Easy, Moderate, or Hard), and the game should begin by calling the appropriate function.

When the user has finished playing the selected game, a message should be displayed indicating whether the user won or lost the game. A menu should then be generated asking if the user wishes to repeat the game just played. If so, the difficulty level menu should be displayed again and the game repeated.

If the user does not wish to repeat the same game, the program should display a menu asking if the user wants to play another game. If so, the game selection menu should be generated again, followed by level selection, etc. If the user does not wish to play another game, the program should display a message saying "Thanks for playing" and terminate.

7. You are to program part of the interface for a simple ATM. When the user inserts their card and types the correct PIN (you do NOT have to write this part of the program), the system will place the users' checking account balance in a variable `CBal` and the users' savings account balance in `SBal`.

   You are to write a function that will accept `SBal` and `CBal` as inputs and return two variables `NewCBal` and `NewSBal` containing the checking and savings balances after the transaction is completed. The function should do the following:

   ■ Display a menu titled "Main Menu" with the following three options.

   • Get cash
   • Get balance
   • Quit

   ■ If "Get cash" is selected, another menu titled "Withdrawal amount" with the following four items is displayed:

   • $20
   • $60
   • $100
   • $200

   ■ After selecting an amount, a menu titled "From which account?" should be displayed showing the following two options:

   • Checking
   • Savings

   ■ At this point, the program should verify that the selected account contains sufficient funds for the requested withdrawal.
   ■ If not, a message should be displayed that says: "Sorry. You do not have sufficient funds in your SSSS account to withdraw $XX" where SSSS is either Savings or Checking and $XX is the selected withdrawal amount.
   ■ If funds are available, the program should call a function named `Disp20(x)`, where x is the number of $20 bills to dispense. (See note about `Disp20` below.) After that, the withdrawal amount should be subtracted from the appropriate balance.
   ■ After processing the "Get cash" request, the program should return to the main menu.

   Note About `Disp20(x)`: The purpose of this function is to dispense the requested number of $20 bills—that is, to shove x bills out of the slot in the ATM machine. This does not really exist, since we do not have an ATM machine to work with. Thus, if you try to run your code, you will get an error ("Undefined function . . .").

   In order to test your program, add the following function to your current path:

   ```
   function[]=Disp20(x)
   fprintf('%.0f $20 bills were dispensed.\n',x)
   ```

   where x is the number of bills to be dispensed.

   This allows you to know if the program reached the proper location in the code. It is fairly common in software development to use a "dummy" function in the place of a real one when the device to be controlled has not been completed or is not available in order to help verify whether the software is reaching the correct places in the code for various situations.

   ■ If Get Balance is selected, another menu titled "Which account?" should appear with the two choices:

   • Checking
   • Savings

   and the program should then display "Your SSSS balance is $bb.bb.," where SSSS is either Savings or Checking and $bb.bb is the balance in the selected account.

- After processing the "Get balance" request, the program should return to the main menu.
- If Quit is selected, the function should return to the calling program with the updated balances in `NewCBal` and `NewSBal`. Note that the new balances will be equal to the original balances if no money was drawn from an account, but they must still be returned in the two new balance variables.

8. For this assignment, you will need the Cincinnati Reds player data file CincinnatiBaseball2010 .mat. The data in this file is a capture of the 2010 season on Baseball-Reference.com (http://www.baseball-reference.com/teams/CIN/2010.shtml).

   In the MATLAB workspace provided, there are three variables of interest:

- `PlayerData`, a 37 × 9 matrix which contains:

  - Column 1: The age of the baseball player
  - Column 2: Games played or pitched
  - Column 3: At bats
  - Column 4: Number of runs scored/allowed
  - Column 5: Singles hit/allowed
  - Column 6: Doubles hit/allowed
  - Column 7: Triples hit/allowed
  - Column 8: Home runs
  - Column 9: Runs batted in

  Note that each row in `PlayerData` represents a different baseball player.

- `PlayerNames`, a 37 × 1 "cell array" that contains the names of each player in the `PlayerData` matrix.
- `PlayerPositions`, a 37 × 1 "cell array" that contains the position abbreviation of each player in the `PlayerData` matrix.

  (a) Create a function that will accept a single input, a cell array of player positions for an entire team—with our data set, this would be a variable like `PlayerPositions`. Your function must create a new cell array that contains only the unique positions, sorted alphabetically.

  (b) Create a program that will display the average age, at bats, home runs, and runs batted in for user-selected positions. In order to allow the user to select the positions, you must use the menu function to allow the user to click positions to include in the analysis, as well as a "Done" button to allow the code to continue to display the output.

  Sample Output: Assuming the user selects all positions

Position Stats for 2010 Cincinnati Reds

Pos	Ave(Age)	Ave(AB)	Ave(HR)	Ave(RBI)
1B	25	288	19	58
2B	29	626	18	59
3B	31	242	8	39
C	32	197	5	32
CF	25	514	22	77
IF	27	3	1	4
LF	29	338	11	52
MI	24	38	1	2
OF	27	123	4	11
P	27	21	0	1
RF	23	509	25	70
SS	31	347	5	34
UT	36	23	2	2

Sample Output: Assuming user presses "3B," "OF," "Done"

Position Stats for 2010 Cincinnati Reds

Pos	Ave(Age)	Ave(AB)	Ave(HR)	Ave(RBI)
3B	31	242	8	39
OF	27	123	4	11

(c) MLB.com is consistently ranked one of the top 500 websites in the United States, bringing in millions of visitors a year. However, recent surveys have shown that MLB.com is one of the slowest top 500 websites and is looking for new solutions to speed up their website. After hearing about the baseball statistics programs we have developed in our class, MLB.com has contracted us to help them create a MATLAB program to help them improve the user experience of their website. Recreate the analysis in part **(b)**, except this time export the output to a Microsoft Excel Workbook.

Sample Output: Assuming the user selects all positions

	A	B	C	D	E	F
1	Position Stats for 2010 Cincinnati Reds					
2	Pos	Ave(Age)	Ave(AB)	Ave(HR)	Ave(RBI)	
3	1B	25	288	19	58	
4	2B	29	626	18	59	
5	3B	31	242	8	39	
6	C	32	197	5	32	
7	CF	25	514	22	77	
8	IF	27	3	1	4	
9	LF	29	338	11	52	
10	MI	24	38	1	2	
11	OF	27	123	4	11	
12	P	27	21	0	1	
13	RF	23	509	25	70	
14	SS	31	347	5	34	
15	UT	36	23	2	2	
16						

Sample Output: Assuming user presses "3B," "OF," "Done"

	A	B	C	D	E	F
1	Position Stats for 2010 Cincinnati Reds					
2	Pos	Ave(Age)	Ave(AB)	Ave(HR)	Ave(RBI)	
3	3B	31	242	8	39	
4	OF	27	123	4	11	
5						

9. Download the weekly retail gasoline and diesel prices Excel workbook associated with this review question and place it in your main MATLAB Current Directory. This data set is based on the data set available from the U.S. Energy Information Administration: http://www.eia.gov/dnav/pet/pet_pri_gnd_dcus_nus_w.htm.

   (a) Calculate the average, minimum, and maximum retail fuel prices for each of the different types of fuel (regular, midgrade, premium, diesel) over the duration of the entire sample set. Your code should work in such a way that if the original Excel file were modified to include more weeks (rows), you would not need to change your MATLAB code. Your code should appear similar to the format below, where the blanks are replaced by the actual calculated values:

   Average Weekly Retail Gasoline and Diesel Prices

	Regular	Midgrade	Premium	Diesel
Min:				
Max:				
Average:				

   Your code should calculate the values shown in the output—you should not hard code the values in the output. Each value you display should appear with two decimal values. You may use any built-in MATLAB function, including functions that find the minimum, maximum, or average values.

   (b) We have decided that we want to modify our previous analysis in part **(a)** to export the computed max, min, and average values to a new Microsoft Excel workbook. The data itself should be exported to a sheet named "Fuel Price Analysis." Your data should appear similar to the worksheet below. Much like part **(a)**, your code should be written in such a way that if the original Excel file were modified to include more weeks (rows), you would not need to change your MATLAB code.

	A	B	C	D	E	F
1	Average Weekly Retail Gasoline and Diesel Prices					
2		Regular	Midgrade	Premium	Diesel	
3	Min					
4	Max					
5	Average					
6						
7						
8						
9						

10. You are an engineer working for M & M / Mars™ Corporation in the M&M plant. For Halloween, M&Ms are produced in "fun size" bags, each containing approximately 17–18 M&Ms for folks to hand out.

    To help with quality control, you create the worksheet below. The factory workers will examine sample bags of M&Ms, and enter the weight of the bag and the individual count of M&Ms contained in the bag.

    Online, you have been given the following data in a Microsoft Excel workbook called **CandyCount.xlsx** with the data stored on a sheet named "M and M data." Only a portion of the actual data is shown.

	A	B	C	D	E	F	G	H
1	Count of M and M in bag by color							
2	Bag Number	Mass of Bag (m) [g]	Red	Orange	Yellow	Green	Blue	Brown
3	1	15.0	3	1	5	4	4	1
4	2	13.9	3	2	3	5	1	2
5	3	14.7	1	1	9	2	3	1
6	4	14.3	2	5	4	2	1	3
7	5	14.7	2	3	6	3	2	2
8	6	13.9	2	4	5	3	1	2
9	7	14.2	3	5	2	2	4	1
10	8	14.7	3	5	2	3	5	0

Write a MATLAB program to read in the Excel data, calculate the average number of red, orange, yellow, green, and brown M&Ms, and write it to a new Microsoft Excel file on a sheet named "Candy_Analysis_USERID." Your sheet should appear as follows, where the highlighted portion is replaced by the values you calculate in your solution. Your program should also use formatted `fprintf` statements to display the table in the command window, filling in the missing information (highlighted).

	A	B	C	D	E	F	G
1	M and M Analysis by Color						
2		Red	Orange	Yellow	Green	Blue	Brown
3	Average per bag:	2.4	3.8	3.5	3.0	2.9	2.0
4	Average mass per bag:	14.7					
5	Average total per bag:	17.6					
6	Average Maximum Count:	5.4					
7	Color that appears the most:	Orange					
8	Color that appears the least:	Brown					

11. The following Microsoft Excel file has been provided online. The file contains energy consumption data by energy source per year in the United States, measured in petaBTUs.

Year	Fossil Fuels	Elec. Net Imports	Nuclear	Renewable
2007	101.605	0.106	8.415	6.830
2006	99.861	0.063	8.214	6.922
2005	100.503	0.084	8.160	6.444
2004	100.351	0.039	8.222	6.261
2003	98.209	0.022	7.959	6.150

(a) Write the MATLAB code necessary to read the Microsoft Excel file and store each column of data into different variables. Create the following:

- `Yr`: a vector of all of the years in the worksheet.
- `FF`: a vector of all of the fossil fuels for each year in Yr.
- `ENI`: a vector of all electric imports for each year in Yr.
- `Nuc`: a vector of all nuclear energy consumption for each year in Yr.
- `Ren`: a vector of all renewable energy consumption for each year in Yr.
- `Hdr`: a cell array of all of the headers in row 1.

You may not hard-code these variables—they should be imported from the Excel file to receive credit.

(b) Create a new variable, **TotalConsumption**, which contains the sum of the four columns of energy consumption data for each year. In other words, since all five of the vectors (`Yr`, `FF`, `ENI`, `Nuc`, and `Ren`) we created in part (a) have the same length, the new variable, **TotalConsumption**, should have the same length. You may assume that you have correctly defined the variables in part (a).

(c) Calculate the average fossil fuel consumption in the entire data set. For this code, you may assume that you have correctly defined **FF**, the variable containing the fossil fuel consumption data, in part (a). Display the result of the calculation in the format shown below, where the number is shown to two decimal places:

The average fossil fuel consumption is _____ petaBTU.

(d) Write the MATLAB code necessary to generate the table below of nuclear energy consumption by year using formatted output in the Command Window. You may assume that you have correctly defined the vector Nuc and cell array Hdr in part **(a)**. Note the nuclear energy consumption should be displayed to two decimal places.

Nuclear Energy Consumption by Year [petaBTU]

2007	2006	2005	2004	2003	... etc
8.42	8.21	8.16	8.22	7.96	

12. An Excel file named Dart Tosses.xlsx, saved in a worksheet named "Darts" contains the measurements in this data set contain the horizontal and vertical distance from the bullseye of a dart board from 20 different tosses of a dart. A portion of the data is shown below. Write the MATLAB code necessary to determine the darts that were the closest and the furthest from the bullseye. The program should tell the user in the command window, using formatted output, the darts that are closest and farthest from the bullseye.

Dart	X	Y
Dart 1	4.04	0.55
Dart 2	2.63	0.35
Dart 3	1.10	2.97
Dart 4	4.89	5.60
Dart 5	4.00	2.07
Dart 6	5.25	0.68
Dart 7	0.05	1.28

UMBRELLA PROJECT

BREAKEVEN ANALYSIS: SMALL PARTS

A "**widget**" is defined by the Oxford dictionary as: **An indefinite name for a gadget or mechanical contrivance, especially a small manufactured item.**

It is used when the actual product being considered is irrelevant to the discussion.

[Photo courtesy of S. Stephan]

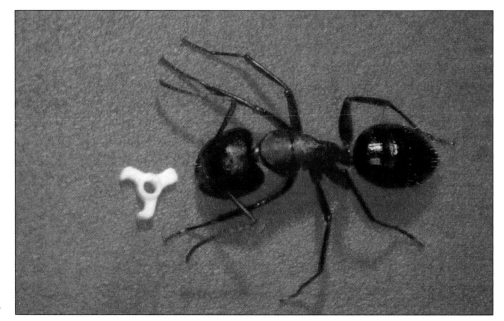

Our engineering team has been asked by upper management to analyze the cost of upgrading an aging machine line that produces widgets. The initial information available from three different machine vendors is outlined below. Perform a breakeven analysis on all three machines and recommend a vendor for purchase.

PROJECT A SPECIFICATIONS

From the initial analysis, the following can be assumed:

- Current material cost is $0.55/widget. With a part redesign, research hopes to reduce the material cost to $0.35/widget, but purchasing is worried because the cost of raw materials may increase the material cost to $0.85/widget. Your analysis should be able to handle fluctuations in material cost. The choice of machine has no effect on the material cost.

■ The current selling price per widget is $3.50. It is expected any increase in material cost will be directly passed on to the customer, so your analysis should be able to handle fluctuations in selling price.

- The current plant operates five days per week.
- The annual sales volume is estimated at four million parts per year.
- Management will not consider any project length greater than two years.

Each machine vendor has supplied the following information:

Machine Company #1: WYSIWYG

Machine cost: $2,750,000

Energy cost: $0.45/widget

Labor cost: $0.25/widget

Maximum capacity per day: 5,500 widgets

Machine Company #2: Thingamabob, LLC

Machine cost: $3,250,000

Energy cost: $0.25/widget

Labor cost: $0.35/widget

Maximum capacity per day: 7,000 widgets

Machine Company #3: GizmosRUs

Machine cost: $4,500,000

Energy cost: $0.15/widget

Labor cost: $0.05/widget

Maximum capacity per day: 6,000 widgets

Special features: This machine is able to run "lights out," extending the time of operation from five to seven days per week.

PROJECT B SPECIFICATIONS

From the initial analysis, the following can be assumed:

■ The current material cost is $0.75/widget. With a part redesign, research hopes to reduce the material cost to $0.65/widget, but purchasing is worried because the cost of raw materials may increase the material cost to $0.85/widget. Your analysis should be able to handle fluctuations in material cost. The choice of machine has no effect on the material cost.

■ The current selling price per widget is $3.00. It is expected any increase in material cost will be directly passed on to the customer, so your analysis should be able to handle fluctuations in selling price.

- The current plant operates five days per week.
- The annual sales volume is estimated at five million parts per year.
- Management will not consider any project length greater than two years.

Each machine vendor has supplied the following information:

Machine Company #1: WidgetsRUs

Machine cost: $2,500,000

Energy cost: $0.40/widget

Labor cost: $0.25/widget

Maximum capacity per day: 5,000 widgets

Machine Company #2: Widgets, Inc.

Machine cost: $3,500,000

Energy cost: $0.15/widget

Labor cost: $0.35/widget

Maximum capacity per day: 7,500 widgets

Machine Company #3: Klein Teil

Machine cost: $5,000,000

Energy cost: $0.15/widget

Labor cost: $0.05/widget

Maximum capacity per day: 6,500 widgets

Special features: This machine is able to run "lights out," extending the time of operation from five to seven days per week.

DATA ANALYSIS

Workbook

Each team must submit a final workbook, including the following components:

- A worksheet that allows an end user to change the initial vendor information and material costs and observe the effect in the graphs.
- A breakeven analysis graph, using the number of widgets produced.
- A breakeven analysis graph, using time in months.

The workbook will be graded for neatness, organization, unit consistency, and proper plots. It should be a proper workbook, containing heading information on every page.

Written Section

Each team must submit a final written communication, including the following components:

- A summary of the analysis performed, including one of the following combinations:
  - One breakeven analysis graph and one summary table
  - Two breakeven analysis graphs
- A final vendor recommendation and supporting facts to justify your recommendation.
- An explanation including supporting facts of WHY the other companies were not chosen.

For the submission format, each team may choose or may be assigned by your instructor one of the following options. The written portion will be graded on neatness, organization, spelling/grammar/mechanics, formatting of embedded graphs, and strength of conclusions.

Written Option One: Memo

A memo template, provided electronically and shown in Section 4.4 in your textbook, must be used.

Written Option Two: Short Report

A report template, provided electronically and shown in Section 4.4 in your textbook, must be used.

Written Option Three: Poster

A poster template, provided electronically and shown in Section 4.4 in your textbook, must be used.

Written Option Four: PowerPoint Presentation

A PowerPoint presentation using voice-over should be developed to address the requirements given above. Information on developing a presentation is given in your textbook in Section 4.1, and the accompanying online slides discuss voice-over.

TRENDLINE ANALYSIS:
HOOKE'S LAW

When Robert Hooke first published his now famous law of elasticity in 1660, he published it in anagram form:

ceiiinosssttuv

Before patents, many scientists used anagrams to release discoveries without giving away details. He published the solution in 1678.

The designer of a museum exhibit of miniature string instruments faces a challenge that the instruments need regular tuning, but the museum budget cannot support someone with a trained ear to perform this function. You have been asked to investigate the possibility of using spring tensioners to allow an untrained tuner to be able to confirm that the string tension is correct without audible tuning. If the desired string tension of the miniature violin in the exhibit is 0.022 newtons (about one ten-thousandth of the normal violin string tension), what combination of springs can confirm this string tension yet stretch less than 1 centimeter? The idea is that a tension of 0.022 newtons will stretch the springs a known amount that can be easily measured.

PREDICTION

Prior to testing, each lab partner(s) should individually create a sketch of the force (ordinate) versus the displacement (abscissa) for the following springs:

- Spring #1: Spring constant of 1 newton per meter [N/m].
- Spring #2: Stiffer than Spring #1.
- Spring #1 and #2 in series.
- Spring #1 and #2 in parallel.

Discuss the plots of each lab partner as a group. Settle on the plot that is the "best" of the group. Each person in the group should sketch this "best" set of curves as a second set of lines on their original plot using a different line type and/or color to distinguish them.

DATA COLLECTION

Create an Excel workbook to collect the data. Place all data on a single worksheet.

Single Springs

- Connect one of the springs to an eye hook at one end.
- Measure the original distance between the bottom hook on the spring and the bottom edge of the board.

- Gently hook one mass on the bottom of the spring.
- Measure the new distance between the bottom hook of the spring and the bottom edge of the board.
- Record the displacement and the corresponding mass used in the worksheet.
- Repeat for a minimum of five different masses for this spring. Note that the masses have a hook on the top and an eyelet on the bottom so they can be easily hooked together.
- Repeat for each spring provided. Please note you do NOT need to use the same masses for each spring. Choose masses that are appropriate for the spring being tested.

For simplification, label the springs in order from the stiffest to the least stiff.

Springs in Combination

Repeat Part II for sets of two springs in parallel and then for two springs in series, forming all combinations you can with the available springs. Please note you do NOT need to use the same masses for each spring combination. Choose masses that are appropriate for the springs being tested.

With two springs you can form two combinations: one in parallel and one in series. With three springs of unique stiffness, you can form eight combinations using the following configurations:

- All three in parallel.
- All three in series.
- Two parallel springs in series with the third (three ways).
- Two series springs in parallel with the third (three ways).

DATA ANALYSIS

Workbook

Each team must submit a final workbook, including the following components:

- All data collected.
- Complete the necessary unit conversions within the worksheet. Be sure to use consistent units. Remember, mass and weight (which is a force) are NOT the same! Most spring constants are reported in units of newton per meter; complete any necessary conversions within the worksheet BEFORE creating the graph.
- Create a graph or graphs containing the data sets, with appropriate trendlines:
  - The force, position data set for the springs individually
  - The force, position data set for the springs in parallel
  - The force, position data set for the springs in series

The workbook will be graded for neatness, organization, unit consistency, and proper plots. It should be a proper workbook, containing heading information on every page.

Written Section

Each team must submit a final written communication, including the following components:

- List the spring constants in units of newton per meter for each spring and each spring combination. Sort the list in the order of stiffness, with the stiffest listed first.

- A summary of the analysis performed, including one of the following combinations:
  - One graph and one summary table
  - Two graphs

- Find a spring combination that stretches less than 1 centimeter under a force of 0.022 newtons. Note that the springs provided may not be sufficient to accomplish this task, in which case you should determine a configuration using as few springs as possible and only considering the stiffnesses provided.

For the submission format, each team may choose or may be assigned by your instructor one of the following options. The written portion will be graded on neatness, organization, spelling/grammar/mechanics, formatting of embedded graphs, and strength of conclusions.

Written Option One: Memo

A memo template, provided electronically and shown in Section 4.4 in your textbook, must be used.

Written Option Two: Short Report

A report template, provided electronically and shown in Section 4.4 in your textbook, must be used.

Written Option Three: Poster

A poster template, provided electronically and shown in Section 4.4 in your textbook, must be used.

Written Option Four: PowerPoint Presentation

A PowerPoint presentation using voice-over should be developed to address the requirements given above. Information on developing a presentation is given in your textbook in Section 4.1, and the accompanying online slides discuss voice-over.

TRENDLINE ANALYSIS:
PENDULUMS

The direction of a Foucault Pendulum's swing moves in relation to the rotation of the Earth.

It is named for Léon Foucault, a French physicist, who is credited with the experiment in 1851, the first dynamic proof of the Earth's rotation.

[Photo courtesy of E. Stephan]

You work for a clock-manufacturing company. You have been asked to explore the best combination of mass and length for a pendulum to put into a grandfather clock. If the desired pendulum swing is 30 times per minute, with the motion of the pendulum to the right and back to the left counting as one full swing, what is the best mass and length of pendulum to choose?

DATA COLLECTION

Effect of Length

Hang a 50-gram mass from the loop at the end of a long piece of fishing line. Wrap the other end of the line around a rod so the distance from the mass to the rod is about 10 inches. Carefully measure and record in a worksheet the distance from the center of the rod to the center of the mass. One person should release the pendulum and the other should use a stopwatch to time how long is required for five full swings, with the motion of the pendulum to the right and back to the left counting as one full swing. The person releasing the pendulum should pull the mass to the side about 15 degrees from the vertical, keeping the body of the mass in line with the fishing line.

After release, the student with the stopwatch should wait for at least one full swing and then start the stopwatch when the mass is at one side or the other and just beginning to reverse direction. Count five full swings and then stop the watch when the mass returns the fifth time to the point where the watch was started. Record the time in the worksheet. Keeping the length constant, repeat this procedure for a total of three trials.

Repeat the entire procedure above for lengths of approximately 15, 20, 25, and 30 inches. Remember to carefully measure and record the actual distance from the center of the rod to the center of the mass.

Effect of Mass

Hang a 200-gram mass from the loop at the end of one of the pieces of fishing line. Wrap the other end of the line around a rod so the distance from the mass to the rod is about 20 inches. One person should release the pendulum and the other should use the stopwatch to time how long is required for five full swings, with the motion of the pendulum to the right and back to the left counting as one full swing. The person releasing the pendulum should pull the mass to the side about 15 degrees from the vertical, keeping the body of the mass in line with the fishing line. After release, the student with the stopwatch should wait for at least one full swing and then start the stopwatch when the mass is at one side or the other and just beginning to reverse direction. Count five full swings and then stop the watch when the mass returns the fifth time to the point where the watch was started. Record the time in the worksheet. Keeping the mass constant, repeat this procedure for a total of three trials.

Repeat the entire procedure above for masses of 100, 50, 20, and 10 grams.

Since the masses are different sizes, you will have to adjust the length from the center of the rod to the center of the mass each time you change masses. **This length should be the same for all five masses**.

DATA ANALYSIS

Analysis Option One: Workbook

Each team must submit a final workbook, including the following components:

- All data collected.
  - **Calculate** the average of the three times and record the "Average Time." Divide each average time by five (you timed five complete swings, not just one) to obtain the value for the period corresponding to each condition.

- **Model for Effect of Length:**
  - Draw a proper plot of period and length.
  - Fit an appropriate trendline model.
- **Model for Effect of Mass:**
  - Draw a proper plot of period and mass.
  - Fit an appropriate trendline model.

Do not be tempted to ascribe undue significance to small amounts of experimental error. You may see this effect in this experiment. In both trendline models, be sure to consider if the data has an offset, and include any corrections necessary.

The workbook will be graded for neatness, organization, unit consistency, and proper plots. It should be a proper workbook, containing heading information on every page.

Analysis Option Two: MATLAB

Program

Each team must write a MATLAB program that will analyze the data you collected in the lab and present your results using formatted output and plots. At a minimum, your program should:

- **Ask** the user to enter data similar to the data collected in the lab. Be sure to give the user good instructions as to the type, format, and units of the input required for each part of the experiment.
- For both the effect of length data and the effect of mass data, the program should **calculate** the average of the three times for each data set. Divide each average time by five (you timed five complete swings, not just one) to obtain the value for the period corresponding to each condition.
- **Create** a set of graphs:
  - Draw a proper plot of period and length.
  - Draw a proper plot of period and mass.
  - Fit appropriate trendline models to each graph using polyfit.

All plots should appear in separate plot figures.

- Do not be tempted to ascribe undue significance to small amounts of experimental error. You may see this effect in this experiment. In both trendline models, be sure to consider if the data has an offset, and include any corrections necessary.
- In the command window, your program should **display using formatted output** the trendline equations for the different springs analyzed in this project; an example is shown in Tables A and B, below. All trendline values must be displayed with a reasonable number of decimal places using output from the polyfit function. For all tables, you do not need to show the table instructions "Using the trendlines . . .," the gridlines or the table headings in bold font or with a background color—you should duplicate the column format only. Note the values shown are EXAMPLES only—these are not the experimental results you will obtain.

Table A: Using the trendline on the period, length graph, for each data set, determine the type of expression shown by the trendline (linear, power, or exponential). Then, based upon the model chosen, list the value and units of the parameters m and b.

For linear:	m = slope	b = intercept
For power:	m = exponent	b = constant
For exponential:	m = exponent	b = constant

Model Type	Value and Units of "m"		Value and Units of "b"	
Linear	10	m / s	30	s

Table B: Using the trendline on the period, mass graph, for each data set, determine the type of expression shown by the trendline (linear, power, or exponential). Then, based upon the model chosen, list the value and units of the parameters m and b.

For linear:	m = slope	b = intercept
For power:	m = exponent	b = constant
For exponential:	m = exponent	b = constant

Model Type	Value and Units of "m"		Value and Units of "b"	
Linear	10	m / s	30	s

All MATLAB codes should include a proper header, including all your team member names, course/section, date, problem statement, documentation on all variables used in the program/function, and any assumptions made in the design of your algorithm. All MATLAB codes should contain an adequate number of comments throughout the code.

Written Section

Each team must submit a final written communication, including the following components:

- Which trendline type did you choose to determine the effect of each variable? Justify the choice in terms of the physical meaning of the trendline parameters, such as the physical meaning of the slope and the intercept if the trendline is linear.
- Explain why you did not choose the other trendline types in terms of their physical meaning.
- Determine the optimum mass and length of the pendulum to achieve 30 swings per minute.

For the submission format, each team may choose or may be assigned by your instructor one of the following options. The written portion will be graded on neatness, organization, spelling/grammar/mechanics, formatting of embedded graphs, and strength of conclusions.

Written Option One: Memo

A memo template, provided electronically and shown in Section 4.4 in your textbook, must be used.

Written Option Two: Short Report

A report template, provided electronically and shown in Section 4.4 in your textbook, must be used.

Written Option Three: Poster

A poster template, provided electronically and shown in Section 4.4 in your textbook, must be used.

Written Option Four: PowerPoint Presentation

A PowerPoint presentation using voice-over should be developed to address the requirements given above. Information on developing a presentation is given in your textbook in Section 4.1, and the accompanying online slides discuss voice-over.

TRENDLINE ANALYSIS:
BOUNCING SPRINGS

A mass bouncing on a spring is an example of simple harmonic motion. In this process, energy is transformed between potential energy and kinetic energy.

Other examples of simple harmonic motion include pendulums and molecular vibrations.

Various devices have been designed and patented to agitate a fishing line to make lures more attractive to fish. A group of friends wants to be more successful in fishing, but does not want to go to the trouble and expense of purchasing one of these devices. They ask you as an engineer if you can design one from scrap materials.

BASIC INFORMATION

NOTE

To model the oscillating mass-spring system mathematically, differential equations are required. You will learn how to solve such equations in your future courses.

This experiment investigates the oscillation frequency of a mass suspended from a spring. When a mass is suspended from a spring and allowed to come to rest, the spring will be stretched so that the upward spring force balances the downward force (weight) of the mass. We will call the location of the mass suspended in this way the rest position.

If the mass is raised above the rest position, the spring is less stretched and the spring force on the mass is less than the weight of the mass. If the mass is then released, it will move downward due to the imbalance in the forces. When the mass reaches the location of the rest position, its velocity is not zero. In order to slow the mass to a stop, there must be an upward force exerted on the mass. As the mass descends beyond the rest position, the spring is stretched more, increasing the upward spring force so that it is greater than the force (weight) of the mass, causing the mass to decelerate (accelerate upward).

Eventually, the mass slows to a stop, but since the mass is below the rest position there is a net upward force that causes the mass to begin rising again. This continues over and over with the mass oscillating up and down with the rest position in the middle of the oscillation. Eventually, frictional forces will bring the mass back to the rest position at a complete stop.

DATA COLLECTION

Initial Setup

1. Clamp a wooden block to a table securely, with a hook on the underside of the block, and positioned so that the hook is well beyond the edge of the table.

2. Hang the spring provided from the hook.
3. Measure the distance from the underside of the block to the lowest point on the spring. Record the initial distance from the block to the bottom of the spring.
4. Read this entire step before doing it!

   - For each of the three masses provided, CAREFULLY hang the mass on the bottom of the spring, and then slowly and gently lower it to the rest position.
   - DO NOT simply let go of the mass or you may damage the spring.
   - With each mass in the rest position, measure the distance from the underside of the block to the lowest point of the spring.
   - Record the values of the masses and the corresponding distances to the rest position in an Excel workbook.

5. Repeat for each spring provided. Please note you do NOT need to use the same masses for each spring. Choose masses that are appropriate for the spring being tested.

Effect of Mass

1. Attach the smallest mass to the spring and carefully lower it to the rest position. Remeasure the rest position and verify that it agrees with the value recorded earlier.
2. Raise the mass vertically until the spring is almost, but not quite, back to the unstretched length. The distance from the underside of the block to the lowest point on the spring will be slightly more than the initial spring position. Determine the amount you raised the spring from the rest position. Record the amount raised.
3. Read this entire step before doing it!

   - Reset the stopwatch.
   - Carefully release the mass from the raised position so that it oscillates up and down with no horizontal motion.
   - Wait for three oscillations, and then, while watching the mass, start the stopwatch when the mass reaches its lowest point.
   - Count 20 full oscillations, and when the mass reaches the lowest point on the twentieth cycle, stop the watch.

4. Record the time for 20 oscillations in the table below.
5. Repeat Step 3 for the remaining masses, raising each the same amount in Step 2. Please note you do NOT need to use the same masses for each spring. Choose masses that are appropriate for the spring being tested.
6. Within the worksheet, calculate the period and frequency for each mass.

Effect of Release Position

1. Choose one spring. Attach the 50-gram mass to the spring, and carefully lower it to the rest position.
2. Raise the mass ¾ inch vertically.
3. Read this entire step before doing it!

   - Reset the stopwatch.
   - Carefully release the mass from the raised position so that it oscillates up and down with no horizontal motion.
   - Wait for three oscillations, and then, while watching the mass, start the stopwatch when the mass reaches its lowest point.
   - Count 20 full oscillations, and when the mass reaches the lowest point on the twentieth cycle, stop the watch.
   - Record the time for 20 oscillations in an Excel workbook.

4. Repeat Step 3 by raising the mass a distance of 1, 1¼, 1½, and 1¾ inches. Note: If the stretch amount at rest position is less than one or more of these increments, come as close as you can without going beyond that value.

5. Within the worksheet, calculate the period and frequency for each mass.

DATA ANALYSIS

Analysis Option One: Workbook

Each team must submit a final workbook, including the following components:

- All data collected.
  - **Calculate** how much the spring was stretched by each mass (distance from block to rest position minus initial distance from block to bottom of spring) and record the spring extensions.
- **Model for Effect of Length:**
  - **Calculate** the period and frequency for each mass.
    - The period is the length of time required for one oscillation.
    - The frequency is the number of oscillations per second, which is simply the inverse of the period.
  - Draw a proper plot of period and length.
  - Fit an appropriate trendline model.
- **Model for Effect of Initial Displacement:**
  - **Calculate** the period and frequency for each mass.
    - The period is the length of time required for one oscillation.
    - The frequency is the number of oscillations per second, which is simply the inverse of the period.
  - Draw a proper plot of frequency and initial displacement.
  - Fit an appropriate trendline model.

In both models, be sure to consider if the data has an offset, and include any corrections necessary.

The workbook will be graded for neatness, organization, unit consistency, and proper plots. It should be a proper workbook, containing heading information on every page.

Analysis Option Two: MATLAB

Program

Each team must write a MATLAB program that will analyze the data you collected in the lab and present your results using formatted output and plots.

At a minimum, your program should:

- Ask the user to enter data similar to the data collected in the lab. Be sure to give the user good instructions as to the type, format, and units of the input required for each part of the experiment.
- **Calculate** how much the spring was stretched by each mass (distance from block to rest position minus initial distance from block to bottom of spring) and record the spring extensions.

- **Calculate** the period and frequency for each mass.
  - The period is the length of time required for one oscillation.
  - The frequency is the number of oscillations per second, which is simply the inverse of the period.
- **Calculate** the period and frequency for each release position.
  - The period is the length of time required for one oscillation.
  - The frequency is the number of oscillations per second, which is simply the inverse of the period.
- **Determination of Spring Constants:**
  - Using MATLAB (not by hand), determine the force applied to each spring in the Initial Setup.
  - Draw a proper plot of force (ordinate) and elongation. The graph should contain three data sets, one for each spring tested.
  - Fit an appropriate trendline model using the output from the `polyfit` function.

- **Model for Effect of Mass:**
  - Draw a proper plot of frequency (ordinate) and mass. The graph should contain three data sets, one for each spring tested.
  - Fit an appropriate trendline model using the output from the `polyfit` function.

- **Model for Effect of Initial Displacement:**
  - Draw a proper plot of frequency (ordinate) and initial displacement.
  - Fit an appropriate trendline model using the output from the `polyfit` function.

All plots should appear in separate plot figures.

- In the command window, your program should display using formatted output the trendline equations for the different springs analyzed in this project; an example is shown below. All trendline values must be displayed with a reasonable number of decimal places using output from the polyfit function. For all tables, you do not need to show the table instructions "Using the trendlines . . .," the gridlines; or the table headings in bold font or with a background color—you should duplicate the column format only. Note the values shown are EXAMPLES only—these are not the experimental results you will obtain.

Table A: Using the trendlines on the force, elongation graph, determine the stiffness (k value) of each spring. As a formatted output table, list the spring name, trendline equation generated from polyfit, and the spring stiffness value and units similar to the following table.

Spring Name	Trendline Equation	Spring Stiffness (k) [value and units]
Alpha	$F = 10 \times$	10 N / m
Beta	$F = 32 \times$	32 N / m

Table B: Using the trendlines on the frequency, mass graph, for each data set, determine the type of expression shown by the trendline (linear, power, or exponential). Then, based upon the model chosen, list the value and units of the parameters m and b.

For linear:	m = slope	b = intercept
For power:	m = exponent	b = constant
For exponential:	m = exponent	b = constant

Spring Name	Model Type	Value and Units of "m"		Value and Units of "b"	
Alpha	Linear	10	m / s	30	s
Beta		25		50	

Table C: Using the trendline on the frequency, displacement graph, for the data set, determine the type of expression shown by the trendline (linear, power, or exponential). Then, based upon the model chosen, list the value and units of the parameters m and b.

For linear:	m = slope	b = intercept
For power:	m = exponent	b = constant
For exponential:	m = exponent	b = constant

Model Type	Value and Units of "m"		Value and Units of "b"	
Exponential	20	m	30	m / s

All MATLAB codes should include a proper header, including all your team member names, course/section, date, problem statement, documentation on all variables used in the program/function, and any assumptions made in the design of your algorithm. All MATLAB codes should contain an adequate number of comments throughout the code.

Written Section

Each team must submit a final written communication, including the following components:

- Which trendline type did you choose to determine the effect of each variable? Justify the choice in terms of the physical meaning of the trendline.
- Explain why you did not choose the other trendline types in terms of their physical meaning.
- Determine a mass and initial displacement required to create a complete oscillation every two seconds. Make sure the initial displacement and mass chosen will not permanently damage the spring and make sure the mass is a size commonly available. This will provide a starting point for the next phase of water testing. (This will not be part of our experiment.)

For the submission format, each team may choose or may be assigned by your instructor one of the following options. The written portion will be graded on neatness, organization, spelling/grammar/mechanics, formatting of embedded graphs, and strength of conclusions.

Written Option One: Memo

A memo template, provided electronically and shown in Section 4.4 in your textbook, must be used.

Written Option Two: Short Report

A report template, provided electronically and shown in Section 4.4 in your textbook, must be used.

Written Option Three: Poster

A poster template, provided electronically and shown in Section 4.4 in your textbook, must be used.

Written Option Four: PowerPoint Presentation

A PowerPoint presentation using voice-over should be developed to address the requirements given above. Information on developing a presentation is given in your textbook in Section 4.1, and the accompanying online slides discuss voice-over.

UMBRELLA PROJECT

MATHEMATICAL MODEL ANALYSIS:
CANTILEVER BEAMS AND CLEAN WATER

The New River Gorge Bridge in West Virginia, shown here, is the world's third longest single arch bridge. The bridge is 3,030 feet long, with an arch of 1,700 feet. The bridge is 876 feet above the New River, the fifth highest vehicular bridge in the world.

[Photo courtesy of E. Stephan]

Our engineering firm has been contacted by a biosystems engineering company researching a device to measure the amount of suspended sediment in rivers. The instrumentation is hanging from a long metal bar over the water. As water flows by the instruments, the resulting drag force produced makes the bar bend.

Your design team must conduct tests on the metal bar to determine a mathematical model that can be used in the field to determine the optimal instrumentation configuration.

NAE GRAND CHALLENGE FOR ENGINEERING: PROVIDE ACCESS TO CLEAN WATER

One of the 14 National Academy of Engineering Grand Challenges is **Provide Access to Clean Water.** According to the NAE website (www.engineeringchallenges.org):

> "Lack of clean water is responsible for more deaths in the world than war. About 1 out of every 6 people living today do not have adequate access to water, and more than double that number lack basic sanitation, for which water is needed. In some countries, half the population does not have access to safe drinking water, and hence is afflicted with poor health. By some estimates, each day nearly 5,000 children worldwide die from diarrhea-related diseases, a toll that would drop dramatically if sufficient water for sanitation was available."

The goal of this textbook project is to develop a model of a cantilever beam frame that will allow various water quality test instruments to be suspended over, and partially immersed in, the water below. This frame is part of a wastewater treatment facility being constructed for Engineers Without Borders (EWB), a nonprofit group that designs sustainable engineering projects for developing communities around the world. Your team must analyze the type of beam that will be used to suspend instruments over a mangrove swamp on the edge of the property set aside for the project. These instruments are especially vital during the rainy season, when runoff flows through the swamp, creating unpredictable sediment and flow conditions. You will make a recommendation as to the optimal beam configuration that allows all the instruments to be installed.

THEORY BACKGROUND

In studying springs and circuits, we established a general expression regarding how an applied force causes a change and meets with a resistance. The deflection (Δ) of a cantilever beam under a load (P) fits this same linear model.

$$P = K\Delta$$

When a "system"...	... is acted upon by a "force"...	... to change a "parameter"...	... the "system" opposes the change by a "resistance"	Equation
Physical object	External push or pull (F)	Acceleration (a)	Object mass (m)	$F = ma$
Beam	Vertical force or load (P)	Tip deflection (Δ)	Beam stiffness (K)	$P = K\Delta$

Previously, we discussed the spring stiffness (k) as a property of the spring. In reality, this one property summarizes a variety of factors that determine how the spring responds to a force. In the case of a cantilever beam, the beam stiffness (K) is a function of the beam length (L), the beam width (W), the beam thickness (T), a material property of the beam (E, the beam's modulus of elasticity or Young's modulus), and a geometrical constant (c).

We wish to study the behavior of a beam to determine the relationship between the beam stiffness and the other variables. To begin with, we assume a power relationship of the variables that make up K, but do not know the power values (m, n, p, q).

$$K = cE^m W^n T^p L^q$$

We will begin by holding several parameters constant, and varying a single parameter to determine the effect on the stiffness. For example, the beam shape (c), material (E), width (W), and thickness (T) will be kept constant. At different beam lengths (L), several masses will be applied and the resulting deflection measured.

PROJECT SPECIFICATIONS

Data Collection

Data will be collected in class according to the specific instructions provided for you by your professor, or sample data will be provided in an Excel workbook.

Model Development

The mathematical model to predict the load or deflection of a cantilever beam must be recorded on the Model Form (Appendix A, Part I). The exponent values ($m, n, p,$ and q) and corresponding graphs necessary to complete the calculations (outlined under Data Analysis) and the shape constant "c" must be completed and verified by the instructor before proceeding with the model verification.

Model Verification

Each team will test their mathematical model. To make your prediction, you must use your worksheet or your program (outlined under Data Analysis) to predict actual field conditions.

Once it has been verified your worksheet or program is calculating correctly, you will use it to complete Appendix A, Part II:

- Given the known load, predict the deflection, or
- Given the maximum deflection, predict the maximum load.

Each team will be given up to six beams at random and asked to predict the outcome of a given experimental scenario. Once the team has predicted the outcome, the team will test the actual beams or be given simulated results. The prediction and results will be compared in the final report.

Model Application

The model will be used to design a cantilever beam component of a frame that will allow various water quality test instruments to be suspended over, and partially immersed in, the water below. This frame is part of a wastewater treatment facility being constructed for Engineers Without Borders (EWB), a nonprofit group that designs sustainable engineering projects for developing communities around the world. Your team must analyze the type of beam that will be used to suspend instruments over a mangrove swamp on the edge of the property set aside for the project.

You will make a recommendation as to the optimal beam configuration that allows all the instruments to be installed, taking into consideration the following:

- Cost of the beams available, minimizing costs wherever possible. Remember that EWB operates solely on donations and funds raised by volunteers.
- More than one frame can be constructed, so more than one beam can be recommended.
- No more than three instruments can be mounted on a single beam.
- Only the support beam for the instruments must be recommended; the frame components are not considered part of the design. In order to prevent failure of the beam itself, no more than 1 inch of deflection is allowed.

Your recommendations should include a short discussion justifying the selection of the beams, summary table with the beams selected, and the instruments to be mounted on each one.

Table 1 contains a list of the instruments to be installed. A portion of each one will be suspended below the water line, creating a drag force. The table includes the estimated drag force of each instrument. Assume all instruments are mounted at the end of the beam, and that up to three instruments can be mounted at the same location using specially designed mounting brackets.

Table 2 contains the beams available for use. All beams must be clamped at the top of the support frame with no beam length extending above the top of the frame. The beams may not be cut or made shorter than the length shown in the table. Your team only needs to recommend the beams to be used to mount the instruments, not the entire support frame. Only one beam of each type is available.

Table 1 Instruments used to test water quality and estimated forces each will apply when mounted in slow-moving water

Instrument	Estimated Drag Force [pound-force]
Turbidity meter	3
Portable sampler	1
Velocity flow meter	4
Nutrient meter (ammonium and orthophosphate)	2
pH electrode	0.5

Table 2 List of materials available to construct the cantilever beam segments of the frames that will support water quality test instruments

Beam Type	Cross Section [in × in]	Length [in]	Cost
Aluminum	0.5 × 0.25	18	$10
Aluminum	0.25 × 0.25	30	$4
Aluminum	1.0 × 0.125	18	$13
Stainless steel	0.5 × 0.125	24	$15
Stainless steel	0.5 × 0.25	24	$20
Stainless steel	0.25 × 0.25	18	$10

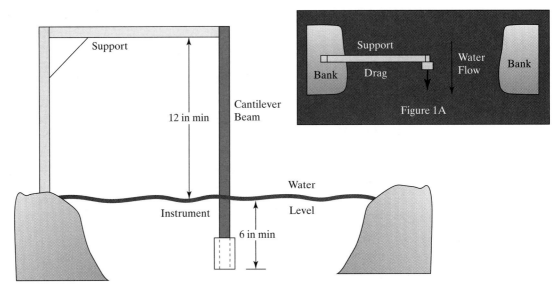

Figure 1 View from the side of a support frame with cantilever beam and mounted water quality test instrument. Drag forces will push the beam perpendicular to the bank of the swamp, in the direction of the water flow. Figure 1A: View from above—the frame, beam, and instrument.

DATA ANALYSIS

Analysis Option One: Workbook

Examine the data taken in the lab. You should reorder your data into the following groups. Note that some beams may appear in more than one group . . . and one beam should appear in EVERY group.

- Length data, testing 1 inch wide by ⅛ inch thick aluminum beams.
- Width data, testing ⅛ inch thick by 18 inches long aluminum beams.
- Thickness data, testing 1 inch wide by 18 inches long aluminum beams.
- Modulus of elasticity data, testing 1 inch wide by 18 inches long by ⅛ inch thick beams.

Create two graphs for each parameter (eight total), using the data above to determine the relationship between the stiffness and the parameter. Each graph should be displayed as an independent worksheet and not embedded within a worksheet.

- Length: (A) force versus deflection and   (B) stiffness versus L
- Width: (A) force versus deflection and   (B) stiffness versus W
- Thickness: (A) force versus deflection and   (B) stiffness versus T
- Modulus: (A) force versus deflection and   (B) stiffness versus E

For parameters that are not linear, a graph must be included using semilog or log–log coordinates making the data appear linear. For all graphs, the appropriate trendline equation and R^2 value must be given.

Create one worksheet showing the determination of shape constant "c" for each stiffness trendline generated (four total), the average value for "c", the standard deviation, and units where appropriate.

Create one worksheet, allowing the user to predict actual field conditions according to the following specifications:

- Use Conditional Statements, Data Validation, and LOOKUP functions to:
  - Create a list to allow the user to choose the input calculation units (AES, Metric, English).
    - AES units should ask for calculations in units of inches, pounds-mass, pounds-force, and pound-force per square inch.
    - Metric units should ask for calculations in units of centimeters, grams, newtons, and pascals.
    - English units should ask for calculations in units of feet, slugs, pounds-force, and pound-force per square foot.
  - Create a list to allow the user to choose the beam material.
  - Request correct input parameters in the selected unit system from the user.
  - Create an error message if the data entered are outside the allowable range for the parameters as tested during the experiment. If the input parameters are outside the range of those tested—for example, if the user enters a beam length of 50 feet—a warning should appear to tell the user the parameters exceed the acceptable range.
- The user should be able to choose to enter the load or deflection. Then:
  - If the load is entered, the program will predict the deflection.
  - If the maximum deflection is entered, the program will predict the maximum load.

The workbook will be graded for neatness, organization, unit consistency, and proper plots. It should be a proper workbook, containing heading information on every page.

Analysis Option Two: MATLAB

MATLAB Code and Excel Workbook

Examine the data taken in the lab in the Excel workbook. You should reorder your data into the following groups. You may reorder this data manually in Excel, or within your code after importing the data. Note that some beams may appear in more than one group . . . and one beam should appear in EVERY group.

- Length data, testing 1 inch wide by ⅛ inch thick aluminum beams.
- Width data, testing ⅛ inch thick by 18 inches long aluminum beams.
- Thickness data, testing 1 inch wide by 18 inches long aluminum beams.
- Modulus of elasticity data, testing 1 inch wide by 18 inches long by ⅛ inch thick beams.

Import the data into MATLAB using `xlsread`. Create `polyfit` solutions for each parameter, using the data to determine the relationship between the stiffness and the parameter.

- Length: (A) force versus deflection and   (B) stiffness versus L
- Width: (A) force versus deflection and   (B) stiffness versus W
- Thickness: (A) force versus deflection and   (B) stiffness versus T
- Modulus: (A) force versus deflection and   (B) stiffness versus E

A menu should ask the user to choose one parameter to display graphically. The chosen parameter should have two corresponding graphs appear, listed as (A) and (B) above. The graphs should be proper plots, and show both experimental data and the appropriate trendline(s). Once displayed, the program should ask the user if they wish to see another parameter displayed.

The code should determine the shape constant "c" for each beam parameter (four total) and then determine the average value for "c."

The code should create a GUI, allowing the user to predict actual field conditions.

- Create a list to allow the user to choose the input calculation units (AES, Metric, English) as a pop-up menu:
  - AES units should ask for calculations in units of inches, pounds-mass, pounds-force, and pound-force per square inch.
  - Metric units should ask for calculations in units of centimeters, grams, newtons, and pascals.
  - English units should ask for calculations in units of feet, slugs, pounds-force, and pound-force per square foot.
- Create a list to allow the user to choose the beam material as a pop-up menu.
- Request correct input parameters in the selected unit system from the user as edit text boxes.
- Create an error message if the data entered are outside the allowable range. (If the input parameters are outside the range of those tested—for example, if the user enters a beam length of 50 feet—a warning should appear to tell the user the parameters exceed the acceptable range.)
- The user should be able to choose to enter the load or deflection as a pop-up menu. Then:
  - If the load is entered, the program will predict the deflection.
  - If the maximum deflection is entered, the program will predict the maximum load.

It is suggested, for efficiency, you examine ways to use functions throughout your code.

Written Section

Each team must submit a final written communication, including the following components:

- Discussion of the method used to determine the mathematical model:

$$K = cE^m W^n T^p L^q$$

including a set of sample figures to illustrate the method used to determine the exponent for one parameter (one plot of force, deflection; one plot of stiffness, chosen parameter illustrating logarithmic axis).
- Table showing the resulting exponents m, n, p, and q, and the shape constant c.
- A summary table that compares predicted load or deflection, actual results, and error calculation.
- Recommendation as to the optimal beam configuration and total cost that allows all the instruments to be installed for the EWB project. Be sure to justify your selection and explain WHY other choices are not better.

For the submission format, each team may choose or may be assigned by your instructor one of the following options. The written portion will be graded on neatness, organization, spelling/grammar/mechanics, formatting of embedded graphs, and strength of conclusions.

Written Option One: Short Report

A report template, provided electronically and shown in Section 4.4 in your textbook, must be used.

Written Option Two: Poster

A poster template, provided electronically and shown in Section 4.4 in your textbook, must be used.

Written Option Three: PowerPoint Presentation

A PowerPoint presentation using voice-over should be developed to address the requirements given above. Information on developing a presentation is given in your textbook in Section 4.1, and the accompanying online slides discuss voice-over.

APPENDIX A

Part I: Model Development

Model: The model should take the form of $K = cE^m W^n T^p L^q$

Model Parameter	Value	Verified by/on
Exponent m on (E) parameter		
Exponent n on (W) parameter		
Exponent p on (T) parameter		
Exponent q on (L) parameter		
Value and units of constant c		

Part II: Model Verification

Testing Results: Fill in the chart below during actual testing. You will be given a maximum of six beams, forming a combination of beams to test under various loading conditions and beams to test under various deflection conditions.

Test	Beam Parameters			Deflection Achieved	
Load Applied	Length	Width	Thickness	Prediction	Actual
1					
2					
3					

Test	Beam Parameters			Load Required	
Deflection	Length	Width	Thickness	Prediction	Actual
1					
2					
3					

Prediction Verified by: _____

Actual Results Verified by: _____

STATISTICAL ANALYSIS: CONFIGURATION MATTERS

The Cooper River Bridge connects Charleston and Mount Pleasant, SC. The cable-stayed bridge has a main span of 1,546 feet, the second longest cable stay bridge in the Western Hemisphere.

[Photos courtesy of E. Stephan]

Our engineering firm has been contacted to test prototypes of bridge struts that will be part of a new bridge design. The design utilizes prefabricated metal–carbon composite struts that are bonded into the desired design configuration on site. The goal of this project is to test three different configurations and recommend a design for a new bridge.

GRAND CHALLENGE FOR ENGINEERING: RESTORE AND IMPROVE URBAN INFRASTRUCTURE

One of the 14 National Academy of Engineering Grand Challenges is **Restore and Improve Urban Infrastructure.** According to the NAE website (www.engineeringchallenges.org):

> "It is no secret that America's infrastructure, along with those of many other countries, is aging and failing, and that funding has been insufficient to repair and replace it. Engineers of the 21st century face the formidable challenge of modernizing the fundamental structures that support civilization.
>
> The problem is particularly acute in urban areas, where growing populations stress society's support systems, and natural disasters, accidents, and terrorist attacks threaten infrastructure safety and security . . . solutions to these problems must be designed for sustainability, giving proper attention to environmental and energy-use considerations . . . along with concern for the aesthetic elements that contribute to the quality of life."
>
> "Rebuilding and enhancing urban infrastructure faces problems beyond the search for engineering solutions. Various policies and political barriers must be addressed and overcome. Funding for infrastructure projects has been hopelessly inadequate in many areas, as the American Society of Civil Engineers' "report card" documented. And the practice of letting infrastructure wear out before replacing it, rather than incorporating technological improvements during its lifetime, only exacerbates the problems."

The goal of this textbook project is to test three different configurations, and recommend a design for a new bridge.

STRUT DESIGN AND CONSTRUCTION

Data will be collected in class according to the specific instructions provided for you by your professor, or sample data will be provided in an Excel workbook.

All prototypes of the bridge struts must be constructed utilizing the following specifications:

- Only the glue provided may be used. No additional materials, such as adhesive, tape, and caulk may be used.
- Five prototypes of three different bridge strut designs are to be built according to the designs shown below. Other strut design prototypes will not be included.
- The bridge strut spans an open distance of 1 foot; the total length of the bridge strut may not be less than 1 foot, 6 inches, and may not exceed 2 feet.
- The bridge strut ends will be clamped to the supporting surfaces.
- The prototypes should be allowed to dry for at least 48 hours prior to testing.
- The bridge must accommodate a bucket with a wire handle for testing.

The bridge strut prototypes utilize the following testing method:

- The magnitude of applied load at failure will be determined by measuring the mass in the container hung on the bridge, using a scale, recorded in units of grams.
- A strut will be considered to fail when its deflection from horizontal reaches 1 inch or when the strut cracks or breaks.

Flush:

Staggered:

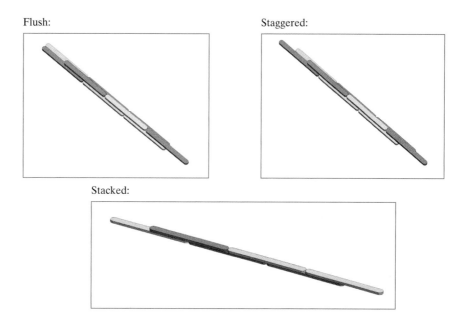

Stacked:

DATA ANALYSIS

Workbook

Each team must submit a final workbook, including the following components:

- Data analysis on each design compiled for all given data for all three designs:
  - Raw data for each trial
  - Average
  - Median
  - Standard deviation
  - Minimum and maximum

- Determine the maximum and minimum values that define the range of two and three standard deviations from the average (plus and minus).
- Use unique conditional formatting on all three designs for the following conditions:
  - For cells outside of two but less than three standard deviations on the low side of the average.
  - For cells outside of two but less than three standard deviations on the high side of the average.
  - For cells outside of three standard deviations on either side of the average.

- For each design, write a conditional statement that determines the following:
  - If the average for the team is within two standard deviations on either side of the mean, leave the cell blank.
  - If the average for the team is outside of two but less than three standard deviations on the low side of the average, show the word "Low."
  - If the average for the team is outside of but less than three standard deviations on the high side of the average, show the word "High."
  - If the average for the team is outside of three standard deviations on either side of the average, show the words "WAY OFF!"

- Three separate worksheets, one for each design, of the compiled data shown as a:
  - Histogram
  - Cumulative Distribution Function (CDF)

The workbook will be graded for neatness, organization, unit consistency, and proper plots. It should be a proper workbook, containing heading information on every page.

Written Section

Each team must submit a final written communication, including the following components:

- A discussion of:
  - Any outliers (two standard deviations away from the average).
  - The upper load limit for each design.
  - Stick count per bridge and cost comparison; assume each stick represents a strut made of metal-carbon composite, costing $500 each, and the cost of the glue, which represents the bonding agent for the composite struts, is negligible.
  - Final design choice and justification.
- Determine the optimal strut design to recommend, answering the following questions:
  - What is the recommended maximum load for each design?
  - Which designs can withstand a load imposed by a mass of 4 pounds-mass 90% of the time?
  - At what load would you expect 20% of the bridge struts for each design to fail?
  - Which design has the highest load to cost ratio?

For the submission format, each team may choose or may be assigned by your instructor one of the following options. The written portion will be graded on neatness, organization, spelling/grammar/mechanics, formatting of embedded graphs, and strength of conclusions.

Written Option One: Short Report

A report template, provided electronically and shown in Section 4.4 in your textbook, must be used.

Written Option Two: Poster

A poster template, provided electronically and shown in Section 4.4 in your textbook, must be used.

Written Option Three: PowerPoint Presentation

A PowerPoint presentation using voice-over should be developed to address the requirements given above. Information on developing a presentation is given in your textbook in Section 4.1, and the accompanying online slides discuss voice-over.

MATLAB MINI-PROJECT: ADVANCED PERSONALIZED LEARNING

Many textbooks used in engineering disciplines use online assessment tools—including those produced by our publisher, Pearson, that have research-backed data that show that students perform better on examinations, work more efficiently, have better evidence of problem solving transfer—and even cheat less on their homework—when using online systems for completing algorithmic homework and assignments that provide rich, interactive feedback to students.

[Photo courtesy of E. Stephan]

One of the Grand Challenges for Engineering, according to the National Academy of Engineering, is **Advance Personalized Learning.** According to a study published by the American Society for Engineering Education, undergraduate enrollment in engineering programs has increased 14% in the last 10 years. However, with more students interested in engineering, many engineering departments and programs that serve those students have not grown proportionally, so class sizes and student-to-faculty ratios have both increased. In addition, research on education has shown that even in a single classroom, many students have different learning styles, so the traditional one-size-fits-all approach to teaching and learning is quickly being replaced with more "personalized" learning systems. Multidisciplinary research in fields like neuroscience, intelligent systems, and education will help steer the future of developing future personalized learning systems.

PROBLEM STATEMENT

In this project, you will build a program to provide personalized feedback based on student performance and participation on different components in a first-year engineering course. In the provided Microsoft Excel workbook, there are two different sheets—"Students" contains a list of all students and their grades on different assignments, and "Assignments" contains a list of each assignment, the textbook section that assignment covered, as well as a list of recommended activities for the student.

We want to write a program that will read the data from the Excel workbook and allow the instructor to provide personalized feedback by analyzing their responses. There should be two modes of operation. First, your program should give the user the menu option to look up a specific student by typing their username as a string and displaying their personalized feedback in MATLAB's Command Window, or reporting an error message if the student is not found in the grade book. In addition to the Personalized Feedback Criteria listed below, the report should also contain the students' name, username, and overall grade.

The second menu option is to process the entire class roster and write a new Excel file that contains only students who have a calculated overall grade less than 70. The Excel workbook should contain a sheet, "Flagged Students," that has three columns of information: the students' name, username, and their overall numerical grade so far. Sample output and a screenshot of the resulting "intervention" file are included below.

Personalized Feedback Criteria:

- **Participation:** If the student has missed more than three days of class, your program should display something in the command window that recommends they attend class more often. Note that the attendance information is stored in columns labeled Day 1–Day 26 and absences are marked with 0's; any number greater than 0 is considered present or excused.
- **Assignment Performance:** If any score on an assignment is less than 70, your code should look up the section coverage and display a recommendation for review for every assignment completed that meets the criteria. Note that assignment grades are located in columns labeled A1–A13.
- **Exam Performance:** For the first exam in the course, the scores on each individual question are recorded in columns EQ1–EQ16, where EQ1 corresponds to Exam Question 1, EQ2 corresponds to Exam Question 2, etc. The overall weighted exam score is in the "Exam Grade" column. If the student missed more than half of the

points available on each question, your program should look up and display the recommended items for review.

Questions 1–8, 12–13, and 15 were worth 5 points each; all remaining questions were worth 20 points each.

- **Overall Grade:** The overall grade should be determined using the following rubric, and scaled out of the current points possible:
  - Assignments are worth 1% each
  - Exams are worth 25% each

For example, if 5 assignments and one exam had been completed, the grade would be the total points earned divided by 30, displayed as a percentage.

HELPFUL FUNCTIONS

Comparing Strings: `strcmp` and `strcmpi`

In order to compare strings, you will need to use the string comparison function, `strcmp`, or the case-insensitive (e.g., dbowman and DBOWMAN are equivalent) string comparison function `strcmpi`. The syntax is simple:

```
A='dbowman'; B='DBOWMAN';
if strcmp(A,B)==1
% this is FALSE
        fprintf('These variables are equivalent')
end
```

```
A='dbowman'; B='DBOWMAN';
if strcmpi(A,B)==1
% this is TRUE
        fprintf('These variables are equivalent')
end
```

For more information: http://www.mathworks.com/help/techdoc/ref/strcmp.html.

All MATLAB codes should include a proper header, including your name, course/section, date, problem statement, documentation on all variables used in the program/function, and any assumptions made in the design of your algorithm. All MATLAB codes should contain an adequate number of comments throughout the code.

SAMPLE OUTPUT

The results shown below are a SAMPLE ONLY and are not necessarily the result you will achieve by correctly conducting the analysis.

```
Type the student's username: custu1

Student Name:  LAST1 FIRST1
Student Username:  CUSTU1

Days Missed:
Total Absences:  0
```

```
Recommended review exam question 8:  Grade: 0 out of 5 points
- Ch 18.1 - 18.2:              Read: pages 519 - 523
- Ch 18.1 - 18.2:              Watch: online materials for 18.1 - 18.2
- Ch 18.1 - 18.2:              Recommend: p. 567: 14, 15

Recommended review exam question 13: Grade: 0 out of 5 points
- Ch 18.1 - 18.2:              Read: pages 519 - 523
- Ch 18.1 - 18.2:              Watch: online materials for 18.1 - 18.2
- Ch 18.1 - 18.2:              Recommend: p. 567: 14, 15
** OVERALL GRADE:             91.3
```

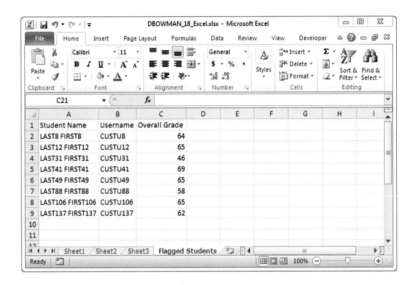

MATLAB MINI-PROJECT:
SECURING CYBERSPACE

Creating secure cyber systems is essential since so much of our critical utilities, finances, and even entertainment relies on secure networks. In 2011, Sony's PlayStation Network, a gaming and digital media service used by devices like the PlayStation 3 and other Sony media devices, suffered a long-term outage due to an illegal intrusion by hackers who gained access to personal information of Sony's customers, impacting over 77 million registered accounts—the largest security breach in history. According to Sony, the cost of the security breach and network outage was over $171 million dollars.

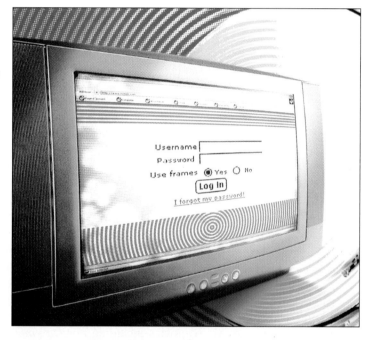

BACKGROUND

One of the 14 NAE Grand Challenges is **Advance Health Informatics**. Medical records today are plagued by mixtures of old technologies like paper and new technologies like computers and Tablet PCs. Computerized records are often incompatible, using different programs for different kinds of data, even within the same hospital! Sharing information over networks complicates things even further because of differences in computer systems, data recording rules, and network security concerns for protecting patient privacy.

In addition to Advancing Health Informatics, another one of the 14 NAE Grand Challenges is **Secure Cyberspace**. Modern communication devices like smartphones and computers, as well as communication protocols like web, phone, text messaging, e-mail, and instant messaging are all heavily used by the average consumer to do work and keep in touch with friends and family—every day. In addition, many companies and government agencies rely on these cyber utilities to conduct everyday work, so protection of cyberspace will prevent or reduce disruptions of essential services and minimize problems with stolen data like identity theft.

With secure networks and standards for electronic medical records, work to address these two Grand Challenges will greatly advance the practice of medicine globally and will help to improve the quality of living for everyone.

PROJECT A: ROTATION ENCODING AND DECODING

The purpose of encoding data is to hide the words from plain text by shifting the letters by a certain number of characters. This type of encoding is known as a Caesar cipher, named after Julius Caesar who used this trick to communicate coded information to his generals. For example, if we shifted (or sometimes called rotated) the characters up by three in the word "dog," the "d" would become "g," "o" would become "r," and "g" would become "j," or "grj." Once a character hits the end of the alphabet, it will wrap around to the beginning of the sequence—for example, "hazy" with a rotation of three would be "kdcb." To decode a word coded with a Ceasar cypher, you need to rotate the opposite direction. For example, "wljhu" decoded with a rotation of three would be "tiger."

For this assignment, your encoder and decoder functions must accept two arguments—the string and the rotation number, and return the encoded or decoded string. For storing encoded data in an Excel workbook, you might want to consider storing the encoded text and the rotation number in separate cells.

PROJECT B: STEGANOGRAPHIC ENCRYPTION AND DECRYPTION

Steganography is the practice of hiding messages in such a way that no one can detect the original message apart from the original sender and the intended recipient and requires some "key" to interpret the original message. Many different forms of steganography have existed before computers, including concealing messages in Morse code on knitting yarn during World War II in order to pass secret messages without being detected by enemies. Many movies, including *The Da Vinci Code*, have referenced the act of hiding encrypted information within different mediums, including painted pictures. Steganography is one of the original ideas for implementing secure text transfer and storage between computers. Assume we want to encrypt the word "tiger" using a steganographic encryption scheme—the result is one of many possibilities:

Encrypted Text: T3gctkWwb4iV846gsBoe99yr

Key: [5 11 16 20 24]

Note that the 5th, 11th, 16th, 20th, and 24th character in the encrypted text spell the word "tiger" and that there would be no possible way to know that information without the key to unlock the text.

Consider the following:

Encrypted Text: bm2dUo61Lo1L52sb9KSe

Key: [2 6 10 15 20]

If you pull out the 2nd, 6th, 10th, 15th, and 20th characters, you will discover the hidden word is "moose".

Here is another example:

Encrypted Text: 26Hy2t52x9bi77rSHgVeYLu2rr7R DypTcs9uEWVb

Key: [6 12 18 20 25 29 34 37 41]

This encrypted text contains the words "tiger cub." Note that between each character there are between 1 and 10 random uppercase, lowercase, or numbers separating the important text. The space between tiger and cub is given as the space character in the encrypted text, and other special characters (like dashes or commas) should be included in the same way.

For this assignment, your encryption function must accept one argument—the string to be hidden, and return two arguments—the encrypted text and the "key" for the text—for each line of text required. For example, for this assignment, there would be six encoded strings (for the name, gender, etc.) and six keys, so you would need to call the encryption function six times to generate both. For the decryption function, your function must accept two arguments—the encrypted text and the "key," and return one argument—the decoded text. For storing encrypted data in an Excel workbook, you might want to consider storing the encrypted text and the key in two or more cells in the worksheet.

PROBLEM STATEMENT

You work for a biomedical technology firm developing a standardized suite of software for use at every hospital across the world. As part of the software suite, your job is to develop the interface and secure file format in order to store medical information in an efficient, but safe manner.

To do this, your team is required to develop a program to store or read medical data in a Microsoft Excel workbook. Each patient must have their medical data stored as a row in a single Microsoft Excel worksheet within the workbook and all of the data must be stored in a "secure" fashion. Your team must also write functions to (a) encode or (b) encrypt the medical data before storing in the workbook, as well as functions to (a) decode or (b) decrypt the medical data to display the patient information when queried by the user. You will be assigned by your professor to complete either Project A (the creation of an encoder and decoder) or Project B (encrypter and decrypter).

DESIGN CRITERIA

Code

Each team must create the following programs/functions in MATLAB:

PGM_TEAM.m—the main interface (program) to allow the user to choose to create a new patient in the command window or read patient information from a Microsoft Excel workbook. The resulting data, from either method, should display in the command window using formatted output. Finally, the program should create a workbook that contains (as (a) encoded or (b) encrypted data):

- Patient name (text)
- Patient gender (text)
- Patients data of birth (YYYY-MM-DD) (text)
- Number of children (number)
- List of allergies (text)
- List of drug prescriptions (text)

- For Project A: Encode/Decode:
  - **Encoder_TEAM.m:** A function that implements a simple character rotation encoding scheme.
  - **Decoder_TEAM.m:** A function that implements a simple character rotation decoding scheme.

- For Project B: Encrypt/Decrypt:
  - **Encrypt_TEAM.m:** A function that implements a simple steganographic encryption algorithm.
  - **Decrypt_TEAM.m:** A function that will decrypt the encrypted text created by the Encrypt_TEAM function.

In all program names, the TEAM stands for your team identifier, as given by your instructor. All MATLAB codes should include a proper header, including all your team member names, course/section, date, problem statement, documentation on all variables used in the program/function, and any assumptions made in the design of your algorithm. All MATLAB codes should contain an adequate number of comments throughout the code.

Microsoft Excel Workbook

In addition to the MATLAB code, your team should design a Microsoft Excel workbook to contain the information required by the program, as well as any other information your code might need to display the user information. Before you submit your code, the workbook your team submits should contain the following patients:

Patient: LUKE SKYWALKER	Patient: LEIA ORGANA	Patient: HAN SOLO
Gender: Male	Gender: Female	Gender: Male
DOB: 1965-11-05	DOB: 1973-10-13	DOB: 1965-12-15
Children: 2	Children: 0	Children: 1
Allergies: Grass, Mold	Allergies: *none*	Allergies: Carbonite, Wookie dander
Prescriptions: Zocor®, Daforce	Prescriptions: *none*	Prescriptions: Cymbalta®

ASCII CHARACTERS

All text and characters stored in MATLAB and other computer programming languages are represented as ASCII (American Standard Code for Information Exchange) characters. Consider the following assignment statement:

```
m='dog'
```

You might have noticed that if you accidentally use a %f control character in an `fprintf` statement when you are trying to display a string, you might see something like this:

```
>>fprintf('The animal is a %.0f\n',m)
The animal is a 100
The animal is a 111
The animal is a 103
```

The reason for this strange output is due to the fact that MATLAB interprets the characters as their ASCII decimal values, rather than their intended character representation:

```
>> fprintf('The animal is a %s\n',m)
The animal is a dog
```

The table below shows each ASCII character and the corresponding decimal value for each letter or number. Note that there are many more characters available in the full set of ASCII values (http://www.asciitable.com/) but for the encoder, we only need to worry about uppercase, lowercase, and numerical values.

Character	A	B	C	D	E	F	G	H	I	J	K	L	M
Value	65	66	67	68	69	70	71	72	73	74	75	76	77

Character	N	O	P	Q	R	S	T	U	V	W	X	Y	Z
Value	78	79	80	81	82	83	84	85	86	87	88	89	90

Character	a	b	c	d	e	f	g	h	i	j	k	l	m
Value	97	98	99	100	101	102	103	104	105	106	107	108	109

Character	n	o	p	q	r	s	t	u	v	w	x	y	z
Value	110	111	112	113	114	115	116	117	118	119	120	121	122

Character	0	1	2	3	4	5	6	7	8	9
Value	48	49	50	51	52	53	54	55	56	57

Note that we actually change a string by adding or replacing values in a string by treating a string character like a vector of numbers. For example, if we want to shift the "o" in the word "dog" stored in variable "m" to create the word "dug":

```
m(2)=m(2)+6
```

Likewise, we could replace a character in a string by typing an ASCII value directly, so if we want to change the current value in "m" to display the word "dig," we could type:

```
m(2)=105
```

HELPFUL FUNCTIONS

- **char:** If you want to convert a decimal value to a corresponding ASCII character, you will want to use the char function. For example:

  ```
  X=[char(97),char(103),char(101)]
  ```

  will store the string "age" in the variable X.

- **randi(x):** If you want to generate a random positive whole/integer number, the randi function will generate a number between 1 and x, the input argument. For example, if you call randi(5), you will randomly generate a 1, 2, 3, 4, or 5. This function might be useful in the encryption function for generating random uppercase, lowercase, or numerical characters.

MATLAB MINI-PROJECT:
DO YOU WANT TO PLAY
A GAME?

Video games have existed in some capacity since 1947 when Thomas T. Goldsmith Jr. and Estle Ray Mann designed a "Cathode Ray Tube Amusement Device" that allowed the player to control a dot on the screen of a television.

[Photo courtesy of E. Stephan]

BACKGROUND

While not directly addressing a specific engineering Grand Challenge, video games actually have some positive effects on video game players, including better hand-eye coordination and visuomotor skills. Video games are often exploited as a tool for education, especially simulation-based games like Sim City, which has been shown to improve strategic thinking and planning skills.

Card games, on the other hand, have existed as early as the ninth century during the Tang Dynasty, where games were played using different leaves. The modern card game with four suits originated in France in approximately 1480, and the contents of the deck of cards have only had a few changes since then, including the addition of the "Ace" and "Joker" cards.

DESIGN CRITERIA

For this project, your team is required to implement any card game using MATLAB.

- All selections ("Hit," "Stand," "Do you want to play again?," etc.) should be done using the menu function. Any numerical or text input may be done with input functions. All inputs (using either menu or input functions) must be validated using `while` loops.
- Your game must implement a real card game. You may choose any **reasonable** card game except overly trivial games like **War** or **52 pick-up**. Acceptable games allow the player to discard and/or draw cards from the deck. For a comprehensive list of card games, visit http://www.pagat.com/alpha.html. Your code should clearly explain to the player which game is being played and any special "house" rules that apply.
- Your card game should only involve one human player. The person running your code may play against computer players, generated by your code. Your instructor will require one or more of the following options required for your project:
  - **Option 1:** Your game does NOT need to display cards graphically (output to the Command Window is fine—e.g., "You drew the Queen of Hearts. Cards in your hand: three of Spades, Queen of Hearts, Ace of Clubs"), but you will earn bonus points if you figure out a way to allow the user to select/see their cards graphically.
  - **Option 2:** Your game should include a GUI to allow the user to press cards to select/discard/etc. The following link contains card images that are free for you to use in your code: http://www.jfitz.com/cards/
- As soon as the game is over, the user must be able to select whether they want to play again using a menu. If the player chooses to play again, the Command Window should clear and the game should start again.
- *Hint*: If you need to generate a random number in your program, look up the `randi` built-in function.

Code

Each team must create the following programs/functions in MATLAB:

- **MP3_TEAM.m**—the main interface (program) to allow the user to play your card game, where TEAM is your team identifier as assigned by your instructor.

You may create any additional functions, but you must include your TEAM identifier in each function name.

All MATLAB codes should include a proper header, including all your team member names, course/section, date, problem statement, documentation on all variables used in the program/function, and any assumptions. All MATLAB codes should contain an adequate number of comments throughout the code.

Grade Scaling

The difficulty of the game your team attempts will be accounted for in the grading of your project.

- 1.0 multiplier: Straightforward games that calculate scores based on card faces like Blackjack.
- 1.1 multiplier: Games that require your code to calculate hands for two or more "computer" players like Euchre.
- 1.2 multiplier: Games that require calculating scores using hand ranking like Poker (flush, straight, pairs, three of a kind, etc.).

This multiplier will be weighted on your final project grade—if your calculated score is 80 and you implemented a Poker game, your overall score would be 80*1.2 = 96.

MATLAB PROJECT:
IMAGE PROCESSING

Image-processing applications have enabled computers to recognize differences and segments in images, websites like Facebook to recognize people in pictures, and cell phones to be able to serve as bar code or QR scanners. While some of these actions seem trivial to the human eye, there's a significant amount of computations required in the background for these simple actions to occur.

[Photo courtesy of E. Stephan]

Your firm has been contacted by a fastener company that needs a device to count the number of nuts and bolts on an assembly line. Cameras are located above the assembly line and take pictures periodically of the contents within the view of the camera.

PROBLEM STATEMENT

Develop a MATLAB function to count the number of nuts and bolts within the current view of a camera on an assembly line. You are given test images to use for the development of your function.

- You must create four MATLAB functions to count the number of nuts and bolts using:
  - Function 1: MATLAB built-in functions
  - Function 2: No built-in function to perform image erosion
  - Function 3: No built-in function to perform image thresholding
  - Function 4: No built-in function to perform image erosion or thresholding
- Using the four functions, determine which function performs best on a single laptop. Record the time it takes for each function to execute.
- Choose the best performing function and execute it on each team member's laptop to determine the best performing laptop. Record the execution time, laptop model, processor information, and RAM size for each test.
- Choose the best performing laptop. Using the best performing function on the best performing laptop, record the execution time for the following:
  - Different resolution images (three images)
  - Different nut/bolt arrangements (three images)
- Propose a solution to the customer based on efficiency of the software and the company's design specification.

DESIGN CRITERIA

Code

Save each image-processing function as a separate file with the following naming scheme:

- Using MATLAB built-in functions (saved as nutsbolts_TEAM.m)
- No built-in function to perform image erosion (saved as nutsbolts_no_e_TEAM.m)
- No built-in function to perform image thresholding (saved as nutsbolts_no_t_TEAM.m)
- No built-in function to perform image erosion or thresholding (saved as nutsbolts_no_e_or_t_TEAM.m)

where TEAM is your team identifier as assigned by your instructor.

Written Section

Each team must submit a final written communication, including the following components:

- Discussion of determination of the most efficient function and hardware configuration.
- Sample figures to illustrate the method used to determine the optimum configuration.
- Total cost analysis for each function developed for the client.
- Discussion of the performance of your function as more nuts and bolts are within the view of the camera, including a proper plot of the relationship between execution time the number of nuts and bolts in the image.
- Discussion of the performance of your function using an image with a higher resolution including a proper plot of the relationship between the execution time and the total number of pixels in an image.
- Discussion of the algorithm developed to determine whether an object is a nut or bolt.
- Your instructor will provide a list of speed and cost requirements. You will recommend the best configurations to meet the specifications required.
- Function cost and performance analysis.
- Computer performance analysis.

For the submission format each team may choose or may be assigned by your instructor one of the following options. The written portion will be graded on neatness, organization, spelling/grammar/mechanics, formatting of embedded graphs, and strength of conclusions.

Written Option One: Short Report

A report template, provided electronically and shown in Section 4.4 in your textbook, must be used.

Written Option Two: Poster

A poster template, provided electronically and shown in Section 4.4 in your textbook, must be used.

Written Option Three: PowerPoint Presentation

A PowerPoint presentation using voice-over should be developed to address the requirements given above. Information on developing a presentation is given in your textbook in Section 4.1, and the accompanying online slides discuss voice-over.

Project Website

Each team must develop and maintain a project website that contains pages dedicated to:

- **A "home" page:** includes a short blurb about the project.
- **Project definition:** a description of your project and the charge your team must accomplish.
- **Team information:** an introduction of each team member (personal) and description of their role within your team (professional).
- **To do list:** a continually updated list that includes an owner, description of the item, due date, and status of whether that item is complete or incomplete.
- **"Daily" updates:** a "blog" update posted (M, W, F) by **ONE** member of the team that documents the events of meetings, who, what, and when every item on the "to do" list is being developed.
- **Weekly updates:** a "blog" update of the big picture of your project development. **These should be INDIVIDUALLY posted by EVERY member of the team by Fridays at 5:00 p.m.** Each team member is required to post a weekly update answering the following questions:

  - What were the major accomplishments for your team last week?
  - What does your team plan to accomplish by this Friday?
  - What were the "road bumps" encountered by your team last week and how did you solve these problems?
  - What problems do you anticipate this week?

- **Calendar:** an embedded calendar of relevant due dates for the project, to-do items, and group meetings.

Each team may use the Google Site Project template or develop your own page from scratch. The site must be secure (only your team + your instructor should have access to your project site) and editable by any member of your team. This security is built in to Google Sites, but you are free to explore alternatives. For more information on how to set up, use, or share Google Sites or use Templates with Google Sites, please refer to the following link: http://sites.google.com/site/projectwikitemplate_en/.

UMBRELLA PROJECT

MATLAB PROJECT:
DMV LINES

Queuing research occurs in diverse fields like computer networking to make digital communication faster, traffic routing to help people move around faster, all the way to the design of theme parks to allow customers to move around the park without standing in too many long lines.

Your firm has been contacted by the Department of Motor Vehicles (DMV) to conduct an analysis of their current customer service performance and recommend changes that would benefit the service time of each customer. To perform the analysis, your team is provided with a list of arrivals to the DMV over the span of three days. Your team must determine the ideal arrangement of the DMV in order to guarantee the smallest average waiting time for each customer.

PROBLEM STATEMENT

You must conduct six analyses of different first come, first serve queuing schemes:

- Two lines, each capable of holding eight people;
- Four lines, each capable of holding four people;
- Eight lines, each capable of holding two people;
- Two lines, each separated by length of service time, holding eight people each;
- Four lines, each separated by length of service time, holding four people each;
- Eight lines, each separated by length of service time, holding two people each.

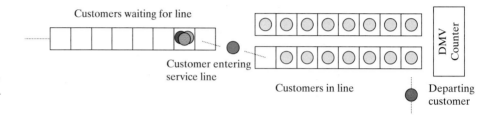

Figure 1 Example of "two lines, each capable of holding eight people."

- The best performing queuing scheme minimizes the average wait time of the customer and serves the most customers. We define wait time of a customer as the sum of the service times of the people in front of that customer when they first enter the line. Note that wait time does not include the customer's own service time.
- If the customer arrives at the DMV and every line is full, the time the customer waits for a line to open up must be included in the total waiting time for that customer. The DMV closes every day at 5:00 p.m., so any customers not yet in one of the queues will be "rejected."
- After selecting the best performing queuing scheme, design a system that can be used at the entrance of the DMV to "ticket" each customer.
- In the service time separated queuing schemes your team will want to try different designs for finding the ideal separation criteria (e.g., all waiting times greater than five minutes go in line 1, everyone else in line 2).

DESIGN CRITERIA

Code

Save each queuing scenario as a separate file with the following naming scheme:

- Two lines, each capable of holding eight people—Save as Lines2_8_TEAM.m
- Four lines, each capable of holding four people—Save as Lines4_4_TEAM.m
- Eight lines, each capable of holding two people—Save as Lines8_2_TEAM.m
- Two lines, each separated by service time—Save as SLines2_8_TEAM.m
- Four lines, each separated by service time—Save as SLines4_4_TEAM.m
- Eight lines, each separated by service time—Save as SLines8_2_TEAM.m
- Ticketing system—Save as Ticketing_TEAM.m

where TEAM is your team identifier as assigned by your instructor.

MATLAB PROJECT: DESIGNING A BETTER VACUUM

Robotic vacuums and other small electronics often rely on embedded systems that are designed to do only a few different tasks. Hardware on embedded systems is usually designed in a manner that reduces the energy usage requirements of the device, but may suffer operational or speed restrictions as a result.

[Photo courtesy of E. Stephan]

Your team is doing consulting work for an electronics manufacturer that is designing a new line of automatic vacuum robots. These robots are unique because they contain sensing capabilities that can map out the floor plan of a room before they begin cleaning. Your team has been charged with developing the algorithms that will control the different models of robotic vacuums.

PROBLEM STATEMENT

■ The manufacturer plans to develop three models of robotic vacuums:

- The economy robot. This robot should clean 100% of the floor in a floor plan, even if the surface has been detected to be clean. The battery life of the economy robot is poor. A robot with "poor" battery life will be able to clean 250 square feet before recharging.
- The regular robot. This robot should clean 100% of the floor in a floor plan, even if the surface has been detected to be clean. The battery life of the regular robot is average. A robot with "average" battery life will be able to clean 350 square feet before recharging.
- The high-end "intelligent" robot. This robot should only clean the dirty parts of the floor. The battery life of the high-end robot is poor. A robot with "poor" battery life will be able to clean 250 square feet before recharging.

You may assume that it takes every robot 60 seconds to recharge. When the robot's battery is empty, it will travel back to the charge station at a rate of 1 pixel per second. After it is fully charged, it will return to where it left off at a rate of 1 pixel per second (without cleaning on the return path).

■ The "sensing capabilities" of the robot will generate a matrix. You may assume that the manufacturer has a separate interface to generate the matrix, so you will not need to develop the sensor—instead, you will write your code assuming that the matrix is provided by some separate mechanism. Test matrices are provided by the manufacturer in a Microsoft Excel workbook. In addition, a blank floor plan is provided in the workbook for creating your own custom floor plan.

■ Your team will need to export movies of your robot cleaning each of the provided floor plans. The provided Microsoft Excel file with floor plan matrices also contains a blank floor plan so that you can create your own custom floor plan for testing. Upon completion, you should have 12 total movies of your three vacuums cleaning the four different floor plans, all due at the same time as the source code.

FLOOR PLAN MATRIX

The manufacturer has developed a sensing mechanism that will generate a 30 × 25 matrix of values that represent different values, as defined below. The algorithms developed should use this matrix as the input to clean the floor.

Floor Plan Matrix Values:

- 0   A wall, couch, or any other surface where the vacuum cannot travel
- 4   Clean floor
- 3   Dirty floor
- 2   Charging station
- 1   The current location of the robot in the image

You may assume that the matrix will always be bordered by a wall and the charging station is always in the same location.

DESIGN CRITERIA

Code

It is suggested, for efficiency that you examine ways to use functions throughout your code.

Save each image-processing function as a separate file with the following naming scheme:

- The economy model (saved as economy_TEAM.m).
- The regular model (saved as regular_TEAM.m).
- The high end model (saved as highend_TEAM.m).

You should create a main GUI that will allow the user to select the model, the floor plan to use, and whether or not the code should export a floor plan movie after your vacuum finishes cleaning the floor plan. Save this program as VacuumProject_TEAM.m, where TEAM is your team identifier as assigned by your instructor.

Each robot should display a summary in the command window (after cleaning a floor plan), including the following:

- The ratio of clean floor to dirty floor. For example, "36% of the floor was covered with dirt."
- The area of the floor plan. You may assume that each "pixel" represents a 1 foot $\times$ 1 foot tile. "Floor plan surface area: 1,200 square feet."
- The amount of time it took the robot to clean the floor. You may assume that it takes five seconds to clean a single dirty "pixel" on the floor.
- The number of times the robot had to recharge. "The robot had to recharge five times."

PowerPoint Presentation/Video

A PowerPoint presentation using voice-over should be developed to address the requirements given above. Information on developing a presentation is given in your textbook in Section 4.1, and the accompanying online slides discuss voice-over.

Each team must submit a final presentation, including the following components:

- Introduction and background on the project.
- Discussion of the algorithm your team designed for each vacuum model.
- Sample figures to illustrate the performance difference in each model.
- Discussion about who should buy each robot-consider housing size (house, apartment, townhome, dorm), power efficiency (how much power does the robot consume relative to the size of each home?), and time (how long will the robot run in each housing unit?).

The presentation will be graded for neatness, organization, presentation quality, formatting of embedded graphs, and strength of conclusions. As an alternative, your team can create a narrated video and submit it in a common video format (AVI, MOV, etc.)—ask your instructor before using a different format. This narrated PowerPoint should last no longer than five minutes.

Project Website

Each team must develop and maintain a project website that contains pages dedicated to:

- **A "home" page:** includes a short blurb about the project.
- **Project definition:** a description of your project and the charge your team must accomplish.
- **Team information:** an introduction of each team member (personal) and description of their role within your team (professional).
- **To do list:** a continually updated list that includes an owner, description of the item, due date, and status of whether that item is complete or incomplete.
- **"Daily" updates:** a "blog" update posted (M, W, F) by **ONE** member of the team that documents the events of meetings, who, what, and when every item on the "to do" list is being developed.
- **Weekly updates:** a "blog" update of the big picture of your project development. **These should be INDIVIDUALLY posted by EVERY member of the team by Fridays at 5:00 p.m.** Each team member is required to post a weekly update answering the following questions:
  - What were the major accomplishments for your team last week?
  - What does your team plan to accomplish by this Friday?
  - What were the "road bumps" encountered by your team last week and how did you solve these problems?
  - What problems do you anticipate this week?
- **Calendar:** an embedded calendar of relevant due dates for the project, to do items, and group meetings.

Each team may use the Google Site Project template or develop your own page from scratch. The site must be secure (only your team + your instructor should have access to your project site) and editable by any member of your team. This security is built in to Google Sites, but you are free to explore alternatives. For more information on how to set up, use, or share Google Sites or use Templates with Google Sites, please refer to the following link: http://sites.google.com/site/projectwikitemplate_en/.

CREATING IMAGES AND VIDEO IN MATLAB

Assume we read in the first floor plan to a matrix A. In an image, we will refer to the "values" in a matrix as pixels.

To plot the matrix A as an image:

```
imagesc(A)        % display the contents of matrix A
colormap(gray)    % convert the image to a scaled grayscale image.
axis('image')     % resize the plot in the figure to the aspect

                  % ratio of the matrix
axis off          % turn off the axis labels
```

The above segment of MATLAB code will create the image shown in the following page. Note that the wall pixels are black, the shades of gray represent the dirt and the charging station, and white represents the floor. *The shades of gray will differ in your image since you will also be drawing the current location of the robot.*

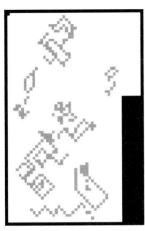

Image of a floor plan showing the walls, dirt, floor, and charging station (top left) without a robot.

Videos in MATLAB

To follow the path of your robot, you will need to use a couple of MATLAB functions to capture each plot and generate the movie.

getframe

getframe will capture the image currently drawn on the figure and store it into a variable in MATLAB's workspace. Typically, this function is used within a for-loop to capture multiple frames.

```
for i = 1:1:30
    % ...
    imagesc(A);        % display the image
    M(i)=getframe;     % capture the figure
    % ...
end
```

movie

The movie function will allow you to play back the frames stored in the workspace. In the example above, the frames are stored in the variable M.

```
movie(M)
```

To control the speed of the video, you can change the frame rate of the video.

```
movie(M,1,30)
```

The above line of code will play the movie M one time (second argument: 1) at a rate of 30 frames per second (third argument: 30).

movie2avi

movie2avi will allow you to export the video stored in MATLAB's workspace to an AVI video file. AVI files can be embedded in PowerPoint presentations and played on computers without any additional software.

```
movie2avi(M,'MyMovie.avi','fps',30)
```

As an example, the above line of code will export the MATLAB video to a file named "MyMovie.avi" with a frame rate of 30 frames per second.

In order to determine the proper frame rate, you should try to make each video between 45 and 60 seconds long.

COMPREHENSION CHECK ANSWERS

CHAPTER 3

CC 3-1

Product	Computer	Automobile	Bookshelf
Inexpensive	Less than $300	Less than 1/5 the median annual U.S. family income	Less than $30
Small	Folds to the size of a DVD case	Two can fit in a standard parking space	Collapses to the size of a briefcase
Easy to assemble	No assembly; just turn on	All parts easily replaceable	Requires only a screwdriver
Aesthetically pleasing	Body color options available	Looks like the Batmobile©	Blends well with any décor
Lightweight	Less than one pound	Less than one ton	Less than 5% of the weight of the books it can hold
Safe	Immune to malware	Receives 5 star rating in NCAP crash tests	Stable even if top-loaded
Durable	Survives the "Frisbee® test"	200,000 mile warranty	Immune to cat claws
Environmentally friendly	Contains no heavy metals	Has an estimated MPG of at least 50	Made from recycled materials Low VOC finish

CC 3-2

- Decrease aerodynamic drag
  - Remove unnecessary protrusions
  - Redesign body based on wind-tunnel testing
- Decrease weight
  - Use carbon-composite materials
  - Remove unnecessary material
- Increase engine efficiency
  - Use ceramic parts (e.g., valves and pistons) so engine can run hotter, thus more efficiently
  - Change to hybrid design with regenerative braking
  - Limit maximum acceleration if consistent with safety
- Manage friction
  - Improved bearings in wheels, engine, etc.
  - Redesign tires for less slippage on road surface
- Encourage fuel-efficient driving
  - Include displays for real-time and cumulative fuel usage (MPG)

CC 3-3

PAT

Hindrance: Tended to slow the team down due to difficulty understanding things.

Helpful: Always available and willing to do assigned tasks. The individuals who helped explain things to Pat probably developed a deeper understanding of the material in the process, thus this was helpful to them personally, though not to the team directly. Probably the second most useful team member.

CHRIS

Hindrance: Often absent, seldom prepared, seldom contributes, offers excuses for poor performance. Definitely the worst team member.

Helpful: Teaches the other team members about the real world, and how to deal with slackards.

TERRY

Hindrance: Not a team player, impatient, not encouraging to others, wants to dominate the team. Despite cleverness, probably the second worst team member.

Helpful: Solves problems quickly.

ROBIN

Hindrance: No obvious hindrance.

Helpful: The real leader of the group, kept things going, encouraging to weaker members. The most useful team member.

CHAPTER 5

CC 5-1

(a) Significant figures: 2   Decimal places: 4
(b) Significant figures: 4   Decimal places: 0
(c) Significant figures: 3   Decimal places: 2
(d) Significant figures: 3   Decimal places: 0
(e) Significant figures: 2   Decimal places: 0 or 3
(f) Significant figures: 5   Decimal places: 0
(g) Significant figures: 7   Decimal places: 0
(h) Significant figures: 5   Decimal places: 3

CC 5-2

(a) -58.9
(b) 247
(c) 2.47
(d) 0.497

CC 5-3

(a) Elevator capacity may require a conservative estimate of 180 pounds.
(b) Serving sizes are usually measured in integer cups, so estimate at 1 cup; but to make sure the bowl is full, round up to 1.3 cups.
(c) If recording, conservative might be to round up to 33 seconds. If it doesn't matter, 30 seconds is fine.

CC 5-4

(a) Scientific notation: 5.809×10^7 Engineering notation: 58.083×10^6
(b) Scientific notation: 4.581×10^{-3} Engineering notation: 4.581×10^{-3}
(c) Scientific notation: 4.268×10^7 Engineering notation: 42.677×10^6

CHAPTER 6

CC 6-1

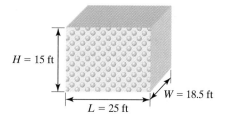

$H = 15$ ft
$W = 18.5$ ft
$L = 25$ ft

CC 6-2

Objective:

Determine the mass of the gravel in units of kilograms.

Observations:

- Assume the container is full, thus the height of the gravel is 15 feet.
- Assume the gravel is flat on the top and not mounded up.
- Neglect the space between the gravel pieces.

CC 6-3

List of variables and constants:

- ρ = density   [=] $\text{lb}_m/\text{ft}^3 = 97\ \text{lb}_m/\text{ft}^3$—given
- V = volume of container [=] ft^3
- m = mass of gravel   [=] kg
- L = length   [=] ft   = 25 ft—given
- W = width   [=] ft   = 18.5 ft—given
- H = height   [=] ft   = 15 ft—assumed

CC 6-4

Equations:

1. Density in terms of
   mass and volume: $\rho = m/V$
2. Volume of a rectangle: $V = L \times W \times H$
3. Unit conversion: $1\ \text{kg} = 2.205\ \text{lb}_m$

CC 6-5

- First, find the volume of the container; use equation (2)
  $V = L \times W \times H$
- Substituting L, W, and H:
  $V = (25\ \text{ft})(18.5\ \text{ft})(15\ \text{ft}) = 6937.5\ \text{ft}^3$
- Second, find the mass of gravel; use equation (1)
  $\rho = m/V$
- Substituting the density and the volume:
  $(97\ \text{bm/ft}^3) = m\ (6937.5\ \text{ft}^3)$ $m = 672937.5\ \text{lb}_m$
- Convert from pound-mass to kilograms:

$$\frac{672{,}937.5\ \text{lb}_m}{} \left| \frac{1\ \text{kg}}{2.205\ \text{lb}_m} \right. = 305{,}881\ \text{kg} \cong 306{,}000\ \text{kg}$$

CHAPTER 7

CC 7-1

(a) $5 = 10^{-9}$ picoliters
(b) $8 = 10^{-14}$ microliters
(c) $3 = 10^{-19}$ liters

CC 7-2

(a) Incorrect: 5 s
(b) Incorrect: 60 mm
(c) Incorrect: 3,800 mL OR 3.8 L

CC 7-3

5.5 mi

CC 7-4

451 in

CC 7-5

328 ft

CC 7-6

2 km/min

CC 7-7

26 flushes

CC 7-8

84 ft^3

CC 7-9

0.16 m^3

CC 7-10

1042 L

CC 7-11

Quantity	Units	Exponents			
		M	L	T	Θ
Geometric quantities					
Area	ft^2	0	2	0	0
Volume	gal	0	3	0	0
	L	0	3	0	0
Rate quantities					
Flowrate	gal/min	0	3	−1	0
Evaporation	kg/h	1	0	−1	0

CC 7-12

$$1 \ \Omega = 1 \ \frac{\text{kg} \, \text{m}^2}{\text{s}^3 \text{A}^2}$$

CC 7-13

$cm^{3/2}$

CC 7-14

(a) 210 Tm
(b) 1.3×10^{11} mi
(c) 1,400 AU
(d) 0.022 light years

CC 7-15

2.72 gal

CHAPTER 8

CC 8-1

13.3 N

CC 8-2

2.2 N

CC 8-3

11 lb$_f$

CC 8-4

$3122 \ \frac{\text{lb}_m}{\text{ft}^2}$

CC 8-5

1.53

CC 8-6

194.5 K

CC 8-7

$0.000357 \ \frac{\text{BTU}}{\text{g K}}$

CC 8-8

221 in Hg

CC 8-9

1.22 atmW

CC 8-10

220 kPa

CC 8-11

2.4 L

CC 8-12

16.9 m

CC 8-13

20 K

CC 8-14

0.83 hr

CC 8-15

61.3 W

CHAPTER 10

CC 10-1

Relative addressing; cell F23 displays 13

CC 10-2

Absolute addressing; cell H26 displays 30

CC 10-3

Mixed addressing; cell D30 displays 30; cell F28 displays 5

CC 10-4

Mixed addressing; cell G30 displays 5; cell J28 displays 30

CC 10-5

A1.	12
A2.	11
A3.	10
A4.	32
A5.	3.14159265
A6.	#NAME? (Excel doesn't understand "PI" without the function parenthesis)
A7.	110
A8.	48
A9.	1.57079633
A10.	1
A11.	0.89399666
A12.	0.78540775
A13.	45
A14.	#NAME? (Excel does not have a function named "cubrt")
Correct expression:	or =27^(1/3) =POWER(27,1/3)

CC 10-6

NO; possible answers given

(a) =IF(B3>B4,B3 - B4,1) OR
=IF((B4 - B3)>=0,1,B3 2 B4)

(b) =IF(B3<=B4,'MAX',B4) OR
=IF((B3 - B4)>0,B4,'MAX')

(c) =IF((B2 + B3 1 B4)<100,
'Too Low','')

CC 10-7

NO; two possible answers given

=IF(B4>0,IF(B4,100<'liquid',
'steam'),'ice')
=IF(B4>=100,'steam',IF
(B4<=0,'ice','liquid'))

CC 10-8

NO; two possible answers given

=IF(B2>=80,IF(B3>=80,IF
(B4>=80,'OK', 'Recycle'),
'Recycle'),'Recycle')
=IF(OR(B2<80,B3<80,B4<80),
'Recycle','OK')

CC 10-9

In cell E7: =VLOOKUP(B7, G8:H17,2,FALSE)

This assumes the table of planetary gravities is in G7:H17 (header in row 7).

The name of the planet is typed into cell B7.

** See video in online materials

CC 10-10

** See video in online materials

CC 10-11

City ascending: Fountain Inn

City descending first, Site Name descending second: Rochester Property

Contaminant ascending first, Site Name ascending second: Sangamo Weston

CHAPTER 11

CC 11-1

- No title
- No axis labels
- Is year 1 1993 or 2000 or . . .?
- Need different line types
- This may be better shown as a column chart instead of a line graph

CC 11-2

- No title
- Poor abscissa scale
- No units or description of axis quantities (what is IC?)
- Need different symbols and line types
- Shouldn't use light colors
- No legend
- Need gridlines if wish to read data

CC 11-3

- Title could be easier to understand
- Percent of what? Unclear
- Symbols too small to distinguish
- Is abscissa value average for decade, or value in that year?
- Colors hard to see on background; need different line types, darker line colors, and white background

CC 11-4

- Title rather cryptic
- No ordinate quantity description
- No vertical axis line
- No gridlines
- No abscissa labels
- Use legend instead of labels on lines
- Different line types needed

CC 11-5

(a) Equal to a constant positive value
(b) Equal to zero
(c) Equal to a constant negative value
(d) Increasing at a constant rate
(e) Increasing at an increasing rate
(f) Increasing at a decreasing rate

CC 11-6

Velocity:

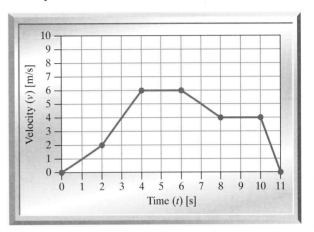

Distance:

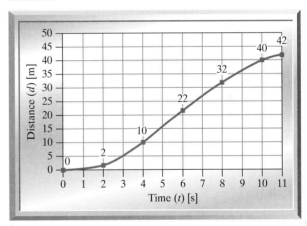

CC 11-7

(a) Machine 2
(b) Machine 2
(c) 1×10^6 feet
(d) Machine 2
(e) Machine 1

CC 11-8

(a) Fixed: $2500; Variable: $50/yr
(b) Approximately $750/yr
(c) It takes approximately 1.1 years to break even

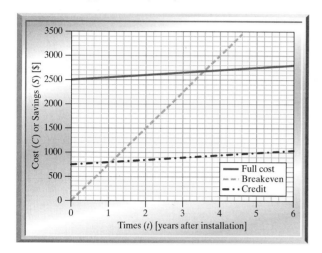

CC 11-9

** See video in online materials

CHAPTER 12

CC 12-1

(a) atm/K   **(b)** 35.6 g   **(c)** 14.1 L

CC 12-2

63.1 lb_m/ft^3

CC 12-3

3.36×10^{-4} $lb_m/(ft\ s)$

CC 12-4

0.455 St

CC 12-5

NOTE: Units on all values are N/m.
Singles: 1, 2, 3
2 in parallel: 3, 4, 5
2 in series: 0.667, 0.75, 1.2
3 in parallel: 6
3 in series: 0.545
Parallel combination of 2 in series with third: 1.5, 1.33, 0.833
Series combination of 2 in parallel with third: 3.67, 2.75, 2.2

CC 12-6

NOTE: Units on all values are ohms [Ω].
Singles: 2, 3
2 in series: 4, 5
2 in parallel: 1, 1.2
3 in series: 7
3 in parallel: 0.75
Series combination of 2 in parallel with third: 1.71, 1.43
Parallel combination of 2 in series with third: 4 (same as 2 + 2 in series), 3.2

CC 12-7

11.4 g

CC 12-8

387 °C

CC 12-9

$C_0 = 35$ g; $k = 1$ h^{-1}

CHAPTER 14

CC 14-1

Mean: 9.3
Median: 9
Variance: 12.9
Standard Deviation: 3.6

CC 14-2

Mean: 102.6
Median: 104
Variance: 295.4
Standard Deviation: 17.2

CC 14-3

(a) Population size has decreased
(b) Variance has increased
(c) Mean has increased

CC 14-4

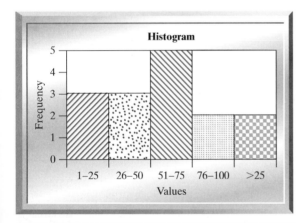

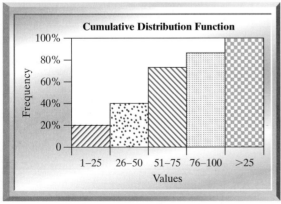

CC 14-5

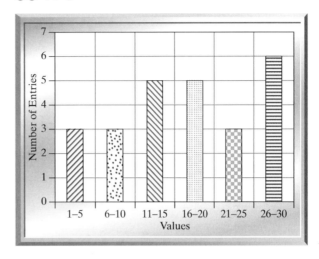

CC 14-6

Not in control

Rule 2—More than nine consecutive points below mean after 26 hr

Rule 3—Downward trend between 25 hr and 37 hr

Rule 5—Two out of three in zone A—several after 29 hr

Rule 6—Four out of five on same side of mean in zone B—start at 33 hr

Rule 8—Eight consecutive points outside of zone C—after 28 hr

CC 14-7

Machine A not in control: Rule 7—Fifteen consecutive points in zone C

Machine B possibly not in control: Possibly Rule 3—Six points with downward trend

CC 14-8

** See video in online materials

CHAPTER 15

CC 15-1

Known:

- The minimum value in the sum will be 2
- The maximum value in the sum will be 20
- Only even numbers will be included

Unknown:
- The sum of the sequence of even numbers

Assumptions:

- [None]

CC 15-2

Known:

- The minimum value in the product of powers of 5 will be 5
- The maximum value in the product of powers of 5 will be 50

Unknown:

- The product of the sequence of powers

Assumptions:

- Only include integer values

Concerning the actual operation to be performed, there are two alternate assumptions, since the wording is intentionally unclear:

1. Multiply all values between 5 and 50 that are an integer power of 5. In other words, the product of 5 and 25.
2. Multiply all integer powers of 5 with a power between 5 and 50 inclusive. In other words $5^5 * 5^6 * 5^7 * \cdots * 5^{49} * 5^{50}$.

CC 15-3

Known:

- Lowest power of 5 in product is 5
- Highest power of 5 in product is 50

Unknown:

- Product of all integer powers of 5 between the lowest and highest powers

Assumptions:

- The result will not exceed the maximum representable number in the computer. Note that by the time the sequence gets to $x = 17$, the product is approximately one googol (10100).

Algorithm:

1. Set product = 1
2. Set x = 5
3. If x is less than or equal to 50
   (a) Multiply product by 5x
   (b) Add 1 to x
   (c) Return to Step 3 and ask the question again
4. End the process

CC 15-4

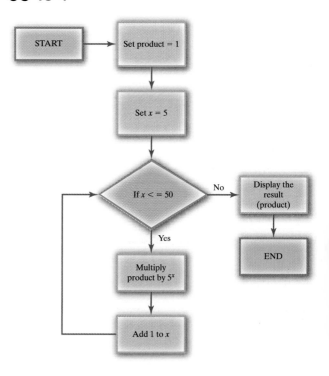

CHAPTER 16

CC 16-1

c, d, e, g

CC 16-2

A = 1
B = 1.2000
C = 1.0000 e − 10
D = 2.0000 + 7.0000i
E = 2
The semicolon suppresses the output to the screen.

CC 16-3

Command Window	Workspace
Hello	'Hello'
Hello World!	'Hello World!'
It's 3:00 PM	'It's 3:00 PM'

Two consecutive single quotes store a single quote.

CC 16-4

A = [1,2,3]
B = [1,2,3]
C = [2;4;6]
D = [2,3,4,5,6]
E = [1,2,3;4,5,6]
F = [1,2,3;4,5,6]
G = [1,2,3;4,5,6]
H = [1,2,3;4,5,6]

- The colon produces a sequence of consecutive integers from the value on the left of the colon to the value on the right of the colon, inclusive.
- A comma inside square brackets marks the end of one value in a row and the beginning of the next value in that row.
- A blank space does the same thing as the comma. Spaces after a bracket or after a semicolon are ignored.

CC 16-5

314.1953
```
Dogs=areaCircle(10);
```

CHAPTER 17

CC 17-1

```
UserHeight=input('Please  enter  your  height
in inches. ');

CurrentTemp=input('Please enter the current
temperature in degrees F. ');
```

CC 17-2

```
EyeColor=input('Please  enter  the  color  of
your eyes. ','s');

ThisMonth=input('Please  enter  the  name  of
the current month. ','s');
```

CC 17-3

```
Month=menu('The current month is:', 'Jan',
'Feb', 'Mar', 'Apr', 'May', 'Jun', 'Jul',
'Aug', 'Sep', 'Oct', 'Nov', 'Dec');
(all on same line)
C=5
```

CC 17-4

%.2f
%.3f
0.35
0.354

CC 17-5

```
The Past Three Years of the Clemson-USC
Football Rivalry:
```

Year	USC	Clemson
2010	29	7
2009	34	17
2008	14	31

CC 17-6

```
% assumes the low temperatures in array LowT
max_low=max(LowT);
min_low=min(LowT);
range=max_low-min_low;
number_bins=round(sqrt(length(LowT)));
bin_size=range/number_bins;
% assumes center_value and bin_range defined
```

```
N=hist(LowT,center_value)
LTSum=cumsum(N);
CDF=LTSum/max(LTSum)*100;
bar(CDF)
set(gca,'XTickLabel',bin_range);
xlabel('Temperature [T] [deg F]');
ylabel('CDF Value [%]');
title('CDF of Daily Low Temperature in
Clemson, SC in October 2006')
```

CHAPTER 18

CC 18-1

Red	Yellow	Green	Action
0	0	0	Power out—Caution
0	0	1	Continue through intersection
0	1	0	Caution—Decide to stop or go
0	1	1	—
1	0	0	Stop
1	0	1	—
1	1	0	—
1	1	1	—

T < 0°C	T > 100°C	State
0	0	Liquid
0	1	Steam
1	0	Ice
1	1	—

Score < 60	Score < 70	Score < 80	Score < 90	Grade
0	0	0	0	A
0	0	0	1	B
0	0	1	0	C
0	0	1	1	C
0	1	0	0	D
0	1	0	1	D
0	1	1	0	D
0	1	1	1	D
1	0	0	0	F
1	0	0	1	F
1	0	1	0	F
1	0	1	1	F
1	1	0	0	F
1	1	0	1	F
1	1	1	0	F
1	1	1	1	F

CC 18-2

```
5 <= t && t < 10
-30 < M && M <= 20
Y <= 100
```

CC 18-3

There is not a logical expression after the "else" because you know (or the program "knows") that the statement following the else is only executed if the conditional statement in the if statement was false. Since Speed > 65 is false, this implies that your speed is 65 or less—you do not need to ask the question—you already know the answer.

CC 18-4

```
if ChargeLevel==discharged
DeviceCheck(battery)
DeviceCheck(alternator)
if BatteryStatus==defective
DeviceReplace=battery
elseif AlternatorStatus==defective
DeviceReplace=alternator
end
end
```

CC 18-5

```
DeviceCheck(battery)
DeviceCheck(alternator)
if ChargeLevel==discharged &&
BatteryStatus==defective
DeviceReplace=battery
end
if ChargeLevel==discharged &&
AlternatorStatus==defective
DeviceReplace=alternator
end
```

CHAPTER 19

CC 19-1

```
for evens=2:2:20
       disp(evens)
end
```

CC 19-2

```
for fives=5:5:50
       disp(fives)
end
```

CC 19-3

```
for odds=13:-2:-11
       disp(odds)
end
```

CC 19-4

Formulae:

A(k)=2*k	B(k)=2*k-1	C(k)=k^2
D(1)=1	for k>1, D(k)=D(k-1)+k	E(k)=7+k/4
F(k)=11-k	G(k)=-1^(k+1)	H(k)=2^k
I(1)=1	for k>1, I(k)=I(k-1)*k	J(k)=1/k

Complete Program:

```
% Print Header row
fprintf('\tk\tA\tB\tC\tD\tE\tF\tG\tH\tI\
   tJ\n')
% Calculate one row at a time and print
   that
```

```
% row
for k = 1 : 10
    A(k)  = 2 * k;
    B(k)  = 2 * k - 1;
    C(k)  = k ^ 2;
    E(k)  = 7 + k/4;
    F(k)  = 11 - k;
    G(k)  = -1 ^ (k+1);
    H(k)  = 2 ^ k;
    J(k)  = 1/k;
    % Handle initialization of first
    % elements of D and I
    if k == 1
        D(k) = 1;
        I(k) = 1;
    else
        D(k) = D(k - 1) + k;
        I(k) = I(k - 1) * k;
    end
    % Print the row
    fprintf('\t%d\t%d\t%d\t%d\t%d\t%.2f\
       t%d\t%d\t%d\t%d\t%.2f\n',...
       k,A(k),B(k),C(k),D(k),E(k),F(k),G(k),
       H(k), I(k),J(k))
end
```

CC 19-5

```
UserNumb=-1;
while UserNumb < 0
      UserNumb=input('Please enter a
      non-negative number. ')
end
```

INDEX